S0-DMW-973

Crop Improvement Utilizing Biotechnology

Authors

Rup Lal, Ph.D.
Department of Zoology
Sri Venkateswara College
Dhaula Kuan, New Delhi

Sukanya Lal, Ph.D.
Department of Zoology
University of Delhi
Delhi, India

CRC Press, Inc.
Boca Raton, Florida

Library of Congress Cataloging-in-Publication Data

Rup Lal, 1953—
Crop improvement utilizing biotechnology / authors, Rup Lal,
Sukanya Lal.
p. cm.
Bibliography: p.
Includes index.
ISBN 0-8493-5082-4
1. Plant genetic engineering. 2. Plant biotechnology. 3. Crops — Development.
I. Sukanya Lal. II. Title.
SB123.57.R87 1990
631.5—dc19 89-31207
CIP

This book represents information obtained from authentic and highly regarded sources. Reprinted material is quoted with permission, and sources are indicated. A wide variety of references are listed. Every reasonable effort has been made to give reliable data and information, but the author and the publisher cannot assume responsibility for the validity of all materials or for the consequences of their use.

Direct all inquiries to CRC Press, Inc., 2000 Corporate Blvd., N.W., Boca Raton, Florida 33431.

International Standard Book Number 0-8493-5082-4

Library of Congress Card Number 89-31207
Printed in the United States

PREFACE

During the 1980s, several research achievements of considerable impact on plant breeding have been witnessed. Among these, the most relevant are detection and utilization of somaclonal variation, cloning plants *in vitro,* isolation of protoplasts and their regeneration into functional plants, regeneration of somatic hybrid plants, development of gene vehicles which allow the insertion of alien genes into plants, and control of frost damage to plants by genetically engineered bacteria.

Researchers have directed increasing effort toward the isolation of mutants from plant tissue cultures without applying any selection. These efforts have been based on the observation that the process of growing plant cells in culture yields a high frequency of stable heritable variants which express traits useful in crop improvement. The variation has recently been termed as somaclonal variation. This subject is being extensively investigated and several variants showing resistance to pathogens and herbicides have been regenerated from cell culture. Tissue culture techniques have already proved useful for the mass propagation of species for which there is no other adequate way of producing large numbers of plants in a short time. Meristem culture for rapid propagation and virus elimination in planting stock are currently the most practical applications of tissue culture available to less developed countries. In the past, it has been typical for yields in potato and cassava crops to be greatly reduced by several potato viruses. Virus free potato and cassava plants have now been regenerated through tissue culture.

One recent innovation is the introduction of methods for freezing plant cells and calli, while retaining cell viability. Several laboratories have devised specific successful freezing regimes to preserve these materials. Another application for this technology would be to freeze germ plasm or plant embryos of rare mutants for later use in breeding, tissue culture, or molecular biology programs. Frozen embryo banks might replace or supplement seed storage facilties.

The ability of isolated protoplasts to regenerate into entire plant offers unique opportunities for genetic manipulation of plants. Protoplast to plant systems are now available for several crop species and protoplasts are proving more receptive to added DNA. Many transformation, selection, and regeneration procedures have been demonstrated to obtain plants resistant to pathogens and certain environment stresses. This success with protoplast regeneration has encouraged work on protoplast fusion leading to the recovery of somatic hybrids. Recent developments have led to the unidirectional transfer of cytogenes through chloroplasts and mitochondria. These cytogenes confer tolerance to herbicides and pathogens. This area therefore will be the most active one for industrial research, and considerable research progress will be made in the next 10 years.

Some success has been obtained in the selection of cells for qualitative traits such as resistance to antimetabolites, herbicides, pathotoxins, etc. Extensive work is being done on selecting useful quantitative traits such as resistance to salt, heat, cold, soil pH, and flooding or drought in several laboratories. The crop species for which regeneration from cells can be accomplished routinely include tomatoes, potatoes, rape seeds, cabbages, citrus, sunflowers, carrots, cassava, alfalfa, millet, clove, and endive. Until recently, it was difficult to regenerate corn, wheat, millet, and soybeans through tissue culture. Methods have now been developed to regenerate plants in large numbers from single cells or protoplasts in these species. The next step will be the exploitation of these species for selection of improved plants or introduction of alien genes into the cells and their subsequent regeneration into complete plants.

The field of recombinant DNA technology or genetic engineering of plants have grown from work initially carried out on microorganisms. This technology allows for detailed manipulation of individual genes. The importance of such development has been highlighted recently with the investment of large sums of venture capital in small companies whose products or potential products are based on the application of recombinant DNA and tissue culture techniques. Significant progress since 1980 has made it possible to attempt cellular and molecular genetic manipulation and improvement of plant species of economic importance.

Remarkable progress has been made in perfecting vectors for delivery of novel genetic information to plants. Most of the research during the past few years has focused on understanding how certain plant pathogens transfer their DNA into a host plant. Emphasis has now been shifted to the transfer of genes through *Agrobacterium* into plant cells and their subsequent regeneration into complete plants. Successful introduction and expression of several genes in regenerated plants have been obtained for several species. In a few cases, foreign genes have been shown to be in a heritable form in the mature plant. The transformation of genes which have insecticidal property from *Bacillus thuringiensis* to tobacco plants and their inheritance from parents to progeny is a remarkable success. Vector systems for agriculturally important monocots have also been developed. The first authorized field trial of genetically engineered ice deletion bacteria to protect plants from frost damage was already conducted in 1987, making it a watershed year of biotechnology.

We would like to express our gratitude to Dr. J. Khurana, MSU-DOE Plant Research Laboratory, Michigan State University, East Lansing, Michigan, who has contributed his comments and criticism and has been generous enough to keep us abrcast of the latest developments in the field. We are thankful to Dr. B. S. Attri for his critical suggestions during the preparation of the book. We are especially indebted to Dr. V. Krishna Moorthy for his continuous encouragement during the project. We must also express our thanks to Drs. Surendar Kumar, P. S. Dhanaraj, V. V. S. N. Rao, Jai Shree, and Sarath Chandran who read every chapter carefully. Our thanks are also due to Mr. C. L. Kaushal for his help and enthusiastic participation in the project. Many thanks are due to Vasu Deva Sastry for his excellent secretarial help. Thanks are also due to M. L. Varghese, T. S. R. Murthy, and A. S. V. N. Rao for typing the manuscript.

THE EDITORS

Rup Lal, Ph.D., has been working in the Department of Zoology, Sri Venkateswara College, University of Delhi, since 1979. Born on September 27, 1953, in Kanoh, Himachal Pradesh, India, he obtained his B.Sc. degree from D.A.V. College, Jullundur in 1973. He obtained his M.Sc. degree from Kurukshetra University, and his Ph.D. degree in 1980 from the University of Delhi. Since then he has been actively engaged in research and teaching.

Dr. Rup Lal has been the recepient of many research grants from the goverment of India. At present he is the Principal Investigator of a project "Effects of Pesticides on *Rhizobium*-legume Associations", funded by the Department of Environment, Government of India. A recepient of Alexander Von Humboldt Fellowship, he is currently working on microbial degradation of xenobiotics at the University of Bielefeld, West Germany.

Dr. Rup Lal has been recognized by the University of Delhi as a supervisor for Ph.D. students in Zoology, being the first lecturer in a college under the university to get this honor. He has authored 2 books and over 50 research papers in the area of pesticide microbiology. His current interest includes gene cloning, plasmid isolation, and microbial detoxification of xenobiotics.

Sukanya Lal, Ph.D., is a Research Associate in the Department of Zoology, University of Delhi, Delhi. She was born on October 28, 1957, in Delhi, India. She graduated from Government College Hamirpur, in 1975. She received her M.Sc. and Ph.D. degrees in Zoology from the University of Delhi, India. She has co-authored 1 book and over 15 research papers in the area of pesticide microbiology.

TABLE OF CONTENTS

Chapter 1
Somaclonal Variation in Crop Improvement 1
I. Introduction 1
II. Terminology 1
III. Methodological and Experimental Approach 2
IV. Survey of Major Crops for Somaclonal Variation 9
A. Potato 9
B. Tobacco 11
C. Alfalfa 12
D. Wheat 14
E. Rice 16
F. Maize 17
G. Lettuce 18
H. Sugarcane 18
I. Tomato 19
J. Strawberry 19
K. Celery 20
V. Genetic Bases of Somaclonal Variation 20
A. Changes in Chromosome Number 20
B. Structural Changes in Chromosomes 30
C. Point Mutations 31
1. Spontaneous Mutations 31
2. Induced Mutations 32
3. Gene Expression and Gene Amplification and Deamplification 34
4. Mitotic Crossing-Over and Transposable Elements 35
5. Changes in Cytogenes 36
D. Genetic Analysis 37
VI. Applications 44
A. Strategies for Crop Improvement 46
B. Somatic Cell Selection 50
1. Selection for Ethionine Resistance 50
2. Selection for Salt and Herbicide Tolerance 51
C. Selection for Disease Resistance 52
D. Application in Research 53
VII. Conclusions and Future Prospects 55
References 59

Chapter 2
Role of Tissue Culture in Rapid Clonal Propagation and Production of Pathogen-Free Plants 73
I. Introduction 73
II. Culture System Steps and Methods of Shoot Induction and Proliferation 74
A. Multiplication by Apical and Axillary Shoots 74
B. Multiplication by Adventitious Shoots 76
C. Multiplication through Callus Culture 77
III. Factors Affecting Morphogenesis and Proliferation Rate 80
A. Nature of the Donor Plant and Explant 81

B. Composition of Culture Media ... 83
C. Culture Environment ... 86
D. Nature of Genotype ... 89
IV. Applications ... 90
A. Production of Pathogen-Free Plants ... 90
1. Meristem Culture ... 90
2. Thermotherapy and Chemotherapy ... 92
3. Factors in Disease Elimination ... 96
B. Germplasm Conservation and Exchanges ... 98
C. Small-Scale Cloning, Seed Production, and Mass Propagation ... 99
V. Conclusions and Future Prospects ... 108
References ... 111

Chapter 3
Protoplasts in Crop Improvement ... 117
I. Introduction ... 117
II. Methodology ... 117
A. Protoplast Isolation and Regeneration ... 117
1. Isolation ... 117
a. Factors Affecting Protoplast Isolation ... 119
i. Leaf Mesophyll Protoplasts ... 119
ii. Cell Culture Protoplasts ... 126
iii. Enzymes ... 126
iv. Osmoticum ... 127
v. Miscellaneous Isolation Conditions ... 127
2. Protoplast Purification ... 127
3. Culture of Protoplasts ... 129
a. The Feeder Layer Technique ... 130
i. Protoplast Irradiation ... 130
ii. Plating of Irradiated Protoplasts ... 130
4. Cell Wall Regeneration, Cell Division, Organogenesis, and Whole Plant Regeneration ... 131
B. Protoplast Fusions and Regeneration of Somatic Hybrids ... 131
1. Fusion and Isolation ... 131
a. Complementation Methods ... 133
b. Visual Selection ... 134
c. Microisolation and Fluorescent Cell Sorting ... 135
d. Identification by Electrophoresis ... 137
e. Use of Genetically Transformed Protoplast Cells ... 137
III. Applications ... 138
A. Regeneration of Plants from Protoplasts ... 139
B. Protoplast Transformations by Direct, Chemically Mediated, and Microinjection Methods ... 145
C. Electroporation of Protoplasts and Transfer of Foreign Genes ... 149
D. Selection of Mutants ... 151
E. Protoplasts as a Tool for Quantitative Assay of Phytotoxic Compounds ... 152
F. Regeneration of Complete Plants with Desired Characters from Somatic Hybrids ... 152
1. Interspecific Hybrids ... 152
2. Intergeneric Hybrids ... 158

3. Organelle Transfer by Protoplast Fusions and Cytoplasmic Genetic Traits 161
4. Nuclear Genetic Traits and Their Transfer through Somatic Hybrids 165
IV. Strategies for the Transfer of New Traits and Development of a New Variety 168
A. Transfer of Organelle and Organelle-Controlled Traits 168
1. The Donor-Recipient Protoplast Fusion 168
a. Nature of Donor and Recipient 169
b. Iodoacetate Perfusion Treatment 170
c. Arrest of Nuclear Division in Donor Protoplasts 170
2. Selection, Isolation, and Regeneration 170
3. Identification of Organelle-Controlled Traits 171
a. Plastome Identification 171
b. Chondriome Analysis 172
B. Development of a New Variety 173
V. Conclusions and Future Prospects 176
References 179

Chapter 4
Cell Selection and Long-Term High-Frequency Regeneration of Cereals and Legumes 195
I. Introduction 195
II. Tissue Culture and Long-Term High-Frequency Regeneration of Cereals and Legumes 196
A. Explant Source 196
B. Culture Conditions 200
1. Rice 200
2. Wheat 203
3. Sorghum 205
4. Millet 205
5. Corn 206
6. Oats 207
7. Soybean 207
8. Pigeon Pea 212
III. Cell Selection 214
A. Selection Strategies 214
1. Callus Initiation and Culture 216
2. Screening 216
3. Plant Regeneration 217
4. Testing of Regenerated Plants and Progeny 217
B. Selection Methods 218
1. Callus Selection 218
2. Selection via Suspension Culture 222
3. Selection via Plated Cells 222
IV. Applications 223
A. Salt Tolerance 224
B. Tolerance to Mineral Stress 235
C. Herbicide Resistance 237
D. Disease Resistance 238
V. Conclusions and Future Prospects 243
References 246

Chapter 5

Agrobacteria-Mediated Gene Transformation and Vectors for Gene Cloning in Plants

in Plants 255
I. Introduction 255
II. Plant Gene Vectors 255
A. *Agrobacterium* Ti Plasmid System 255
1. Incorporation, Localization, and Organization of T-DNA 256
2. Vectors for *Agrobacterium*-Mediated Plant Transformation 258
a. Cointegrating Vectors 258
i. Disarmed Vectors 258
ii. Nondisarmed Vectors 259
b. Binary Vectors 260
3. Limitations in Vector Construction and Plant Regeneration 260
4. Selectable Markers for Plant Transformation 261
5. Transformation of Plants 262
a. Cocultivation 262
b. Leaf Disk Transformation 262
B. Caulimovirus System 263
1. Localization and Organization of CaMV 263
2. Use as a Cloning Vector 264
C. Gemini Virus System 267
1. Structural Features 268
2. Use as a Closing Vector 268
III. Applications 270
A. Pest Resistance Transgenic Plants 270
B. Herbicide Resistance Transgenic Plants 273
C. Transgenic Plants with Marker Genes (Preliminary Success) 276
D. Genetic Transformation of Marker Genes through *Agrobacterium* for Selection of Hybrid Seeds 278
E. Vector Design to Facilitate the Expression of Foreign Genes 279
F. Agroinfection 280
1. *Agrobacterium* as an Organism for the Experimental Storage and Transmission of Plant Viruses 281
a. Storage Efficiency and Flexibility 281
b. Viral Genome Release from T-DNA 283
c. T-DNA Transfer and Its Analysis 284
2. Transformation of Plant Cells with Viral Genetic Information 285
a. Transient Expression 285
b. Expression of Viral Genes in Host and Nonhost Plants 285
c. Transgenic Plants Containing Oligomers of Viral Genomes 286
d. Superinfection of Plants Transgenic for Viral Sequences 288
3. Satellite Defenses for Plants 289
Acknowledgments 292
References 292

Chapter 6

Plant Frost Injury and Its Management ... 299

I. Introduction ... 299
II. Supercooling of Water and Ice Formation ... 299
III. Ice Nucleation-Active Bacteria on Plant Surface and Frost Damage ... 300
A. Types of Ice Nuclei-Active Bacteria ... 301
B. Distribution of INA Bacteria on Plant Surfaces ... 303
1. Factors Affecting the Distribution of INA Bacteria ... 303
a. Variation Due to Plant Species and Plant Parts ... 303
b. Seasonal Variations ... 304
C. Activity of INA Bacteria ... 304
1. Variation Among Different INA Bacteria ... 304
2. Variation Due to Temperature ... 305
3. Population Density of INA Bacteria ... 307
4. Miscellaneous Conditions ... 307
D. Quantitative Estimation of INA Bacteria ... 308
IV. Control of Plant Frost Injury ... 309
A. Traditional Methods ... 309
B. Bactericides ... 309
C. Antagonistic Bacteria ... 312
1. Factors Affecting the Activity of Antagonistic Bacteria ... 314
a. Time of Inoculation of Plants with Antagonists ... 314
b. Effect of Inoculum Size of Non-INA Bacteria (Antagonists) ... 315
c. Time of Inoculation in Relation to Host Plant ... 317
D. Construction and Use of Ice$^-$ Bacteria for the Control of INA Bacteria ... 319
1. Construction of Ice Nucleation-Deficient Bacteria ... 319
2. Use of Ice$^-$ Site-Directed Deletion Mutant of *Pseudomonas syringae* for Frost Control ... 323
E. Control of Almond, Pear, Tomato, and Potato ... 334
F. Mechanisms of Antagonism ... 335
G. Molecular Mechanisms of Ice Nucleation ... 337
Acknowledgments ... 338
References ... 338

Index ... 345

Chapter 1

SOMACLONAL VARIATION IN CROP IMPROVEMENT

I. INTRODUCTION

Somaclonal variation refers to the genetic variability in plants from tissue and cell cultures.[1] The agronomic value of the plants can be significantly improved in many traits for which genetic variability is generated through somaclonal variation. Its widespread nature came into light only recently.[1-3]

The idea of application of variations to crop improvement is not new. Plants have been regenerated from cell culture, especially by the micropropagators for nurseries, for several years. They have also noted variations from time to time in the plants raised by them. Until 1975, though such variations in plants were fully recognized, no attempt was made to explain them.[4-6] It is surprising to note that even chromosomal changes were reported to occur in cell cultures and regenerates,[4,5,7,8] but such changes did not cause much concern at that time.

A major achievement in this direction was made by Skirvin and Janick,[9] although their work did not receive much attention at that time. They recovered variant plants from callus cultures of scented geranium (*Pelarogonium* spp.). These plants were different in their oil component, fascination, pubescence, and anthocyanin production. Several workers were impressed from such culture-induced variations in the early 1980s.[10] These variations, including chlorophyll deficiency and enhanced axillary shooting in lettuce cultures, were named as "phenovariants" by Sibi.[11] Interestingly, in all the plants which showed such variations the chromosomal number was normal ($2n = 18$). These studies further aroused interest to look slightly beyond the chromosomal level to understand the phenomenon of variations. Further genetic-based information was gathered from the experiments on anther culture-induced variations.[12-21] These studies only recognize ploidy changes and chromosomal number responsible for such variations. However, it was the critical assessment of such variations by Alhoowalia,[22,23] who for the first time reported gross structural changes (reciprocal, translocations, deletions, and inversions) responsible for the production of variants in *Lolium*.

Until the late 1980s, many authors associated these variations only with leaf protoplasts.[24,25] However, subsequent work on potatoes revealed variations from cultures of leaf protoplasts and other explants including rachis, stem, and petiole.[26-29] It was finally in 1981 that Larkin and Scowcroft[1] named the entire aspect somaclonal variation.

The potential application of somaclonal variation for crop improvement has been suggested frequently.[1,3,6,30-33] From the standpoint of utility, knowledge of fundamental mechanisms and understanding of the origins of somaclonal variation are necessary for their successful utilization in crop improvement. Hence, in this chapter, several aspects of somaclonal variation are described with more emphasis on experimental design, sources of variation, mechanism of variation, and practical application in crop improvement that has emerged from extensive research during these years.

II. TERMINOLOGY

Early variant plants regenerated from cell cultures of geranium were termed "calliclones" by Skirvin and Janick,[9] while plants regenerated from protoplasts of potato were termed "protoclones" by Shepard et al.[33] Larkin and Scowcroft,[1] in their review in 1981 and in several subsequent reviews, have used the more general term "somaclonal variation" for the variation detected in plants derived from any form of cell culture. However, the type of genetic variations recovered in regenerated plants is to a large extent dictated by the genetic constitution of the particular cell population which is regenerated. Thus, it is necessary to distinguish between plants regenerated from somatic and gametic tissues. Therefore, Evans et al.[2] coined another term "gametoclone", which refers to plants regenerated from cell cultures originated from the gametic tissue; and the term "somaclone" is kept exclusively for plants regenerated from cell culture.

For genetic analysis Orton[25] recognized the need for unified nomenclature and introduced the R, R_1, R_2, etc. (respectively, regenerate, first-generation selfed progeny, second-generation selfed progeny, etc.) system which is now in wide use. Variation on this system includes the use of "P" instead of "R"[34] and the use of a subscript to denote the population of primary regenerated plants.[35]

Larkin et al.[36] proposed a new system: SC_1, SC_2, SC_3, etc. equivalent, respectively, to R, R_1, R_2, etc. The SC refers to somaclone and was introduced primarily to avoid confusion with the established R system. The authors argue that passage of cells or tissues through culture often, if not usually, results in the introduction of mutations. If cell and corresponding regenerated plants are diploid, mutations should be heterozygous. Hence, the R designation is numerically inconsistent with the established "filial" nomenclature denoting hybridity and successive self-generations (F_1, F_2, F_3, etc.). Orton[37] pointed out that the "1" subscript denoting primary regenerated plant is potentially misleading because it presumes the existence of heterozygous mutations in plants regenerated from cell and tissue cultures, thus introducing a bias with respect to the presence or absence of genetic variation. Perhaps the strongest argument in favor of the R system is that it is a historical precedent[34] and is already in wide use and broadly accepted.

III. METHODOLOGICAL AND EXPERIMENTAL APPROACH

A need to improve and standardize experimental designs in somaclonal variation so that results are repeatable has been realized recently. In general, most of the studies on somaclonal variation in plants suffer from narrow applicability due to potentially confounding factors or lack of adequate replication. In addition, germplasm and details of experimental conditions are often not described in sufficient detail. For example, tissue from a single plant is often used to represent the general characteristics of the entire populations, varieties, and species. However, plants of a different genotype within a species have been shown to exhibit reproducible differences under culture conditions.[38,39] Hence, extrapolations beyond the individual genotype are difficult, and the lack of a defined genotype or access to identical genotypes makes reconstruction of experiments impossible. Further, explant tissue is generally assumed to be in a state of genetic uniformity, permitting the conclusion that spontaneous genetic instability occurs *in vitro*. Polyploidy[40] and aneuploidy[4,5,41] have been noted in differentiated plant tissue, and appear to be a normal feature of some plants cultured *in vitro*.

Genetic background, explant source, the characteristics of the initiated growth (e.g., cell association, latent differentiation, intrinsic growth rate), medium composition, and overall culture age are the major factors that appear to affect the results on somaclonal variation.[42] Orton[42] proposed that genetic and developmental architecture of explant and the genotype(s) of explant donor(s) are the major factors for variability in cultures. In addition, precise standardization, consideration of all possible invariant factors, and genotype and environment interactions must be considered for accurate assessment of the results. Standardization of culture media, light and temperature regimes, and transfer method can further help in the accurate assessment of somaclonal variation. In most studies explants with reasonably homogenous genotype have not been used. Pure or inbred lines for F_1 hybrid varieties are, in most cases, reasonably homogeneous with respect to genotype; these should be used in preference to undefined populations such as natural collections, open-pollinated outcrossing cultivars, early-generation breeding materials, and hybrid varieties other than strict F_1 types. It is also better to use inbred lines where genetic study is the aim in somaclonal variation.

A major difficulty encountered is the nonavailability of standard procedures for the detection of somaclonal variation. There is also a lack of standard parameters in detection and analysis of variation. In general, variation in a single cell can be attributed to base changes, duplications, deletions, transpositions, translocations, inversions, chromosome loss or gain, and polyploidy. Further, chloroplast and mitochondrial genomes are dependent on nuclear genomes; thus consideration of both nuclear and cytoplasmic genes for such variation is essential. In fact, several characters in plants now appear to be under the control of cytogenes.

The production of variants due to changes in the cytogenes responsible for these characters has been overlooked due to the nonavailability of sophisticated techniques. Since several new methods such as DNA probes, use of restriction endonucleases, blotting, and hybridization have been made available recently to scientists, coming years will witness the discovery of many organelle-based genes responsible for somaclonal variation in several plants.

A summary of the techniques for the detection and measurement of genetic variation in somatic tissue was compiled by Orton.[42] He suggested that the spectrophotometric estimation of cellular DNA content is the crudest method of detection of somaclonal variation. In this technique, tissues are typically fixed, treated with weak hydrolysis, and stained with Feulgen reagent, which contains a DNA-specific dye. A modified light microscope is then used to measure absorbance in reference to a standard of known DNA content. The main disadvantages of the technique are high error components and the inability to detect mutations other than nuclear polyploidy. Errors can be reduced by utilizing isolated nuclei, fluorescent DNA-specific dyes, and fluorescence-activated sorting. Under certain circumstances, it may be desirable to study polyploidy alone in isolation from other potentially confounding classes of mutations, but DNA contents alone can only be of descriptive value in the total phenomenon of somaclonal variability. The main advantages of this technique are the relative ease of sampling and the fact that interphase (not mitotic) cells are used.

Striking variability in chromosome number among cell cultures of *Haplopappus gracilis* tissue was first reported by Mitra and Steward.[43] The change of chromosome number is a more direct and convincing demonstration of variation than DNA content. However, such determinations are reasonably tedious and progressively more difficult as chromosome number increases. In addition, only a limited population of cells in a culture which is in division stage can be sampled. Thus, variation in nondividing cells cannot be predicted. Though changes in chromosome number imply changes in cellular information, this may in certain instances be erroneous, as in the case of Robertsonian chromosome fissions and fusions. Further aneuploidy is a condition that is defined by the presence or absence of a specific chromosome or chromosomes, which cannot be predicted from chromosome number alone.

Evans and Sharp[44] published a description of experiments designed to generate somaclones of tomato and to ascertain the genetic basis of somaclonal variation (Figure 1). Several tomato seeds of a well-characterized, open-pollinated tomato variety UC82B were sown in a greenhouse. Young, fully expanded leaves were removed from the plant from cultured leaf explant.[45] Callus that developed from the culture leaf explant regenerated shoots within 4 weeks.[44] Regenerated shoots were rooted on medium containing naphthalene acetic acid, then transferred to the greenhouse for maturation and fruit collection. Self-fertilized seed was collected from mature fruits on R_0 plants. To evaluate the R_1 generation, seed was sown in the greenhouse where seedling characters, such as chlorophyll deficiency, anthocyanin content, and leaf shape, were monitored. R_1 seedlings were transplanted to the field to classify mature plant characteristics such as pedical type, fruit shape, and fruit and flower

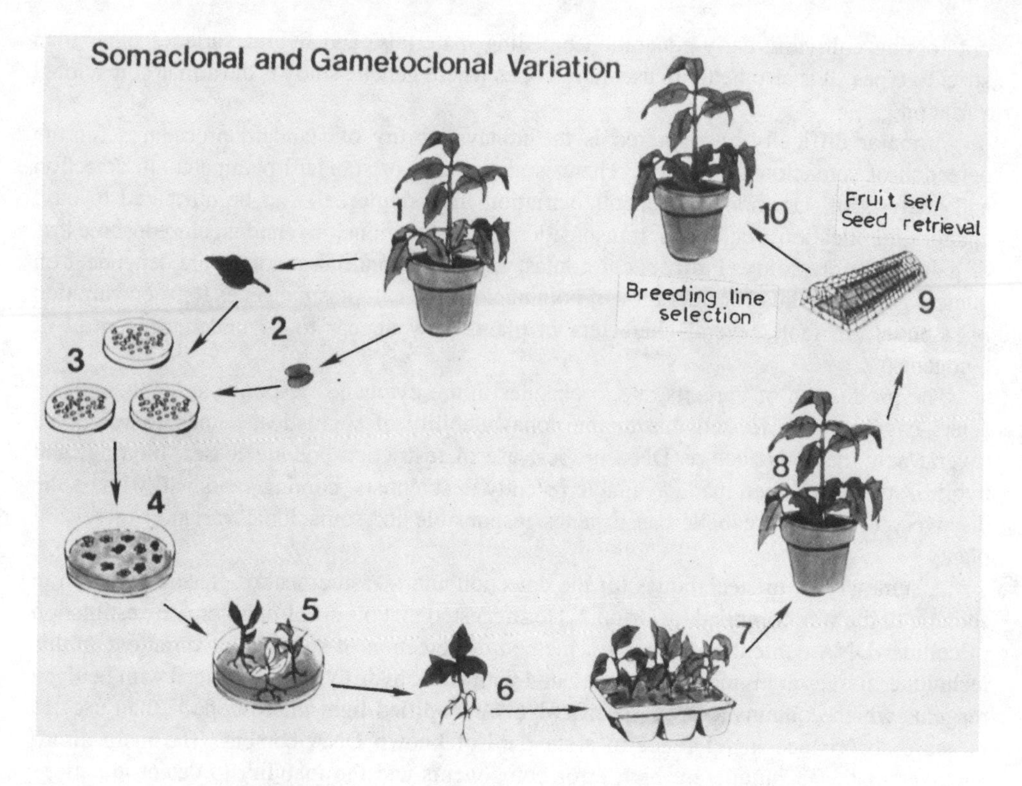

FIGURE 1. Scheme for production of somaclonal variation in tomato. (1) Donor plants are identified, (2) explant removed and (3) transferred to cell culture, (4) callus formation is followed by (5) shoot regeneration, (6) rooted plantlets are (7) transferred to the greenhouse where they are (8) raised to maturity, (9) fruit and self-fertilized seed are collected in the greenhouse, and (10) R_1 progeny are then evaluated in the greenhouse or field. (From Evans, D. A., Sharp, W. R., and Medina-Filho, H. P., *Am. J. Bot.*, 71, 759, 1984. With permission.)

color (Figure 2). In addition, preliminary data on agronomic characters were collected in the R_1 field evaluation. Variation in chromosome number, particularly tetraploidy $2n = 48$, was detected (Figure 3). Sterile aneuploid lines were also detected. However, an emphasis was placed on the analysis of R_1 progeny of regenerated plants, and most aneuploid lines were discarded as little or no R_1 seed was collected. Several R_1 progenies were observed to segregate for morphological characters. Among the first 230 regenerated UC82B plants, 13 plants with discrete nuclear mutation were identified. These included recessive mutations for male sterility, jointless pedicel, tangerine virescent leaf, flower, and fruit color, lethal chlorophyll deficiency, virescene and mottled leaf appearance, and dominant mutation controlling fruit ripening and growth habit. Genetic analysis was completed by evaluating self-fertilized R_2 progeny of selected R_1 plants and, in some cases, by crossing with known mutant lines. This systemic approach also led to the completing of extensive genetic analysis with the tangerine-virescent fruit variant.

It has now been realized that chromosome number alone is too crude to permit the use of this technique as a tool to study somaclonal variation, and many other good, reliable techniques are available now. For instance, Murata and Orton[46] conducted extensive karyological studies on cultured celery cells, and estimated that approximately 40% were diploid ($2n = 22$) based on raw chromosome number, the remainder being hypoaneuploid. However, comparisons of presumed diploid cells with those that have been established as normal revealed that they all had distinct chromosome structural changes. Karyotype analysis, even at high resolution, cannot predict small point mutations which appear to be very common in the production of somaclones from tissue culture. Thus, the precise nature and source of striking structural changes based on karyological observations alone are not enough.

FIGURE 2. View of field trial of R_1 tomatoes recovered from somaclonal variation program. (From Evans, D. A., Sharp, W. R., and Medina-Filho, H. P., *Am. J. Bot.*, 71, 759, 1984. With permission.)

FIGURE 3. Tetraploid tomato seedling (2n = 48) developed by somaclonal variation. (From Evans, D. A., Sharp, W. R., and Medina-Filho, H. P., *Am. J. Bot.*, 71, 759, 1984. With permission.)

The technique of differential chromatin staining is also of general utility in the visualization of specific karyotypic changes, although this technique of differential staining of chromatin has long been an important tool in diagnostic cytogenetics of mammals. Unfortunately, plants have proven to be less suitable in revealing striking patterns of chromosomal changes.[47] Nonetheless, Ashmore and Gould[48] were successful in using the technique of Giemsa banding to pinpoint specific structural changes in altered karyotypes of cultured crown gall tissues of *Crepis capillaris*.

A more precise method to detect somaclonal variation with certainty in a precise manner is the utilization of physiological and biochemical parameters of known genotype which are expressed in plant tissue. Starting with tissue of known genetic constitution at a given locus or set of loci, changes in the expression of the loci can be used as tools to estimate somaclonal variation at the level of the structural gene. Orton[42] suggested that diploid organisms which are heterozygous at a defined locus can be used to visualize all three possible phenotypes in cultured tissues. Large quantities of the cells to identify phenotypes through electrophoretic protein can be raised starting from a single cell. Qualitative changes in phenotypic expression, such as loss of electromorph activity or altered mobilities, can be seen directly and are probably indicative of underlying mutations.

In hexaploid wheat there are three loci controlling the synthesis of constitutive alcohol dehydrogenase (Adh) in seed endosperm. They are located on chromosome 4A, 4B, and 4D, respectively; named Adh-A1, Adh-B1 and Adh-D1; and the monomer units are known as α, β, and δ, respectively.[49] The three Adh subunits dimerize at random to produce six functionally active molecules. Under electrophoresis the dimers migrate to one of the three positions. Wheat endosperm tissue is triploid, and the relative intensity of the active dimers is 1:4:4 for $\alpha\alpha$, $\alpha\beta + \alpha\delta$, and $\beta\beta + \delta\delta + \beta\delta$, respectively, reflecting random subunit dimerization. The relative intensities of three electrophoretic classes can be deduced for homozygous and heterozygous null mutants at each of the Adh-A1, Adh-D1 loci. The relative intensities of the zymogram bands provide the basis for screening SC_2 seed of somaclones regenerated from wheat.[3] Compared to other plant characters, isozymes are quite rapidly screened, are relatively free from environmental variation, and are simply inherited when combined with the availability of good cytogenetic stocks such as wheat. They can become powerful tools in the analysis of genetic variation. The Adh isozymes of wheat were thus used in a study of somaclonal variation by Davies et al.,[50] enabling a detailed description of several translocations to be made. Brettell et al.[51] also used Adh isozymes to isolate somaclonal variants of maize. One variant with altered isozyme mobility was analyzed at the molecular level and found to differ from the parent by a single nucleotide substitution. The utilization of β-amylase isozyme to detect somaclonal variation is also becoming popular.[52] The β-amylase isozyme represents the products of at least four loci on four chromosomes in wheat and it resolves into many bands when separated by isoelectric focusing.

However, this proposal suffers from a lack of conclusiveness with respect to underlying cause or causes. For example, in *Phaseolus vulgaris* changes in isozyme patterns were observed, but it was difficult to distinguish the relative genetic and developmental (i.e., nongenetic) components of variation.[53] This was a consequence of the lack of understanding of the genetic control of the isozyme phenotypes in the organism and the fact that they were developmentally unstable. Orton[54] used the presence or absence of isoenzyme pattern of bands of one constituent parent or another on gel electrophoresis (in plants regenerated from cultures of an interspecific *Hordeum vulgare*) in order to detect the somaclonal variation. Since both cultures and corresponding regenerated plants were karyologically unstable, variation in isoenzyme patterns was speculated to have been a manifestation of aneuploidy.

Changes of allele expression or a complete loss of allele expression in cultured heterozygous tissues could conceivably occur as a consequence of a broad range of both genetic and nongenetic factors. Among the possible genetic causes of allele inactivation are point mutation, insertion of a transposable element, somatic recombination, nondisjunction, deletion, and position effects resulting from translocation or inversion. However, the involvement of epigenetic changes alone seems highly unlikely, since complete, selective, allele inactivation at a given locus has not been demonstrated in plants. However, modulation of Adh allele expression as a function of tissue has been observed in corn.[55] Epigenetic events would not be expected to be stable across developmental transitions in culture, or especially among sexual progenies of regenerates.

Different classes of mutations are expected to give rise to stable, sexually heritable variation with the possible exception of transposable elements.[42] It is also possible to distinguish through heritable variation among different classes of mutations. In point mutations, cells should be normal both karyologically and cytogenetically and corresponding regenerated plants should transmit the resulting null allele in a Mendelian fashion. Loss of allele expression due to nondisjunction associated with monosomy, and trisomy, should be observed among at least certain of the presumptive heterozygous clones. Although deletions and inversions may not be directly discernible, cytological abnormalities and segregation distortions should reveal their existence in regenerates and corresponding progenies.

Genetic changes resembling single gene mutations have now been detected in numerous crops. For example, recessive single-gene mutations are suspected if the variant does not appear in the R_0 plant and if self-fertilized R_1 progeny segregate in an expected 3:1 Mendelian ratio for a morphological trait.[56] To confirm that the trait of interest is indeed a single-gene trait, progeny tests should be completed on selected single R_1 plants to identify segregators and nonsegregators and to ensure continued 3:1 transmission in the R_2 segregating population. When possible, complementation tests should be completed with known mutants for the crop of interest. Genetic analysis of this type has been completed for several tomato somaclones by Evans and Sharp.[56] The information from this analysis was used to map somaclones to specific loci (Figure 4). There is a 3:1 ratio in R_1 plants; the first step in this type of analysis has been reported for maize,[35] *Nicotiana sylvestris,*[57] rice,[58,59] and wheat.[36]

Selective modulation of allele activity by the transposable element Ds at the Adh locus has been reported by Osterman and Swartz.[60] It has also been shown that the insertion occurs at the promoter region of Adh proximal to the actual coding sequence and that reversion to normal occurs concomitantly with the excision of Ds.[61] However, it is not possible to distinguish changes due to transposable elements from point mutation due to insufficient knowledge of the behavior of transposable elements. Recently, Peschke et al.[62] have been able to detect such elements with certainty.

Orton[63] observed loss of allele expression in cultured celery *(Apium graveolens)* tissues, heterozygous for two linked loci, Pgm-2 and Sdh-1. Specifically, cells progressively exhibited loss of the PGM-2^F band over a period of approximately 30 months in culture, while heterozygosity was retained at Sdh-1. The resulting homo/hemizygous phenotype was highly stable and transmitted to regenerates. However, no inheritance studies could be carried out due to abnormal development. High mutation frequencies and the absence of reciprocal clones were taken as strong indications that point mutation and somatic recombination were not involved. Moreover, Pgm-2 and Sdh-1 are linked and no altered Sdh-1 phenotypes were observed concomitantly with loss of PGm-2^F. Thus, it was concluded that aneuploidy alone was responsible for the loss of allele in cultured celery cells.

As mentioned earlier, identification and characterization of somaclonal variants have become easier and more precise with the development and progress in new techniques. For example, restriction enzymes have already been used to demonstrate the occurrence of spontaneous mutational events involving mitochondrial (mt) DNA in cultured plant tissues.[64] Boeshore et al.[65,66] have obtained direct evidence of hybrid composition of the mitochondrial genomes in clonal tissue derived from fused protoplasts from *Petunia* species. The chloroplast genome seems to be relatively more stable than mt DNA in cultured tissues, but further

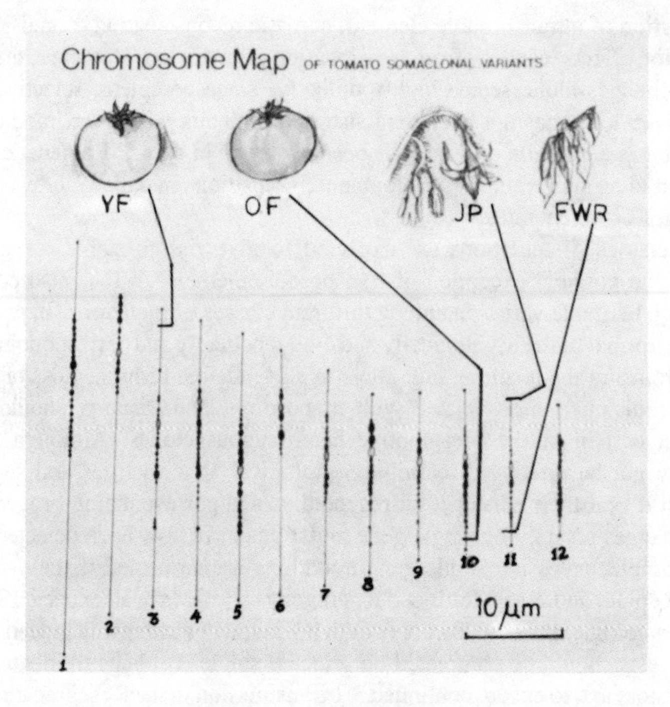

FIGURE 4. Chromosomal map of tomato somaclonal variants (conventional genetic complementation tests were used to map four somaclonal variants to specific sites on the tomato genome). The yellow fruit (YF), orange fruit (OF), and jointless pedicel (JP) are recessive mutations, while the *Fusarium* race 2 wilt resistance (FWR) is a dominant mutation. (From Evans, D. A. and Sharp, W. R., *Biotechnology,* 4, 528, 1986.With permission.)

experimentation is necessary to determine whether chloroplast genomes recombine in somatic tissue or not.

Molecular approaches are providing unprecedented insights into the sources and phenomenology of somaclonal variation with respect to genes encoded in the nucleus.[67] Although the nuclear genome is organized and maintained in a much different and more complicated manner than those of the cytoplasmic organelles, rapid molecular cloning techniques that utilized improved vectors have made it possible to obtain comprehensive stable genomic libraries. Specific constituents of such libraries will be useful as direct genetic or chromosome structural probes in experimental studies. Several cloned probes that reveal restriction site polymorphisms have been described.[67] Molecular polymorphisms can be subjected to standard inheritance and linkage tests. Armed with a set of cloned fragments of known arrangement, one could initiate test cultures and conduct powerful experimental studies on somaclonal variation. Application of various other techniques coupled with the latest technique available in molecular biology can further enhance resolution of chromosome structural mutations and aneuploidy. For example, fluorescence-activated sorting capability is presently being applied to the purification of metaphase chromosome fractions based on DNA content. This permits the direct mapping of library probes to linkage groups, and, potentially, the linear order as well.[68]

Radiolabeled cloned probes of highly repeated ribosomal RNA genes of rye have been successfully hybridized directly to metaphase chromosomes immobilized on glass slides and visualized directly by autoradiography.[42] Thus, there is a theoretical possibility of utilizing directly the library probes which could be used directly as markers for genome architecture via *in situ* hybridization. However, for this to become practicable, it is necessary to develop techniques to permit visual perception of single-copy hybridization events.

Other limited approaches that have been utilized to detect and measure heritable, qualitative changes in nuclear DNA of cloned somatic tissues include CsCl gradient separation, thermal denaturation curves and reassociation kinetics, and resistance to S_1 nuclease digestion.[69] Another approach to the study of somaclonal variation is the analysis of somaclonal variation in regenerates which is a direct reflection of the spontaneous variation in the cell cultures. Since inheritance studies are necessary, the ideal starting materials for such experiments are inbred lines and, presumably, homogeneous explants. The main advantage to this approach is that direct morphological phenotypes are restored as markers for the detection and measurement of variation, thus obviating the need for the tedious, expensive laboratory protocols described above. In addition, direct cytogenetic analysis can be used to detect and quantify certain classes of chromosomal variability,[70] and variation is often observed within as well as among regenerating units.[8,54,71-74]

Further complications may result from viability or fertility losses encountered as a consequence of somaclonal variation. Hence, using morphological or physiological variation in regenerates as a direct measure of somaclonal variation in cultured tissues is extremely tenuous. It should be pointed out that regeneration is necessary to carry out inheritance studies to verify bases of certain forms of variation encountered in somaclonal variants. Recent analysis of *Nor* loci for somaclonal variation in wheat is a good example of utilization of modern techniques to correlate somaclonal changes with the changes at the gene level. Breiman et al.[75] analyzed *Nor* loci (genes for 18S, 5.8S, 26S) among wheat cultivars. For this evaluation they used a cloned probe (pTA71) to assay the spacer length of these loci in sexual progenies of plants regenerated from scutellum of immature wheat embryos. They used restriction endonuclease digestion, blotting, and hybridization to detect changes in the *Nor* loci.

IV. SURVEY OF MAJOR CROPS FOR SOMACLONAL VARIATION

A. POTATO

Few world crops rival the potato, *Solanum tuberosum,* as a target for improvement via somaclonal variation. First, the crop is normally propagated and established vegetatively from tuber pieces. Second, some of the best cultivars presently in use are sexually impotent, and breeding of such lines proceeds entirely by clonal selection. Where classical breeding approaches are used, the enormous levels of heterozygosity encountered in inbreeding depression make the selection of superior lines a difficult task as compared to other crops. The potato is also considered to be autotetraploid, which further complicates inheritance studies and breeding.

The first large group of plants from potato protoplasts was produced and investigated by Shepard and Totten.[76] They cultured protoplasts from cultivar (cv.) Russet Burbank and were confronted with an outburst of variation in the regenerants. Subsequent experiments by Shepard and associates confirmed these observations.[24,33,77-80] Trials with potato cv. Maris Bard,[81,82] cvs. Maris Bard and Fortyfold,[83] cvs. Desiree, Maris Bard, Fortyfold, and Majestic,[84] cv Bintje,[85-89] cvs. Maris Piper, Foxton, and Feltwell,[90] cv Record,[91] and cvs. Desiree and Pentland[92] showed similar results. Further studies on potato cultivars and their dihaploids were published by Tiemann et al.[93]

Extensive variations among plants regenerated from cultured protoplast of cv. Russet Burbank have been described by Secor and Shepard.[78] In their initial somaclone experiments, some 1700 plants were evaluated for overall morphological variations. In two subsequent growing seasons, this population was reduced to 65 clones possessing acceptable vigor and fine tuber characteristics. The clones were then evaluated under replicated field conditions for 35 characters including preplant sprouting, vine quality, maturity, morphology, and yield. Secor and Shepard[78] concluded that cell culture-generated variation provides enough variability for selective improvement of Russet Burbank. Variation among potato somaclones is not, however, a consequence of the protoplast technique. That plants derived from a single callus vary indicates that variation arose during callus growth.[27,82,94] The recovery of variation in potato without protoplast culture is important, because plant regeneration from leaf or other tissue explant is far simpler than that from protoplast culture.

The potato somaclones were also screened for resistance to both late and early blight, to which the parent Russet Burbank is highly susceptible. More than 800 plants tested varied widely in their reaction to late blight *(Phytophthora infestans)*.[33] About 2% of the somaclones displayed enhanced resistance, which was transmitted through subsequent tuber generation. The form of resistance recovered in the somaclones appeared different from the common *R* gene type derived from *S. demissum*. Similarly, several disease-resistant somaclones were recovered from 500 plants screened for resistance to early blight *(Alternaria solani)*. On the basis of toxic assay five plants were initially identified as resistant. Four of them subsequently displayed resistance in the field. Somaclones were also raised by Nelson et al.[95] from *S. brevidens* and by Barsby et al.[96] from *S. brevidens, S. tuberosum,* and *S. fernandezianum*. Jones et al.[84] obtained with this method colonies of cells from 35% of the protoplasts of cvs. Maris Bard and Desiree and 50 to 80% regeneration from these calli. The variations comprised grossly aberrant plants; more or less slight morphological deviations in type of habit, leaf shape, leaf and flower color, and tuber traits (such as shape, texture of the surface, flesh color, uniformity of the tuber nest, and others); physiological variation such as time of flowering and length of variation period; and resistance to *Globodera pallida, A. solani,* and *P. infestans* in the leaves. The number of different deviating morphological traits in a single protoclone varied between 1 to 16.

Early results showed that variation existed among callus clones derived from protoplasts and could be transmitted into corresponding regenerated plants. Large-scale field studies were initiated for the purpose of determining the types and degrees of variation among regenerated potato to select and evaluate any of anticipated economic significance.[33,78,97] Plants regenerated from mesophyll protoplasts of potato showed some variability in chromosomal rearrangements.[76] Lines have been identified that are resistant to diseases (late blight and early blight) to which the parent line Russet Burbank is susceptible.[80]

Variants with resistance to fungal parasites are of special interest.[98] For tests with *A. solani,* Matern et al.[24] isolated two phytotoxins similar to lipids from the culture filtrates. Mixed together and transferred with a needle onto the leaves, they caused typical concentric rings. Mackay[92] reported interesting somaclonal variation in cyst formation and with eelworn phenotype pa-3 in potato. Rehbein[99] treated potato callus pieces grown from a cell cluster of dihaploids with culture filterates of *Erwinia carotovora* var. *carotovara*. The growth of the calli was delayed, but after some weeks some cells from other calli, having withstood the toxin, started a new growth.

Thomas et al.[82] conducted two potato experiments, the first with cv. Maris Bard in which they found considerable variation in ten morphological traits in 20 out of 23 protoplast-derived plants. The traits included the appearance of anthocyanin on leaves (parent has none), prostrate growth, reduced terminal leaf overlapping, increased terminal and lateral leaf fusion, and increased green color of leaves. The experiment with cv. Majestic involved the production of 17 shoots from *in situ* stem callus. These were excised and rooted. No striking variations were observed for the same ten morphological traits examined in the first experiment. Again, because of differences in cultivar and culture conditions, it was not possible to directly compare the two experiments. The absence of variation in the second experiment suggests that the variation phenomenon may be controllable. Thomas et al.[82] also confirmed that the variation observed by van Harten et al.[26] in leaf, rachis, and petiole callus regenerants was not a consequence of protoplast fusion.

Bright et al.[28] reported that morphological variants were recovered from a number of potato cultivars following culture of leaf rachis and petiole pieces. They also compared the frequency of tuber color variants following culture of cv. Desiree from tuber, leaf, stem, and petiole explant. This is in contrast to the results of van Harten et al.[26] who used the same cultivar and who found leaf culture to give the highest frequency of tuber color variants. The observation by Wenzel et al.[100] that plants regenerated from protoplasts of dihaploids failed to exhibit striking variation is circumstantial evidence that mutations may occur and accumulate over time in clones propagated from tuber.[37,101,102] However, variation was observed as the age of these callus cultures was increased. Thus, the culture and regeneration of potato cells and tissues offer an extremely powerful breeding tool, and the economic impact of the technique in the foreseeable future is likely to be significant. The long-term impact of recent efforts to develop sexual breeding techniques which integrate haploidization, diploidization, selection for combining ability, and release and sale of cultured hybrid seed or propagates generated from fused dihaploid protoplasts is likely to be even greater.

B. TOBACCO

Tobacco (and various solanaceous relatives) remains as the undisputed model system of choice in plant somatic cell manipulation. It is also one of the major species in which high levels of somaclonal variation have been reported.[1,103] Somaclonal variation has been observed among tobacco plants regenerated from a broad range of callus tissues (microspores, pith, leaf mesophyll). Characters for which variant phenotypes have been observed include flowering date, leaf shape, plant height, total vegetative yield, alkaloid content, sugar content, chlorophyll content, Co_2 absorption, disease resistance, and male sterility.[57,104-109]

Barbier and Dulieu[29] noted simple and quantitative genetic variation from tobacco cotyledon and leaf protoplast cultures. Their results suggest that some of the variations either preexisted in the cotyledon cells or occurred early in culture. Some of the variations increased with the duration of culture. The same authors have extended their study in a series of well-conceived experiments which are exemplary in exploiting the protoplasts for somaclonal variation.[110] The donor plant was a yellowish-green heterozygote at two loci affecting chlorophyll differentiation (a_1^+/a_1, yg^+/yg) as used in the earlier study. In one experiment 1048 colonies from individual leaf protoplasts were grown and then subdivided into 4 subcolonies which were regenerated into one shoot each. Some (13.1% of the 1048) showed sectoring among the subcolonies for variant green shoots and some (3.8% of the 1048) showed sectoring for variant yellow shoots. In an analogous way,[85] plants from cotyledon protoplasts were regenerated. Some regenerants (18.8%) showed green sectoring and some (2.3%) showed yellow sectoring among the subcolonies. In total there were 199 variants producing colonies: 128 showed one variant, 1 showed four variant to nonvariant shoots. Barbier and Dulieu[110] suggested that the high proportion with two or more variants indicates an appearance or sorting out of the genetic changes during the first or the second divisions of protoplasts. In another experiment, 25 cotyledon protoplasts were cultured and subcloned into four sectors after each of the three successive culture periods. The pedigree was maintained and one shoot regenerated from each of the 1600 final subcolonies. Each protoplast was thus represented by 64 shoots and the experimental design allowed a fluctuating analysis. Remarkably, all of the variants observed (2/25 sectored green, 2/25 sectored yellow) had appeared in the first period of culture. There was no new variation in the later two periods of the culture. Thus, Barbier and Dulieu[110] favored the notion that single-strand lesions which accumulate in somatic cells of the cotyledon or leaf only get the opportunity for expression after one or two divisions of the isolated cells.

By contrast, Lorz and Scowcroft[111] found possible gene conversion-type events to increase with culture duration in experiments with the aurea mutant sulfur, of tobacco. A total of 2222 colonies were cultured from individual leaf protoplasts of heterozygous (Su/

su) donor. A number of shoots were regenerated from each of the 2156 morphogenic colonies. Variants at the sulfur locus were scored as green or yellow shoots. Twenty colonies gave rise only to variant shoots (14 gave all green shoots, 6 gave all yellow shoots) and these were conservatively assumed to represent protoplasts which were already su/su, Su/Su in the leaf. Such preexisting variation is well known in these plants and can appear as yellow or green spots or yellow/green twin spots on the leaves. In addition to this preexisting variation, Lorz and Scowcroft found 59 colonies which gave rise to both variant and non-variant plants. The variation in these genetically heterozygous colonies must have arisen during the culture period. Further, by doubling the duration of callus culture the frequency of heterogeneous colonies increased from 1.4 to 6% on the standard medium and from 2.3 to 3.6% on a more stressful medium giving slower growth.

Tobacco plants have been produced with resistance to herbicides. Mutagenized tobacco cultures were selected for tolerance to picrolam by Chaleff and Parsons.[106] These authors also attempted to regenerate plants from seven surviving colonies. Two of the seven seemed unstable. All the five remaining mutant cultures plants with stable, heritable, picrolam resistance characters. Three were dominant and two semidominant. Two of the dominant genes were linked and the others were unlinked. They also estimated a frequency of 2.7×10^{-6} per selectable unit which appears an admirable microbial genetic number. Larkin,[3] however, pointed out that cautions were required in the interpretation of such frequencies because they were extremely difficult to derive accurately with plant cell cultures. One reason for the difficulty is the density-dependent efficiency of planting. However, the Chaleff group, to their surprise, found three of the five picrolam resistance (PmR) mutants which were also resistant to hydroxylurea (HuR) and the resistance was due to a single dominant nuclear gene which segregated independently of PmR in two of the three cases. This recovery of HuR without selection and without linkage to PmR is remarkable. The HuR and PmR mutants were not cross-selectable.

Genetic interest is also coming to bear on culture-induced variation in tobacco where breeders are beginning to use anther culture to speed up the production of pure breeding lines. Considerable agronomic and morphological variations have been reported from anther culture among double microspore-derived lines. This work sparked a controversy as to whether residual heterozygosity in high inbred donor lines could explain the variations. Arcia et al.[15] argue against residual heterozygosity in the parental material of tobacco. For some heterozygosity to have been maintained through so many selfing generations would require heterozygotes to exhibit significant advantage in the seedling stage prior to transplantation. Such an advantage was not observed by Collins and Legg,[16] who remained unconvinced about the nonexistence of residual heterozygosity whose elimination in the dihaploids resulted in the variability. DePaepe et al.[17] also appeared to discount the explanation based on residual heterozygosity. They found extensive variability amongst dihaploid plants obtained from binucleate microspores. A second cycle of androgenesis from the dihaploids generated even further variation. A line developed from an androgenic tobacco plant containing a new mutation conferring tobacco horn worm resistance has been released.[112] *In vitro* techniques are now routinely used in direct association with established tobacco breeding programs, and it seems reasonable to speculate that somaclonal variation will progressively gain importance as a source of useful mutations in adapted, commercial backgrounds.

C. ALFALFA

Due to self-incompatibility, inbreeding depression, and polyploidy, it is difficult to improve alfalfa through conventional breeding techniques. However, recovery of desirable traits through somaclonal variation may present an alternative to improvement via recurrent backcrossing or hybridization and recurrent mass selection. An important feature of alfalfa is that plants capable of regeneration from tissue cultures occur in essentially all cultivars and experimental populations that have been extensively tested. The frequency of regenerable genotypes in most materials is less than 10%; however, strains with a high frequency of regenerable plants have been bred[113,114] and identified by screening cultivars.[115,116]

Regeneration in alfalfa has been interpreted as occurring via organogenesis and embryogenesis based on the morphology of regenerating structures. However, one line of recent evidence suggests that regeneration is predominantly via embryogenesis. Schenk and Hildebrandt[117] found that somatic embryos of alfalfa contain 11S seed storage protein. The presence of storage protein is interpreted as evidence of embryogenesis and was used to distinguish adventitious shoots from somatic embryos.[118] Plants can also be generated from callus cultures initiated from any tissue or organ, from suspension-cultured cells, and from protoplasts of leaf mesophyll cells. This makes alfalfa suitable for plant tissue culture research.

It is well established that alfalfa cultures contain cells with the somatic chromosome number of the donor and cells with spontaneous doubled chromosome numbers;[114] thus, cultures are heterogeneous for ploidy levels. Consequently, plants are routinely regenerated with either the donor or the spontaneously doubled chromosome number. In addition, cultures in some experiments are known to be genetically heterogeneous due to somaclonal variation, induced mutations, and unstable allele. In spite of this heterogeneity at the cell level, regenerated plants are rarely chimeral and the vast majority have one ploidy level. Therefore, most alfalfa plants regenerated from culture appear to be derived from single cells.

Cultivated alfalfa behaves as an autotetraploid with random bivalent pairing of the four homologous chromosomes and tetrasomic genetic ratios.[119] Haploids ($2n = 2x = 16$) of the cultivated tetraploid alfalfa ($2n = 4x = 32$) are prized for research, including tissue culture studies, because they have the diploid chromosome number, are potentially fertile, and have disomic genetic ratios. Several alfalfa cultivars have been regenerated with attempts to produce the desired haploids by anther culture of cultivated tetraploid alfalfa.[120,121] Ironically, the first plants regenerated from alfalfa anther cultures were not haploids; they had the somatic chromosome number of the donor. In these plants callus was derived from somatic connective tissue of the anther. Nonetheless, haploids were obtained by parthenogenesis associated with 4x-2x crosses.[122] These maternally derived haploids were used to develop seed-reproduced populations of cultivated alfalfa at the diploid level (CADL),[123] which in turn have been a source of somaclonal variants in tissue culture.

Herbage yield in alfalfa is considered a quantitative genetic trait, i.e., it is controlled by many genes, each of which has a small effect on yield. It is also well known that single qualitative genes such as those for dwarfism can override the quantitative system. A field study of herbage yield of 32 diploid and 16 tetraploid regenerates of diploid alfalfa (HG_2) revealed variants for herbage yield.[124] The most outstanding variant for herbage yield (NSI) in alfalfa was regenerated from an unmutagenized cell line.[125] Variant NSI yielded threefold more dry matter (in a clonal evaluation for spaced plants) than 4x HG2 and there was a significant increase in herbage yield. Clonal NSI was tetraploid and its superiority over 4x HG2 was confirmed in a follow-up study to compare the best somaclonal derivative of HG2 with the best sexual derivatives of HG2.[126] The HG2 control for these studies was a typical spontaneously doubled regenerate of HG2 that was selected as a 4x HG2 control soon after HG2 was developed. This 4x HG2 control is considered a valid 4x control clone, because it is indistinguishable from 4x HG2 obtained by colchicine doubling. Autotetraploid 4x HG2 was comparatively self-sterile, but 25 self-progeny were obtained by selfing several thousand flowers. Thus, a sexual population was initiated from 4x HG2. The self-progeny were sib-mated, and selection for thrifty fertile plants was continued for three additional generations. This resulted in a sexual population of plants termed HAG (advanced generation from HG2). Three phenotypically superior (HAG4) plants were selected as the best sexual derivatives of HG2 and cloned by vegetative cuttings. Clone NSI was the superior somaclonal selection from an array of 100 regenerates of HG2 and was also cloned by vegetative cuttings for field studies with HAG clones. In the clonal herbage yield test NSI yielded 2.5 times more herbage than 4x HG2, which agreed with the previous study of Reisch et al.[124] However, the best HAG sexual derivative yielded twice as much as NSI. Hence, the herbage yield ranking was HAG4 $>$ NSI $>$ HG2, and the differences were significant. The ranking was the same for seed production based on hand pollinations in the greenhouse and open pollination in the field. Both HAG4 and NSI were superior to the original parent and donor, respectively, and the superiority was genetically heritable.

The genetic mechanism for the alteration during tissue culture of 2x HG2 and NS1 is not known; however, quantitative traits were altered and the change was probably more complex than a single-gene mutation.[125] Perhaps one or more chromosome substitutions occurred during the polyploidization step similar to the mechanism hypothesized for similar changes in sorghum that involved polyploidization and somatic reduction. Somatic reduction and great variation in nuclear size have been reported in octoploid alfalfa.[127]

The best somaclonal variant and the best sexual derivative of alfalfa were selected from large arrays of plants raised through tissue culture in order to provide at least a preliminary evaluation of the potential of the different methods for improving plants for quantitative traits.[125] About 2 years were required to produce and identify the best somaclonal variant and to complete four to five sexual generations. In the herbage yield test of diploid variants of HG2, 22 variants ranked above HG2 and 11 ranked below.[124] However, only variant R85 yielded significantly more dry matter than HG2. The enhanced yield of R85 probably was due to an induced dominant mutation for improved lateral branching and leafage and was not associated with a quantitative change. Actual herbage yield of R85 was about twofold that of HG2. The R85 dominant mutation has been transferred to the 4x cultivated level by both colchicine doubling and 2n gametes. This mutation has potential value as a genetic marker and for the improvement of alfalfa. However, enhanced leafage is associated with sparse flowering and limited seed production. This negative association between leafage and seed yield may limit the usefulness of R85 mutation as an agronomic trait.

Somaclonal variants in alfalfa were noted from the plants regenerated from callus.[120] Several alfalfa variants have also been found among plants regenerated from cells or protoplasts. Reisch et al.[124] conducted a phenotypic analysis of plants regenerated from alfalfa callus cultures with and without exposure to the mutagen ethyl methane sulfonate (EMS) and the selective agent ethionine (a methionine analog which inhibits feedback-controlled methionine biosynthesis). A number of morphological characters, including some affecting plant yield, were obtained. Curiously, the frequency of variants was clearly enhanced by ethionine selection, an observation which prompted the authors to suggest that ethionine was mutagenic. Alternatively, the compound could be acting as a selective agent for variant cells, in general, as suggested by Larkin and Scowcroft.[103,128] However, a number of variants were obtained including the highest yielding clone, in the absence of EMS and ethionine. Additional reports on inheritance have revealed that most of these (including a leafy/branched mutant which was speculated to enhance effective forage yield) have a simple genetic basis.[25]

D. WHEAT

Wheat is the largest of the cereal crops in terms of both hectares and production and a staple food for over one third of the world population. Wheat contributes more calories and protein to the world diet than any other food crop. It was difficult to use tissue culture for the improvement of wheat and extensive work has been taken up only recently.

It is argued by the breeders of many sexually propagated crops that sufficient genetic variation exists in present germplasm collections to permit significant breeding gains well into the future. What is needed are techniques by which existing variation can be transmitted or selected more efficiently. Enormous stores of wheat germplasm are maintained worldwide.

Why turn to the generation of presumably random mutations generated in culture as an alternative to already preexisting variation? First, there is an immediate advantage in the addition of single, beneficial mutations to existing superior varieties, thus bypassing the need for lengthy backcrossing of existing mutations. Second, it appears that at least part of the spectrum of stable variation produced is inherited in a complex fashion. Further beneficial true-breeding variants for characters such as yield, uniformity, and quality can be obtained in superior varieties directly. Finally, there exists a distinct possibility that beneficial mutations that do not exist in the world germplasm collection will surface. Until basic research can illuminate the distinctions between somaclonal variants and those generated in mutation breeding approaches, and perhaps determine ways in which its manifestations can be controlled, it would appear worthwhile to conduct empirical experiments on somaclonal variation for crop improvement in species with a well-established effective breeding technique such as whcat.

Plants now can be regenerated from tissue culture of wheat with reasonable efficiency.[129] Many cultivars of wheat can be tissue cultured and plants can be regenerated from callus derived from immature embryos.[130] Although more than 200 plants have been recovered from a single embryo, 20 plants from each immature embryo was the average. These initial (SC_1) plants displayed some phenotypic variations, but their progeny revealed the true extent of the genetic changes induced during culture.[36] Progeny of 142 somaclones of the Mexican double dwarf accession Yaqui 50E through two generations segregated for height, maturity, tiller number, presence or absence of awns, glume color, grain color, clavate head shape, and leaf waxiness. The frequency of morphological variants which appears in the analysis of somaclonal progeny of wheat has been found to differ among cultivars.[3] The cv. Yaqui 50E exhibited a greater frequency than that of Millewa and Warigol.

Albino wheat plants regenerated from anther culture had large deletions in the chloroplast genome.[131,132] Gosch-Wackerle et al.[133] and Shimada and Yamada[134] found independently that in bread wheat (*Triticum aestivum)* plant regeneration was obtained most efficiently from scutellar calli derived from immature embryos. This method was subsequently adopted as the standard procedure to regenerate wheat from tissue culture. Different degrees of variation in morphological features, yield components, and biochemical traits (grain-protein, composition, and analysis) were found among wheat regenerants and their progeny.[36,135-137] Recently, Breiman et al.[75] analyzed the progenies of plants regenerated from scutellar callus of bread wheat for changes in rRNA genes located at the sites of the nucleolar organizer region (*Nor* loci). Cvs. Chinese Spring and Miriam were found to be stable in their organization of the *Nor* loci, whereas three progeny plants of ND 7532 showed reduction in the number of rDNA spacers. Individual double haploids from wheat anther culture varied in yield, floret fertility, disease resistance, seed weight, and seed proteins.[138]

Eapan and Rao[139] cultured two genotypes of *T. turgidum* subspecies *dicoccum* and two genotypes of the subspecies *durum*. Among regenerants of the responding *dicoccum* line, there were chlorophyll variants from 50% of the culture. The plants, however, did not reach flowering stage. In a large experiment, Maddock et al.[140] regenerated over a thousand plants from a number of cultivars. Significant differences were recorded between somaclones of one cultivar in plant height, time of flowering, head morphology, and fertility.

Khuspe et al.[141] found somaclone lines from wheat cv. N1917 which displayed considerable variation in the field over six generations. Height, maturation period, number and size of grains, and total yield were variable. The variation was heritable and true-breeding variant lines could be selected in subsequent generations. Replicated field trials showed some of the lines with 15% higher grain yields than controls. Larkin and associates have observed true-breeding variants in wheat primary regenerants.[36,142,143] Some were fully awned from tip awned and some were white grained from red grained.

Gliadins are a class of grain storage protein in wheat coded by a multigene family at a number of macroloci on the group 1 and 6 chromosome. Larkin et al.[36] described heritable variation in the electrophoretic patterns of gliadins in wheat somaclones. Differences between regenerants of the same cultured embryo assure that the variation was from culture and not an outcrossed embryo. No gliadin changes were found in control grain from the donor plants. Cooper et al.[137] and Maddock et al.[136] have also confirmed that wheat tissue culture can generate heritable alterations in the expression of gliadins.

E. RICE

Rice is a diploid monocot of tremendous world importance. It is grown in more than 100 countries and on every continent of the world (except Antarctica) and produces more calories and carbohydrates per hectare than any other cereal. This grain, second only in economic importance to wheat, is the staple food for one half to two thirds of the world population.

It is relatively easy to regenerate rice plant from cell cultures as compared to other members of the family Graminae. Extensive regeneration through callus culture has been recently achieved at the TCCP laboratory[129] Oono[19-21] reported dramatic variations in rice somaclones from a true breeding dihaploid derivative of cv. Norin 1O. Heritable changes affecting chlorophyll development, plant height, heading date, maturity, and grain yield were observed in the progeny of more than 1000 somaclones examined over three generations. Many of the variants bred true within two generations of selfing. Progeny of about 70% of the somaclones differed from the parent at least in one heritable mutation.

In a series of experiments beginning in 1970, regenerant progeny lines were examined from rice anther and embryo cultures by Oono.[18-21] Many traits (both morphological and quantitative) were followed over a number of generations. The donor plant was absolutely homozygous (being a dihaploid derived from the inbred cultivar Norin 8). Seed calli were initiated from 75 seeds and 1121 somaclones were regenerated. The selfed progeny of these were examined over three generations for many characters including plant height, chlorophyll mutations, heading date, panicle length, grain yield, and 1000-g weight. Many of the variations observed in the primary somaclones D_1 segregated in the progeny D_2, and by D_3 many variant lines had been obtained in a pure-breeding state. Among them were some with increased grain number and 1000-g weight. When the D_1 families were analyzed for just five characters, only 28.1% of the somaclones were normal, and 28.0% were variant in two or more characters.

Following the procedure of Oono[19-21] and using a broader base of germplasm, Suenaga et al.[144] also succeeded in obtaining a significant array of mutations among plants regenerated from rice callus cultures. Approximately 54% of regenerated plants were undesirable. Certain mutations such as new dwarfism and altered heading dates may ultimately be useful, but require further studies. Zhao et al.[145] also regenerated plants from callus cultures of 23 rice varieties and found high levels of variation among their selfed progenies. Mutations were obtained which affected floral and foliar morphology, grain shape, heading date, and fertility. Most characters were found to be fixed following selection among selfed progeny of regenerated plants and remained stable through two more cycles of selfing. Further variations were found to be greater in upland than in lowland types. A number of lines with good characters were noted and these are apparently being integrated into ongoing rice pedigree and backcross breeding programs.

Different components of the culture media have been reported to influence the rate of somaclonal variation in rice.[146] Five agronomic characters were assessed over three generations on lines derived from nearly 1200 regenerants of rice seed calli. The cultures were grown on media modified in various components. Increased concentration of $MgSO_4$, NH_4NO_3, and KI was responsible for ,increasing variation, whereas high concentration of $CaCl_2$, KH_2PO_4 and $MnSO_4$ decreased it.

Fukai[58] initiated a callus culture from a single rice seed and maintained it for 2 weeks before regenerating 12 plants. Segregation was observed in the progeny for four presumed independent gene mutations. He observed that each mutation had occurred independently. Among the second-generation (SC_2) progeny of 12 plants derived from the callus of a single seed, four recessive mutations (early heading, albina, short culm, and sterility) were observed. Because of the pedigree relationship of these plants, one genetic event was assumed responsible for each mutation and the mutational sequence was early heading, albina, short culm, and sterility.

Somaclonal variation has been observed in regenerated rice plants in the characters such as leaf blade angle, flage leaf angle, culm angle, leaf color, leaf pubescence, leaf senescence, culm strength, disease resistance, panicle type and size, panicle exertion, sterility, seedling height, mature plant height, sheath color, tiller number, leaf blade length, blade width, days to heading, panicle number, panicle weight, 100-seed weight, grain type, grain quality, yield, apiculus color, and hull color.[147] In a field trial during the 1986 season, the line Labelle 215 out-yielded the parent variety Labelle and the high-yielding commercial variety Lemont. However, grain quality was seriously reduced (head rice) due to breakage of large grain during milling.

Different rice germplasms show different responses as far as development of variation in somaculture is concerned. Somaclonal variation observed in the field with somaclonal lines from Labelle and Lemont was great during both 1985 and 1986 growing seasons.[147] However, cvs. Melrose, Tetep, Taipei 309, and Tadukan showed little or no variation. Somaclones from cv. Zenith showed several types of variation when planted in field in 1986. This included variation in grain type (changes from medium to long grain), hull color (strain hulled to gold hulled), height, resistance to sheath blight, and pubescence.

Anther culture in rice has also been pursued extensively as a mean to obtain homozygous from heterozygous (F_1 hybrid) plants, thus reducing the period from original hybridization to final testing and release.[148] It was estimated that roughly 100,000 ha in China is used to grow rice varieties developed via anther culture.[146] It is generally presumed that all variation observed among dihaploids obtained from microspores is a reflection of the assortment of alleles contributed by the original parents of hybrid. However, striking new variants are sometimes observed, the most conspicuous example being albinism. Up to 90% plantlets regenerated from rice microspores may be albina.[149] Schaeffer[150] has obtained evidence of mutations other than albinism among androgenic rice plants. Although it is impossible to determine the precise frequency of mutational events, several independent cases of dwarf mutations were associated with a reduction in leaf width. Other inherited characters observed among dwarf plants were reduced number of grains per plant, reduced awn length, reduced seed weight, increased tillering, and the tendency for open hulls. These characters were observed among androgenic plants from a recognized pure line, cv. Calrose 76. Similar mutations arising among dihaploids regenerated from microspores of multiple heterozygotes could certainly have been lost in the sea of segregants. Hence, it appears that new mutations arising during the culture process may contribute to the pool of genetic variation, both desirable and undesirable, observed among the androgenic rice plants produced for breeding purposes.

F. MAIZE

Edallo et al.[35] identified 17 classes of defective endosperm or seedling mutants among progeny of 77 plants regenerated from maize tissue cultures. Classification of a somaclone progeny variant as a mutant was based on its confirmation to the Mendelian segregation ratio. The somaclones analyzed were derived from two different genotypes and progeny analysis revealed that each somaclone carried an average of one inherited variation. Somaclonal variants in height, node number, ear arrangement, and stalk number from immature maize embryo culture were reported by Green and co-workers.[151,152] Eight maize regenerants segregated for recessive kernal mutations, and one for mutations that cause premature wilting among progeny of 51 initial (SC_1) regenerants.[153] Segregation for some mutants was not observed until the third (SC_3) generation, suggesting that the male and female influorescence on the progenitor SC_1 plants differed genetically. Alternatively, an event during the culture phase that causes somaclonal variation may persist after regeneration.

A report by Beckert et al.[154] has confirmed in maize what has been suspected of plants regenerated from cultured cells in general: that quantitatively inherited as well as simply inherited characters are affected. Later, the same group[155] conducted an analysis of quantitative and qualitative variants arising among plants regenerated from scutellum-derived callus cultures of several maize inbreds. Prominent among simply inherited mutations were dwarf and chlorophyll deficiencies. No variants, however, were considered useful. Recently, Peschke et al.[62] have reported transposable activity among the progeny of tissue culture-derived plants of maize.

G. LETTUCE

Lettuce was among the first crop species in which somaclonal variation was observed and characterized. Sibi[34] noted three qualitative (flat, glaucous leaves, yellow leaves, pale midveins) and numerous quantitative variants among a population of plants regenerated from cotyledon-derived callus cultures of the cv. Val d'Orge. The "flat, glaucous leaves" variant was shown to be maternally inherited, while the other two qualitative characters diminished in intensity over selfed generations and could be environmentally corrected. Interestingly, it has been observed that hybrids between sexual descendants of regenerates and seed-propagated Val d'Orge, or among sexual descendants of regenerates, exhibited hybrid vigor in the form of increased leaf area as compared to self-pollinated progenies. This heteroitic effect was observed regardless of plant used as female parent. Sibi hypothesized that this hybrid vigor was a consequence of the mixture of wild-type and culture-induced mutant organelles during gamete fusion, resulting in the zygote having new, stable, heterogenous organelle genotypes. However, the genetic studies to verify this hypothesis were not reported. Subsequently, Sibi[34] filed a patent application pertaining to the use of plant tissue culture as a tool for obtaining useful germplasm (No. 4003156), issued on January 18, 1977.

Out of nine variant characters observed among selfed progeny of lettuce plants (regenerated from mesophyll protoplast-derived calli of the cv. Climax), seven were attributed to nuclear recessive mutations.[156-158] All of these variants were considered as horticulturally undesirable. The remaining two variants increased or decreased seedling size, but did not segregate upon selfing and could therefore have been caused by homozygous nuclear or cytoplasmic mutations. The increased seedling-size mutation was considered as potentially undesirable from the standpoint of rapid establishment, although the effect was diminished during later development stages. No crosses were conducted to determine whether regenerated plants contained new mutations which might contribute to hybrid vigor.

H. SUGARCANE

Plants regenerated from sugarcane cell cultures expressed variation for stalk, leaf morphology, sugarcane yield, and disease resistance.[159] While several useful clones with increased sucrose yield have been isolated, progress has been most striking in recovery of plants with resistance to several diseases of sugarcane. These include resistance to Fiji virus disease,[160] downy mildew,[159] and culmi colus smut.[161] In most cases this resistance has been expressed in field conditions and has been transmitted asexually to cuttings of the resistant variants. Larkin and Scowcroft[128] have demonstrated that some clones isolated for tolerance to eyespot toxin either reverted to normal or were unstable. On the other hand, field evaluations of somaclones for resistance to Fiji and downy mildew diseases grown in four locations over 5 years have demonstrated stability of the resistant phenotypes.[162] Based on the work with sugarcane in several laboratories, some general suggestions have been published and appear to be useful for nearly all crop species.[2] First, important sugarcane varieties could be used as donor tissue and somaclonal variation can be directed toward recovery of specific incremental improvements not yet achieved using conventional breeding approaches. Second, it has been suggested that a second characteristic could be produced using a second culture cycle.

I. TOMATO

Several somaclonal variants in *Lycopersicon peruvianum*, *L. chilense*, and *L. cheesmanii* var. *minor* have recently been reported.[163] In R_1 plants of *L. chilense* pseudo-self-compatibility was observed. Self-compatibility in these plants was thought to be due to tetraploidization (2n = 28). The plants of *L. chilense*, however, were diploid and it was suggested that a change in S allele had occurred. However, variations were passed to next generation and one of the autotetraploid R_2 plants of *L. peruvianum* had pedicels that were altered. Instead of jointed pedicels in normal plants, pedicels were jointless in this species. The wild *L. cheesmanii* (var. *minor*) species is known for its difficult germination of seeds due to hardness of the seed coat. Several variants of R_3 seeds of this species germinated easily with no preliminary processing.

Leaf disk tomato line (*L. esculentum*, Mill.) from a fully tobacco mosaic virus (TMV)-susceptible (+/+) isogenic line (GCRI-26) was used to regenerate somaclones by Barden et al.[164] Out of 370 somaclones inoculated with TMV-Flavrum, six were eventually selected as virus-free and putatively resistant. R_1 progeny from self-pollination of the six somaclones showed TMV resistance for varying periods ranging from 28 to 55 d, while seedlings of the source plant usually became infected in 10 to 20 d. In resistant plants, no symptoms were visible and TMV could not be detected either by enzymatically linked immunosorbent assay (ELISA) or by back inoculations to *Nicotiana glutinosa*. Further attempts to develop TMV-resistant plants for a prolonged period were successful and Smith and Murakishi[165] reported resistance in somaclones even up to 150 d. Somaclones appeared normal in morphology, pollen viability, and chromosomal number. In reciprocal crosses, when somaclonal plant lines were used as female parent, the offspring showed a high level of resistance; whereas in plant lines, when somaclonal variant was used as the pollen parent, low resistance was evident indicating the role of cytoplasmic factor in deciding resistance. In addition, F_2 plants were more resistant than F_1.

Evans and associates have characterized several somaclonal variation in tomato.[2,56] They have been successful in the production of tetraploid tomato seedling by somaclonal variation. Tomato plants with normal fruit, but higher pigment and lycopene concentration, deficient in chlorophyll, and bearing fruits of orange and yellow color were also raised from variant cell lines.

Evans and Sharp[44] have also demonstrated a genetic basis to somaclonal variation in tomato. Several somaclonal variation and some 13 different nuclear mutations have been recovered among the progeny of only 230 regenerants of tomato. The mutations, which had simple Mendelian inheritance, included fruit color, pedicel jointedness, male fertility, chlorophyll formation, and growth habit. The authors stated that autotetraploidy was frequent among primary regenerants. Selfed seed was obtained from 230 fertile diploid "R" plants and large replicated field plots were examined over 2 years. Variants were carefully tested to determine their mutational origin and potential allelism to known markers. Thirteen independent mutations were apparently present in the primary regenerates, none of which were observed in the seed-propagated pure line UC82B.

J. STRAWBERRY

In strawberry, where meristem culture is routinely used for clonal propagations, varietal differences occur in the frequency of off-type plants.[166] Among plants derived by adventitious shoot formation from leaf explant in *Begonia* × *hiemalis*, Roest et al.[167] found that in one variety, 43% of regenerants were variant (color, size, and form of leaves and flowers), whereas for another variety only 7% were variant. Some varieties of strawberry are known to be inherently unstable through conventional cutting propagation. Such cultivars produced higher frequencies of variants from callus culture.[9,168]

K. CELERY

Celery denotes the domesticated form of the species of *Apium graveolens* L., characterized by enlarged, succulent petioles on a single rosette. A number of reports have been published which provide a description of one or more populations of regenerated plants. Because few or no genetic studies were conducted on the derivative culture or ensuing generations, the relationship of this variation to genetic instability *in vitro* cannot be established. For example, Williams and Collins[169] observed little or no phenotypic variation among regenerated celery plants and speculated that the derivative culture and, indeed, the celery in general was highly stable. This conclusion was premature, as subsequent work has documented instability *in vitro* and the transmission of this variation to regenerated plants. Moreover, regenerates may have contained new masked recessive mutations, which are sometimes present at high frequencies in regenerated plants.[25,170] Fujii[171] conducted extensive field trials of plants regenerated from suspension cultures derived from axial buds of the celery cv. "Tall Utah 52-70 R improved". In one such trial only 54.0% were deemed marketable, the remainder having one or more major deficiencies. In the other trial in which quantitative data were generated, 62.0% of the plants were marketable. Common characteristics in the trials were abnormal petiole morphology, stunting and premature flowering, and multiple growth centers. Rappaport et al.[172] initiated a project to attempt to tap into somaclonal variation to rapidly obtain heritable resistance to the casual pathogen of *Fusarium* yellow in a commercial celery variety. Approximately 400 plants were regenerated from suspension cultures of cv. Tall Utah 52-70 R and challenged directly with pathogen. Twelve plants were identified which appeared to be resistant. Unfortunately, these plants were either escapes or the resistance was transient, since selfed progeny were all susceptible.

V. GENETIC BASES OF SOMACLONAL VARIATION

The events which give rise to somaclonal variants must be critically analyzed at the genetic, cytogenetic, and molecular levels for effective use of somaclonal variation. The possible causes of variations observed in plants raised through tissue culture have been discussed extensively by several authors.[173-178] Several genetic bases such as karyotypic changes, factor mutations in the genome and plasmon, translocations, deletions, inversions, cryptic chromosome changes, gene rearrangements, and nonconventional mutations comprising gene amplifications and transposable elements have been reported to be responsible for the appearance of different somaclones in plants (Table 1).

A. CHANGES IN CHROMOSOME NUMBER

The outset of chromosome instability has been well characterized in many plants, such as *Datura carola*[209] and *Haplopappus gracilis*.[210] The most frequently reported variation has been polyploidy, which is, in part, dependent upon the concentration of cytokinin used in the medium.

Callus and suspension cultures derived from immature ovules of cultivated barley *(Hordeum vulgare)*, an allotetraploid relative *(H. jubatum)*, and interspecific hybrid (HV × HJ) all exhibited polyploidy, aneuploidy and chromosomal rearrangements.[54] Plants regenerated from HV × HJ cultures also showed variation with respect to chromosomal number and

TABLE 1
Genetic Changes in Plants Showing Somaclonal Variation

Species	Genetic alterations	Ref.
Allium sativum	Fragments, rings, bridges, breakage, and fusion	179
Apium graveolens	Large chromosome, fusions	170
	Monosomics, trisomics	42, 54, 102
	Chromosomal loss or fusions, accessory chromosomes, hypodiploidy, and tetraploidy	170
	Hypodiploidy, tetraploidy	169
	Aneuploids, polyploids, intermediates	74
	Increase in chromosomal number, aneuploidy	46
	Changes in chromosomal number, structure, and marker enzymes	96
Avena sativa	Monosomics, trisomics, bivalents, multivalents, telocentrics, deletions	38
Haworthia	Monosomics, trisomics	73
Hordeum jubatum	Polyploids, aneuploids	72
H. setata	Multivalents, bridges, fragments, interchanges, inversions	73
H. vulgare	Polyploids, aneuploids, chromosomal fragments, bridges, and fusions	180
H. vulgare × *H. jubatum*	Multivalents, mitotic multicentrics, interchanges	54
Lolium multiflorum × *L. perenne*	Acrocentrics, telocentrics, multivalents, bridges + fragments, deletions, inversions, translocations	22
Lycospersicon esculentum	Polyploidy	44
L. peruvianum	Mutation in S-allele	181
Medicago sativa	Polyploidy	182
	Increase in GS m-RNA level	183
	Translocations	184
Nicotiana tabacum	Mutation for chlorophyll deficiency	103, 185
	Mutation conferring resistance to chlorsulfuron, polyploidy, chromosomal rearrangements	186—191
N. sylvestris	Activation of mutator gene	103, 188
Pelargonium zonale	Polyploidy	168
Saccharum officinarum	Chromosomal mosaics	160, 192
Solanum tuberosum	Multivalents, bridges, heterochromatin fragments, heteromorphic bivalents, telocentrics, isochromosomes, interchanges, deletions	193
	Large and small chromosomes, translocations, and deletions	194
	Changes in mitochondrial DNA and reduction in rDNA genes	195, 196
	Monosomics and trisomics	83, 194
	Aneuploidy, polyploidy	68, 85
	Increase in chromosomal number, tetraploids, aneuploids	85—87
Stylosanthes guianensis	Tetraploid, polyploid, and aneuploid	197
Triticum aestivum	Reduction in rDNA spacers	75
	Mitotic bridges, inversions, multivalents, interchanges, heteromorphic bivalents, iso-ring, isochromosomes, deletions	193, 198, 199
	Changes in Adh isoenzyme	51, 200
	Multivalents	201, 202
	Chromatid exchange, deletion in chloroplast genome	131
Triticale	Deletions, interchanges, heterochromatization, gene amplifications	203
	Interchanges, dicentrics, C-banding changes, DNA-sequence depletion	204
	Monosomics, trisomics	205, 206

TABLE 1 (continued)
Genetic Changes in Plants Showing Somaclonal Variation

Species	Genetic alterations	Ref.
	Repeat sequence of DNA spacer region	196
Triticum crassum × *Hordeum vulgare*	Multivalents, interchanges, or pair control changes	201
T. tanschii × *Secale cereale*	Multivalents, interchanges, or pair control changes	202
Zea mays	Transposable elements	62
	Heteromorphic bivalents, deletions, inversions	152, 207, 208
	Variation in Adh loci	51

From Evans, D. A., Sharp, W. R., and Medina-Filho, H. P., *Am. J. Bot.*, 71, 759, 1984. With permission.

isozyme expression. The extent to which each class of chromosomal variability was present in culture was dependent upon differentiated state, age, and history of the culture. Studies on the variability of chromosomes in *Stylosanthes guianensis* plants regenerated from callus cultures revealed several changes in chromosomal number.[19] Regenerated plants from callus culture (using leaf and hypocotyl explants) in *S. guianensis* (CIAT 2243) in SC_1 showed 24 plants (out of total 114, SC_1 plants) as tetraploid ($2n = 2x = 40$). Subculture further affected the ploidy changes of the regenerated plants and the frequency of tetraploids was higher among plants regenerated from 60- and 90-d-old cultures (Figure 5). This was reflected in wide variation in SC_1 plants in vegetative characters such as leaf size, flower size, pubescence, and reaction to inoculation with anthracnose cultures (Figure 6).

In order to assess if variability observed in the SC_1 plants can be transmitted through seeds in the field and to determine the extent of such variability, several characters of the SC_2 plants have been evaluated by scientists in the CIAT project.[197] These characters are vigor of plants, later growth, internode distances, leaf size, reaction to anthracnose attack, flowering, seed production, nitrogen content and digestability, and plant fresh and dry weight (Figure 7). SC_2 plants were grouped into ploidy levels (67 diploid clones and 10 tetraploid clones) and displayed wide variability in these characters. Some somaclones were rated with highest number of stems and with vigor as high as the controls.

In alfalfa tissue culture chromosome doubling appears to be very common. Chromosomally doubled and near-doubled plants are easily recognized because of their larger leaves, flowers, and general appearance.[211] Doubled plants also possess simple or complex genetic changes. When the first plants were regenerated from a tetraploid clone of ''Saranac'' alfalfa (S-4), about 5% of the 200 plants studied were spontaneous octoploids.[120] Clone S-4 was used in several experiments, each of which yielded a few octoploids.[113,120] A study of 50 octoploids indicated that four plants were variant for characteristics not merely associated with chromosome doubling. Two plants had altered intensity of flower pigmentation, one had abnormal leaves, and one was variant for flower color and had markedly increased susceptibility to leaf spot.[211]

Chromosome doubling is also common in suspension culture of diploid alfalfa clone HG2.[114] The proportions of chromosomally doubled plants regenerated after 7, 21, 42, and 56 d in suspension were 10, 30, 40, and 55%, respectively. Most of the doubled plants were tetraploids, but a few presumed octoploids were also produced. The presumed octoploids may well have been near hexaploids, because plants from doubling and redoubling HG2 in callus or suspension culture in other experiments have all proved to be hexaploids.[182] Aneuploidy has also been reported in cell cultures at low frequency. In addition to numerical chromosome changes, mitotic abnormalities, such as anaphase bridges and fragments[212] and chromosome rearrangements,[54] are also common.

Detailed cytological analysis of two *S. guianensis* cell cultures revealed wide variability

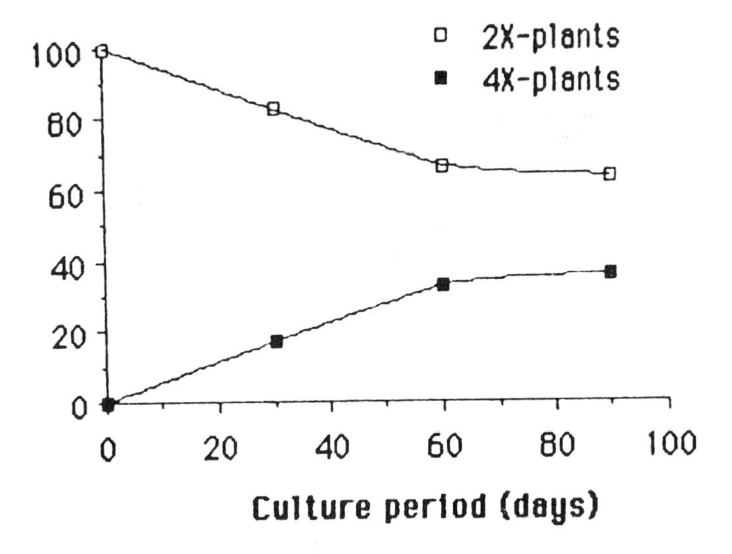

FIGURE 5. Somaclonal variation in *Stylosanthes;* frequency of diploid and tetraploid plants regenerated from 30-, 60-, and 90-d-old callus cultures (*S. guianensis* CIAT 2243). (From Roca, W. M., Szabados, L., and Hussain, A., Biotechnology Research Unit, Annual Report, CIAT, Colombia, 1987. With permission.)

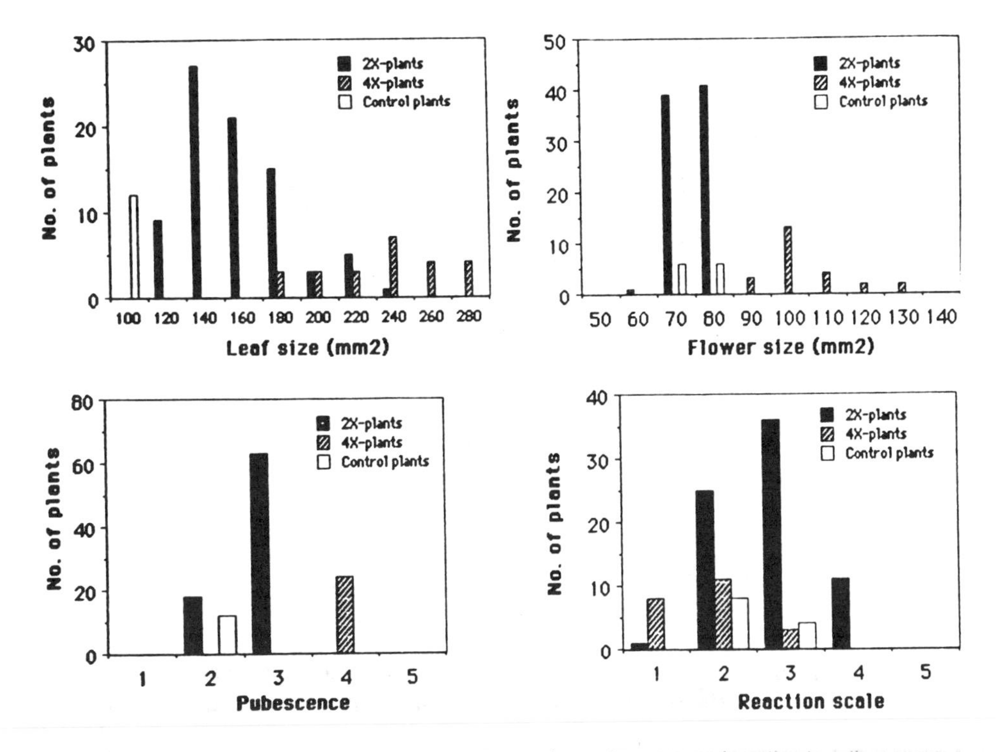

FIGURE 6. Frequency distribution of leaf size, flower size, pubescence, and reaction to anthracnose inoculation of regenerated *S. guianensis* (CIAT 2243), diploid (2 × plants), and tetraploid (4 × plants) variants and parental (control) plants. (From Roca, W. M., Szabados, L., and Hussain, A., Biotechnology Research Unit, Annual Report, CIAT, Colombia, 1987. With permission.)

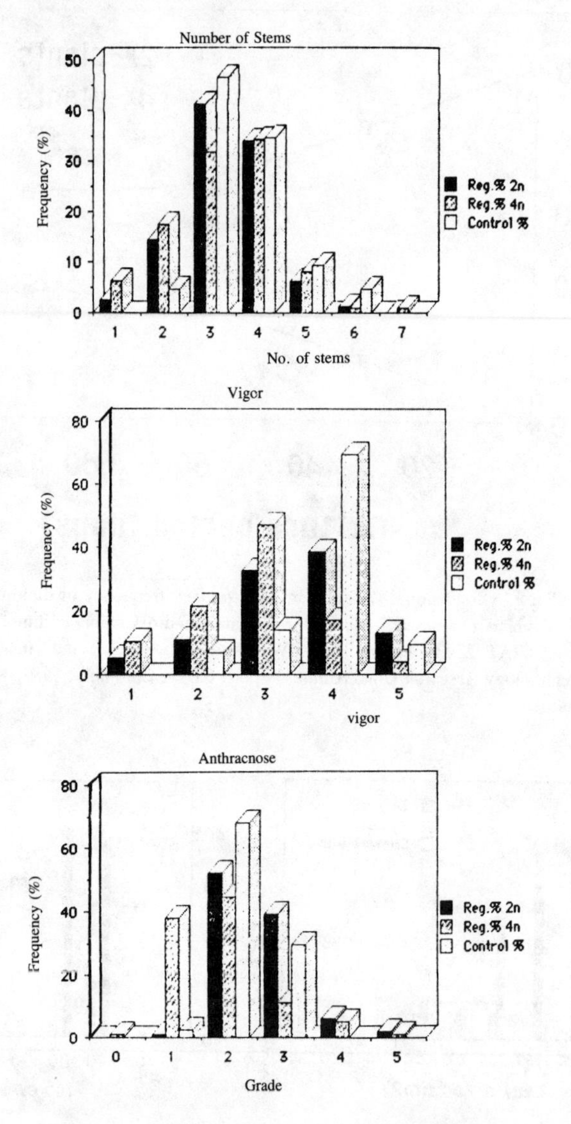

FIGURE 7. Somaclonal variation in progenies (SC_2) of regenerated plant of *S. guianensis* (CIAT 2243) grown in the field, in Santander de Quilichao. Frequency percent of plants with variation in number of stems per plant, vigor (1 = less vigorous, 5 = very vigorous), and reaction to anthracnose (0 = plants without anthracnose symptoms; 5 = dead plants due to anthracnose). (From Roca, W. M., Szabados, L., and Hussain, A., Biotechnology Research Unit, Annual Report, CIAT, Colombia, 1987. With permission.)

in chromosome number. Besides diploid (2n = 20) chromosome number, cells with tetraploid, polyploid, and aneuploid chromosome numbers have been found with high frequency.[197] Morphogenic cell cultures of cv. CIAT 136 had chromosome numbers between 20 and 100, the majority of cells having 26 to 40 chromosomes (Figure 8). Nonmorphogenic, habituated cell cultures of cv. CIAT 2243 (2243 H) had chromosome numbers in an even wider range, i.e., varying from 18 to 116 (Figure 8).

Instability at the chromosome level of callus cells of *Stylosanthes* may be partly responsible for the morphological variability observed in regenerated plants.[197] Variability of

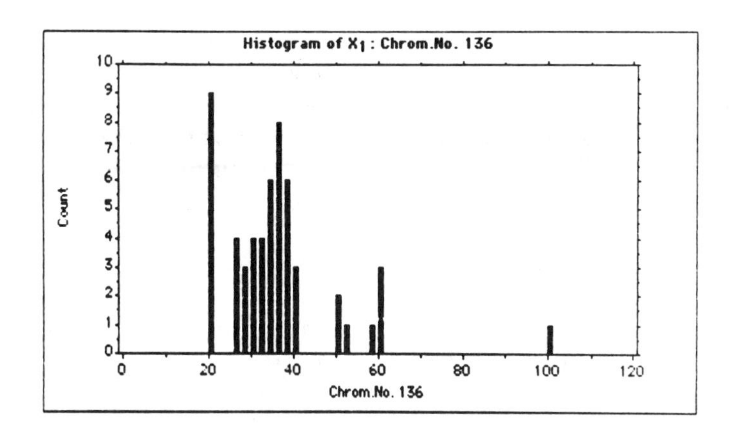

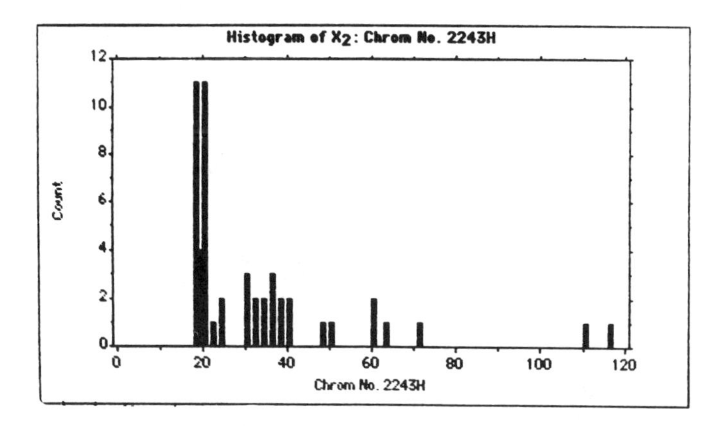

FIGURE 8. Variability in chromosome number of *S. guianensis* morphogenic cell suspension cultures of CIAT 136 and nonmorphogenic habituated, cell suspension cultures of CIAT 2243 (2243 H). Even though chromosome frequencies of $2\times = 20$ prevailed, there are cells with $4\times$ polyploid and aneuploid numbers. (From Roca, W. M., Szabados, L., and Hussain, A., Biotechnology Research Unit, Annual Report, CIAT, Colombia, 1987. With permission.)

this type suggests that genetic instability does exist in these cultures, generating a wide range of genetic changes. No plant of *Stylosanthes* with aneuploid chromosome number has been found up to now, indicating aberrant cells. Indeed, this is an advantage of somaclonal variation over mutant induction, since somaclones can be stabilized immediately in most cases, in contrast to mutants which require several backcross generations.

Time in culture evidently influences chromosome doubling more than any chemical treatments evaluated so far. More than 1000 plants have been regenerated from diploid HG2 alfalfa in several experiments spanning over a period of 10 years.[114,182] A survey of the chromosome doubling in these experiments, some of which involved a 3-week suspension culture and various chemical treatments, indicated that doubling was between 20 and 30% regardless of the treatment.

Chromosome doubling in culture could be advantageous when a diploid form of cultivated alfalfa is used in experiments to produce variability. Spontaneously doubled tetraploid variants could be crossed directly with cultivated tetraploids. In this way a wide range of such tetraploid variants has been regenerated from HG2 alfalfa.[182] Chromosome doubling in culture when the donor cells are from cultivated tetraploids is usually undesirable. This is because potentially useful genetic traits in doubled plants have to be scaled down to the tetraploid level for use. In research, to select salt-tolerant variants of alfalfa, plants regenerated after 32-month-old cultures were likely doubled or redoubled.[213] However, these plants were sterile and could not be used in breeding. A problem of doubling associated with a long selection regime for disease resistance was overcome by modifying the selection regime.[214] In initial experiments, to select for resistance to culture filtrate of *F. oxysporum* fungus, tetraploid alfalfa callus was subcultured for more than a year on media containing increasing concentrations of fungal filtrate. After this length of time in culture, only hexaploid and octoploid plants were regenerated. The selection regime was modified to greatly shorten the time in culture and tetraploid plants with the desired resistance were recovered.

Several workers have been successful in raising alfalfa plants from protoplasts.[184,215] Chromosomal doubling and partial doubling is a prominent form of variation in such plants.[216,217] Octoploids were common in a study where chromosome numbers were determined for about 100 plants regenerated from mesophyll protoplasts of two tetraploid clones of ''Regen-S''.[184] About half of the regenerated plants of one clone and about one third of the other were octoploids. Many octoploids were variant for forage yield.

Protoplasts of strain ''Europe'' of alfalfa have been regenerated into normal and variant protoclones.[216] The variant protoclones were all doubled or partially doubled and reported to be more resistant to verticillium wilt than normal protoclones. Five variants for which chromosome numbers were determined included one hexaploid, one near heptaploid, and three near octoploids. The mechanisms by which hexaploids arise in the culture of diploid and tetraploid donor cells are not known. Hexaploid may arise from octoploid cells by gradual loss of chromosomes or from fragmented spindles during restitution divisions. In the case where the donor is diploid, two mitotic divisions of the nucleus followed by one cytokinesis could set three nuclei apart from the fourth and result in a euploid hexaploid cell and a diploid cell. All hypothetical mechanisms of hexaploidy from either diploid or tetraploid somatic cells involve abnormal events. The fact that hexaploids have occurred in several independent experiments may suggest that there is a selective advantage of hexaploidy over octoploidy in some cases. The chromosome number of callus from a tetraploid donor has been observed to double and redouble the 16x level,[116] and yet hexaploids and octoploids are the highest ploidy levels regenerated. Perhaps these are the upper limits of autopolyploidization in alfalfa.

Apart from alfalfa, the cytogenetics of several regenerated plants has not been clearly elucidated. In some cases, normal diploid plants have been regenerated despite the presence of abnormal cells in the cultures for regeneration.[218,219] This probably indicates that at least in some species a selective advantage exists for the organogenesis of diploid cells. On the other hand, polyploid and aneuploid plants have been regenerated *in vitro* from a large number of plant species. Polyploid plants have been recovered in many commercially important plant species including *Pelargonium zonale*,[9] *Nicotiana tabacum*,[187] *Lycospersicon esculentum*,[44] and *Medicago sativa*.[182] Aneuploids have been reported more often from plants of polyploid or hybrid origin such as *N. tabacum*[8] and *Saccharum hybrida*,[6] where loss or addition of a few chromosomes can be tolerated. Plants with chromosomal mosaics have also been recovered from cell culture.[220]

Studies by Sree Ramulu et al.[85-87] presented an interesting picture of the cytological events in potato calli of cv. Bintje. The variation in chromosome number began with the first division of the protoplasts. The cell colonies and calli contained a mixture of cells with chromosomes ranging from 24 to 200. In further growth stages of the calli there was a trend to increasing numbers of cells with tetraploid chromosome number and aneuploid number around 48 chromosomes. Other cells were continuously sorted out. Regenerants of the same callus, being segregants of chimeras, delivered different phenotypes. Of the regenerated plants, 63.2% were normal looking and euploid and tetraploids; 20% varied in few characters with the optimum chromosome numbers between 39 and 47. The rest of the somaclones showed variation in several characters with chromosome numbers between 50 to 96. The calli differed in giving rise to only normal or only variant plants or mixtures. Several variations in chromosome number in potato have been reported by Creissen and Karp[194] and Wheeler et al.[221] For an autotetraploid, such as potato, a surplus of chromosomes need not necessarily imply detrimental consequences for the phenotype. Of the normal looking Maris Bard protoclones, 20% were aneuploid. Work in the Max-Planck Institute[221] also showed that crosses of octoploid forms of *Solanum acaule* and *S. stoloniferum*, as well as of the hexaploid *S. demissum* with *S. tuberosum* and backcrosses, produced useful hybrids which could be selected with chromosome number up to at least 2n = 60.

Somaclonal variation among dihaploid potato plants derived from protoplasts is comparatively minor. Only 67% of a group of interdihaploids responded to protoplast culture at all and only a few of these profusely.[100,222-224] All regenerants doubled their chromosome number. Of the petri plate culture only 4% were aneuploid (being abnormal and sterile). A group of 30 uniform-looking somaclones were screened by polyacrylamide gel-electrophoresis of the total proteins which were identical.[223] Similarly, Binding et al.[225] found less karyological and morphological variability in protoplast cultures of dihaploids.

Williams and Collins[169] examined celery plants regenerated from 9- to 15-month-old callus cultures of the VCV "Lathom Blanching" for chromosomal instability. Chromosome counts conducted in root tips revealed departures from the normal chromosome number at an appreciable frequency. Among the cytological variants hypodiploidy (27.3% of cells of regenerants) and tetraploidy (2.3%) were very common. Surprisingly, all the plants were morphologically indistinguishable from normal. Reduced chromosome numbers were probably a consequence of end-to-end fusions and not bonafide loss. Such end-to-end fusions have been observed previously in celery.[226] Based on these observations, Williams and Collins[169] concluded that the process of culture *in vitro* of celery is genetically stable.

Browers and Orton[74] also conducted a study to compare chromosomal variability *in vitro* and in corresponding regenerated celery plants. Suspension cultures were established from petiole callus tissues of two genotypes, PI 169001 and PI 171500. At least 70% of the PI 171500 culture cells were nondiploid, and only abnormal regeneration was observed. The PI 169001 culture consisted of approximately 80% of cells indistinguishable from diploid, and exhibited a strong regeneration response. Comparisons of the culture and root tip meristem cells of corresponding regenerated plants revealed the presence of aneuploid cells in regenerates which were always present in "mixtures" with diploid cells. However, polyploid and intermediate (2n = 23 to 43, 45 to 87, etc.) cells were not observed in regenerates, whereas they were present in the culture.

Different celery cultures which may appear similar *in vitro* frequently can be distinguished based on regeneration responses. In experiments conducted by Browers and Orton,[74] one culture regenerated only abnormal globular embryos, while the other regenerated normal plantlets quite profusely. Intermediate responses have been noted, as embryos appear to develop normally during early stages and exhibit a breakdown later. Such a case was reported by Orton[102] where attempts were made to regenerate subclones characterized by differing patterns of *Pom-2* phenotype and its karyological constitution.[63] Regeneration responses ranged from null to the recovery of embryos which "germinated" into structures that failed to develop into plants. Typically, the root axis was indistinguishable from normal, but the shoot never developed beyond a club-shaped, abnormal, cotyledon-like structure (Figure 9). A comparison of the karyological makeup of the culture and root tip meristem cells of these abnormal regenerates revealed the following for two separate subclones: (1) variability was dramatically attenuated from culture to regenerate, and (2) all abnormal regenerates were aneuploid, and some showed striking structural changes (Figure 10). It was concluded that most of the cells in the callus tissues of these subclones were genetically impaired with respect to regeneration capacity, but that one or more minor components were able to develop into an abnormal plantlet.

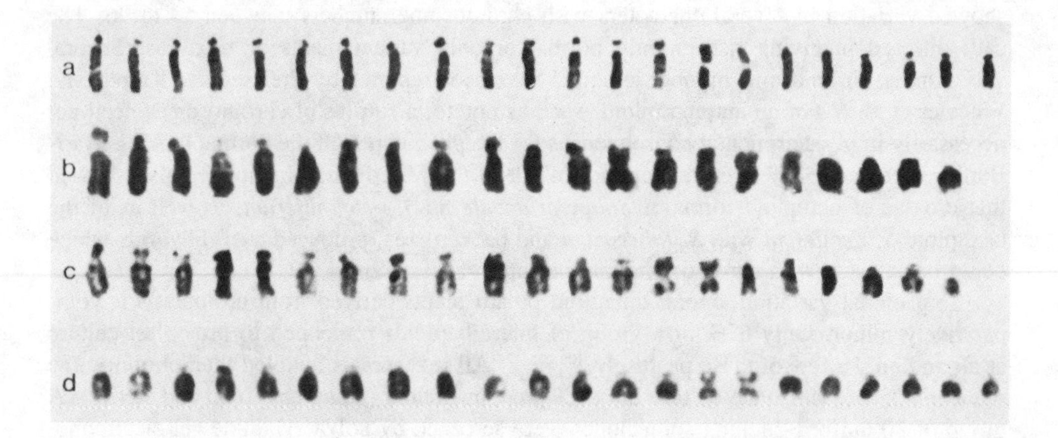

FIGURE 9. Karyotypes of plants (celery) regenerated from a PI 169001 × Tall Utah 52-70 R culture: (a) normal, (b) Pgm-2^{FF} clone with hypodiploidy and structural changes, (c) Pgm-2^{S} variant clone with hypodiploidy and minor structural rearrangement, and (d) developmentally normal regenerate with apparently normal karyotype. (From Orton, T. J., *Cell Cult. Somatic Cell Genet. Plants,* 3, 345, 1986. With permission.)

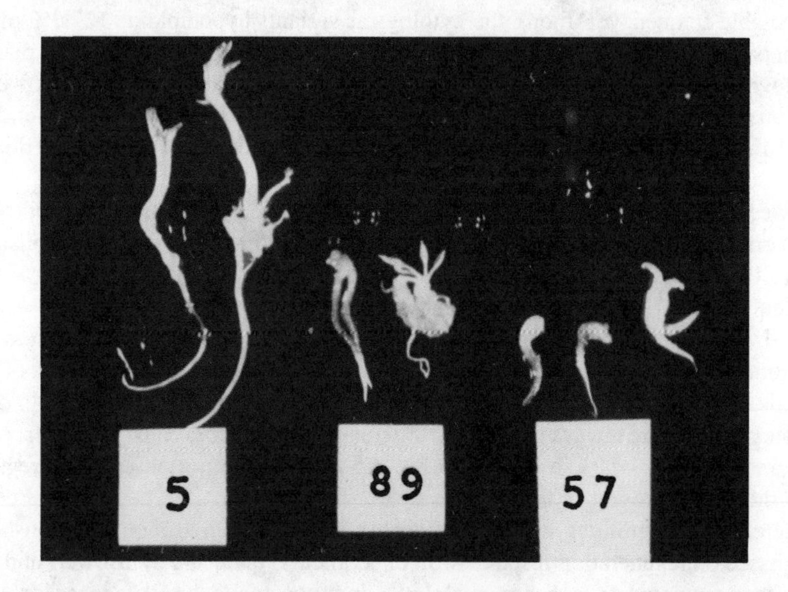

FIGURE 10. Morphology of normal (5) and abnormal (89 and 57) regenerates. (From Orton, T. J., *Cell Cult. Somatic Cell Genet. Plants,* 3, 345, 1986. With permission.)

In the experiment in which genetically marked tissues were examined in celery culture,[170] plants were regenerated and genetic and cytogenetic comparisons were made with the 6-month-old culture. About 84% of cells of this culture were indistinguishable from diploid, the remainder being hypodiploids and cells bearing chromosome fusions. Cytogenetic studies of regenerated plants were conducted on microsporocytes. Since this is a rather late developmental stage, variation present in the original population of regenerates might have been lost; for example, by sectoring out within plants and poor competitive ability of variants within the mixture.

In vitro culture of plant cell can result in aneuploid plants (somaclones).[70,227,228] The aneuploidy includes simple monosomics (2n − 1) and trisomics (2n + 1). Chromosome number variability may arise early in culture, but its frequency increases with time. Examples of culture-generated monosomics and trisomic plants include oats,[153] ryegrass,[206] *Haworthia*,[73] celery,[63] wheat,[193] triticale,[205] potato,[83,194] and alfalfa.[125]

Commercial sugarcane varieties are chromosomal mosaics.[192] Hence, it is not surprising that plants regenerated from somatic tissue are highly variable for chromosome number. In fact, plants regenerated from cell culture express greater variation in chromosome number than donor tissue.[192] Chromosomal variation has been cited as a genetic basis for most of the variation in potato somaclone's resistance to early and late blight.[33,79] However, detailed genetic analysis has not been reported by several authors for potato varieties.[27,85,229] Whenever detailed cytogenetic analysis has been completed, aneuploidy was found to be prevalent among regenerated plants.[83] Aneuploids have also been identified among plants regenerated in alfalfa. Groose and Bingham[230] studied aneuploidy in a tetraploid alfalfa donor that was heterozygous for four heritable traits. Loss of chromosomes carrying dominant alleles could be identified by phenotypic expression of recessive traits among regenerated plants. Heterozygous traits included flower color (loss of a dominant allele would change flower color from purple to white); multifoliolate leaves (loss of dominant alleles would shift the plant from trifoliolate to multifoliolate); ability to regenerate (loss of a dominant allele would reduce or prevent regeneration); and nuclear restoration of cytoplasmic male sterility (loss of dominant alleles would shift plant from male fertile to sterile). Further, the variants for each marked trait were identified and many of the variants were found to be associated with a chromosome loss. At least 11% of 116 regenerates lost one or more chromosomes with aneuploid chromosome numbers ranging from 2n = 28 to 31.

Aneuploid plants of alfalfa with 28 chromosomes were recultured to test for shifts toward euploidy at either the triploid (2n = 24) or tetraploid (2n = 32) level.[125] Of 113 plants regenerated, 87 appeared to be unaltered clones of the donor, and the 26 others appeared to be chromosomally doubled. While there appeared to be selective advantage for doubling of the 28 chromosome in aneuploid (only 2 of 116 plants were doubled in the main experiment), its donor-type regenerates appeared more uniform than regenerates of the original tetraploid.

Several plants were regenerated from protoplasts of two donor alfalfa clones of Regen-S by Johnson et al.[184] Both donors produced about the same frequency (12 to 15%) of aneuploids with 2n = 31. This was not the case for aneuploids with 2n = 33; one donor produced 6%, while the other donor produced none. One donor in this study and the donors in the study using genetic markers[230] did not produce aneuploids with 2n = 33. The donor genotype influences the type and frequency of aneuploidy. In a somatic cell selection experiment in alfalfa diploid clone HG2 (2n = 16) gave rise to a low frequency of aneuploids with 2n = 31 (2%) and 2n = 33 (2%), along with 65% diploids, 20% tetraploids, and 1% hexaploids. It is interesting that no aneuploids near 16 chromosome number were produced.[125] Alfalfa aneuploids with 2n = 17 and 18 have been produced by 3x-2x crosses and were thrifty and essentially indistinguishable from diploids.[231] Similarly, aneuploids near the tetraploid level found in populations or produced by interploid crosses were indistinguishable from tetraploids.[232] Hence, aneuploid regenerates that differ greatly from the donor in tissue culture experiments may be suspected of possessing variant characteristics not associated with aneuploidy as such.

Aneuploidy resulting from culture is more likely to result in genic imbalance in diploids as compared to polyploids. For example, only one aneuploid out of 110 regenerants was found in maize (2n = 2x = 20).[35] Even in polyploids chromosome number variants are extremely unlikely to contribute to new cultivars. The possible exception might be sugarcane, which is a highly polyploid hybrid.

D'Amato[227] has suggested that due to the nuclear genetic difference between differentiated cells and the genetic phenomenon occurring during the first few mitotic divisions of the callus formation, an explant comprises a heterogeneous cell population from as early as the first day of culture. During growth of callus of liquid cell cultures additional chromosomal changes appear to occur in high frequency. Hence, it is not surprising that chromosome number variation has often been reported for established plant cell cultures.[70] This variation in chromosome number appears to be influenced by several factors: (1) preexisting chromosome variation in plants used for culture initiation;[233,234] (2) nuclear fragmentation associated with the first cell division of callus initiation;[235] (3) endoreduplication or endomitosis occurring during culture initiation;[227] and (4) abnormalities of the mitotic process resulting in aneuploid cells. Evidence from several laboratories suggests that chromosome stability can be achieved by selection and use of appropriate culture medium[236] and by using short intervals between subcultures.[190] In this way, liquid cultures of Su/Su genotype of *N. tabacum* ($2n = 48$) maintained stable chromosome for 6 years.

B. STRUCTURAL CHANGES IN CHROMOSOMES

Though much work has not been done on the structural rearrangement, gross structural rearrangements rather than the alteration in the chromosome whole number appear to be the major cause of somaclonal variation.[70] Major chromosomal rearrangements which have resulted in somaclonal variation are deletions, inversions, duplications, and translocations.[3] In different regenerants of genus *Triticum* and its various regenerated hybrid, multivalents[193,201,202,237] and meiotic bridges[198] have been reported. Genetic variability in somaclones of *Zea mays* has been traced to the formation of heteromorphic bivalents and multivalents.[151,152,207] Correlation between somaclones and chromosomal structural modifications such as fragments, rings, and bridges in *Allium sativum*,[179] large chromosomes in *Apium graveolens*,[170] fragments, bridges, and multivalents in *Hordeum vulgare*,[72,180] acrocentrics, telocentrics, dicentrics, rDNA spacers, and mitotic C-banding in *Triticale*,[200,203,205] multivalents, bridges, heteromorphic bivalents, telocentrics, and isochromosomes in *Solanum tuberosum*,[193,194] translocations in *M. sativa*,[184] heteromorphic bivalents, multivalents, and telocentrics in *Avena sativa*,[38] and acrocentrics, telocentrics, and bridges in *Lolium*[22,206] have been reported. Late replicating heterochromatin has also been proposed to be involved.[38,207,238]

Structural changes have important implications where tissue and cell cultures are being used for cloning and as vehicles for the generation of variability.[219,239] Dicentric chromosomes and other structural changes have often been observed among cultured cells of many plant species.[72,240] Changes relative to the native karyotype in the mean ratio of longest and shortest chromosomes have been used as a rough indicator of the presence and extent of chromosomal rearrangement.[72] Detailed karyotypic analysis is limited to plant species with a small number of large chromosomes such as *Crepis capillaris*[238] and *Haplopappus gracilis*.[212] Extensive karyotypic analysis of cultured celery cells (*Apium graveolens*) revealed a net loss of long acrocentrics concomitant with a net increase in the number of short acrocentrics.[46] Several cells exhibited clear karyotypic changes. Orton and associates revealed several structural changes in celery by using genetically marked tissues.[25,63,96,102]

Chromosome structural changes have been reported in only one alfalfa study so far; however, the full range of abnormalities found in other plants may likely be found in alfalfa as research progresses. Johnson et al.[184] found 11 protoclones with chromosome interchanges among about 100 protoclones evaluated. Somaclonal variants with undetected chromosome structural changes have probably occurred in some studies. At the tetraploid level only major chromosome structural changes (e.g., a translocation that significantly increases chromosome length) are detectable in mitosis, because the morphology of all the eight chromosomes is similar. Meiotic analysis of tetraploids is complicated, because multivalents due to translocations must be distinguished from those sometimes occurring due to autotetraploidy.

Alhoowalia[198] examined meiosis in regenerated wheat plants and observed bridges and fragments at anaphase, suggesting inversions and deletions. Ryschka[199] reported trivalents and quadrivalents during meiosis in primary somaclones, suggesting translocations. Using both mitotic and meiotic studies, Karp and Maddock[193] observed no structure alterations in the wheat regenerants. Larkin[241] reported several structural changes in the chromosomes in wheat. A number of isochromosomes, non-N-banded chromosomes, telosomics, and translocations were observed. The translocations in wheat culture varied in size from large to small. Telocentrics were reported from wheat and rye regenerants.[204] Lapitan and Ryan[203] used C-banding to examine octoploid primary *Triticales* produced by crossing wheat and rye. Ten somaclonal amphidiploids were examined with C-banding. Deletions, translocations, and DNA amplifications were found to be very common. These authors also concluded that cultures showing extraordinary frequencies of translocations can be used for introgressing alien genes into wheat.

C. POINT MUTATIONS

1. Spontaneous Mutations

In addition to changes in chromosome number and structure, single gene mutations also appear to be common in somaclones of different crop plants.[200] Cooper et al.[137] have confirmed that wheat tissue culture can generate heritable alterations in the expressions of grain storage protein in wheat. An example of an apparent point mutation from culture is the variant gene conferring chlorsulfuron resistance in tobacco isolated following selection *in vitro*.[186,242] It appears that at least one of the mutants, the target enzyme for this herbicide, acetoacetate synthase, is not produced in greater amount, but has a greater tolerance to the chlorsulfuron than the wild-type enzyme.[242] Apparent activation of the previously silent element (mutator genes) has been implicated following plant cell culture. Examples of unstable (mutable) variants include a protoclone of *N. tabacum*[111] and a somatic hybrid of *N. tabacum* and *N. sylvestris*.[188]

Mutations in the heterozygous state are not so easy to detect as ploidy and detailed investigations with protoclones have not been done so far. However, Prat[57] and Lorz and Scowcroft[111] were able to detect mutations even in heterozygotes with somaclones of *N. sylvestris* and *N. tabacum*, respectively. Sacristan[243] reported that in *Phoma lingam* (resistant/tolerant somaclones) factor mutations and mitotic segregation can account for a large part of the somaclonal variation. With regard to the genetic markers in celery, clones isolated from the 6-month-old callus retained the original heterozygous phenotype at the five co-dominant isozyme markers.[102] All but one of the 96 regenerated plants were phenotypically indistinguishable from the original donor plant with respect to markers and overall morphology. The single variant plant died before flowering and could not be verified genetically. Selfed progeny of 55 of these 96 regenerates were examined for any segregation distortions at the disease resistance and flowering behavior locus, and for any new recessive mutations. Fertility was depressed among the remaining 41 plants, thus precluding genetic studies. Reduction in fertility has also been reported among lettuce plants regenerated from leaf mesophyll protoplasts.[158] No segregation distortions involving the seven marker systems were noted among the 55 selfed families. A new recessive mutation conditioning abnormal leaf morphology was recovered in one family, and this bred true to the next selfed generation (Figure 11). Similar mutations were also expected in selfed progeny of regenerates analyzed by Rappaport et al.[172]

Several reports have appeared in which variation was observed among regenerated celery plants. Because few or no genetic studies were conducted on the derivative culture or ensuing generations, the relationship of this variation to genetic instability *in vitro* cannot be established. For example, Williams and Collins[169] observed little or no phenotypic variation among regenerated celery plants, and speculated that the derivative culture, and indeed of celery in general, was highly stable. This conclusion was the transmission of this variation to regenerated plants. Moreover, regenerates may have contained new masked recessive mutations, which are sometimes present at high frequencies in regenerated plants.[96]

FIGURE 11. Abnormal leaf morphology mutant recovered in a plant regenerated from a celery callus culture: (a) normal and (b) mutant. (From Orton, T. J., *Cell Cult. Somatic Cell Genet. Plants,* 3, 345, 1986. With permission.)

Since growth of cells in culture results in single gene mutations,[44] cultures of microspores could result in recessive mutations that are visible in the Ro generation. Hence, it is likely that gametoclonal variants can be detected in the Ro generation; however, evaluation of R_1 progeny must be completed to detect the full spectrum of somaclonal variation. It is also likely that mutant cells do not regenerate as well as wild-type cells do. Hence, the mutation spectrum and frequency obtained from regenerated haploids may be strikingly different from diploid tissue, in which recessive mutations in heterozygous conditions may have no effect on regeneration capacity. Oono[21,146] has used gametoclonal variation to recover several mutants in rice as homozygous diploid plants. These included variation for heading date, seed fertility, plant height, morphology, and chlorophyll content.

It is also possible that certain variants may not be recovered using conventional mutagenesis, but may be recovered using gametoclonal variation. Using conventional mutagenesis or somaclonal variation, no new variants for the S locus governing self-incompatibility in *Lycospersicon peruvianum* could be detected.[181] However, S allele changes were detected for 17 of 53 gametoclonal regenerants of two genotypes. These changes were heritable and one new type of S allele was detected. This result suggests that the mutation spectrum obtained by gametoclonal variation may differ from that obtained by somaclonal variation.

2. Induced Mutations

The frequent occurrence of spontaneous variants in culture has minimized the need to increase variability by mutagenesis. When cells in culture are exposed to a mutagen, the background frequency of spontaneous variation makes it difficult to determine whether a given variant is spontaneous or induced. However, on a population basis the effect of mutagenesis can be established when the frequency of variants is significantly higher than in control cultures. Evidence of variant origin may be established if a certain type of variant occurs only after mutagenesis.

As the entire aspect of somaclonal variation is quite recent and frequency of variability even with any chemical treatment or induction is very high, little work has been done on the induction of variants. However, variants have been recently induced by chemicals in alfalfa which are described here in detail. Cells from only one diploid alfalfa genotype, the clone HG2, are known to have been exposed to chemical mutagens and plants regenerated.[125] The frequencies of phenotypic variants have been compared for four treatments: controls, methane sulfonic acid ethyl ester (EMS), ethionine (ETH), and sequential treatment with EMS and ETH. Spontaneously doubled plants without other apparent phenotypic changes were not considered variants in these studies. The frequency of phenotypic variants in the control experiments varied from 1% of 194 plants to 10% of 30 plants. Treatment with EMS alone did not increase the variant frequencies. A potentially important result is that the only dominant mutations studied so far (R85 and R121) were obtained after sequential treatment with EMS and ETH.[182]

Whereas EMS was used as a chemical mutagen in the experiment reported by Reisch et al.[124] and Reisch and Bingham,[182] ETH was used as an analog of methionine to select for cells that overproduced methionine. However, it is known that ETH can cause mutations by transethylating nucleic acids[244,245] and altering enzymes involved with DNA replication.[246] Hence, many cultures were exposed to two potential mutagens. A wide array of morphological variants of donor HG2 was regenerated after the combined treatment.[182] Leaf size of individual variants ranged from petite to normal; leaf morphology of variants ranged from two leaflets to many multifoliolate leaves. Plant stature varied from dwarf to taller and lateral branching was less uniform than the donor plant. Some variants produced significantly less adventitious rooting of cuttings than the donor.[182] Subsequent research has established that most variants tended to be less fertile than donor HG2 and that some were sterile. Variant R105 was an example of a variant with unusual leaf shape and morphology, but it was sterile.

Variant R85 is particularly striking in the celery plant with extra leafiness and more uniform side branching than donor HG2.[125] The R85 trait was expressed in about 50% of F_1 progeny in crosses with a normal clone, which indicated that the trait was under dominant control. In two generations of backcrossing the R85 trait to cultivated diploids, segregation was one mutant to one normal in each generation as expected for a dominant trait. Intercrossing backcross derivatives assumed to be heterozygous for the dominant allele produced about three mutants to one normal.

Variant R121 in celery was also very attractive and leafy in appearance, but was a semidwarf.[125] It generated as a tetraploid, which flowered profusely and was fully fertile. Crosses to tetraploid produced mainly R121-type semidwarf progeny in a ratio of five mutants to one normal. This indicates that the R121 trait is dominant and that the dominant allele is in the duplex condition (e.g., DDdd). The mutation from normal to semidwarf evidently took place in the diploid donor cells prior to spontaneous doubling (dd-Dd-DDdd). The R121 trait was backcrossed twice to normal cultivars and the trait was expressed in each generation at about one mutant to one normal as expected for a dominant in simplex condition (Dddd). A group of backcross-derived semidwarfs and six tetraploid F_1 progeny was obtained. In the F_2 generation produced by selfing, one of the six F_2 families segregated for purple and white flowers. To determine the locus at which the mutant allele occurred, the white-flowered F_2 plants were crossed with basic color factor-recessive stocks C1 and C2, and with flower pigment-recessive stock p. The allelism test indicated that the new mutant was allelic to C2. As in the study of Groose and Bingham,[230] a dominant C2 allele mutated during tissue culture. The C2 mutation in the latter case frequently reverted to the functional C2 state when returned to culture. It is thus tempting to speculate that C2 alleles are relatively mutable in tissue culture.

Certain other recessive traits induced by chemicals in alfalfa are either trivial, as in the case of albinos, or difficult to study.[125] Several variants expressed altered morphology, recemes, or flowers and reduced fertility. These altered characters were not expressed in the F_1, indicating that they were recessive or not heritable. The classification of these traits was difficult in the F_2 generation, as the F_2 generation of the control HG2 was quite variable with a broad range of morphology and fertility. These factors made it difficult to conclude that certain variant F_2 generations were different from the control. Although it had been easy to distinguish variants of donor HG2 against the uniform clonal background of HG2, it was difficult to distinguish them in a variable F_2 generation. Thus, genetic analyses were conclusive only for most unique and qualitative variants.

The mechanisms for induction for somaclonal variation are largely unexplored.[247] In wheat, variation in height, awn, head shape, and heading date appeared in induced mutants and were also common among the somaclonal mutants.[3] However, this apparent similarity is just a reflection of the fact that there are many loci in the wheat genome which can affect these morphologies. Sree Ramulu[181] documented that extensive mutagenic experiments in many laboratories have failed to generate any new S alleles governing gametophytic incompatibility. Brock[248] summarized the induced mutagenesis in higher plants. In these species which were adequately marked genetically (peas, barley, tomato, and *Arabidopsis*), the indications were that induced mutagenesis in most cases affected groups of adjacent genes. Although direction of induced mutagenesis from dominant to recessive is more common, Konzac[249] has described a radiation-induced dwarf mutant in the wheat line Karcag 522 which was dominant.

3. Gene Expression and Gene Amplification and Deamplification

Reports have now begun to appear in which molecular approaches have been used to analyze somaclones. Brettell et al.[200] used a cloned gene of Adh to evaluate an electrophoretic variant at the Adh1 locus. They were able to demonstrate that the somaclonal variant was the result of an alteration of a single base pair resulting in change of a single amino acid in the polypeptide sequence. Landsmann and Uhrig[196] used random cloned fragments of potato DNA to probe 12 regenerated potato plants. Two of the 12 plants had alterations resulting in deficiencies in ribosomal RNA genes, although this was not associated with a change in plant morphology. Such deficiencies in ribosomal DNA (rDNA deamplifications) have been previously reported for flax grown under stress conditions.[200] Hence, tissue culture stress might be influencing the genome of regenerating plants in a similar fashion to flax plants grown in stress environments. Brettell et al.[200] examined ribosomal rDNA genes and detected a reduction in rDNA spacer sequences in one family of regenerated plants. These and other reports confirm that somaclonal variation causes genetic changes that range from simple single base pair changes (mutations) to more complicated chromosomal changes such as deletions, translocations, and changes in chromosome number.

Transient amplifications of certain fractions of DNA, particularly heterochromatin, have been implicated in the differentiation process of cell culture, which may accentuate the lateness of chromosome breakage in culture.[25] Lapitan et al.[250] found a number of translocations and deletions in wheat and rye somaclones. Davies et al.[50] reported interesting chromosomes, an isochromosome, and a number of translocations and amplification and deamplification of genes. There are evidences of transient amplification of heterochromatin sequences[25] in cultured cells. In contrast, 70% reduction in 25S-rDNA was found in potato somaclones.[196] Repeat sequence of the rDNA spacer region was reported in triticale somaclones.[200] Gene amplification in response to selection has been demonstrated by Donn et al.[183] Alfalfa cell cultures were challenged with increasing levels of the nonselective herbicide L-phosphinothricin, which is a competitive inhibitor of glutamine synthase (GS). A selected line showing more than 20 times of the tolerance was stable at least for 12 months without selection. This line showed 6 to 7 times the GS activity, 8 times GS mRNA level, and 4 to 11 times the number of GS gene sequence present.

An extensive analysis of dihaploid plants of *N. sylvestris* derived by consecutive rounds of androgenesis was used to identify both heritable quantitative and qualitative changes in the nuclear DNA.[69] An increasing proportion of a 1.703-g/cm^3 satellite signaled an increase in GC-rich sequences or, alternatively, an increase in the degree of methylation. An increase in AT-rich sequences and progressive increase of inverted repeat sequences were also noticed. Two potato somaclones were shown to be deficient for certain repeat sequences.[196] In triticale, repeat sequence of rDNA spacer region has been observed.[251]

Brettell et al.[200] isolated DNA from 192 lines, restricted with Taq I and probed with Southern blots with a 2.7-kb Taq I rDNA spacer region clone. This probe represented the 12 × 136-bp repeat sequence of the rDNA spacer region. The cultivar currency gave four bands of 3.2, 3.1, 2.8, and 2.5 kb. One somaclone of currency displayed 80% reduction in the 2.5-kb (IR) band relative to the other bands. This depleted IR DNA variant was heritable.

The genes for 18S, 5.8S, and 26S cytoplasmic ribosomal RNA called *Nor* loci in eukaryotes are organized in tandem arrays of repeats in the nucleolar organizer regions on one or several chromosomes. Since *Nor* loci are variable among wheat *(Triticum aestivum)* cultivars, but uniform within each cultivar, Breiman et al.[75] analyzed these loci in order to evaluate somaclonal variation in this species. Thirty-eight progeny plants of the cvs. Chines Spring and Miriam were found to be stable in their organization of the *Nor* loci. On the other hand, three progeny plants of ND 7532 showed reduction in the number of rDNA spacers. Grain glutenin and gliadin profiles of sexual progeny of the plants derived from scutellar calli of Chines Spring, Miriam, and ND 7532 were identical to control, indicating low somaclonal variation in these grain proteins of the plants.

4. Mitotic Crossing-Over and Transposable Elements

Several other minute changes are responsible for the range of variation detected in regenerated plants.[37,130] The most frequently cited changes are small chromosomal rearrangements. While large changes like translocations, deletions, and inversions have all been detected, it is also likely that less dramatic structural changes that cannot be detected under light microscope also occur frequently. Small changes in chromosome structure will also alter expression and genetic transmission of specific genes. These changes include deletion of one copy (or copies) of a gene, duplication of one copy (copies) of a gene, or gene conversion during the repair process. In addition, recombination or chromosomal breakage may occur in preferential regions or ''hot spots'' of particular chromosomes, thereby affecting some regions of the genome in a disproportionately higher frequency. With all the phenomena occurring simultaneously the altered segregation ratios as detected by Lorz and Scowcroft[111] are not unexpected.

In addition, mitotic crossing-over could also account for some of the variation detected in regenerated plants. This could include both symmetric and asymmetric recombination. Mitotic crossing-over may also account for the recovery of homozygous recessive single gene mutations in some regenerated plants.[44] As breeders previously had access only to variation that is normally transmitted through meiosis, the recovery of the products of mitotic crossing-over may constitute a unique source of new genetic variation.

Mobile element transposition (transposable elements) can create heritable genetic changes by number of means particularly due to insertional disruption of the integrity of the gene, an element acting as a promotor moving to or from an affected gene; activation of silent genes; piggy-backing of genes to positions with different regulation; and the transposition of a nonautonomous element that may subjugate a gene under the control of a master element elsewhere.[3,142] In fact, mobilization of transposable element is one response to ''genomic shock''.[252] Apparent activation of previously silent elements (mutator genes) has been implicated following plant cell culture by several authors. Variation in the insertion of plasmid-like DNA found in mitochondria of cms-corn has been detected in corn cell cultures.[253] Heterozygous light-green (Su/su) somaclones with a high frequency of colored spots on the leaf surface have been detected for a clone of both *N. tabacum*[111] and *N. tabacum* + *N. sylvestris* somatic hybrids.[44] The somatic hybrids had an unstable pattern of inheritance that would be consistent with an unstable gene.

A qualitative change in gene expression was noted in an experiment designed to monitor loss of genetically marked chromosomes during tissue culture in alfalfa.[117] A white-flowered variant was regenerated from a purple donor carrying one functional allele for anthocyanin synthesis.[230,254] Chromosome counts revealed that anthocyanin synthesis switched off during culture without loss of the chromosome carrying the functional allele for synthesis. The functional allele had mutated to a nonfunctional state. When the white-flowered mutant (WFM) was recultured, about 40% of the regenerates of WFM unexpectedly reverted to normal function and were purple flowered. Basic color factor allele ''C'', which is necessary for anthocyanin synthesis in all parts of the plants, is known to be involved in the unstable system. Test crosses of WFM and purple and white regenerates of WFM indicated that purple regenerates carried one dominant C allele (Cccc) and the WFM and white regenerates were homozygous recessive (cccc). Regeneration of white and purple plants could be explained by the donor being a periclinal chimera; however, a test of genetic transmission of the unstable allele proved that genetic reversion takes place. To test genetic transmission, WFM was crossed to a white-flowered tester and all progeny were white flowered. Then, several progeny were challenged to regenerate and revertants were observed among regenerates of two of five progeny that regenerated. The unstable condition was, therefore, transferred through a single cell (zygote) to sexual progeny that reverted in culture, indicating genetic reversion. The reversion in plants was 2.22% based on four pigmented shoots among 180 shoots from crown buds and only 0.12% based on 9 flowers with revertant sectors among about 7500 white flowers. Average reversion of WFM *in vitro* was 23%. Groose and Bingham[230] thus concluded that the mutant allele was much more unstable *in vitro*. They also proposed that genetic phenomena in the alfalfa flower system were typical of the interaction between a transposable element and a gene.

The example described above, however, lacks the definitive molecular demonstration that DNA transposition is involved. A more satisfying demonstration has been recently reported in maize by Peschke et al.[62] A high frequency of structurally altered chromosomes in maize (*Z. mays* L.) plants regenerated from tissue culture led to the prediction that newly activated transposable elements could be detected in regenerated plants.[62] Test cross of 1200 progeny from 301 regenerated maize plants confirmed that 10 regenerated plants from two independent embryo cell lines contained an active Ac transposable element. No active Ac elements were present in explant sources. Recovery of transposable element activity in regenerated plants indicated that the same tissue culture-derived genetic variability might be the result of insertion or excision of transposable elements, or both.

5. Changes in Cytogenes

The structural and functional organization of higher plant chloroplast genomes (cpDNA) is now fairly well understood.[255,256] Mitochondrial genome (mtDNA) has been more difficult to isolate and is generally larger than cpDNA.[257,258] The detailed information on cpDNA and mtDNA which has been obtained only during recent years made it possible to gain tremendous insight on the role of cytogenes in somaclonal variation.

Frequency of nuclear as well as cytoplasmic encoded variation increases when somatic cells or tissues are passed through *in vitro* culture as compared to conventional seed or vegetative propagation. The role of nuclear encoded functions in somaclonal variation is becoming clear, but cytoplasmic encoded variation is less clear. Sibi[34] observed morphological variation in plants regenerated from lettuce. The variation was markedly inherited.[259-262]

In addition to single nuclear gene changes, cytoplasmic genetic changes have also been detected in somaclonal variants. The most detailed work on the evaluation of cytoplasmic genetic via somaclonal variation has been done on Gengenbach and colleagues by evaluating plants for two cytoplasmic traits.[263] Sensitivity to host-specific toxin of *Drechslera maydis* race T (the causative agent of Southern corn leaf blight) is associated with all genotypes containing Texas male sterility (cms-T) cytoplasm. In seed-derived plants these two characters are tightly associated. Gengenbach et al.[263] selected cell lines for resistance to toxin and regenerated resistant plants with the aim to recover resistant cms breeding lines. However, resistance was associated with a concomitant reversion to male fertility. The restriction endonuclease pattern of mitochondrial DNA (mtDNA) revealed[195] significant changes in mitochondrial DNA of plants derived from cell culture.

A change in chlorophyll pigmentation was observed in tissue culture of alfalfa progeny of WFM.[125] A variant with chlorophyll-deficient sectors (CDS) was regenerated along with 25 normal green plants. Interestingly, the white-flowered donor did not yield regenerates that were revertant for flower color (flower color being stable white). Some shoots developing from the crown of the CDS plant are pure chlorophyll deficient. Flowers in pure chlorophyll-deficient sectors have been used as pollen and seed parents to study the transmission of CDS. Male and female transmission of CDS was studied in reciprocal crosses over several generations and the findings were unexpected. The chlorophyll-deficient sector was transmitted biparentally as a cytoplasmic trait and the transmission was higher through the pollen.[264-267]

No changes were observed in the restriction patterns of mitochondrial DNAs of 20 alfalfa protoclones when digested with five restriction endonucleases.[184,217,268] The restriction pattern of chloroplast DNAs from 23 protoclones was identical to the parents following digestion with four endonucleases. However, digestion with Xba I identified an 8.4 kb band in the parents and one protoclone that was absent in 22 other protoclones.[268] Rose et al.[268] and Johnson et al.[217] concluded that in alfalfa major cpDNA or mtDNA changes do not occur as a result of protoplast culture. The authors recognized that smaller changes not detectable by their system could occur, or that examination of additional protoclones may identify major rearrangements.

D. GENETIC ANALYSIS

Genetic analysis of variants has revealed several alterations that occur at the gene level, most of them being single gene mutations. Recessive single gene mutations are suspected if the variant does not appear in R_0 plant and self-fertilized R_1 progeny segregate in an expected 3:1 Mendelian ratio for the trait under consideration. To confirm that trait of interest is indeed a single gene trait, progeny tests should be completed on R_1 plants to identify segregators and nonsegregators and to ensure continued 3:1 transmission in R_2 segregating populations. When possible, complementation tests should be completed with known mutants for the crop of interest. This type of detailed genetic analysis to confirm single gene traits has been completed for several tomato somaclones[44] and has been used to map somaclones to specific loci (Figure 4). A 3:1 ratio in R_1 plants, the first step in this type of analysis, has been reported for maize,[35] *N. sylvestris*,[57] and rice.[58,59]

In some cases use of specific genetically marked strains has aided in evaluation of plants regenerated from cell culture. Dulieu and Barbier[185] and Lorz and Scowcroft[111] have regenerated plants from *N. tabacum* with specific chlorophyll deficiency markers present in heterozygous condition. By regenerating plants from heterozygous tissue, genetic changes could be detected by appearance of albino or dark-green regenerated plants. Dulieu and Barbier[185] have reported a high frequency (9.6%) of variant regenerates at the al and y loci and have ascribed these genetic changes to deletion and mitotic recombinations. Lorz and Scowcroft[111] detected genetic changes in 3.7% of morphogenetic cell colonies using the Su locus. When light-green regenerated shoots (Su/su) were self-fertilized, up to 37% of the regenerated plants had segregation ratios that were distorted from a 1:2:1 ratio of dark-green/light-green/albino. These authors did not speculate on the genetic basis of altered R_1 segregation ratios. The value of using genetically marked heterozygous donor material was evident in each of these studies, as the authors were able to demonstrate that much of the reported variation was proportional to the duration of culture.

Somaclonal variation also depends on the occurrence and recovery of regenerated plants with Mendelian and non-Mendelian genetic variation from cell cultures. The genetic variation in these plants appears to result from both preexisting variation in the explant donor tissue as well as from the variation induced by cell culture. These changes in the integrity of the genome are attributed to mutant induction, mitotic crossing-over, organelle mutation, and sorting. Evans et al.[2] proposed two important selection steps. These steps of the procedure serve as sieves to recover a population of R_1 plants that are most suitable for a breeding program: (1) the culture medium provides a sieve for singling out cells from the foundation cell population which possesses genome competence for plantlet regeneration; and (2) greenhouse selection identifies those Ro plants with normal development that are capable of undergoing flower and fruit formation and seed setting. This permits elimination of Ro plants with deleterious genetic backgrounds. The Ro plants are selfed and the resulting R_1 seed is used for field trial evaluation and selection of breeding lines.

Genetic analysis of somaclonal variation has been helpful in revealing the genetic architecture of several plants. For instance, in alfalfa (which is an outcrossing species) null alleles and/or cryptic deletions have been found to contribute to inbreeding depression and outcrossing.[269] A spontaneous recessive mutation for simple leaves and rudimentary flowers occurred in a diploid clone (CADLSh-3) of alfalfa that had undergone selection for several generations. The trait was designated MCB and morphology of two phenotypically MCB mutations was identical in tetraploid alfalfa. One of the tetraploid mutations was induced by X-rays, and the other was discovered in a population.[270,271] A simple-leaf trait that was studied at Wisconsin in 1969 had normal flowers and therefore was different from the MCB trait.[272] The diploid MCB mutation evidently arose in callus culture of CADLSh-3. Evidence for this is the fact that the F_2 generation of progeny involving a plant regenerated from CADLSh-3 segregated three normal to one MCB. This MCB trait in diploid alfalfa and the ones studied in tetraploid alfalfa by Murray and Craig[271] behaved as single-recessive genes.

A spontaneous diploid variant (C49) with abnormal flowers arose in a control culture of HG2 alfalfa.[125] This floral trait was expressed directly in the diploid genome of C49, although it behaved as a recessive trait when crossed to several different normals in genetic analysis (HG2 and all of its variants are self-sterile and must be crossed to normals for genetic analysis). An F_2 generation was produced by sib-mating F plants to minimize inbreeding. The F_2 derived from C49 segregated 59 normal to 13 mutant, suggesting the 3:1 segregation ratio expected for a single gene with a deficiency in the mutant class (expected 54:18). Examination of F_2 segregations from individual F_1 plants showed that about half segregated three normal to one mutant, and half were deficient for the mutant class. This accounted for the deficiency of recessives in the F_2 and suggested that C49 may be carrying only one recessive allele and that the oppositional allele may be a null allele in the diploid genome of HG2.

Genetic analysis was completed by evaluating self-fertilized R_2 progeny of selected R_1 and in some cases by crossing with known mutant lines of tomato.[2] The most extensive genetic analysis has been completed with the tangerine-virescent (tv-tcl) fruit variant and the jointless pedicel (Figure 12). This character is a single recessive allele that results in orange flowers and fruit and yellow virescent leaves. The R_1 segregation data observed in the field (30 red to 6 tangerine fruit) first suggested control of the trait by a single recessive allele. Fruit was collected from eight individual self-fertilized plants. Six of these R_1 plants had red fruit, while two had tangerine fruit. The two tangerine fruits contained seed that

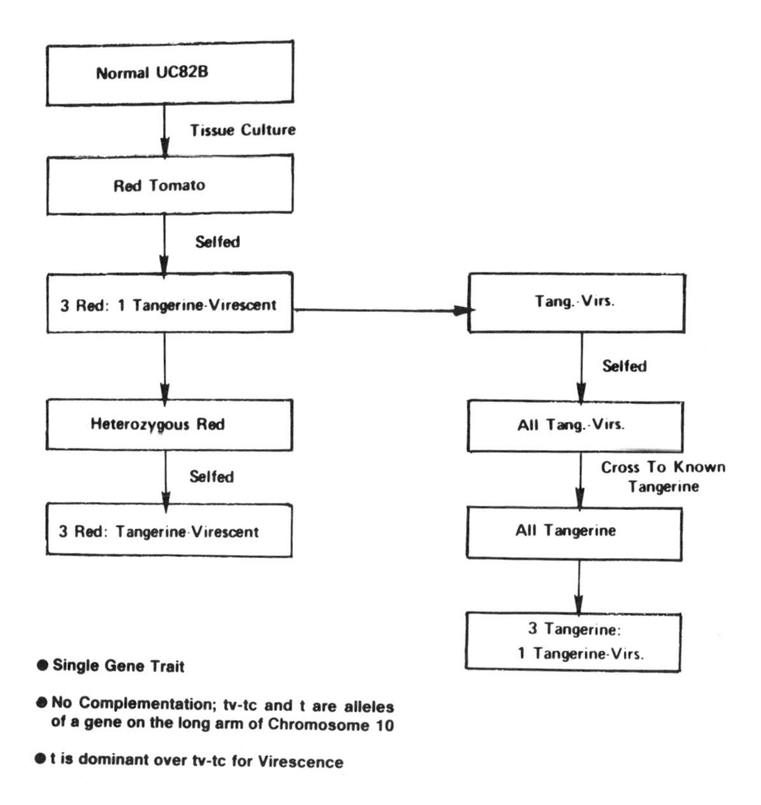

FIGURE 12. Scheme for the procedure used for genetic analysis of tangerine-virescent somaclonal variant. (From Evans, D. A., Sharp, W. R., and Medina-Filho, H. P., *Am. J. Bot.*, 71, 759, 1984. With permission.)

bred true for tangerine-virescent phenotype. Of the six red-fruited plants, three bred true for red fruit, and these segregated for red vs. tangerine-virescent. The pooled progeny of the three segregating plants showed a 3:1 ratio. In addition, among the segregating R_2 progeny the flower, fruit, and leaf color defects cosegregated suggesting control of two pigments, carotenoids (flower and fruit) and chlorophyll a (leaf) by pleiotropic genes. Such pleiotropic genes controlling carotenoids and chlorophyll (two compounds found in plastids) have been reported previously. Evans et al.[2] crossed a single mutant gene R_1 plant to a known fruit variant, tangerine, and all hybrid progeny had tangerine fruit. Using this complementation test, it was concluded that the new somaclone (tv-tcl) was a mutant for a new allele in a previously known gene at position 95 on the long arm of chromosome 10. In addition, by evaluating the self-fertilized progeny of the hybrids between the tv-tcl and the earlier tangerine mutant for the virescence character,[44] it has been possible to conclude that the locus contains two elements, one controlling chlorophyll synthesis and the other controlling carotenoid synthesis, that mutate independently. The new tv allele recovered by somaclonal variation is recessive to the allele for the virescent phenotype.

A well-characterized recessive mutant trait of tomato is jointless pedicel. Plants with this character are desirable for mechanical harvesting as the harvested fruit has no stem attached. Two mutants with jointless pedicel were identified in somaclones of UC82B.[2] The first variant was normal and recessive which was detected in about 25% of the R_1 progeny. However, the original regenerated plant of the second variant (j-tc2) already expressed the mutant trait. This jointless Ro plant bred true in both the R_1 and R_2 generations. Based on crosses with a known jointless mutant and with a normal jointed tomato, it was ascertained that the original Ro plant was homozygous recessive for the jointless mutant.[2] This new

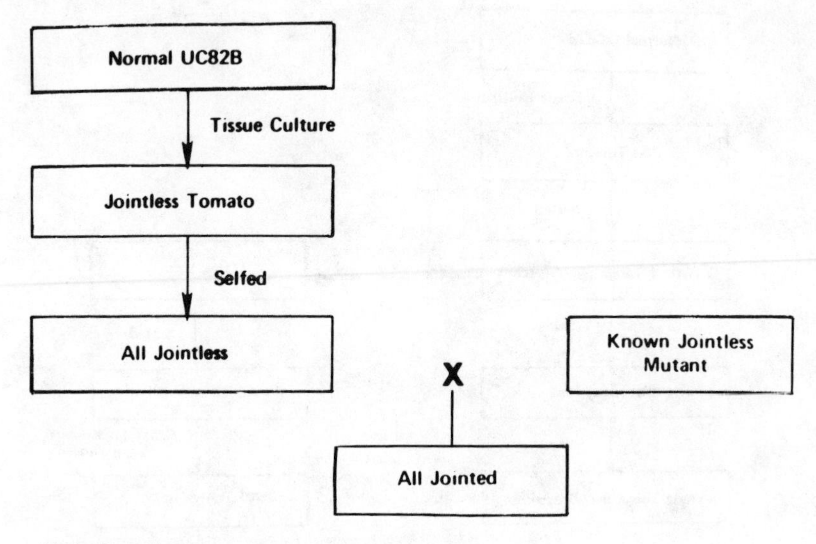

FIGURE 13. Scheme for the procedure used for genetic analysis of the jointless pedicel somaclonal variant. (From Evans, D. A., Sharp, W. R., and Medina-Filho, H. P., *Am. J. Bot.*, 71, 759, 1984. With permission.)

mutation complements the known mutation so that it is not encoded in the same gene (Figure 13) and may represent a new mutation in an undescribed gene encoding for jointless pedicel.

The chlorophyll-deficient mutant in tomato was identified in the R_1 generation in the greenhouse by Evans et al.[2] This mottled mutation (m-tcl) uncovered among somaclones of UC82B is depicted in Figure 14. Selected mottled R_1 plants breed true for the mottled appearance (Figure 15). The variegation is somewhat similar to the previously reported plastome mutant. When chlorophyll-deficient sectors of plastome mutants are placed under *in vitro* conditions, it is possible to establish shoot cultures of pure chlorophyll-deficient tissue.[273] However, the mottled phenotype of m-tcl is quite stable and is maintained in shoots regenerated from leaf explants of mottled plants. In addition, it is also possible to discern orange-red mottling on fruits of homozygous (m-tcl/m-tcl) plants. Once again, as with tv-tcl, the m-tcl mutant has an effect on both the chlorophyll and carotenoid pigments. This mutant appears to be distinct from hundreds of previously reported tomato mutants.

Several other single gene mutations have been identified in R_1 progeny tests of regenerated UC82B tomato cultivars including a semidominant allele controlling an electrophoretic variant of Adh and a dominant allele conferring resistance to *F. oxysporium* race 2. In addition, somaclones have been identified with agriculturally important traits. For instance, new lines have been developed with higher pigment (Figure 16).

Morphological variants have also been detected among regenerated plants of UC82B that breed true in R_1 generation.[2] In many cases the altered somaclones have larger leaf size, darker leaf color, or reduced fruit set. In some cases these R_1 plants have been evaluated and preliminary evidence obtained by restriction enzyme analysis isolated chloroplast DNA suggests that genetic changes have occurred in the chloroplast DNA.[2] Chloroplast genes are inherited maternally in tomato; hence, one would expect the variant to breed true in R_1 progeny. It is not surprising that chloroplast DNA variants are uncovered, as the number of plastids in a developing shoot apex is much smaller than in a mature cell.[274,275] Hence, if a

FIGURE 14. Mottled leaf somaclonal variant (R_1 progeny test). (From Evans, D. A., Sharp, W. R., and Medina-Filho, H. P., *Am. J. Bot.*, 71, 759, 1984. With permission.)

FIGURE 15. R_2 progeny test from selected, self-fertilized, mottled R_1 plant breeding true for mottled leaf type. (From Evans, D. A., Sharp, W. R., and Medina-Filho, H. P., *Am. J. Bot.*, 71, 759, 1984. With permission.)

FIGURE 16. Fruit of normal UC82B (left) and somaclone (right) with higher pigment. The concentration of lycopene is significantly higher in this variant. (From Evans, D. A., Sharp, W. R., and Medina-Filho, H. P., *Am. J. Bot.*, 71, 759, 1984. With permission.)

mutation occurs in chloroplast DNA, it is more likely to become dominant plastid type during sorting out if the mutant occurs in 1 of 10 plastids than in 1 of 100 plastids.

In addition, several other tomato varieties have been regenerated using the procedure developed for UC82B. Single gene mutations have been recovered in several of these varieties including several male sterile and chlorophyll-deficient mutations.[2] A recessive mutant for chlorophyll deficiency (Figure 17) was identified in one tomato breeding line. This new mutant appears distinct from other known chlorophyll-deficient mutations. New fruit color mutants have also been detected including dominant orange mutation in C38 and a recessive yellow mutation in C40 (Figure 18). C38 and C40 are both Campbell Soup Company commercial tomato varieties. The orange mutation appears to be pleiotropic as it is associated with bushy foliage and alters flower color. The yellow mutation, on the other hand, affects only fruit color.

A substantial number of somaclonal variants have been shown to transmit the altered phenotype to sexually produced progeny. Examples exist in oats, *Hordeum* spp., *Lolium* spp., rice, wheat, maize, rape, lettuce, celery, tomato, alfalfa, and *Nicotiana*.[130] Genetic analysis of a number of these variants has revealed that progeny segregation patterns confirm to Mendelian expectations.[25] Recessive mutants occur more frequently, but dominant and codominant mutations also appear. Heterozygosity in the original explant could inflate the observed frequency of somaclonal mutants.[110] However, evidence from the analysis of somaclonal variation among dihaploids derived from anther cultures[17,57] and among somaclones derived by protoplast culture[111] dispels residual heterozygosity as a major cause of

FIGURE 17. Chlorophyll-deficient somaclone with green stripe on cotyledon. (From Evans, D. A., Sharp, W. R., and Medina-Filho, H. P., *Am. J. Bot.*, 71, 759, 1984. With permission.)

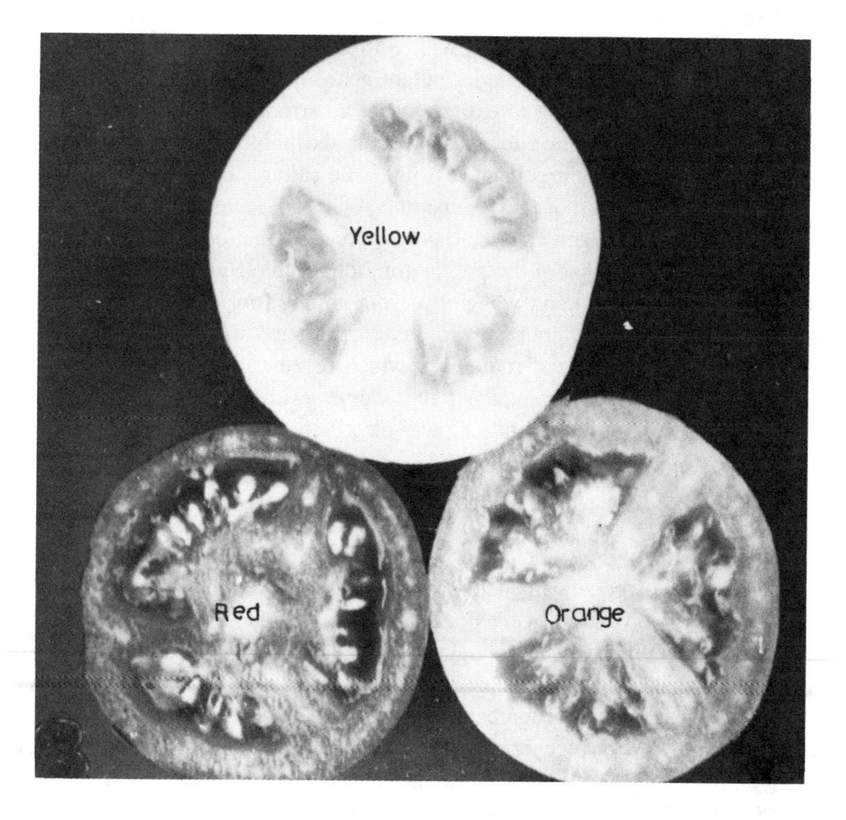

FIGURE 18. Red (control), orange, and yellow (somaclones) fruit. Orange is controlled by a single dominant allele. Yellow is controlled by a single recessive gene. (From Evans, D. A., Sharp, W. R., and Medina-Filho, H. P., *Am. J. Bot.*, 71, 759, 1984. With permission.)

variation. Segregation of aberrations, such as translocations, during meiosis in the initial regenerant can then lead to additional genetic change.

Yaqui 50E variants of wheat were examined in backcross and testcross using both mature plant height and seedling responses to gibberellic acid.[241] In all these crosses shortness was dominant in F_1 plants; gibberellin responsiveness was generally associated with talls and unresponsiveness was associated with short variants. F_2 generation values did not fall into classes indicating involvement of single genes. Similarly, the variant alleles conferring awnlessness appeared to be dominant to the donor tip awned condition.[274] However, F_2 data from other crosses were significantly nonconforming to a 3:1 pattern. Larkin et al.[274] also described heritable heading date variants in somaclonal lines of Yaqui 50E. Extremely SC_2 individuals gave early SC_3 progeny and late SC_2 individuals gave late SC_3 progeny. When two brown glumed Yaqui 50E were crossed back to Yaqui 50E, both groups of F_1 plants were brown glumed indicating dominance or semidominance of the variant trait. Most of the variants for grain color appeared to be segregating in the SC_2 generation.

With development of sophisticated techniques, results on genetic analysis at the molecular level are beginning to appear,[51] SC_1 regenerants of the wheat cv. Millewa have been screened to SC_2 families for variation in the expression of NAD-dependent Adh-coded genes on group 4 chromosomes.[50] Seventeen regenerants gave rise to progeny having altered Adh1 zymograms. Progeny with altered zymograms in 13 of these regenerants were aneuploid. The remaining four regenerants gave rise to euploid progeny with altered Adh1 zymograms. The genetic of three of these somaclonal mutants revealed that regenerants possessed a 4Aα isochromosome, a 3BS/4Aα-translocation, and a 7BS/4Aα-translocation. Ryan and Scowcroft[52] recently regenerated 149 plants from tissue culture of wheat and found one heterozygous regenerant for a variant pattern characterized by the presence of at least five new β-amylase isozyme bands. The F_2 homozygous variant crossed back to the parent segregated in an approximate 3:1 ratio. It was, however, not established whether this represented a dominant or codominant single mutant gene. It was concluded that this variant represented a rare mutation leading to expression of a currently unexpressed locus.

Analysis of variants at defined loci also reveals the nature of variation and possible mechanisms responsible. Plants regenerated from tissue cultures of maize were screened for variants of Adh1 and Adh2. Out of 645 regenerants, one stable mutant of Adh1 was detected by Brettell et al.[51] The mutant gene (Adh1-Usv) produced a functional enzyme with a slower electrophoretic mobility than that of the proginator Adh1-S allele and was stably transmitted to the progeny. Because the mutant was not present among four other plants derived from the same immature embryo, the mutant arose as a consequence of culture procedure. The gene of Adh1-Usv was cloned and sequenced which revealed a single base change. This gene could translate a polypeptide sequence to a valine residue, substituting for a glutamic acid residue resulting in a protein with a slower electrophoretic mobility.

VI. APPLICATIONS

In the beginning many considered the advantages of tissue culture to be its ability to produce large numbers of genetically identical plants. Later studies revealed that this was not really so and several variants were produced with the plants regenerated from single cells. These variants were, however, thought to be undesirable in the beginning.[2] Probably the first work recognizing the potential of somaclonal variation, a source of variability for crop improvement, was done with sugarcane and later with potatoes.[4,33,276,277] In these studies researchers found that plants regenerated from callus and protoplast cultures varied in a number of traits, including morphological characteristics, maturity date, yield, and response to pathogens (Table 2). This work which began with sexually propagated crops was later extended to many plant species. Several useful characters have been obtained in various crops including tobacco,[1,103,173,282,312] corn,[1,103,173,313] barley,[1,103] sorghum,[1] wheat,[36,103,173]

TABLE 2
Useful Somaclonal Variation in Different Plants

Crop	Mode of propagation	Useful somaclonal variants/mutations	Ref.
Alfalfa	Seed	Yield, disease resistance	197
		Multifoliate leaves, plant habit, plant height, dry matter yield	143
		Resistance to phosphinothricin	198
		Resistance to *Fusarium oxysporum* f. sp. *medicaginis*	229, 278, 279
		Resistance to *Verticillium albotrum*	280
Barley	Seed	Haploidy introgression	54
		Plant height, tillering, and fertility	281, 282
Carrot	Seed	Resistance to glyphosate	283
Cassava	Seed	SC_2 plants showed changes in plant vigor, lateral growth internode distance, leaf size, flowering date, seed production, nitrogen content, and plant fresh and dry weight	197
		SC_1 plants showed changes in leaf shape, leaf size, flower size, pubescence, internode distance, seed size, number of seeds per plant, deep green leaves, larger flowers, plant height, and seed yield	197
Celery	Seed	Disease resistance	172, 284
		Leaf shape, size, and form, flower morphology, plant height, fascination, pubescence, anthocyanin pigmentation, essential oil composition	281
		Resistance to *Cercospora apii, Septoria apiicola, Pseudomonas cichorii, F. oxysporum* f. sp. *apii*	285
		Resistance to *F. oxysporum* f. sp. *apii*	286
Geranium	Vegetative (cutting)	Morphological characteristics (floral, vegetative), essential oil composition, anthocyanin pigmentation	9, 287, 288
Lettuce	Seed	Seedling vigor, heterosis	260, 289
		Leaf weight, length, width, flatness, color, and bud number	281
Lotus	Vegetative	Resistance to 2,4-D	290
Maize	Seed	Pollen fertility	281
		Endosperm and seedling mutants; *Drechslera maydis* race T toxin resistance, mt-DNA sequence rearrangement	22, 263
		Resistance to *Helminthosporium maydis* race T and HmT-toxin	208, 291, 292
		Height, disease resistance, reversion to male fertility	64, 155, 263
		Resistance to *H. maydis*	101
Oat	Seed	Plant height, heading date, leaf striping, twin culms, awn morphology	281
		Height, heading date	92
		Height, heading date, leaf striping, awns resistance to *H. oryzae*	129, 293
Onion	Seed	Bulb size, shape	179
		Bulb size and shape, clone number, aerial bulbil germination	281
Orange	Cutting	Salt tolerance	294
Potato	Vegetative (tuber)	Yield, quality, uniformity, disease resistance	24, 33, 78, 80
		Resistance to *Phytophthora infestans*	79, 295—297
		Resistance to *Alternaria solani*	24
		Tuber shape, yield, maturity date, plant habit, stem, leaf, and flower morphology, early and late blight resistance	130
Rape	Seed	Glucosinolate content	98, 298
		Flowering time, growth habit, waxiness, glucosinolate content	130, 218
		Phoma lingam tolerance	243, 299
		Seed color	300
		Alteration in flowering time, glucosinolate content, growth habit	281

TABLE 2 (continued)
Useful Somaclonal Variation in Different Plants

Crop	Mode of propagation	Useful somaclonal variants/mutations	Ref.
Rice	Seed	Yield components, height, heading date, salt tolerance	19, 20, 144, 150
		Resistance to *H. oryzae*	301
		Tiller number, panicle size, seed fertility, flowering date, plant height	130
Sorghum	seed	Fertility, leaf morphology, growth habit	281
Sugarcane	Vegetative (tiller)	Resistance to *Sclerospora sacchari*	159, 302, 303
		Resistance to *H. sacchari*	1, 159
		Resistance to *Puccinia melanocephala*	304
		Eye spot, Fiji virus, downy mildew, culmicolus spot disease, auricle length, estrase isoenzyme, sugar yield	130
		Resistance to Fiji disease	159, 302
		Yield, sugar content, disease resistance	4—6, 128
Tobacco	Seed	Plant height, leaf size, yield, alkaloids, reducing sugars, specific leaf chlorophyll loss	130
		Leaf color	185
		Leaf spots	2
		Resistance to *Pseudomonas tabaci*	305
		Resistance to *P. syringae*	306
		Resistance to *A. alternaria*	306
		Resistance to paraquat	307
		Resistance to sulfonylureas	186, 240
		Resistance to glyphosate	308
		Resistance to amitrol	308
		Resistance to picrolam	34, 106
		Resistance to hydroxylurea	108, 309
		Resistance to *Peronospora tabacina*	310
		Flowering rate, height, yield, alkaloid content, disease resistance, male sterility, insect resistance, herbicide resistance	12—14, 106—109
Tomato	Seed	Jointless pedicels, male sterility	44
		Male sterility, jointless pedicels, fruit color, indeterminate growth	130
		Male sterility, jointless pedicel, tangerine varescent leaf, flower, and fruit color, lethal chlorophyll deficiency, virescence, mottled leaf appearance, fruit ripening	2
		Resistance to *F. oxysporum* f. sp. *lycopersici* race 2	307, 311
Wheat	Seed	Grain color, height, tiller number, seed storage protein	128, 198
		Plant height, spike shape, awns, maturity, tillering, leaf wax	130
		Albino wheat plants	131

clover,[103,173] carrot,[1] pineapple,[1] lettuce,[314] garlic,[1,103] geranium,[1] *Brassica*,[1,103,282] alfalfa,[125,216] celery,[286,287] tomato,[2,44,56] strawberry,[310,315] and sweet potato.[316]

In recent years, researchers have directed increasing effort toward the isolation of mutants from plant tissue cultures without applying any selection. These efforts have been based on the observations that the process of growing plant cells in culture yields a high frequency of stable, heritable variants which express traits useful in crop improvement.[317-323]

A. STRATEGIES FOR CROP IMPROVEMENT

Out of all the crops investigated, somaclonal variation has proved quite successful in the improvement of crops such as potato, tomato, celery, alfalfa, tobacco, and sugarcane.

TABLE 3
Comparison of Procedures for Variety Development

	Somaclonal variation	Gametoclonal variation	Mutation breeding	Backcross program
Source of variation	Spontaneous and induced	Spontaneous and induced	Induced	Natural populations
Likelihood of success	Undirected variation	Some direction, high percentage of success	Undirected variation	Guaranteed except where linkages not broken
Alteration of qualitative traits	Possible	Possible	Possible	Rarely successful
Rate of progress	More than one trait per generation	More than one trait per generation	More than one trait per generation	One trait in five to seven sexual generations
Chimerism	None or low frequency	None or low frequency	Major problem	None
Species limits	In all species that can be regenerated	In all species that can be regenerated from anthers	All species	Only sexually propagated crops
Time for breeding line development	One generation	One generation	Three generations	Up to six generations

From Evans, D. A., Sharp, W. R., and Medina-Filho, H. P., *Am. J. Bot.*, 71, 759, 1984. With permission.

Evans and associates have done extensive work on somaclonal variation in tomato.[2,56] They have published a description of experiments designed to generate somaclones of tomato and to ascertain the genetic basis of somaclonal variation. As the work done with tomato by Evans and co-workers can be considered as a model for the utilization of somaclonal variation in other crops, it has been discussed in detail

The development of a new crop variety by somaclonal variation has been compared to more conventional approaches such as mutation breeding and backcross breeding (Table 3). It is evident that this new technique is far more superior to the conventional breeding. Based on the work described earlier, several strategies for the use of somaclonal and gametoclonal variation have been worked out recently by Evans and associates.[2] As single gene mutations and organelle gene mutations have been produced by somaclonal variation, one obvious strategy is to introduce the best available varieties into cell culture to select for improvement of a specific character.[2] Hence, somaclonal variation could be used to uncover new variants that retain all the favorable qualities of an existing variety while adding one additional trait, such as disease resistance or herbicide resistance. Work with sugarcane and tomato has already suggested that this approach is feasible. A simple procedure as given by Evans et al.[2] is outlined in Figure 19. Once new R_1 variants are identified, these should be field tested in replicated plots to ascertain genetic stability. Seed should be increased at the same time to permit rapid variety development of promising lines. Reciprocal crosses between desirable R_1 and seed-derived controls can be used to gain an understanding of the genetic basis of somaclonal variants. New promising breeding lines can be reintroduced into cell culture to add an additional character or to improve agronomic performance of a selected somaclonal variant. By using this approach, it is possible to produce new breeding lines with desirable traits in a short period of time (Table 4).

It is also desirable to obtain true breeding homozygous diploid plants at many steps during a conventional breeding program. Hence, the value of gametoclonal variation for a rapid production of doubled lines has long been recognized.[2] One method for rapidly generating new breeding lines using gametoclonal variation is outlined in Figure 20. Two plants each with desirable characteristics can be crossed to produce a hybrid that expresses all the

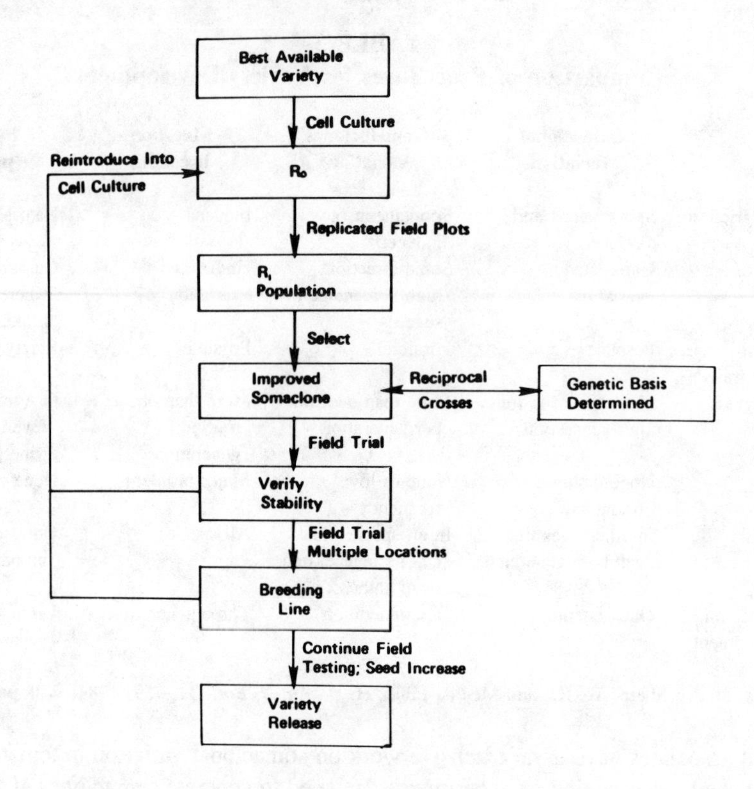

FIGURE 19. Breeding strategy for the use of somaclonal variation for the development of new plant varieties. (From Evans, D. A., Sharp, W. R., and Medina-Filho, H. P., *Am. J. Bot.*, 71, 759, 1984. With permission.)

TABLE 4
Procedures for New Variety Development Using Four Breeding Approaches for an Annual Crop Species

Somaclonal variation	Gametoclonal variation	Mutation breeding	Backcross program
Regenerate plants	Regenerate plants	Mutagenize seed raise M_1 plants	Hybridize with wild species
Self-fertilize	Double chromosomes	—	Backcross F_1
Collect seed	Select best R_0's	Collect seed	Evaluate and backcross BC_1
Evaluate and select R_1's in field	Evalue R_1's	Screen M_2 plants, self-fertilize	Select best BC_2 and backcross
Evaluate R_2's	Evaluate R_2 and cross with second line	Evaluate stability in M_3 generation	Select best BC_3 and backcross
Seed increase and evaluate R_3's	Evaluate and select best hybrids	Evaluate stability in M_4 generation	Select best BC_4 and backcross
Seed increase and evaluate R_4's	Seed increase and evaluate hybrids	Evaluate stability in M_5 generation	Select best BC_5 and backcross
Seed increase, variety trial	Seed increase, variety trial	Cross with existing variety	Select best BC_6 and backcross
Variety release in year 4	Seed increase of commercial F_1 hybrid	Select and cross	Seed increase and evaluate BC_7
	Hybrid variety release in year 5	Seed increase and variety trial	Seed increase and variety trial
		Variety release in year 6	Variety release in year 7—8

From Evans, D. A., Sharp, W. R., and Medina-Filho, H. P., *Am. J. Bot.*, 71, 759, 1984. With permission.

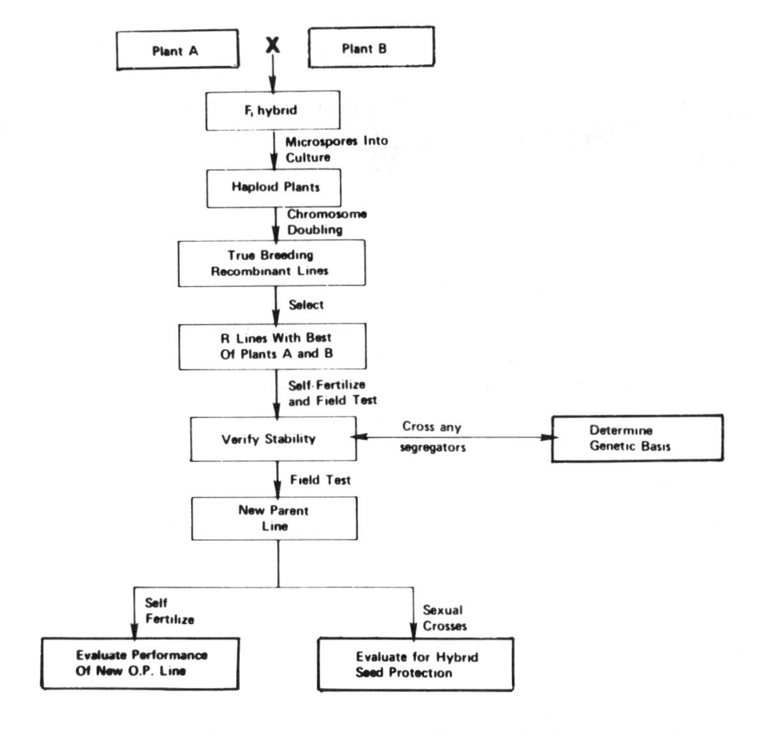

FIGURE 20. Breeding strategy for the use of gametoclonal variation, using a hybrid sorting technique, for the development of new plant varieties. (From Evans, D. A., Sharp, W. R., and Medina-Filho, H. P., *Am. J. Bot.*, 71, 759, 1984. With permission.)

desirable characters of plants A and B, and thus could also include commercially available F_1 hybrids. Anthers from hybrids are cultured to produce haploids, then doubled haploids. The plants with expression of the best characteristics of plants A and B are selected from among the doubled haploids and used to generate new plant lines. The lines can then be self-fertilized to generate new open-pollinated breeding lines or can be used in a breeding program to produce new hybrid seed. In most cases, the variation recovered using this approach is the result of meiotic recombination occurring during sexual reproduction of the F_1 hybrid.

To the extent that desirable somaclones can be identified in the test tube rather than in the field trials, somaclonal variation can be more efficient and cost effective. This requires a correlation between the cellular and whole plant response to specific chemicals used as selective agents. Promising results have already been obtained by selecting for resistance to host-specific pathotoxins[263,301] and for herbicide resistance[106] and resistance to aluminum.[324] The value of cellular selections of desirable genetic modifications has been elucidated also.[108] It is likely that this approach will be more valuable as a tool for correlation between cell selection and whole plant performance.

The value of gametoclonal variation is evident from the reports of new variety development by Chinese workers.[312,325] For wheat and rice, anthers of F_1 hybrid plants were introduced into culture and the recombinant microspores produced new doubled haploid lines that contained genetic information from both parents. Anther culture has been used for the recovery of recombinant plants of an F_1 hybrid between Xian nog 5675 (a variety with white glumes, top awn, clavate spike, and short stalk) and Jili (a variety with red glumes, awn fusiform spike, and tall stalk). Several doubled haploid plants were recovered that expressed mixed characters of two parents. Hence, gametoclonal variation is valuable for hybrid sorting, i.e., development of homozygous diploid recombinant lines from interspecies or intraspecies F_1 hybrids.

A number of categories of *in vitro* selection can be conceived. In all cases, the ideal situation is first to validate the selection strategy. For instance, during selection for resistance to a fungal disease by using a culture filterate in the culture medium, it is desirable to confirm that the culture filterate discriminates between cultures of a known resistant genotype and a known susceptible genotype. It is even better if the discrimination is also between a pair of isogenic-resistant and -susceptible varieties, where the only genetic difference is the presence or absence of the gene(s) for resistance. Such an ideal experimental system is rarely available. Indeed, for the most pressing agricultural problems, there is often no known genotype displaying the trail required. In such cases it may be possible to confirm that the selection strategy discriminates between different types of species.

Somaclonal variation is so prevalent that variants for a desirable trait also may possess undesirable traits. This may be the case for plants regenerated after somatic cell selection and genetic transformaiton. How can the desirable traits be separated from the undesirable? A proven method is to use the selected regenerated plant as a donor parent and backcross the desired trait into a seed-derived elite stock. This eliminates the unwanted background effects of somaclonal variation. In the case of a desired dominant trait, the backcrossing operation requires no more time than the initial somatic cell selection experiment. In addition, it is quite essential that controls be included with proper consideration in methodology.[3]

B. SOMATIC CELL SELECTION

Most of the cell line resistance to pathogenes has been selected from spontaneous variants appearing in culture. However, recent developments in this field indicate that several new varieties of crop can be generated from cells subjected to stress. Somaclonal variation provides the variability. Additional stress, when applied to these cultured cells, increases the rate of production of variants. Research has just begun in this direction and the recent development in producing a stress-resistant plant through induced variation is described briefly.

1. Selection for Ethionine Resistance

When ETH, an analogue of methionine, was incorporated in the culture medium on which plant cells were growing, it caused feedback inhibition of methionine synthesis, thus inhibiting cell growth.[326] Only variant cells resistant to the inhibitory effect of ETH are capable of growth. Plant cell lines that are resistant to growth inhibition by an amino acid analogue often produce increased amounts of the corresponding amino acid.[325] This phenomenon was the basis of the selection strategy using ETH as an analogue of methionine to select alfalfa cell lines that are overproducers of methionine.[124] The experiment was undertaken for two reasons: (1) as a model experiment in alfalfa tissue culture and (2) to improve alfalfa protein, which is slightly low in methionine in terms of amino acid requirements for human and other monogestric animal nutrition.

Reisch et al.[124] selected cell lines of diploid alfalfa HG2 that were resistant to ETH. Suspension-cultured cells were mutagenized with EMS, washed, and then plated on solid medium containing ETH. Typically, only about one to two resistant colonies formed per plate containing ETH, whereas control plates produced 20 to 50 times more colonies. Cell lines were grown for 6 months on media without ETH while plants were being regenerated and then the original cell lines were retested for ETH resistance. Of 110 cell lines retested 15 retained ETH resistance and grew significantly better than HG2 on ETH media. Amino acid analysis of resistant cell lines revealed that some lines contained high concentrations of methionine, cysteine, cystathionine, and glutathione.[124] Of 124 cell lines recovered from plates containing ETH, 91 cell lines regenerated plants. A sample of 25 regenerated plants from ETH-resistant cell lines was retested for resistance to ETH. This sample included plants from cell lines containing high levels of methionine and plants regenerated from ETH-resistant cell lines with control levels of methionine. Seven of the 25 plants retested were resistant to ETH inhibition.[125]

Out of the most resistant variants — R32, R59, and R64 — of alfalfa, only the R32 variant had a tenfold increase in soluble methionine, which was the greatest overproduction found among the variants.[125] Though this cell line (R32) was extremely low in regeneration ability, a plant was eventually regenerated. However, variant R32 was very slow growing, unthrifty, multifoliate, and never flowered over a period of 2 years. The chromosome number of R32 was 2n = 40, but the polyploid chromosome number as such would not account for its abnormalities, because plants with 2n = 40 usually resemble normal alfalfa with 2n = 32. Thus, it was suggested that R32 likely possessed genetic mutations, a very unbalanced 2n = 40 set of chromosomes, or both.[125] Unfortunately, the leaves of variants R32, R59, R64, and several other variants of alfalfa did not contain enhanced levels of unbound or protein-bound methionine. Thus, somatic cell selection was effective in isolating cell lines with ETH resistance and enhanced levels of methionine, yet regenerated plants from the cell lines did not contain enhanced levels of methionine in their leaves.

2. Selection for Salt and Herbicide Tolerance

Preliminary results of Smith and McComb[265] indicated potential for cellular selection for salt tolerance in alfalfa. Four alfalfa cultivars were screened at the whole plant and cellular levels, including one cultivar, W75RS (Regen-S), that demonstrated *in vitro* tolerance (control level of growth at 62.5 m*M* NaCl). However, following selection of a NaCl-tolerant cell line capable of plant regeneration, Smith and McComb[265] found that the regenerated plants were as salt sensitive as the initial plants.

Studies with the production of salt-tolerant alfalfa cell lines appeared to be complete,[125,265,327] but salt-tolerant alfalfa plants have not been regenerated. First, the salt-tolerant alfalfa cell line[327] indicated that a halophytic type of salt tolerance was suitable because the selected cell line required salt for optimal growth. Plant regeneration was possible only after the selected cell lines were maintained *in vitro* for several years. The somaclones were so stunted that whole plant tolerance could not be determined.[213]

The variable results in response to salt may be due to the involvement of several factors responsible for whole plant salt tolerance. At least seven major avoidance and/or tolerance mechanisms for salt tolerance have been proposed by Levitt.[328] Some of these mechanisms or factors may operate at the cellular level only, and thus expressed *in vitro*, while other mechanisms may operate only at the organism level. The negative results of Smith and McComb[265] can not be explained based on the multiple mechanism concept of salt tolerance. Additional selection efforts may result in the selection of a cellular tolerance mechanism that correlates with whole plant tolerance. Preliminary results have demonstrated that alfalfa cell lines HG2 and several Regen-S clones with stable resistance to 1.0% NaCl can be selected in a period of 6 months.[125]

In addition to cellular selection, other *in vitro* approaches are potentially available for obtaining salt-tolerant alfalfa plants. One method would be to screen many somaclones for differential salt tolerance. There is currently no information on the use of somaclone variation for salt tolerance. Another approach would be to incorporate the salt tolerance present in other *Medicago* species by somatic hybridization or sexual hybridization and embryo rescue. Somatic hybrids between *M. sativa* and *M. falcata* have been reported and embryo rescue techniques that allow recovery of *Medicago* interspecific hybrids have been identified.[329]

Through *in vitro* selection several cell lines resistant to herbicides have been obtained and some have been regenerated into complete plants. Prominent among them are tobacco resistant to picrolam,[34,106] amitrol,[308] glyphosate,[308,330,331] sulfonylurea,[186] hydroxyurea,[108,309] paraquat;[332] lotus resistant to 2,4-D;[290] carrot resistant to glyphosate;[283] and alfalfa resistant to phosphinothricin.[183] Other types of *in vitro* selection have been reviewed by Maliga[325] and Negrutiu et al.[333]

C. SELECTION FOR DISEASE RESISTANCE

There are several current applications of research on somaclonal variation in crop improvement. Some of these include the direct selection of disease-resistant plants by selecting cell lines resistant to toxin(s) produced by the pathogen,[214] the isolation of disease-resistant plants by screening large populations of somaclones,[184,216,280] the investigation of host-pathogen interactions,[334] and the evaluation of disease resistance.[335] In order to obtain plants with resistance to toxin(s) produced by the pathogen, the toxin must be involved in pathogenesis. There should be a precise correlation between cellular resistance to the toxin and whole plant resistance to the pathogen. Finally, whole plants must be regenerated from the resistant cell lines.

Development of resistance to disease or pathogens has been the major contribution of somaclonal variation. In these studies large populations of plants were regenerated from cultured cells at the whole plant level for resistance. A number of these studies are at a preliminary stage and only the initial regenerants have been tested; others showed that the resistance was passed on to progeny and inherited in a stable manner.

In alfalfa, plants resistant to *Fusarium oxysporum* were regenerated from cell lines selected for resistance to *Fusarium* culture filtrates.[214] The selection scheme was dependent on the spontaneous occurrence of resistant variant cells in the population of susceptible cells. Initial experiments utilized a step-up concentration procedure whereby resistant cell lines were first selected at low toxin concentration followed by selection at a higher concentration. This involved a fairly long-term (15 months) culture period, and the resistant regenerated plants were all of elevated ploidy levels (hexaploids and octoploids). A subsequent experiment conducted over 7 months (at high concentration only) resulted in resistant regenerated alfalfa plants with the normal tetraploid chromosome number. Based on preliminary inheritance data a spontaneously occurring dominant gene had been selected.[278] This dominant gene appeared to confer resistance to the toxic filtrate at the cellular level and resistance to the pathogen at the whole plant level. Screening of alfalfa protoclones resulted in plants that were resistant to *Verticillium albotrum*[216] and a plant that appeared to be resistant to a crown rot fungal complex.[214] In the latter case, one protoclone was observed with putative resistance to a complex of fungi including *F. solani* and/or *F. roseum* and *Rhizoctonia solani* based on tolerance to a natural epidemic involving these fungi.

Of interest was the finding that all verticillium wilt-resistant clones identified by Latunde-Dada and Lucas[216] showed elevated ploidy levels, similar to the initial fusarium-resistant selection of Hartman et al.[214] Chromosome counts on five verticillium wilt-resistant clones showed one plant to be hexaploid, one plant at or near the heptaploid level, and three plants near the octoploid level. The authors hypothesized that tolerance might be due to gene dosage effects associated with increased ploidy.

Scientists in the TCCP[147] project at Colorado State University grew about 3000 somaclonal lines of rice in the field in 1986. About 1100 lines were grown for the first time. In 1985 two Labelle lines that were resistant to sheath blight were identified. These lines were also resistant in 1986, although one line appeared to be segregating for resistance. In 1986 two more Labelle somaclones and one Zenith somaclone with apparent increased resistance to sheath blight were identified. Many Labelle lines appeared to vary in resistance to the rice blast disease when exposed to natural infection.

Daub[112] has reported protoplast-derived clones of tobacco cultivars for increases in resistance to several major tobacco pathogens including TMV, *Meloidogyne incognita*, and *Phytophthora parasitica* var. *nicotianae* (casual agent of tobacco black shank). Daub[112] has tested over 2500 seed progeny of 64 somaclones derived from two high-quality cultivars that are susceptible to black shank. He was able to identify three somaclones whose progeny showed elevated levels of black shank resistance in both field and greenhouse tests, but could not observe any somaclone with increased resistance to TMV or *M. incognita*. Several other investigators have been able to raise resistant plants from somaclones in the absence of selection.[24,278,279,336,337]

To date at least two cultivars have been released that were derived from somaclonal variation and identified by screening at the whole plant level. The sugarcane cv. Ono was developed following the work of Krishnamurthi and Tlaskal.[160] In this case they isolated subclones of sugarcane resistant to Fiji disease from the susceptible cv. Pindar. The cv. Ono yielded more than the parent cultivar, carried increased resistance, and has been grown to a limited extent in Fiji. Also, a cultivar of sweet potato isolated from meristem tip culture-derived clones has already been released.[316] This cv. Scarlet has yielded disease-resistance characteristics similar to those of the parent cultivar, but has a darker and more stable skin color, a desired quality trait. In both the cases lines were isolated in the absence of *in vitro* selections and were identified by screening the regenerated plants.

Several cell lines by the use of toxin(s) and culture filterates from pathogens to select for plants with toxin and disease tolerance have been selected and in certain cases plants have been regenerated. Prominent among them are maize resistance to *Helmithosporium maydis*,[101,263] tobacco to *Pseudomonas tabaci*,[305] sugarcane to *H. sacchar*,[111,159] potato to *Phytophthora infestans*,[296,297] rapeseed to *Phoma lingam*,[299] tobacco to *Pseudomonas syringae* and *Alternaria alternaria*,[306] alfalfa to *F. oxysporium*,[214,278] rice to *H. oryzae*,[301] and oats to *H. victoriae*.[293]

D. APPLICATION IN RESEARCH

Several research applications are likely to emerge through somaclonal variation in the near future. Strange genetic behavior with distorted segregation patterns of somaclones also becoming popular as tools to obtain true somaclonal mutants among primary regenerants. The somaclonal variation and variant selection are uniquely suited for studies where it is desirable to study normal and mutant gene expression in a common genetic background. For example, a single gene change in a variant represents a near-isogenic line of the donor clone that cannot be produced in some plants like alfalfa, using sexual reproduction because of inbreeding depression.

Somaclonal variation is a result of multiple interaction of genes which is responsible for producing several variants. For example, chromosome instability associated with somaclonal variation may be important in the direct expression of recessive mutations in diploid donor cells.[38] These mutations cannot be detected in diploid cultures unless both the alleles are mutated, or wild-type allele is removed. The latter requirement may be satisfied by the frequent loss of whole or partial chromosomes during culture. Such direct expression of a recessive mutation was observed in the diploid alfalfa variant C49 discussed earlier by Bingham and McCoy.[125] The fact that recessive mutation can be expressed directly, potentially enhances the efficiency of somatic cell selection.

Additional research on the efficiency of somaclonal vs.induced variation is needed in plants as well as other crop species in order to make an objective judgment about methods of producing useful variation. The numbers and types of somaclonal variants obtained for a given amount of input should be compared with those obtained by mutagenesis of seeds, plants, and cell in culture. Mutagenized seeds and plants could be studied in conventional sexual generation as well as used as donors for culture and regeneration. In plants such as alfalfa that regenerate via embryogenesis, the strategy of regenerating plants from mutagenized tissue would eliminate the problem of sectoring. Such a research could best be performed using cultivated alfalfa at the diploid level, which is capable of regeneration from cell cultures and is self-fertile. Diploid stocks that regenerated were developed from tetraploid alfalfa via haploidy and breeding.[114,123] A limited number of diploid regenerators that are self-fertile have been developed by further breeding and selection.[125]

Larkin,[3] from the survey of literature on somaclonal variation, observed that occurrence of dominants is greater in induced variants than the spontaneously occurring variants. This incidence of dominants may make somaclonal variation particularly attractive for plant breeders and scientists alike. The widespread incidence of chromosome breakage and fusion in plant cells should enable more efficient introgression of alien genes by dramatically enhancing recombination between the crop and the alien chromosomes.

In addition to being useful in establishing cytogenetic stocks, the chromosomal instability associated with tissue culture may be extremely useful in interspecific or intergenetic gene transfer. The chromosome breakage and interchanges that occur as a result of tissue culture of barley,[72] oats,[38] wheat,[193] triticale,[250] and *Haworthia*[73] may be mechanism for gene transfer to interspecific hybrids. In interspecific hybrids where chromosome pairing is limited, the possibility of recombination between cultivated and wild genomes may be remote. Tissue cultures of such sterile F_1 hybrids between a cultivated species and a wild species with desirable traits may allow for the incorporation of the desirable traits into chromosomes of the cultivated species by exchanges *in vitro*. Interchanges between wheat and rye chromosomes have been observed in somaclones derived from wheat × rye hybrids.[250] Orton[54] cultured *Hordeum vulgare* × *H. jubatum* hybrid ovary tissue and obtained some *H. vulgare*-like regenerants. Although these were haploids and had lost most *jubatum* features and GOT isozyme, two of them showed *jubatum*-specific esterases.

Interface with protoplast fusion of somaclones was first characterized among plants regenerated from protoplasts. It is, therefore, not surprising that somaclonal variation has been detected in plants regenerated from fused protoplasts.[190] As fusion products are the sum of genetic information from two parent lines, it is possible to recover from a single fusion experiment an array of breeding lines containing various mixtures of genetic information in a single cell (i.e., chloroplast, mitochondria, and nuclei) and each parent line. The units may fuse or segregate. If the units fuse, they may undergo extensive recombination and subsequent segregation producing new, unique units. Nuclear recombination has been detected in products of protoplast fusion as mitotic recombination.[188] Similarly, mitochondrial recombinations have been observed in several laboratories.[338] Most recently, chloroplast recombination was detected.[339] It is possible to obtain nuclear hybrid, cytoplasmic hybrids (cybrids), and plants with recombinant organelles from the same fusion experiment.[340]

For gene transfer purposes it is possible to incorporate chemicals known to cause chromosome elimination such as carbamate herbicide, CIPC.[341] Treatment of protoplast cultures with such chemicals would produce an array of plants containing chromosomes different from each of the two parent lines. This could be used for more selective gene transfer.

Currently, genetic transformation in plants has been limited to a small number of genes, crop species, and transformed plants. All gene transfer techniques require plant regeneration from transformed cell or protoplasts. The major hurdle to progress as described by Evans and Sharp[56] comes from the insufficient knowledge of the make-up of the genomes of major crops. Efforts need to be committed to chromosome mapping of those genes, key to the important value of the added traits. Transformation research has centered on model systems such as tobacco and petunia using the Ti-plasmid of *Agrobacterium tumefaciens*. To date, examination of transformed plants has been limited only to the expression of the transferred gene in regenerated plants. Single gene transformation of T-DNA sequences has been detected in progeny of transformed plants.[342] As the list of transformed species increases, greater attention will undoubtedly be directed to the phenomenon of somaclonal variation. It has already been noted that the following transformation of tomato with *Agrobacterium* cells with a very wide array of chromosome numerical and structural variation makes its appearance.[343] The chromosome variation is substantially greater than in comparable untransformed cell cultures. The promise of molecular biology to transfer a single gene at a time to existing crop varieties without attention to the phenomenon of somaclonal variation may remain unfulfilled.

Results of somaclonal variation may also have an impact on gene transfer experiments using recombinant DNA, since knowledge gained on medium components that increase or decrease somaclonal variation should lead to the development of optimum regeneration protocols for use in transformation experiments. Another possibility would be to transform the embryo or shoot precursor mother cells. This is currently practicable for a number of crops in which the embryo mother cell can be identified, separated, and manipulated. It is also possible that somaclones could be used to develop more suitable variants for uptake, integration, and stable expression of T-DNA.

As somaclonal variation results in a large number of simple genetic changes, new somaclones are in many cases isogenic stocks of the original parental variety. Each somaclone differs from the parent by only a single or small number of genetic changes. Thus, these clones could be a valuable source of germplasm to isolate specific genes of interest. Somaclones and their parent lines could be compared to identify distinct mRNAs correlated with a particular phenotype. These mRNAs could be used to prepare clones that could be tested for genetic linkage with a specific agronomic character. Ultimately, results would be particularly attractive for those crop plants where only a limited gene map exists (e.g., soybean, sugarcane, coffee).

Although no work has been done on the utilization of the chromosomal loss for gene mapping, the apparent tandem loss or interchange of chromosomal segments as a result of tissue culture may be used to map genes to chromosomes in the near future. Aneuploidy due to loss of whole chromosome from tetraploid cultures is relatively common in several plants.[3,253] Bingham and McCoy[125] proposed a strategy to utilize aneuploidy for gene mapping. If a known heterozygote for an unmapped recessive gene is placed in culture, then plants could be regenerated at monthly intervals. Regenerated plants expressing the recessive gene could be cytologically analyzed. If a single chromosomal segment is altered, then one could infer the gene maps to the deficient chromosome or chromosome segment. Additional genetic tests should be conducted to test the inference. The strategy becomes more efficient as the number of recessive heterozygotes increases. Hence, the system will accommodate as many unmapped recessives as can be bred into the stock used in tissue culture.

VII. CONCLUSIONS AND FUTURE PROSPECTS

The first hint that useful variation arises spontaneously among cultured plant cells emerged in the late 1960s in the pioneering work of Heinz et al.[5] with sugarcane. Until early 1980 such a variation was considered by many to be artificial and probably unstable. Since then, evidence has quickly mounted to confirm that this source of genetic variation (somaclonal variation) is stable, has high potential utility, and probably reflects natural mechanisms which maintain and generate variation in somatic tissues.

Somaclonal variation presents itself as a method for generating useful variation without outcrossing. The last few years of analysis have considerably fortified this prospect. This variation has already been exploited to great advantage in crops such as sugarcane, tobacco, potato, rice, maize, tomato, alfalfa, and *Pelargonium*. In sugarcane, for example, calliclones from tissue cultures produce plant lines which differ significantly in their response to Fiji disease (caused by a leaf-hopper-transmitted virus), downy mildew, and *Helminthosporium sacchari*. This wide variation in disease response also includes classes of plants which are highly tolerant and in some cases resistant to a particular disease.[159] Use of a special bioassay, similar to one developed to quantify the sensitivity of leaves to fungal toxins administered at a standardized concentration known to induce leaf damage, enabled a large number of somaclones of Australian varieties of sugarcane to be screened. Many of the somaclones proved resistant or essentially immune to the effects of the toxin. However, most important, the resistant somaclones retained their resistance through subsequent sugarcane generations, although there were some individual plants which expressed susceptibility in some later generation of somaclones.[28] Several beneficial selections have been obtained concurrently in sugar. The independent nature of these results indicates that cell culture should now become an integral part of sugarcane improvement programs. Not only is tissue culture cloning of value to improving sugar yield, but it may also produce novel types of sugarcane, especially useful for animal feedstock or for substrates in alcohol fermentations. In other crops like tobacco, somaclonal variation in pollen and protoplast-derived plants is providing extended opportunities for breeders of air-cured, flue-cured, and burley types of tobacco. Dihaploids of all three types of tobacco derived from anther culture display extensive variations in leaf yield, days to flowering, plant height, leaf number, leaf length, leaf width, total alkaloids, and reducing sugar content. An example of somaclonal variation in this crop can be seen in the leaf-shape and plant-height phenotypes displayed in *Nicotiana tabacum* plants regenerated from protoclones obtained from protoplasts of a single dihaploid leaf. In the case of potato, Shepard et al.[33] screened over 1000 protoclones produced from leaf protoplasts of the established variety Russet Burbank and found significant, yet quite stable, variation in compactness of growth habit, maturity date, tuber uniformity, tuber skin color, photoperiod requirement, and fruit production. Some of these characteristics, e.g., greater tuber uniformity and early onset of tuberization, were far better than in the parent variety itself. Of even greater significance is the fact that some protoclones recovered were more resistant to early blight (caused by *Alternaria solani*) and that resistance was retained through several vegetative generations. Examples of these types of extremely valuable somaclonal variations can be found in an increasing number of crops extending through fruit, tuber, ornamental, forage, and cereal crops (Table 2).

There is the lack of control by the investigator or breeder over the characters being varied. Classical mutagenesis has proved not to be entirely random. It remains to be seen to what extent somaclonal variation will be random and whether we can accept all agronomic characters to be modified by the phenomenon. Another question not yet satisfactorily answered relates to the extent of variation within an individual somaclone. Some of the sugarcane variants chosen for amelioration of one trait have also been reevaluated in field trials for variation in the other agronomic parameters. In at least some cases general performance and quality were as good as the parental cultivars.[159,161] An independent reevaluation of the yield of Fiji disease-resistant variants of sugarcane cultivars did not show any decline. Secor and Shepard examined 65 potato somaclones in replicated field trials analyzing 35 parameters. Three clones varied in only one character from the parent, one clone varied in 17 of the parameters, and the mode class (15 clones) varied in only 4 parameters.

Several mechanisms may be responsible for induction of somaclonal variation. These include the gross karyotypic changes which accompany *in vitro* culture via calli as cryptic chromosomal rearrangement, somatic crossing-over with sister chromatid exchange, transposable elements, gene amplification or diminution, or perhaps various combinations of these processes. Different types of cryptic chromosomal arrangements are well-described aspects of the meiotic behavior of chromosome in plants regenerated from cell cultures. They include reciprocal translocations, deletions, inversions, nonhomologous translocations, and acentric and centric fragment formations. Such rearrangements probably cause losses of genetic material or at the very least realignment and transportation of chromosomal material. This can lead to expression of previously silent genes, especially where loss or switching off of a dominant allele has occurred.

Somaclonal variation in plants is due to either preexisting cellular genetic differences or tissue culture-induced variability.[3] However, there is no experimental evidence to distinguish accurately between these two factors. Plants regenerated from the callus of five cultivars of *Pelargonium* spp. were systematically compared by Skirvin.[32] Plants obtained from *in vivo* root and petiole cuttings and plants regenerated from callus were quite variable, whereas plants obtained from the geranium stem cuttings *in vivo* were uniform. This suggests that at least some variability is preexisting in the donor explants. Changes in organ size, leaf and flower morphology, essential oil constituents, fascination, pubescence, and anthocyanin pigmentation were observed in regenerated plants. It is likely that both preexisting genetic and tissue culture-induced variability have been recovered in these geranium lines. Somaclonal variations, particularly abnormal floral morphology, in plants regenerated from long-term cell cultures were observed by Syono and Furuya.[344] Tissue culture-induced variability with regard to chromosome number in both callus and in plants regenerated from callus cultures for a long duration has been reported.[70,228,238] This results in regeneration of aneuploid plants that are commercially useless in sexually propagated species. However, aneuploidy does not interfere with productivity of sugarcane and potato, which are asexually propagated. The tissue culture environment may enhance the frequency of somatic crossing-over, and if a proportion of such exchanges were asymmetric or between nonhomologous chromosomes, then genetic variants could be generated as a consequence. It is known that the frequency of sister chromatid exchange in plants is quite high (20% per cell division observed in barley). Transposable elements may also be responsible for certain types of genetic instability in cell cultures and it may well be that such elements contribute significantly to somaclonal variation. Several plant cells like those of other eukaryotypes can increase or decrease the quantity of a specific gene product by differential gene amplication or diminution, respectively. For instance, ribosomal RNA gene amplification and diminution are now known to be widespread in wheat, rye, and tobacco, and in flax ribosomal DNA is known to alter directly in response to environmental and cultural pressure.[75,345]

Somaclonal variation represents a new tool for plant breeder, and several breeding strategies as suggested by Evans et al.[2] have been discussed. Somaclonal variation also has the advantage of being accessible in most crop species. Many possible applications can be envisaged at present and many more will emerge in time to come, as our understanding of the phenomena improves. A crop cultivar may be improved dramatically if appropriate selection can be imposed during the culture phase. Some systems lend themselves readily to selection strategies such as pathotoxins, herbicides, heavy metal toxicity, and salinity. Other systems will require more ingenuity in order to devise appropriate selection, such as those for recovering tobacco plants which support only slow TMV development or the use of lysine plus threonine inhibition for selection of variants with potentially increased nutritional value.[284] One of the most exciting and useful prospects of integrating somaclonal variation into existing breeding programs will be by the tissue culture of wide hybrids. A significant proportion of modern plant breeding is aimed at the introgression of alien genes or gene complexes into crop cultivars. Since tissue culture has been strongly implicated in causing chromosomal rearrangements and exchanges, it may aid in the nutritional process. Somaclonal variation is a novel way of generating potentially useful genetic variability. Research with major crop species has already shown that many somaclonal variants have agronomic value. It must be combined with efficient *in vivo* selection and plant breeding programs. Perhaps one of the greatest benefits of somaclonal variation will be the broadening of the genetic base of our crop plants.

With the realization of the importance of somaclonal variation in crop improvement, standardization of techniques and utilization of more sophisticated techniques for detecting changes at gene level have become necessary. It is important (sometimes imperative) to maintain vegetative propagules of the seed-derived plant as the best controls in the identification and comparison of somaclonal variants. In addition, the seed-derived plant may be useful as a recurrent parent into which a single desired mutation or transformed gene may be backcrossed to eliminate undesirable somaclonal variation and regain the original donor genotype. In this regard, it is believed that it may be desirable in many cases to utilize regenerated plants as gene donors for research and plant improvement.

The limits of detection of somaclonal variation are expected to be broadened in the future by biochemical analysis (e.g., electrophoresis) advanced cytogenetic techniques (*in situ* hybridization and improved banding techniques), and additional use of molecular tools (use of restriction endonucleases, blotting, hybridization, and molecular probes). The combination of improved detection methods and proper genetic strategies to separate desired from undesired variation will enhance the usefulness of somaclonal variants.

Based on their work on the improvement of tomato through somaclonal variation, Evans et al.[2] concluded that: (1) variation in number of chromosomes can be recovered in regenerated plants; (2) several single gene mutations have been recovered in tomato varieties; (3) somaclones includc dominant, semidominant, and recessive nuclear mutations; (4) the frequency of single gene mutation is one mutant in every 20 to 25 regenerated plants; (5) new single gene mutants not previously reported using conventional mutagenesis have been recovered using somaclonal variation; (6) the occurrence of 3:1 ratio for single mutants in R_1 plants suggests that mutants are of clonal origin and that the mutation occurred prior to short regeneration (e.g., no mosaics are detected); (7) mitotic recombination may account for some somaclonal variation; and (8) mutations in chloroplast DNA can also be recovered.

At first reading the prefix ''soma'' appears to be inappropriate for cells derived from gametes, for example, the microspores in anther culture. Yet the reprogrammed microspores and in the sporophytic sense the somaclonal term might be appropriate for anther-derived tissues as well. Tissue and anther culture-derived variability, even though genotype dependent, is pervasive, and is adequately documented in many crops such as sugarcane,[6,30,303] potato,[24,33] tobacco,[12,14] maize,[263] barley,[346] *Brassica*,[347] *Pelarogonium*,[9] carrot,[348] chrysanthemum,[349] lilly,[350,351] carnation,[351] clover,[352] and sorghum.[353] This aspect has been named as gametoclonal variation by Evans et al.[2]

The types of genetic changes that are recovered in plants regenerated from cell culture are dependent upon the donor material. In tomato Evans et al.[2] used diploid inbred variety; hence, most regenerated plants were heterozygous resulting in segregation of new mutations in the R_1 generation. Larkin and Scowcroft[103] used heterozygous material so that mutations could be visually detected in regenerated plants. Among these somaclones several plants had distorted segregation ratios in the R_1 generation. Since the generation of variants will ultimately be used for breeding and crop improvement, it is important to distinguish somatic-derived somaclones and gametic-derived gametoclones. The gametes are products of meiosis, governed by Mendel's laws of segregation and independent assortment. These genetic differences which were provided by Evans et al.[2] point out that gametoclonal and somaclonal variations are distinct in the following ways: (1) both dominant and recessive mutants will be expressed directly in haploid-regenerated plants as only a single copy of each gene is present. Hence, regenerated gametoclones (R) can be analyzed directly to identify new variants; (2) recombinational events that are recovered in gametoclones would be the result of meiotic and not mitotic crossing-over; and (3) to use gametoclones, chromosome number must be doubled.

REFERENCES

1. **Larkin, P. J. and Scowcroft, W. R.,** Somaclonal variation — a novel source of variability from cell culture for plant improvement, *Theor. Appl. Genet.*, 60, 197, 1981.
2. **Evans, D. A., Sharp, W. R., and Medina-Filho, H. P.,** Somaclonal and gametoclonal variation, *Am. J. Bot.*, 71, 759, 1984.
3. **Larkin, P. J.,** Somaclonal variation: history, method and meaning, *Iowa State J. Res.*, 61, 393, 1987.
4. **Heinz, D. J. and Mee, G. W. P.,** Plant differentiation from callus tissue of *Saccharum* species, *Crop Sci.*, 9, 346, 1969.
5. **Heinz, D. J., Mee, G. W. P., and Nickell, L. G.,** Chromosome number of some *Saccharum* species hybrids and their cell suspension cultures, *Am. J. Bot.*, 56, 450, 1969.
6. **Heinz, D. J. and Mee, G. W. P.,** Morphologic, cytogenetic, and enzymatic variation in *Saccharum* species hybrid clones derived from callus tissue, *Am. J. Bot.*, 58, 257, 1971.
7. **Murashige, T. and Nakano, R.,** Chromosome complement as a determinant of the morphogenic potential of tobacco cells, *Am. J. Bot.*, 54, 963, 1967.
8. **Sacristan, M. D. and Melchers, G.,** The karyological analysis of plants regenerated from tumorous and other callus cultures of tobacco, *Mol. Gen. Genet.*, 105, 317, 1969.
9. **Skirvin, R. M. and Janick, J.,** Tissue culture-induced variation in scented *Pelargonium* spp., *J. Am. Soc. Hortic. Sci.*, 101, 281, 1976.
10. **Pelletier, G. and Pelletier, A.,** Culture in vitro de tissus de Trefle blanc *(Trifolium repens)*; variabilite des plantes regenerees, *Ann. Amelior. Plant.*, 21, 221, 1971.
11. **Sibi, M.,** La notion de programme genetique chez les vegetaux superieurs. II. Aspect experimental: obtention do variants par culture de tissue in vitro sur *Lactuca sativa* L. Apparition de vigeur chez les croisements, *Ann. Amelior. Plant.*, 26, 523, 1976.
12. **Devreux, M. and Laneri, V.,** Anther culture, haploid plants, isogenetic line and breeding research in *Nicotiana tabacum* L., in FAO/IAEA, Polyploidy and Induced Mutations in Plant Breeding, IAEA, Vienna, 1974, 101.
13. **Oinuma, T. and Yoshida, T.,** Genetic variation among doubled haploid lines of barley tobacco varieties, *Jpn. J. Breed.*, 24, 211, 1974.
14. **Burk, L. G. and Matzinger, D. F.,** Variation among anther-derived doubled haploids from an inbred line of tobacco, *J. Hered.*, 67, 381, 1976.
15. **Arcia, M. A., Wernsman, E. A., and Burk, L. G.,** Performance of anther derived dihaploids and their conventionally inbred parent as lines, in F_1 hybrids and in F_2 generations, *Crop Sci.*, 18, 413, 1978.
16. **Collins, G. B. and Legg, P. D.,** Recent advances in the genetic applications of haploidy in *Nicotiana*, in *The Plant Genome*, Davies, D. R. and Hopwood, D. A., Eds., John Innes Charity, Norwich, England, 1980, 197.
17. **DePaepe, R., Bleton, E., and Gnangbe, F.,** Basis and extent of genetic variability among doubled haploid plants obtained by pollen culture in *Nicotiana sylvestris*, *Theor. Appl. Genet.*, 59, 177, 1981.
18. **Oono, K.,** Production of haploid plants of rice *(Oryza sativa)* by anther culture and their use for breeding, *Bull. Natl. Inst. Agric. Sci.*, 26, 139, 1975.
19. **Oono, K.,** Test tube breeding of rice by tissue culture, *Trop. Agric. Res. Ser.*, 11, 109, 1978.
20. **Oono, K.,** High frequency mutations in rice plants regenerated from seed callus, in 4th Int. Congr. Plant Tissue Cell Culture, Calgary, Canada, Abstr., 1978, 52.
21. **Oono, K.,** *In vitro* methods applied to rice, in *Plant Tissue Culture*, Thorpe, T. A., Ed., Academic Press, New York, 1981, 273.
22. **Alhoowalia, B. S.,** Chromosome changes in parasexually produced ryegrass, in *Current Chromosome Research*, Jones, K. and Brandham, P., Eds., Elsevier, Amsterdam, 1976, 115.
23. **Alhoowalia, B. S.,** Novel ryegrass genotypes regenerated from embryo callus culture, in 4th Int. Congr. Plant Tissue Cell Culture, Abstracts, Calgary, Canada, 1978, 162.
24. **Matern, U., Strobel, G., and Shepard, J.,** Reaction to phytotoxins in a potato population derived from mesophyll protoplasts, *Proc. Natl. Acad. Sci. U.S.A.*, 75, 4935, 1978.
25. **Orton, T. J.,** Somaclonal variation: theoretical and practical considerations, in *Gene Manipulation in Plant Improvement*, Gustafson, J. P., Ed., Plenum Press, New York, 1984, 427.
26. **van Harten, A. M., Bouter, H., and Broertjes, C.,** In vitro adventitious bud techniques for vegetative propagation and mutation breeding of potato *(Solanum tuberosum L.)*. II. Significance for mutation breeding, *Euphytica*, 30, 1, 1981.
27. **Thomas, E., Bright, S. W. J., Franklin, J., Lancaster, V. A., Miflin, B. J., and Gibson, R.,** Variation amongst protoplast derived potato plants *(Solanum tuberosum* cv, 'Maris Bard'), *Theor. Appl. Genet.*, 62, 65, 1982.
28. **Bright, S. W. J., Nelson, R., Creissen, G., Karp, A., Franklin, J., Norbury, P., Kueh, J., Rognes, S., and Miflin, B.,** Modification of agronomic traits using in vitro technology, in *Plant Biotechnology*, Mantell, S. H. and Smith, H., Eds., Cambridge University Press, Cambridge, 1983, 251.

29. **Barbier, M. and Dulieu, H. L.,** Effects genetiques observes sur des plantes de Tabac regenerees de cotyledons par culture in vitro, *Ann. Amelior. Plant.,* 30, 321, 1980.
30. **Nickell, L. G. and Heinz, D. J.,** Potential of cell and tissue culture techniques as aids in economic plant improvement, in *Enzymes and Populations,* Srb, A. M., Ed., Plenum Press, New York, 1973, 109.
31. **Liu, M. C. and Chen, W. H.,** Tissue and cell culture as aids to sugarcane breeding. I. Creation of genetic variation through callus culture, *Euphytica,* 25, 393, 1976.
32. **Skirvin, R. M.,** Natural and induced variation in tissue culture, *Euphytica,* 27, 241, 1978.
33. **Shepard, J. F., Bidney, D., and Shahin, E.,** Potato protoplasts in crop improvement, *Science,* 208, 17, 1980.
34. **Sibi, M.,** Expression of cryptic genetic factors in vivo and in vitro, in *Broadening the Genetic Base of Crops,* Van Harten, A. M. and Zeven, C. C., Eds., Pudoc, Wageningen, Netherlands, 1979, 339.
35. **Edallo, S., Zuccinali, C., Perenzin, M., and Salamini, F.,** Chromosomal variation and frequency of spontaneous mutation associated with in vitro culture and plant regeneration in maize, *Maydica,* 26, 39, 1981.
36. **Larkin, P. J., Ryan, S. A., Brettell, R. I. S., and Scowcroft, W. R.,** Heritable somaclonal variation in wheat, *Theor. Appl. Genet.,* 67, 443, 1984.
37. **Orton, T. J.,** Genetic variation in somatic tissues, method or madness?, *Adv. Plant Pathol.,* 2, 153, 1983.
38. **McCoy, T. J., Phillips, R. L., and Rines, H. W.,** Cytogenetic analysis of plants regenerated from oat *(Avena sativa)* tissue culture: high frequency of partial chromosome loss, *Can. J. Genet. Cytol.,* 24, 37, 1982.
39. **Browers, M. A. and Orton, T. J.,** A factorial study of chromosomal variability in callus cultures of celery (*Apium graveolens* L.), *Plant Sci. Lett.,* 26, 65, 1982.
40. **D'Amato, F.,** Polyploidy in the differentiation and function of tissues and cells in plants, a critical examination of the literature, *Caryologia,* 4, 311, 1952.
41. **Murashige, T. and Nakano, R.,** Tissue culture as a potential tool in obtaining polyploid plants, *J. Hered.,* 57, 115, 1966.
42. **Orton, T. J.,** Experimental approaches to the study of somaclonal variation, *Plant Mol. Biol. Rep.,* 1, 67, 1983.
43. **Mitra, J. and Steward, F. C.,** Growth induction in cultures of *Haplopappus gracilis.* II. The behavior of the nucleus, *Am. J. Bot.,* 48, 358, 1983.
44. **Evans, D. A. and Sharp, W. R.,** Single gene mutations in tomato plants regenerated from tissue culture, *Science,* 221, 949, 1983.
45. **Padmanabhan, V., Paddock, E. F., and Sharp, W. R.,** Plantlet formation from *Lycopersicon esculentum* leaf callus, *Can. J. Bot.,* 52, 1429, 1972.
46. **Murata, M. and Orton, T. J.,** Chromosome structural changes in cultured celery cells, *In Vitro,* 19, 83, 1983.
47. **Greilhuber, J.,** Why plant chromosomes do not show G-bands, *Theor. Appl. Genet.,* 50, 121, 1977.
48. **Ashmore, S. E. and Gould, A. R.,** Karyotype evolution in a tumour-derived plant tissue culture analyzed by Giemsa C-banding, *Protoplasma,* 106, 297, 1981.
49. **Hart, G. E.,** Evidence for triplicate genes for dehydrogenase in hexaploid wheat, *Proc. Natl. Acad. Sci. U.S.A.,* 66, 1136, 1970.
50. **Davies, P. A., Pallotta, M. A., Ryan, S. A., Scowcroft, W. R., and Larkin, P. J.,** Somaclonal variation in wheat: genetic and cytogenetic characterization of alcohol dehydrogenase 1 mutants, *Theor. Appl. Genet.,* 72, 644, 1986.
51. **Brettell, R. I. S., Dennis, E. S., Scowcroft, W. R., and Peacock, W. J.,** Molecular analysis of somaclonal mutant of maize alcohol dehydrogenase, *Mol. Gen. Genet.,* 202, 235, 1986.
52. **Ryan, S. A. and Scowcroft, W. R.,** A somaclonal variant of wheat with additional β-amylase isozymes, *Theor. Appl. Genet.,* 73, 459, 1987.
53. **Arnison, P. G. and Boll, W. G.,** Isoenzymes in cell cultures of bush been *(Phaseolus vulgaris* cv. Contender): isoenzymatic differences between stock suspension cultures derived from a single seedling, *Can. J. Bot.,* 53, 261, 1975.
54. **Orton, T. J.,** Haploid barley regenerated from callus cultures of *Hordeum vulgare* × *jubatum, J. Hered.,* 71, 780, 1980.
55. **Freeling, M.,** Further studies on the balance between Adh-1 and Adh-2 in maize: gene competitive programs, *Genetics,* 81, 641, 1975.
56. **Evans, D. A. and Sharp, W. R.,** Application of somaclonal variation, *Biotechnology,* 4, 528, 1986.
57. **Prat, D.,** Genetic variability induced in *Nicotiana sylvestris* by protoplast culture, *Theor. Appl. Genet.,* 64, 223, 1983.
58. **Fukai, K.,** Sequential occurrence of mutations in a growing rice callus, *Theor. Appl. Genet.,* 65, 225, 1983.
59. **Sun, Z. X., Zhao, C. Z., Zheng, K. L., Qi, X. F., and Fu, Y. P.,** Somaclonal genetic of rice, *Oryza sativa* L., *Theor. Appl. Genet.,* 67, 67, 1983.

60. **Osterman, J. C. and Swartz, D.,** Analysis of a controlling element mutation at the Adh locus of maize, *Genetics,* 99, 267, 1981.
61. **Dauglas, M. W., Perani, L., Radke, S., and Bossert, M.,** The application of recombinant DNA technology toward crop improvement, *Physiol. Plant.,* 68, 560, 1986.
62. **Peschke, V. M., Phillips, R. L., and Gengenbach, B. G.,** Discovery of transposable element activity among progeny of tissue cultures, *Science,* 238, 804, 1987.
63. **Orton, T. J.,** Case histories of genetic variability *in vitro:* celery, *Cell Cult. Somatic Cell Genet. Plants,* 3, 345, 1986.
64. **Pring, D. R., Conde, M. F., and Gengenbach, B. G.,** Cytoplasmic genome variability in tissue culture-derived plants, *Environ. Exp. Bot.,* 21, 369, 1981.
65. **Boeshore, M. L., Hanson, M. R., and Izhar, S.,** A variant mitochondrial DNA arrangement specific to *Petunia* stable sterile somatic hybrids, *Plant Mol. Biol.,* 4, 125, 1985.
66. **Boeshore, M. L., Lifshitz, I., Hanson, M. R., and Izhar, S.,** Novel composition of mitochondrial genomes in *Petunia* somatic hybrids derived from cytoplasmic male sterile and fertile plants, *Mol. Gen. Genet.,* 190, 459, 1983.
67. **Rivin, C. J., Zimmer, E. A., Cullis, C. A., Walbot, V., Huynh, T., and Davis, R. W.,** Evaluation of genomic variability at the nucleic acid level, *Plant Mol. Biol. Rep.,* 1, 9, 1983.
68. **Kao, F. T., Hartz, J. A., Law, M. L., and Davidson, J. N.,** Isolation and chromosomal localization of unique DNA sequences from a human genomic library, *Proc. Natl. Acad. Sci. U.S.A.,* 79, 865, 1982.
69. **De Paepe, R., Prat, D., and Huguet, T.,** Heritable nuclear DNA changes in doubled haploid plants obtained by pollen culture of *Nicotiana sylvestris, Plant Sci. Lett.,* 28, 11, 1982.
70. **Bayliss, M. W.,** Chromosomal variation in plant tissue in culture, *Int. Rev. Cytol. Suppl.,* 11A, 113, 1980.
71. **Ogura, H.,** Genetic control of chromosomal chimeras found in a regenerate from tobacco callus, *Jpn. J. Genet.,* 53, 77, 1978.
72. **Orton, T. J.,** Chromosomal variability in tissue cultures and regenerated plants of *Hordeum, Theor. Appl. Genet.,* 56, 101, 1980.
73. **Ogihara, Y.,** Tissue culture in *Haworthia.* IV. Genetic characterization of plants regenerated from callus, *Theor. Appl. Genet.,* 60, 353, 1981.
74. **Browers, M. A. and Orton, T. J.,** Transmission of gross chromosomal variability from suspension cultures into regenerated plants, *J. Hered.,* 73, 159, 1982.
75. **Breiman, A., Felsenburg, T., and Galun, E.,** *Nor* loci analysis in progenies of plants regenerated from the scutellar callus of bread-wheat, *Theor. Appl. Genet.,* 73, 827, 1987.
76. **Shepard, J. F. and Totten, R. E.,** Mesophyll cell protoplasts of potato, *Plant Physiol.,* 60, 313, 1977.
77. **Gunn, R. E. and Shepard, J. F.,** Regeneration of plants from mesophyll derived protoplasts of British potato *(Solanum tuberosum* L.) cultivars, *Plant Sci. Lett.,* 22, 97, 1981.
78. **Secor, G. A. and Shepard, J. F.,** Variability of protoplast derived potato clones, *Crop Sci.,* 21, 102, 1981.
79. **Shepard, J. F.,** Protoplasts as sources of disease resistance in plants, *Annu. Rev. Phytopathol.,* 19, 145, 1981.
80. **Shepard, J. F.,** The regeneration of potato plants from leaf-cell protoplasts, *Sci. Am.,* 246, 154, 1982.
81. **Thomas, E.,** Plant regeneration from shoot culture-derived protoplasts of tetraploid potato *(Solanum tuberosum* cv. Maris Bard), *Plant Sci. Lett.,* 23, 81, 1981.
82. **Thomas, E., Bright, S. W. J., Franklin, J., Lancaster, V. A., Miflin, J. J., and Gibson, R.,** Variation amongst protoplast derived potato plants *(Solanum tuberosum* cv. "Maris Bard"), *Theor. Appl. Genet.,* 62, 65, 1982.
83. **Karp, A., Nelson, R. S., Thomas, E., and Bright, S. W. J.,** Chromosome variation in protoplast-derived potato plants, *Theor. Appl. Genet.,* 63, 265, 1982.
84. **Jones, M. G. K., Bright, S. W. J., Nelson, R. S., Foulger, D., Creissen, G. P., Karp, A., and Ooms, G.,** Variations in plants regenerated from protoplasts and complex explants of potato, in Proc. 6th Int. Protoplast Symp., Ciba-Geigy Poster Proc., Basel, 1983, 150.
85. **Sree Ramulu, K., Dijkhuis, P., and Roest, S.,** Phenotypic variation and ploidy level of plants regenerated from protoplasts of tetraploid potato *(Solanum tuberosum* L. cv. Bintje), *Theor. Appl. Genet.,* 65, 329, 1983.
86. **Sree Ramulu, K., Dijkhuis, P., Roest, S., Bokelmann, G. S., and DeGroot, B.,** Early occurrence of genetic instability in protoplast cultures of potato, *Plant Sci. Lett.,* 36, 79, 1984.
87. **Sree Ramulu, K., Dijkhuis, P., and Roest, S.,** Genetic variability in protoclones of potato *(Solanum tuberosum* L. cv. Bintje), new types of variation after vegetative propagation, *Theor. Appl. Genet.,* 68, 515, 1984.
88. **Bokelmann, S. J. and Roest, S.,** Plant regeneration from protoplasts of potato *(Solanum tuberosum* cv. Bintje), *Z. Pflanzenphysiol.,* 109, 259, 1983.
89. **Carlberg, I., Glimelius, K., and Eriksson, K.,** Improved culture ability of potato protoplasts by use of activated charcoal, *Plant Cell Rep.,* 2, 223, 1983.

90. **Jellis, G. J., Gunn, R. E., and Boultion, R. E.,** Variation in disease resistance among potato protoclones, in Abstr. 9th Triennial Conf. European Assoc. Potato Research, Interlaken, Switzerland, 1984, 380.
91. **Bright, S. W. J., Wheeler, J. A., and Kirkman, M. A.,** Variation in record clones from tissue cultures, in Abstr. 9th Triennial Conf. European Assoc. Potato Research, Interlaken, Switzerland, 1984, 62.
92. **Mackay, G. R.,** Potato breeding, in 3rd Annu. Rep. Scott. Crop. Research Institute, Pentlandfield, 1984, 73.
93. **Tiemann, U., Pett, S. B., and Koblitz, H.,** Studies on breeding characters of mesophyll protoplast derived potato plants *(Solanum tuberosum* L.), in Proc. Int. Symp. Plant Tissue and Cell Culture Applied Crop Improvement, Olmouc, Czechoslovakia, 1984, 443.
94. **Thomas, E., King, P. J., and Potrykus, I.,** Improvement of crop plants via single cells in vitro — an assessment, *Z. Pflanzenzuecht.*, 82, 1, 1979.
95. **Nelson, R. S., Creisson, G. P., and Bright, S. W. J.,** Plant regeneration from protoplasts of *Solanum brevidens, Plant Sci. Lett.*, 30, 355, 1983.
96. **Barsby, T. L., Shepard, J. F., Kemble, R. J., and Wong, R.,** Somatic hybridization in the genus *Solanum: Solanum tuberosum* and *Solanum brevidens, Plant Cell Rep.*, 3, 165, 1984.
97. **Bidney, D. L. and Shepard, J. F.,** Phenotypic variation in plants regenerated from protoplasts: the potato system, *Biotechnol. Bioeng.*, 23, 2691, 1981.
98. **Wenzel, G.,** Strategies in unconventional breeding for disease resistance, *Annu. Rev. Phytopathol.*, 23, 149, 1985.
99. **Rehbein, E.,** Selection of *Erwinia* — Resistenz bei kartoffeln mittels Gewebckultur, Dissertation, University of Bonn, Bonn, West Germany, 1983.
100. **Wenzel, G., Schieder, O., Przewozny, T., Sopory, S. K., and Melchers, G.,** Comparison of single cell culture derived *Solanum tuberosum* L. plants and a model for their application in breeding programs, *Theor. Appl. Genet.*, 55, 49, 1979.
101. **Brettell, R. I. S. and Ingram, D. S.,** Tissue culture in the production of novel disease-resistant crop plants, *Biol. Rev.*, 54, 329, 1979.
102. **Orton, T. J.,** Spontaneous electrophoretic and chromosomal variability in callus cultures and regenerated plants of celery, *Theor. Appl. Genet.*, 67, 18, 1983.
103. **Larkin, P. J. and Scowcroft, W. R.,** Somaclonal variation and crop improvement, in *Genetic Engineering of Plants (an Agricultural Perspective)*, Kosuge, T., Meredith, C. P., and Hollaender, A., Eds., Plenum Press, New York, 1983, 289.
104. **Mousseau, J.,** Fluctuations induites par la neoformation de bourgeons in vitro, *Strasbourg Coll. Intern. CNRS*, 293, 234, 1970.
105. **Fraser, R. S. S., Loughlin, S. A. R., and Conner, J. G.,** Resistance to tobacco mosaic virus in tomato: effects of the Tm-1 gene on symptom formation and multiplication of virus strain, *J. Gen. Virol.*, 50, 221, 1980.
106. **Chaleff, R. S. and Parsons, M. F.,** Direct selection *in vitro* for herbicide-resistance mutants of *Nicotiana tabacum, Proc. Natl. Acad. Sci. U.S.A.*, 75, 5104, 1978.
107. **Burk, L. G., Chaplin, J. F., Gooding, G. V., and Powell, N. T.,** Quantity production of anther-derived haploids from a multiple disease resistant tobacco hybrid. I. Frequency of plants with resistance or susceptibility to tobacco mosaic virus (TMV), potato virus (PYV) and root knot (RK), *Euphytica*, 28, 201, 1979.
108. **Chaleff, R. S. and Keil, R. L.,** Genetic and physiological variability among cultured cells and regenerated plants of *Nicotiana tabacum, Mol. Gen. Genet.*, 181, 254, 1981.
109. **Deaton, W. R., Legg, P. D., and Collins, G. B.,** A comparison of barley tobacco doubled-haploid lines with their source inbred cultivars, *Theor. Appl. Genet.*, 62, 69, 1982.
110. **Barbier, M. and Dulieu, H. L.,** Early occurrence of genetic variants in protoplast cultures, *Plant Sci. Lett.*, 29, 201, 1983.
111. **Lorz, H. and Scowcroft, W. R.,** Variability among plants and their progeny regenerated from protoplasts of Su/su heterozygotes of *Nicotiana tabacum, Theor. Appl. Genet.*, 66, 67, 1983.
112. **Daub, M. E.,** Tissue culture and the selection of resistance to pathogens, *Annu. Rev. Phytopathol.*, 24, 159, 1986.
113. **Bingham, E. T., Murley, L. V., Kaatz, D. M., and Saunders, J. W.,** Breeding alfalfa which regenerates from callus tissue in culture, *Crop Sci.*, 15, 719, 1975.
114. **McCoy, T. J. and Bingham, E. T.,** Regeneration of diploid alfalfa plants from cells grown in suspension culture, *Plant Sci. Lett.*, 10, 59, 1977.
115. **Mitten, D. H., Sato, S. J., and Skokut, T. A.,** In vitro regenerative potential of alfalfa germplasm sources, *Crop Sci.*, 24, 943, 1984.
116. **Atanassov, A. and Brown, D. C. W.,** Plant regeneration from suspension culture and mesophyll protoplasts of *Medicago sativa* L., *Plant Cell Tissue Organ Cult.*, 3, 149, 1984.
117. **Schenk, R. V. and Hildebrandt, A. C.,** Medium and technique for the induction of growth of monocotyledonous and dicotyledonous plant cell cultures, *Can. J. Bot.*, 50, 199, 1972.

118. **Crouch, M. L.,** Non zygotic embryos of *Brassica napus* L. contain embryo specific storage proteins, *Planta,* 156, 520, 1982.
119. **Standord, E. H., Clement, W. M., and Bingham, E. T.,** Cytology and evolution of alfalfa, in *Alfalfa Science and Technology,* Hanson, C. H., Ed., American Society of Agronomy, Madison, WI, 1972, 87.
120. **Saunders, J. W. and Bingham, E. T.,** Production of alfalfa plants from callus tissue, *Crop Sci.,* 12, 804, 1972.
121. **Saunders, J. W. and Bingham, E. T.,** Growth regulator effects on bud initiation in callus cultures of *Medicago sativa, Am. J. Bot.,* 62, 850, 1975.
122. **Bingham, E. T.,** Isolation of haploids of tetraploid alfalfa, *Crop Sci.,* 11, 433, 1971.
123. **Bingham, E. T. and McCoy, T. J.,** Cultivated alfalfa at the diploid level, origin, reproductive stability, and yield of seed and forage, *Crop Sci.,* 19, 97, 1979.
124. **Reisch, B., Duke, S. H., and Bingham, T.,** Selection and characterization of ethionine-resistant alfalfa *(Medicago sativa)* cell lines, *Theor. Appl. Genet.,* 59, 89, 1986.
125. **Bingham, E. T. and McCoy, T. J.,** Somaclonal variation in alfalfa, *Plant Breed. Rev.,* 4, 123, 1986.
126. **Pfeiffer, T. W. and Bingham, E. T.,** Comparisons of alfalfa somaclonal and sexual derivatives from the same genetic source, *Theor. Appl. Genet.,* 67, 263, 1984.
127. **Sadasivaiah, R. S. and Lesins, K.,** Reduction of chromosome number in root tip cells of *Medicago, Can. J. Genet. Cytol.,* 16, 219, 1974.
128. **Larkin, P. J. and Scowcroft, W. R.,** Somaclonal variation and eyespot toxin tolerance in surgarcane, *Plant Cell Tissue Organ Cult.,* 2, 111, 1983.
129. TCCP: Tissue Culture for Crops Project, Progress Report, Colorado State University, Fort Collins, 1987.
130. **Scowcroft, W. R., Ryan, S. A., Brettel, R. I. S., and Larkin, P. J.,** Somaclonal variation in crop improvement, in *Biotechnology in International Agricultural Research,* International Rice Research Institute, Manila, Philippines, 1985, 100.
131. **Day, A. and Ellis, T. H. N.,** Chloroplast DNA deletions associated with wheat plants regenerated from pollen: possible basis for maternal inheritance of chloroplasts, *Cell,* 39, 359, 1984.
132. **Day, P. R.,** Plant genetics: increasing crop yield, *Science,* 197, 1334, 1977.
133. **Gosch-Wackerle, G., Aviv, L., and Galun, E.,** Induction, culture and differentiation of callus from immature rachises, seeds, and embryos of *Triticum, Z. Pflanzenphysiol.,* 91, 267, 1979.
134. **Shimada, T. and Yamada, Y.,** Wheat plants regenerated from embryo cell cultures, *Jpn. J. Genet.,* 54, 379, 1979.
135. **Alhoowalia, B. S. and Sherington, J.,** Transmission of somaclonal variation in wheat, *Euphytica,* 34, 525, 1985.
136. **Maddock, S. W., Risiott, R., Parmar, S., Jones, M. G. K., and Shewry, P. R.,** Somaclonal variation in the gliadin patterns of grains of regenerated plants, *J. Exp. Bot.,* 36, 1976, 1984.
137. **Cooper, D. B., Sears, R. G., Lockhart, G. L., and Jones, B. L.,** Heritable somaclonal variation in gliadin proteins of wheat plants derived from embryo callus culture, *Theor. Appl. Genet.,* 71, 784, 1986.
138. **Baenziger, P. S., Wesenberg, D. M., Schaeffer, G. W., Galun, E., and Feldman, M.,** Variation among anther culture derived doubled haploids of 'kitt' wheat, in Proc. 6th Int. Wheat Genetics Symp., Kyoto, Japan, 1983, 575.
139. **Eapan, S. and Rao, P. S.,** Plant regenerations from callus cultures of durum and emmer wheat, *Plant Cell Rep.,* 1, 215, 1982.
140. **Maddock, S. E., Lancaster, V. A., Risiott, R., and Franklin, J.,** Plant regeneration from cultured immature embryos and influorescences of 25 cultivars of wheat *(Triticum aestivum), J. Exp. Bot.,* 34, 925, 1983.
141. **Khuspe, S. S., Agarwal, D. C., and Mascarenhas, A. F.,** Plant tissue culture in wheat crop improvement, in Proc. 15th Int. Congr. Genetics Abstr., 1984, 408.
142. **Larkin, P. J.,** *In vitro* culture and cereal breeding, in *Cereal Tissue and Cell Culture,* Bright, S. W. J. and Jones, M. G. K., Eds., Nighoff/Junk, Dordrecht, 1985, 273.
143. **Orton, T. J. and Browers, M. A.,** Segregation of genetic markers among plants regenerated from cultured anther of broccoli (*Brassica oleracea* var. 'talica'), *Theor. Appl. Genet.,* 69, 637, 1985.
144. **Suenaga, K., Abrigo, E. M., and Yoshida, S.,** Seed-derived callus culture for selecting salt-tolerant rices. I. Callus induction, plant regeneration, and variations in visible plant traits, *I.R.R.I. Res. Pap. Ser.,* 79, 1, 1982.
145. **Zhao, C. Z., Zheng, K. L., Qi, X. F., Sun, Z. X., and Fu, Y. P.,** Characteristics of rice plants derived from somatic-tissue and their progenies, *Acta Genet. Sin.,* 9, 320, 1982.
146. **Oono, K.,** Regulation of somatic mutation in rice tissue culture, *Int. Symp. Genet. Manipulation Crops, (Beijing, China),* 5(Abstr.), 1984.
147. TCCP: Tissue Culture for Crops Project, Newsl. No. 7, Colorado State University, Fort Collins, July 1987.
148. **Zhen-Hua, Z.,** Application of anther culture techniques to rice breeding, in Rice Tissue Culture Planning Conf., Int. Rice Research Institute, Los Banos, Laguna, Philippines, 1982, 55.

149. **Chih-Ching, C.,** Anther culture of rice and its significance in distant hybridization, in Rice Tissue Culture Planning Conf., Int. Rice Research Institute, Los Banos, Laguna, Phillipines, 1982, 47.
150. **Schaeffer, G. W.,** Recovery of heritable variability in anther derived doubled-haploid rice, *Crop Sci.,* 22, 1160, 1982.
151. **Green, C. E.,** Prospects for crop improvement in the field of cell culture, *Hortic. Sci.,* 12, 7, 1977.
152. **Green, C. E., Phillips, R. L., and Wang, A. S.,** Cytological analysis of plants regenerated from maize cultures, *Maize Genet. Coop. Newsl.,* 51, 53, 1977.
153. **McCoy, T. J. and Phillips, R. L.,** Chromosome stability in maize *(Zea mays)* tissue cultures and sectoring in some regenerated plants, *Can. J. Genet. Cytol.,* 24, 559, 1982.
154. **Beckert, M., Pollacsek, M., and Cao, M. Q.,** Analysis of genetic variability in regenerated inbred lines, *Maize Genet. Coop. Newsl.,* 56, 38, 1982.
155. **Beckert, M., Pollacsek, M., and Caenen, M.,** Etude de la variabilite genetique obtenue chez le mais apres callogenes et regeneration de plantes *in vitro, Agronomie,* 3, 9, 1983.
156. **Engler, D. E. and Grogan, R. G.,** *In vitro* selection of potential disease resistant somaclonal variants of lettuce from regenerated protoplasts, *Phytopathology,* 72 (Abstr.), 1003, 1982.
157. **Engler, D. E. and Grogan, R. G.,** Isolation, culture and regeneration of lettuce leaf mesophyll protoplasts, *Plant Sci. Lett.,* 28, 223, 1983.
158. **Engler, D. E. and Grogan, R. G.,** Variation in lettuce plants regenerated from protoplasts, *J. Hered.,* 75, 426, 1984.
159. **Heinz, D. J., Krishnamurthi, M., Nickell, L. G., and Maretzki, A.,** Cell tissue and organ culture in sugarcane improvement, in *Applied and Fundamental Aspects of Plant, Cell, Tissue and Organ Culture,* Reinert, J. and Bajaj, Y. P. S., Eds., Springer-Verlag, Berlin, 1977, 3.
160. **Krishnamurthi, M. and Tlaskal, J.,** Fiji disease resistant *Saccharum officinarum* var. Pindar subclones from tissue cultures, *Proc. Int. Soc. Sugar Cane Technol.,* 15, 130, 1974.
161. **Liu, M. C.,** *In vitro* methods applied to sugarcane improvement, in *Plant Tissue Culture,* Thorpe, T. A., Ed., Academic Press, New York, 1981, 299.
162. **Krishnamurthi, M.,** Disease resistance in sugarcane developed through tissue culture, in *Plant Tissue Culture,* Fujiwara, A., Ed., Japanese Association Plant Tissue Culture, Tokyo, 1982, 769.
163. **Vulkova-Atachkova, Z., Zagorska, N., and Tchiporikova, E.,** Somaclonal changes of wild species from the genus *Lycopersicon* and *Solanum lycopersicoides,* in *Tomato Genetics Cooperative Vol. 37,* Cornell University, Ithaca, NY, 1987, 77.
164. **Barden, K. A., Smith, S. S., and Murakishi, H. H.,** Regeneration and screening of tomato somaclones for resistance to tobacco mosaic virus, *Plant Sci.,* 45, 209, 1986.
165. **Smith, S. S., and Murakishi, H. H.,** Inheritance of resistance to tomato mosaic virus (ToMV-O) in tomato somaclones, in *Tomato Genetics Cooperative Vol. 37,* Cornell University, Ithaca, NY, 1987, 65.
166. **Schwartz, H. J., Galletta, G. J., and Zimmerman, R. H.,** Field performance and phenotypic stability of tissue culture-propagated strawberries, *J. Am. Soc. Hortic. Sci.,* 106, 667, 1981.
167. **Roest, S., Van Berkel, M., Bokelmann, G., and Broertjes, C.,** The use of an in vitro adventitious bud technique for mutation breeding of *Begonia* × *hiemalis, Euphytica,* 30, 381, 1981.
168. **Skirvin, R. M.,** 'Velvet Rose' *Pelargonium,* a scented geranium, *Hortic. Sci.,* 11, 61, 1976.
169. **Williams, L. and Collins, H. A.,** Growth and cytology of celery plants derived from tissue cultures, *Ann. Bot. (London) (N.S.),* 40, 333, 1976.
170. **Orton, T. J.,** Genetic instability during embryogenic cloning of celery, *Plant Cell Tissue Organ Cult.,* 4, 159, 1985.
171. **Fujii, D. S.,** In Vitro Propagation of Celery (*Apium graveolens* L.), M.S. thesis, University of California, Davis, 1982.
172. **Rappaport, L., Banks-Izen, M., Fujii, D. S., Orton, T. J., Stuart, D., Pullman, G., Matlin, S. A., and Thompson, R. H.,** Celery cell and tissue culture: toward induced variation and selection for resistance to *Fusarium oxysporum* f. sp. appi, in California Celery Research Program — 1981/1982, Annual Report, Batkin, T., Ed., California Celery Research Advisory Board, Dinuba, 1982, 117.
173. **Scowcroft, W. R. and Larkin, P. J.,** Somaclonal variation and genetic improvement of crop plants, in *Better Crops for Food,* Nugent, J. and D'Cornor, M., Eds., Pitman Publishing, London, 1983, 177.
174. **Duncan, D. R. and Widholm, J. M.,** Cell selection for crop improvement, *Plant Breed. Rex.,* 4, 153, 1980.
175. **Scowcroft, W. R.,** Somaclonal variation: the myth of clonal uniformity, in *Genetic Flux in Plants,* Hohn, B. and Dennis, E. S., Eds., Springer-Wien, New York, 1985, 217.
176. **Miins, F.,** Heritable variation in plant cell culture, *Annu. Rev. Plant Physiol.,* 34, 327, 1983.
177. **Zeng, J. Z.,** Application of anther culture technique to crop improvement in China, in *Plant Cell Culture in Crop Improvement,* Sen, S. K. and Giles, K. L., Eds., Plenum Press, New York, 1983, 351.
178. **Karp, A. and Bright, S. W. J.,** On the causes and origins of somaclonal variation, in *Surveys of Plant Molecular and Cell Biology,* Vol. 2, Miflin, B. J., Ed., Oxford University Press, Oxford, 1985, 199.

179. **Novak, F.**, Phenotype and cytological status of plants regenerated from callus cultures of *Allium sativum* L., *Z. Pflanzenzuecht.*, 84, 250, 1980.
180. **Mix, G., Wilson, G. M., and Foroughi-Wehr, B.**, The cytological status of plants of *Hordeum vulgare* L. regenerated from microspore callus, *Z. Pflanzenzuecht.*, 80, 89, 1978.
181. **Sree Ramulu, K.**, Genetic instability at the S-locus of *Lycopersicon peruvianum* Plants regenerated from in vitro culture of anthers: generation of new S-specificities and S-allele reversions, *Heredity*, 49, 319, 1982.
182. **Reisch, B. and Bingham, E. T.**, Plants from ethionine resistant alfalfa tissue cultures: variation in growth and morphological characteristics, *Crop Sci.*, 21, 783, 1981.
183. **Donn, G., Tischer, E., Smith, J. A., and Goodman, H. M.**, Herbicide-resistant alfalfa cells, an example of gene amplification in plants, *J. Mol. Appl. Genet.*, 2, 621, 1984.
184. **Johnson, L. B., Rose, R. J., Schlarbaum, S. E., and Skinner, D. Z.**, Variation in phenotype and chromosome number in alfalfa protoclones regenerated from nonmutagenized calli, *Crop Sci.*, 24, 948, 1984.
185. **Dulieu, H. and Barbier, M.**, High frequencies of genetic variant plants regenerated from cotyledons of tobacco, in *Variability in Plants Regenerated from Tissue Culture*, Earle, L. and Demarly, Y., Eds., Praeger Press, New York, 1982, 211.
186. **Chaleff, R. S. and Ray, T. B.**, Herbicide-resistant mutants from tobacco cell cultures, *Science*, 223, 1184, 1984.
187. **Brossard, D.**, The influence of kinetin on formation and ploidy levels of buds arising from *Nicotiana tabacum* pith tissue grown in vitro, *Z. Pflanzenphysiol.*, 78, 323, 1976.
188. **Evans, D. A., Bravo, J. E., Kut, S. A., and Flick, C. E.**, Genetic behavior of somatic hybrids in the genus *Nicotiana*. *N. otophora* + *N. tabacum* and *N. sylvestris* + *N. tabacum*, *Theor. Appl. Genet.*, 65, 93, 1983.
189. **Evans, D. A., Flick, C. E., Kut, S. A., and Reed, S. M.**, Comparison of *Nicotiana tabacum* and *Nicotiana nesophila* hybrids produced by ovule culture and protoplast fusion, *Theor. Appl. Genet.*, 62, 193, 1982.
190. **Evans, D. A. and Gamborg, O. L.**, Chromosome stability of cell suspension cultures of *Nicotiana* spp., *Plant Cell Rep.*, 1, 104, 1982.
191. **Evans, I. T.**, Raising the yield potential: by selection or design?, in *Genetic Engineering in Plants*, Kosuge, T., Meredith, C. P., and Hollaender, A., Eds., Plenum Press, New York, 1983, 371.
192. **Krishnamurthi, M.**, Sugarcane improvement through tissue culture and review of progress, in *Tissue Culture of Economically Important Plants*, Rao, A. N., Eds., Asian Network for Biological Science, Singapore, 1981, 70.
193. **Karp, A. and Maddock, S. E.**, Chromosome variation in wheat plants regenerated from cultured immature embryos, *Theor. Appl. Genet.*, 67, 249, 1984.
194. **Creissen, G. P. and Karp, A.**, Karyotypic changes in potato plants regenerated from protoplasts, *Plant Cell Tissue Organ Cultr.*, 4, 171, 1985.
195. **Kemble, R. J., Brettell, R. I. S., and Flavell, R. B.**, Mitochondrial DNA analyses of fertile and sterile maize plants from tissue culture with Texas male sterile cytoplasm, *Theor. Appl. Genet.*, 62, 213, 1982.
196. **Landsmann, J. and Uhrig, H.**, Somaclonal variation in *Solanum tuberosum* detected at the molecular level, *Theor. Appl. Genet.*, 71, 500, 1985.
197. **Roca, W. M., Szabados, L., and Hussain, A.**, Biotechnology Research Unit, Annual Report, CIAT, 1987, 64.
198. **Alhoowalia, B. S.**, Plant regeneration from callus culture in wheat, *Crop Sci.*, 22, 405, 1982.
199. **Ryschka, U.**, Morphogenase bei kalluskulturen des Weizen, *Tagungsber. Akad. Landwirtschaftwiss D.D.R. (Berlin)*, 207, 15, 1983.
200. **Brettell, R. I. S., Pallotta, M. A., Gustafson, J. P., and Appels, R.**, Variation at the *Nor* loci in triticale derived from tissue culture, *Theor. Appl. Genet.*, 71, 637, 1986.
201. **Nakamura, C., Keller, W. A., and Fedak, K.**, In vitro propagation and chromosome doubling of a *Triticum crassum* × *Hordeum vulgare* intergeneric hybrid, *Theor. Appl. Genet.*, 60, 89, 1981.
202. **Fedak, G.**, Cytogenetics of tissue culture regenerated hybrids of *Triticum tauschii* × *Secale cereale*, *Can. J. Genet. Cytol.*, 26, 382, 1984.
203. **Lapitan, N. L. V. and Ryan, S. A.**, Somaclonal variation and eyespot toxin tolerance in sugarcane, *Plant Cell Tissue Organ Cult.*, 2, 111, 1983.
204. **Armstrong, K. C., Nakamura, C., and Keller, W. A.**, Karyotype instability in tissue culture regenerants of triticale (× *Triticosocale* Wittmack) cv. "Welsh" from 6-month-old callus cultures, *Z. Pflanzenzuecht.*, 91, 233, 1983.
205. **Nakamura, C. and Keller, W. A.**, Callus proliferation and plant regeneration from immature embryos of hexaploid triticale, *Z. Pflanzenzuecht.*, 88, 137, 1982.
206. **Alhoowalia, B. S.**, Spectrum of variation in somaclones of triploid ryegrass, *Crop. Sci.*, 23, 1141, 1983.

207. **Benzion, G., Phillips, R. L., and Rines, H. W.,** Case histories of genetic variability in vitro: oats and maize, in *Plant Regeneration and Genetic Variability,* Vasil, I. K., Eds., Academic Press, New York, 1986.
208. **Brettell, R. I. S. and Thomas, E.,** Reversion of Texas male-sterile cytoplasm of maize in culture to give fertile, T-toxin resistant plants, *Theor. Appl. Genet.,* 58, 55, 1980.
209. **Smith, S. M. and Street, H. E.,** The decline of embryogenic potential as callus and suspension cultures of carrot (*Daucus carota* L.) are serially subcultured, *Ann. Bot.,* 38, 233, 1974.
210. **Singh, B. D., Kao, K. N., and Miller, R. A.,** Karyotypic changes and selection pressure in *Haplopappus gracilis* suspension cultures, *Can. J. Genet. Cytol.,* 17,109, 1975.
211. **Bingham, E. T. and Saunders, J. W.,** Chromosome manipulations in alfalfa: scaling the cultivated tetraploid to seven ploidy levels, *Crop Sci.,* 14, 474, 1974.
212. **Singh, B. D. and Harvey, B. L.,** Does 2,4-D induce mitotic irregularities in plant tissue culture, *Experientia,* 32, 785, 1975.
213. **Stavarck, S. J. and Rains, D. W.,** Cell culture techniques: selection and physiological studies of salt tolerance, in *Salinity Tolerance in Plants: Strategies for Crop Improvement,* Staples, R. C. and Toenniessen, G. H., Eds., John Wiley & Sons, New York, 1984, 321.
214. **Hartman, C. L., McCoy, T. J., and Knous, T. R.,** Selection of alfalfa *(Medicago sativa)* cell lines and regeneration of plants resistant to the toxin(s) produced by *Fusarium oxysporum* of sp. *medicaginis, Plant Sci. Lett.,* 34, 183, 1984.
215. **Mroginski, L. A. and Kartha, K. K.,** Tissue culture of legumes for crop improvement, *Plant Breed. Rev.,* 2, 215, 1984.
216. **Latunde-Dada, A. O. and Lucas, J. A.,** Somaclonal variation and reaction to verticillium wilt in *Medicago sativa* L. plants regenerated from protoplasts, *Plant Sci. Lett.,* 32, 205, 1983.
217. **Johnson, L. B., Rose, R. J., Scwarbaum, S. E., and Stuteville, D. L.,** Effects of protocloning in 'Regen-S' alfalfa on somatic chromosome and on mitochondrial and chloroplast DNAs, report of the 29th Alfalfa Improvement Congr., Lethbridge, Alberta, Canada, 1984.
218. **D'Amato, F.,** Cytogenetics of differentiation in tissue and cell culture, in *Applied and Fundamental Aspects of Plant Cell, Tissue and Organ Cultures,* Reinert, J. and Bajaj, Y. P. S., Eds., Springer-Verlag, Berlin, 1977, 343.
219. **D'Amato, F.,** The problem of genetic stability in plant tissue and cell cultures, in *Crop Resources for Today and Tomorrow,* Frankel, O. and Hawkes, J. G., Eds., Cambridge University Press, Cambridge, 1975, 333.
220. **Ogura, H.,** The cytological chimeras in original regenerates from tobacco tissue cultures and in their offspring, *Jpn. J. Genet.,* 51, 161, 1976.
221. **Wheeler, V. A., Evans, N. E., Foulger, D., Judith Webb, K., Karp, A., Franklin, J., and Bright, S. W. J.,** Shoot formation from explant cultures of fourteen potato cultivars and studies of the cytology and morphology of regenerated plants, *Ann. Bot.,* 55, 309, 1985.
222. **Wenzel, G.,** Neue Wege in der Kartoffelzuchtung. I. Vom Shortenkl zur Haploiden, *Kartoffelbau,* 30, 126, 1979.
223. **Wenzel, G. and Foroughi-Wehr, B.,** Anther culture of *Solanum tuberosum,* in *Cell Culture and Somatic Cell Culture Genetics of Plants,* Vol. 1, Vasil, J. K., Ed., Academic Press, New York, 1989, 293.
224. **Wenzel, G. and Uhrig, H.,** Breeding for nematode and virus resistance in potato via anther culture, *Theor. Appl. Genet.,* 59, 333, 1981.
225. **Binding, H., Nehls, R., Schieder, O., Sopory, S. K., and Wenzel, G.,** Regeneration of mesophyll protoplasts isolated from dihaploid clones of *Solanum tuberosum, Physiol. Plant.,* 43, 52, 1978.
226. **Marks, G. E.,** The consequences of an unusual Robertsonian translocation in celery (*Apium graveolens* var. *dulce), Chromosoma,* 69, 211, 1978.
227. **D'Amato, F.,** Chromosome number variation in cultured cells and regenerated plants, in *Frontiers of Plant Tissue Culture,* Thorpe, T. A., Ed., IAPTC/University of Calgary, Calgary, Canada, 1978, 287.
228. **Constantin, M. J.,** Chromosome instability in cell and tissue cultures and regenerated plants, *Environ. Exp. Bot.,* 21, 359, 1981.
229. **Austin, S. and Cassells, A. C.,** Variation between plants regenerated from individual calli produced from separated potato stem callus cells, *Plant Sci. Lett.,* 31, 107, 1983.
230. **Groose, R. W. and Bingham, E. T.,** Variation in plants regenerated from tissue cultures of tetraploid alfalfa heterozygous for several traits, *Crop Sci.,* 24, 655, 1984.
231. **Binek, A. and Bingham, E. T.,** Cytology and crossing behavior of triploid alfalfa, *Crop Sci.,* 10, 303, 1970.
232. **Bingham, E. T.,** Aneuploids in seedling populations of tetraploid alfalfa, *Medicago sativa, Crop Sci.,* 8, 571, 1968.
233. **D'Amato, F.,** Endopolyploidy as a factor in plant tissue development, *Caryologia,* 17, 41, 1964.
234. **D'Amato, F.,** The problem of genetic stability in plant tissue and cell cultures in crop resources for today and tomorrow, Hawkes, J. G. E., Ed., Cambridge University Press, Cambridge, 1975, 333.

235. **Cionini, P. G., Bennici, A., and D'Amato, F.,** Nuclear cytology of callus induction and development in vitro. I. Callus from *Vicia faba* cotyledons, *Protoplasma,* 96, 101, 1978.
236. **Nandi, S., Fridborg, G., and Eriksson, T.,** Effects of 6-(3-methyl-2-buten-1-ylamino)-purine and α-naphthalene acetic acid on root formation and cytology of root tips and callus in tissue cultures of *Allium cepa* var. *proliferum, Hereditas,* 85, 57, 1977.
237. **Ryschka, U.,** Morphogenas bei/Kalluskulturen des Weizen, *Tagungsber. Akad. Landwirtschaftwiss. D.D.R. (Berlin),* 207, 15, 1983.
238. **Sacristan, M. D.,** Karyotypic changes in callus cultures from haploid and dihaploid plants of *Crepis capillaris* (L) Wallr, *Chromosoma,* 33, 273, 1971.
239. **Sunderland, N.,** Nuclear cytology, in *Plant Cell and Tissue Culture,* Vol. 2, Street, H. E., Ed., University of California Press, Berkeley, 1977, 177.
240. **Kao, K. N., Miller, R. A., Gamborg, O. L., and Harvey, B. L.,** Variations in chromosome number and structure in plant cells grown in suspension cultures, *Can. J. Genet. Cytol.,* 12, 297, 1970.
241. **Larkin, P. J.,** Case histories of genetic variability in vitro: wheat and triticale, *Cell Cult. Somatic Cell Genet. Plants,* 3, 367, 1986.
242. **Chaleff, R. S. and Mauvais, C. J.,** Acetolactate synthase in the site of action of two sulfonylurea herbicides in higher plants, *Science,* 224, 1443, 1984.
243. **Sacristan, M. D.,** Selection for disease resistance in *Brassica* cultures, *Hereditas (Suppl.),* 3, 57, 1985.
244. **Talmud, P. J. and Lewis, D.,** The mutagenecity of amino acid analogues in *Coprinus lagopus, Genet. Res.,* 23, 47, 1974.
245. **Razin, A. and Riggs, A. D.,** DNA methylation and gene function, *Science,* 210, 604, 1980.
246. **Davies, P. J. and Perry, J. M.,** The modification of induced genetic change in yeast by an amino acid analogue, *Mol. Gen. Genet.,* 162, 184, 1978.
247. **Morrison, R. A. and Evans, D. A.,** Gametoclonal variation, *Plant Breed. Rev.,* 5, 359, 1987.
248. **Brock, R. D.,** Mutagenesis and crop improvement, in *The Biology of Crop Productivity,* Carlson, P., Ed., Academic Press, New York, 1980, 384.
249. **Konzac, C. F.,** Induced mutations for genetic analysis and improvement of wheat, in *Induced Mutations — a Tool in Plant Research,* IAEA, Vienna, 1981, 469.
250. **Lapitan, N. L. V., Sears, R. G., and Gill, B. S.,** Translocations and other karyotypic structural changes in wheat × rye hybrids regenerated from tissue culture, *Theor. Appl. Genet.,* 68, 547, 1984.
251. **Appels, R.,** Chromosome structure in cereals: the analysis of regions containing repeated sequence DNA and its application to the detection of alien chromosomes introduced into wheat, in *Genetic Engineering in Plants,* Kosuge, T., Meredith, C. P., and Hollaender, A., Eds., Plenum Press, New York, 1983, 229.
252. **McClintock, B.,** The significance of responses of the genome to challenge, *Science,* 226, 792, 1984.
253. **Chourey, P. S. and Kemble, R. J.,** Transposition event in tissue cultured cells of S-cms genotype of maize, *Maize Genet. Coop. Newsl.,* 56, 70, 1982.
254. **Peschke, V. M., Phillips, R. L., and Gengenbach, B. G.,** Transposable element activity in progeny of regenerated maize plants, in Proc. 6th Int. Congr. Plant Tissue and Cell Culture, IAPTC, Minneapolis, 1986, 285.
255. **Fromm, H., Edelman, M., Koller, B., Goloubinoff, P., and Galun, E.,** The enigma of gene coding for ribosomal protein S12 in the chloroplasts of *Nicotiana, Nucl. Acids Res.,* 14, 883, 1986.
256. **Kollar, B., Fromm, H., Galun, E., and Edelman, M.,** Evidence for in vivo trans splicing of pre-mRNAs in tobacco chloroplasts, *Cell,* 48, 111, 1987.
257. **Quetier, F. and Vedel, F.,** Heterogenous population of mitochondrial DNA molecules in higher plants, *Nature,* 268, 365, 1977.
258. **Spruill, W. M., Jr., Levings, C. S., and Sederoff, R. R.,** Recombinant DNA analysis indicates that the multiple chromosome of maize mitochondria contains different sequences, *Dev. Genet.,* 1, 362, 1980.
259. **Sibi, M.,** Multiplication conforme, non conforme, *Selectionneur Fr.,* 26, 9, 1978.
260. **Teolule, E.,** Somatic hybridization between *Medicago sativa* L. and *Medicago falcata* L., *C. R. Acad. Sci. (Paris) Ser. C,* 297, 13, 1983.
261. **Sibi, M.,** Variants epigeniques et cultures in vitro chez *(Lycopersicon esculentum* L.), in *Application de la Culture in Vitro a L'Amelioration des Plantes Potageres,* Dove, C., Ed., Institut National de la Recherche Agronomique, Versailles, France, 1981, 179.
262. **Sibi, M., Biglay, M., and Demarly, Y.,** Increase in the rate of recombinants in tomato *(Lycopersicum esculentum* L.) after in vitro regeneration, *Theor. Appl. Genet.,* 68, 317, 1984.
263. **Gengenbach, B. G., Green, C. E., and Donovan, C. M.,** Inheritance of selected pathotoxin resistance in maize plants regenerated from cell cultures, *Proc. Natl. Acad. Sci. U.S.A.,* 74, 5113, 1977.
264. **Smith, S. E., Bingham, E. T., and Fulton, R. W.,** Transmission of chlorophyll deficiencies provides evidence for biparental inheritance of plastids in *Medicago sativa, Heredity,* 77, 35, 1986.
265. **Smith, M. K. and McComb, J. A.,** Selection for NaCl tolerance in cell cultures of *Medicago sativa* and recovery of plants from a NaCl tolerant cell line, *Plant Cell Rep.,* 2, 126, 1983.
266. **Smith, M. K. and McComb, J. A.,** Use of callus cultures to detect NaCl tolerance in cultivars of three species of pasture legumes, *Aust. J. Plant Physiol.,* 8, 437, 1981.

267. **Smith, S. H. and Oglevee-O'Donovan, W. A.,** Meristem-tip culture from virus-infected plant material and commercial implications, in *Plant Cell and Tissue Culture: Principles and Applications,* Sharp, W. R., Larsen, P. O., Paddock, E. F., and Raghavan, V., Eds., Ohio State University Press, Columbus, 1979, 453.
268. **Rose, R. J., Johnson, L. B., and Kemble, R. J.,** Restriction endonuclease studies on the chloroplast and mitochondrial DNAs of alfalfa *(Medicago sativa* L.) protoclones, *Curr. Top. Plant Biochem. Physiol.,* 3, 178, 1984.
269. **Mulcahy, D. L. and Mulcahy, G. B.,** Gametophytic self-incompatibility reexamined, *Science,* 220, 1247, 1983.
270. **Bayly, I. K. and Craig, I. L.,** A morphological study of the X-ray induced cauliflower head and single-leaf mutation in *Medicago sativa* L., *Can. J. Genet. Cytol.,* 4, 386, 1962.
271. **Murray, B. E. and Craig, I. L.,** A cytogenetic study of X-ray induced cauliflower-head and single leaf mutation in *Medicago sativa* L., *Can. J. Genet. Cytol.,* 4, 379, 1962.
272. **Bingham, E. T.,** Morphology and petiole vasculature of five heritable leaf forms in *Medicago sativa* L., *Bot. Gaz.,* 127, 221, 1966.
273. **Gleba, Y. Y.,** Nonchromosomal inheritance in higher plants as studied by somatic cell hydridization, in *Plant Cell and Tissue Culture, Principles and Applications,* Sharp, S. R., Larsen, P. O., Paddock, E. F., and Raghavan, V., Eds., Ohio State University Press, Columbus, 1979, 775.
274. **Larkin, P. J., Ryan, S. A., Brettell, R. I. S., and Scowcroft, W. R.,** Heritable somaclonal variation in wheat, *Theor. Appl. Genet.,* 67, 443, 1984.
275. **Bendich, A. J. and Gauriloff, L. P.,** Morphometric analysis of cucurbit mitochondria: the relationship between chondriome volume and DBA content, *Protoplasma,* 119, 1, 1984.
276. **Heinz, D. J.,** Sugarcane improvement through induced mutations using vegetable propagules and cell culture techniques, in Proc. of a Panel 1972, IAEA, Vienna, 1973, 53.
277. **Nickell, L. G.,** Crop improvements in sugarcane: studies using in vitro methods, *Crop. Sci.,* 17, 717, 1977.
278. **Ross, H.,** Potato breeding-problems and perspectives, in *Advances in Plant Breeding* (suppl. to *J. Plant Breed.),* Verlag, Paul Parley, Berlin, 1986.
279. **Hartman, C. L., Knous, T. R., and McCoy, T. J.,** Field testing and preliminary progeny evaluation of alfalfa regenerated from cell lines resistant to the toxins produced by *Fusarium oxysporum* f. sp. *medicaginis, Phytopathology,* 74(Abstr.), 818, 1984.
280. **Latunde-Dada, A. O. and Lucas, J. A.,** Somaclonal variation and reaction to *Verticillium* wilt in *Medicago sativa* plants regenerated from protoplasts, *Plant Sci. Lett.,* 32, 205, 1983.
281. **Scowcroft, W. R. and Larkin, P. J.,** Somaclonal variation: a new option for plant improvement, in *Plant Improvement and Somatic Cell Genetics,* Vasil, K., Scowcroft, W. R., and Frey, K. J., Eds., Academic Press, New York, 1982, 159.
282. **Scowcroft, W. R., Larkin, P. J., and Brettell, R. I. S.,** Genetic variation from tissue culture, in *Use of Tissue Culture and Protoplasts in Plant Pathology,* Helgeson, J. P. and Deverall, B. J., Eds., Academic Press, New York, 1983, 139.
283. **Nafziger, E. D., Widholm, J. M., Sterinrucken, H. C., and Killmer, J. L.,** Selection and characterization of a carrot cell line tolerant to glyphosate, *Plant Physiol.,* 76, 571, 1984.
284. **Murakishi, H. H. and Carlson, P. S.,** In vitro selection of *Nicotiana sylvestris* variants with limited resistance to TMV, *Plant Cell Rep.,* 1, 94, 1982.
285. **Wight, J. C. and Lacy, M. L.,** Somaclonal variation occurring in disease response of regenerated celery plants from cell suspension and callus cultures, *Phytopathology,* 75(Abstr.), 967, 1985.
286. **Pullman, G. S. and Rappaport, L.,** Tissue culture-induced variation in celery of *Fusarium* yellows, *Phytopathology,* 73(Abstr.), 818, 1983.
287. **Pullman, G., Rappaport, L., and Heath-Pagliuseo, S.,** Somaclonal variation in celery towards selection for resistance to *Fusarium* wilt, *Hortic. Sci.,* 19(Abstr.), 589, 1984.
288. **Abo El-Nil, M. M. and Hildebrandt, A. C.,** Morphological changes in geranium plants differentiated from anther cultures, *In Vitro,* 7, 258, 1972.
289. **Engler, D. E. and Grogan, R. G.,** Variation in lettuce plants regenerated from protoplasts, *J. Hered.,* 75, 420, 1984.
290. **Swanson, E. B. and Tomes, D. T.,** Evaluation of birdsfoot trefoil regenerated plants and their progeny after in vitro selection for 2,4-dichlorophenoxyacetic acid tolerance, *Plant Sci. Lett.,* 29, 19, 1983.
291. **Umbeck, P. F. and Gengenbach, B. G.,** Reversion of male sterile-T cytoplasm maize *(Zea mays)* to male fertility in tissue culture, *Crop Sci.,* 23, 584, 1983.
292. **Cummings, D. P., Green, C. E., and Stuthman, D. D.,** Callus induction and plant regeneration in oats, *Crop Sci.,* 16, 465, 1976.
293. **Rines, H. W. and Luke, H. H.,** Selection and regeneration of toxin-insensitive plants from tissue cultures of oats *(Avena sativa)* susceptible to *Helminthosporium victoriae, Theor. Appl. Genet.,* 71, 16, 1985.

294. **Kochba, J., Ben-Hayyim, G., Spiegel-Roy, P., Saad, S., and Neumann, H.,** Selection of stable salt-tolerant callus cell lines and embryos in *Citrus sinensis* and *C. aurantium, Z. Pflanzenphysiol.*, 106, 111, 1982.
295. **Behnke, M.,** Kulturen isolierter von einigen dehaploiden *Solanum tuberosum* — Klonen und ihre regeneration, *Z. Pflanzenphysiol.*, 78, 177, 1976.
296. **Behnke, M.,** Selection of potato callus for resistance to culture filtrates of *Phytophthora infestans* and regeneration of resistant plants, *Theor. Appl. Genet.*, 55, 69, 1979.
297. **Behnke, M.,** General resistance to late blight of *Solanum tuberosum* plants regenerated from callus resistant to culture filtrates of *Phytophthora infestans, Theor. Appl. Genet.*, 56, 151, 1980.
298. **Hoffmann, F.,** Mutation and selection of haploid cell culture system of rape and rye, in *Production of Natural Compounds by Cell Culture Methods,* Alfermann, A. W. and Reinhard, E., Eds., Gesellschaft fur Strahlen and Umweltforschung, Munich, 1978, 319.
299. **Sacristan, M. D.,** Resistance responses to *Phoma lingam* of plants regenerated from selected cell and embryogenic cultures of haploid *Brassica napus, Theor. Appl. Genet.*, 61, 193, 1982.
300. **George, L. and Rao, P. S,** Yellow seeded variants in *in vitro* regenerants of mustard *(Brassica juncea* coss var. RAI-5), *Plant Sci. Lett.*, 30, 327, 1983.
301. **Ling, D. H., Vidhyaseharan, P., Borromeo, E. S., Zapata, F. J., and Mew, T. W.,** In vitro screening of rice germplasm for resistance to brown spot disease using phytotoxin, *Theor. Appl. Genet.*, 71, 133, 1985.
302. **Liu, M. C. and Chen, W. H.,** Improvement in sugarcane by using tissue culture methods, *4th Int. Congr. Plant Tissue Cell Cult. (Calgary, Canada),* 163(Abstr.), 1978.
303. **Liu, M. C., Shang, K. C., Chen, W. H., and Shih, S. C.,** Tissue and cell culture as aids to sugarcane breeding. Aneuploid cells and plants induced by treatment of cell suspension cultures with colchicine, *Proc. 16th Congr. Int. Soc. Sugar Cane Technol.*, 1, 29, 1977.
304. **Liu, L. J., Marquez, E. R., and Biascoechea, M. L.,** Variation in degree of rust resistance among plantlets derived from callus cultures of sugarcane in Puerto Rico, *Phytopathology,* 73 (Abstr.), 797, 1983.
305. **Carlson, P.,** Methionine sulfoximine-resistant mutants of tobacco, *Science,* 180, 1366, 1973.
306. **Thanutong, P., Furusawa, I., and Yamamoto, M.,** Resistant tobacco plants from protoplast-derived calluses selected for their resistance to *Pseudomonas* and *Alternaria* toxins, *Theor. Appl. Genet.*, 66, 209, 1983.
307. **Miller, O. K. and Hughes, K. W.,** Selection of paraquat-resistant variants of tobacco from cell cultures, *In Vitro,* 16, 1085, 1980.
308. **Singer, S. and McDaniel, C. M.,** Selection of glyphosate tolerant tobacco calli and the expression of this tolerance in regenerated plants, *Plant Physiol.*, 78, 411, 1985.
309. **Keil, R. L. and Chaleff, R. S.,** Genetic characterization of hudroxyurea-resistant mutants obtained from cell cultures of *Nicotiana tabacum, Mol. Gen. Genet.*, 192, 218, 1983.
310. **Zagorska, N. and Atanassov, A.,** Somaclonal variation in tobacco and sugar beet breeding, in *Tissue Culture in Forestry and Agriculture,* Abstract, Henke, R. R., Hughes, K. W., Constanin, M. J., and Hollaender, A., Eds., Plenum Press, New York, 1985, 371.
311. **Miller, S. A., Williams, G. R., Medina-Filho, H., and Evans, D. A.,** A somaclonal variant of tomato resistant to race 2 of *Fusarium oxysporum* f. sp. lycopersici, *Phytopathology,* 75(Abstr.), 1354, 1985.
312. **Zong-Xiu, S., Cheng-Zhang, Z., Kangle, Z., Xiu-fand, Q., and Ya-ping, F.,** Somaclonal genetics of rice, *Oryza sativa* L., *Theor. Appl. Genet.*, 67, 67, 1983.
313. **Earle, E. D. and Gracen, V. E.,** Somaclonal variation in progeny of plants from corn tissue cultures, in *Tissue Culture in Forestry and Agriculture,* Henke, R. R., Hughes, K. W., Constanin, M. J., and Hollaender, A., Eds., Academic Press, New York, 1985, 139.
314. **Earle, E. D., Gracen, V. E., Yorder, D. C., and Gemmill, K. P.,** Cytoplasm-specific effects of *Helminthosporium maydis* race T toxin on survival of corn mesophyll protoplasts, *Plant Physiol.*, 61, 420, 1978.
315. **Shoemaker, N. P., Swartz, J. J., and Galletta, G. J.,** Cultivar dependent variation in pathogen resistance due to tissue culture-propagation of strawberries, *Hortic. Sci.*, 20, 253, 1985.
316. **Moyer, J. W. and Collins, W. W.,** 'Scarlet' sweet potato, *Hortic. Sci.*, 18, 111, 1983.
317. **Chaleff, R. S.,** Isolation of agronomically useful mutants from plant cell cultures, *Science,* 219, 676, 1983.
318. **Groose, R. W.,** An Unstable Anthocyanin Mutation in Alfalfa (*Medicago sativa* L.) Which Reverts at High Frequency in Tissue Culture, Ph.D. thesis, University of Wisconsin, Madison, 1985.
319. **Henke, R. R., Mansur, M. A., and Constantin, M. J.,** Organogenesis and plantlet formation from organ- and seedling-derived calli of rice *(Oryza sativa), Physiol. Plant.*, 44, 11, 1978.
320. **Krikorian, A. D., O'Connor, S. A., and Fitter, M. S.,** Chromosome number variation and karyotype stability in cultures and culture-derived plants, in *Handbook of Plant Cell Culture,* Vol. 1, Evans, D. A., Sharp, W. R., Ammiratto, P. V., and Yamada, Y., Eds., Macmillan, New York, 1983, 541.
321. **Lesins, K. A. and Lesins, I.,** *Genus Medicago (Leguminosae), Taxogenetic Study,* Dr. W. Junk Publishers, Hange, Boston, 1979.

322. **Pavek, J. J. and Corsini, D. L.,** Field performance of clones from regenerated protoplasts of Russet Burbank, *Am. Potato J.*, 59(Abstr.), 482, 1982.
323. **Popchristov, V. D. and Zaganska, N. A.,** Study of the seed progeny of regenerated plants obtained by the tissue culture of tobacco, in *Use of Plant Tissue Culture in Breeding,* Ustav Experimentalni Botaniky CSAV, Prague, 1977, 209.
324. **Conner, A. J. and Meredith, C. P.,** Large-scale selection of aluminium-resistant mutants from plant cell culture: expression and inheritance in seedlings, *Theor. Appl. Genet.*, 71, 159, 1985.
325. **Maliga, P.,** Resistant mutants and their use in genetic manipulation, in *Frontiers of Plant Tissue Culture,* Thorpe, T. A., Ed., Calgary University Press, Calgary, Canada, 1978, 381.
326. **Widholm, J. M.,** Selection and characterization of biochemical mutants, in *Plant Tissue Culture and Its Bio-Technological Applications,* Barz, W., Reinhard, E., and Zenk, M. H., Eds., Springer-Verlag, Berlin, 1977, 112.
327. **Croughan, T. P., Stavarek, S. J., and Rains, D. W.,** Selection of a NaCl tolerant line of cultured alfalfa cells, *Crop Sci.*, 18, 959, 1978.
328. **Levitt, J.,** Water, radiation, salt and other stresses, in *Responses of Plants to Environmental Stresses,* Vol. 2, Academic Press, New York, 1980.
329. **McCoy, T. J.,** Interspecific hybridization of *Medicago sativa* L. and *M. rupestris* M. B. using ovule-embryo culture, *Can. J. Genet. Cytol.*, 27, 238, 1985.
330. **Singer, S. and Janick, J.,** Tissue culture-induced variation in scented *Pelargonium* spp., *J. Am. Soc. Hortic. Sci.*, 101, 281, 1976.
331. **Lorz, H.,** Variability in tissue culture derived plants, in *Genetic Manipulation: Impact on Man and Society,* Arber, W., Ed., Cambridge University Press, London, 1984, 103.
332. **Thomas, E. and Pratt, D.,** Isolation of paraquat tolerant mutants from tomato cell cultures, *Theor. Appl. Genet.*, 63, 169, 1982.
333. **Negrutiu, I., Jacobs, M., and Caboche, M. I.,** Advances in somatic cell genetics of higher plants — the protoplasts approach in basic studies on mutagenesis and isolation of biochemical mutants, *Theor. Appl. Genet.*, 67, 289, 1984.
334. **Helgesen, J. P., and Deverall, B. J.,** *Use of Tissue Cultures and Protoplasts in Plant Pathology,* Academic Press, New York, 1983.
335. **Miller, S. A., Davidse, L. C., and Maxwell, D. P.,** Expression of genetic susceptibility, host resistance, and nonhost resistance in alfalfa callus tissue inoculated with *Phytophthora megasperma, Phytopathology,* 74, 345, 1984.
336. **Hartman, C. L.,** Use of Alfalfa *(Medicago sativa* L.) Cell Culture to Produce Plants Resistant to *Fusarium* Wilt Caused by *Fusarium oxysporum* f. sp. (medicaginis), M.S. thesis, University of Nevada, 1983.
337. **Marton, L., Manh Dung, T., Mendel, R. R., and Maliga, P.,** Nitrate reductase deficient cell lines from haploid protoplast cultures of *Nicotiana plumbaginifolia, Mol. Gen. Genet.*, 186, 301, 1982.
338. **Belliard, G., Vedel, F., and Pelletier, G.,** Mitochondrial recombination in cytoplasmic hybrids of *Nicotiana tabacum* by protoplast fusion, *Nature,* 281, 401, 1979.
339. **Medgyesy, P., Fejes, E., and Maliga, P.,** Interspecific chloroplast recombination in a *Nictotina* somatic hybrid, *Proc. Natl. Acad. Sci. U.S.A.*, 82, 6960, 1985.
340. **Flick, C. E., Bravo, J. E., and Evans, D. A.,** Organelle segregation following plant protoplast fusion, *Trends Biotechnol.*, 1, 90, 1983.
341. **Roth, E. J. and Lark, K. G.,** Isopropyl-N (3-chlorophenyl) carbamate (CIPC), induced chromosome loss in soybean: a new tool for plant somatic cell genetics, *Theor. Appl. Genet.*, 68, 421, 1984.
342. **Hain, R., Stabel, R., Czernilofsky, A. P., and Steinbiss, H. H.,** Genetic transmission of a selectable chimaeric gene by plant protoplasts, *Mol. Gen. Genet.*, 199, 161, 1985.
343. **Banerjee-Chattpadhyay, S., Schewemmin, A. M., and Schewemmin, D. J.,** A study of karyotypes and their alterations in cultured and *Agrobacterium* transformed roots of *Lycopersicon peruvianum* Mill, *Theor. Appl. Genet.*, 71, 1258, 1985.
344. **Syono, K. and Furuya, T.,** Studies of plant tissue culture. XVIII. Abnormal flower formation of tobacco plants regenerated from callus cultures, *Bot. Mag. (Tokyo),* 85, 273, 1972.
345. **Cullis, C. A. and Goldsborough, P. H.,** *The Plant Genome,* Davies, D. R. and Hopwood, D. A., Eds., John Innes Charity, Norwich, 1980, 91.
346. **Dembrugio, E. and Dale, P. J.,** Effect of 2,4-D on the frequency of regenerated plants in barley and on genetic variability between them, *Cereal Res. Commun.*, 8, 417, 1980.
347. **Grout, B. W. W. and Crisp, P.,** The origin and nature of shoots propagated from cauliflower roots, *J. Hortic. Sci.*, 55, 65, 1980.
348. **Ibrahim, R. K.,** Normal and abnormal plants from carrot root tissue cultures, *Can. J. Bot.*, 47, 825, 1969.
349. **Ben-Jaacov, J. and Langhans, R. W.,** Rapid multiplication of Chrysanthemum plants by stem-tip proliferation, *Hortic. Sci.*, 7, 289, 1972.

350. **Stimart, D. P., Ascher, P. B., and Zagorski, J. S.,** Plants from callus of the interspecific hybrid *Lilium* 'Black Beauty', *Hortic. Sci.*, 15, 313, 1980.
351. **Hackett, W. P. and Anderson, J. M.,** Aseptic multiplication and maintenance of differentiated carnation shoot tissue derived from shoot apices, *Proc. Am. Soc. Hortic. Sci.*, 90, 365, 1967.
352. **Beach, K. H. and Smith, R. R.,** Plant regeneration from callus of red clover and crimson clover, *Plant Sci. Lett.*, 16, 231, 1979.
353. **Gamborg, O. L., Shyluk, J. P., Brar, D. S., and Constabel, F.,** Morphogenesis and plant regeneration from callus of immature embryos of sorghum, *Plant Sci. Lett.*, 10, 67, 1977.

Chapter 2

ROLE OF TISSUE CULTURE IN RAPID CLONAL PROPAGATION AND PRODUCTION OF PATHOGEN-FREE PLANTS

I. INTRODUCTION

The generation of useful genetic and epigenetic-based variation by breeding and selection, maintenance, and bulking of desirable traits by sexual (seed) and asexual (vegetative) propagation methods is an important component of crop improvement. Vegetative means, such as division, cuttings, budding, or grafting, are common methods of propagation in higher plants. In crop species, however, cultivated plants will either be lost or will revert to less desirable forms unless they are propagated under controlled conditions such as directed inbreeding or vegetative propagation. The conventional methods of plant breeding are now proving time consuming and expensive. Tissue culture has gained a significant importance in recent years for propagation of crops *in vitro*. *In vitro* propagation of plants is also necessary which requires the use of true-to-true propagation of several plants. With the progress in methods for *in vitro* propagation the meaning of the word ''clone'' has also undergone a change. The word clone was initially applied to cultivated plants that were propagated vegetatively. This also implies that plants grown from such vegetative parts are not individuals in the ordinary sense, but are simply transplanted parts of the same individuals which are identical. Shull[1] defined the term clone as all groups of genotypically identical individuals which arise by asexual reproduction. Such a definition, however, is extremely difficult to implement in a practical manner. Stout[2] tried to resolve the issue by arguing that a clone should be regarded as an artificial unit and that it should be a collective term for a genetically uniform assemblage of individuals derived originally from a single individual by asexual propagation. This definition includes most widely used methods of artificial propagation as grafting and budding, which are examples of asexual or clonal multiplication, but which are at the same time extreme cases of chimeral development.

Advances in biotechnology and current techniques of single cell can lead to the formation of thousands of plants from a single cell. The products of this rapid vegetative propagation should by definition be considered a single clone. In practical terms such a notion is not acceptable because the products of callus and cell suspension cultures may consist of many abnormal cells. Thus, to consider such products under clones, the uniformity in regenerants of tissue cultures should be clearly established before using an *in vitro* propagation method for clonal propagation. The objective of rapid cloning of plants using *in vitro* methods should be, therefore, to reproduce plants asexually leading to the formation of a more or less uniform plant genotype. Broadly, this individual genotype should include plants from immature zygotic embryos to mature plants, plant parts which may be genotypically different from the rest of the plant (such as bud sprouts and chimeras), primary and secondary somatic embryos, and single cells including protoplasts, immature pollen, and cell suspensions. The genetic instability in suspension, protoplast, and callus cultures has been useful in clonal propagation. Protoplast, callus, and cell suspension materials are now of greater value to the breeding and the selection of cells for crop improvement. The culture techniques are proving particularly useful in the generation of novel types of genetic and epigenetic variation referred to collectively as somaclonal variation. The utilization of such variation is covered in Chapter 1 of Volume I. In this chapter only the current *in vitro* propagation methods and their application in propagation of plants are described.

II. CULTURE SYSTEM STEPS AND METHODS OF SHOOT INDUCTION AND PROLIFERATION

Propagating a plant using tissue culture techniques requires that organs, tissues, or cells be carried through a sequence of steps in which different environmental and cultural conditions are used. These steps have been defined in terms of different stages by Murashige.[3] These stages are

- Stage I — selection and establishment of suitable explants, their sterilization, and transfer to nutrient media
- Stage II — proliferation of shoots on multiplication medium
- Stage III — transfer of shoots to a rooting (or storage) medium followed later by planting into soil or some suitable compost mixture (Stage III is often considered to be the one in which whole plants are regenerated from the proliferated cultures usually by induction of rooting on shoots. A further stage, stage IV is sometimes also included for acclimatizing the newly regenerated plants to greenhouse or field conditions.)

All crop species are not required to be propagated *in vitro* by means of all three stages. However, these stages have been designated to describe the micropropagation process. *In vitro* propagation mainly centers around the formation and multiplication of shoot meristems, each meristem being a potential plant. Cultures for propagation *in vitro* may be started either from existing meristems, the embryo, main shoot or subsequently formed axillary shoots, or from organ explants that are suitable for the induction of adventitious meristems in the form of shoot apices or embryos. When organ explants are used, adventitious structures may be formed either directly within the parent tissue or from intermediate callus or cell suspension cultures. In the following discussion methods of shoot induction and proliferation from various explant sources are discussed.

A. MULTIPLICATION BY APICAL AND AXILLARY SHOOTS

Apical shoots occupy the growing tips of shoots, whereas axillary shoots emerge from their normal position on the plant in the leaf axils. Apical and axillary shoots, depending on the physiological state of the plant, contain a quiescent or active meristem. Leaf axils in vascular plants contain a subsidiary meristem, which is capable of growing into a shoot that is identical to the main axis. However, only a limited number of axillary meristems develop and the majority of them are inhibited by apical dominance. The mechanism of apical dominance is primarily under the control of various growth regulators. In many plants the outgrowth of axillary shoots appears to depend on the supply of cytokinin to their meristem. This situation has been exploited to a great extent for *in vitro* propagation. Shoot tips cultured on basal medium containing no growth regulators typically develop into single seedling-like shoots with strong apical dominance. However, when the shoots of the same explant material are grown on media containing cytokinin, axillary shoots often develop prematurely. This results in precocious branching which leads to the development of secondary, then tertiary, etc. shoots in a proliferating cluster. Such clusters, once developed, can be subdivided into smaller clumps of shoots or separate shoots which will in turn form similar clusters when subcultured on fresh medium. This subdivision process can be continued indefinitely provided the basic nutrient formulation is adequate for normal growth. In this way proliferating shoots of many species have been maintained for several years without any deterioration.[4] The rate of multiplication through axillary shoot proliferation depends upon the relative ability of the genotype to form leaves and vary from species to species. Multiplication rates on a regular 4- to 8-week micropropagation cycle can be achieved by manipulating optimum levels of cytokinin (0.5 to 10 mg/l) under optimal culture conditions.

Extremely impressive rapid clonal propagation levels in the range of 0.1 to 3.0 $\times$ 10^6 within a year have been obtained in this way.

All plants which produce regular axillary shoots and respond to an available cytokinin can be proliferated with this technique. Cultures are started either from excised shoot tips or from *in vitro* adventitious shoots. This excised shoot tip can be dissected from the terminal or lateral shoots and smaller size is preferred to avoid any chance of contamination. However, this has got certain limitations and the larger size of the shoot tips grows faster. The shoot tips are generally placed in a medium containing low levels of cytokinin and auxin. The levels of cytokinin are raised progressively at each subculture until an optimum rate of proliferation is achieved. Whenever there is callus formation at the base, auxin is omitted from the medium. The conditions for maintenance and propagation are to be worked out for each plant and it is not difficult to maintain them throughout the year. In general, the technique of proliferation by axillary shoots is applicable to any plant that produces regular axillary shoots and responds to cytokinins such as BAP (benzylaminopurine), 2iP (2-isopentyl adenine), and zeatin.

Vegetative propagation *in vitro* through apical and axillary shoots is primarily dependent on the juvenility of the plant. Seedlings just after germination are generally in the juvenile phase. There is very much active growth at this stage which is characterized by morphological features such as unique leaf shape and the presence of spines. Explants of axillary and apical buds and shoots are generally taken for culture during this phase of growth. If the material for explant is not available from seedlings and there is no alternative except to obtain it from mature specimens, the rejuvenated stage can be induced by grafting of shoots and seedlings, shoot pruning or spraying with cytokinin, and by maintaining high levels of fertilizers.

The multiplication through apical shoots is direct. Apical shoots (1 to 5 mm in size) are cultured on media containing mixtures of auxin (0.01 to 0.1 mg l^{-1}) and cytokinin (0.05 to 0.5 mg l^{-1}). In subsequent subcultures the level of cytokinin is raised in order to achieve an acceptable rate of proliferation without inducing yellowing or distortion of the shoots. In situations where callus is produced first in the presence of auxin, the possibility of regeneration of shoots from callus is avoided by reducing the levels of auxin.

Cultured material from all sources is often transferred to a rooting medium which contains reduced levels of cytokinin. This is particularly important in many monocotyledons, where root formation and proliferation are concomitant with subculture on a medium containing no cytokinin. This type of *in vitro* micropropagation through the shoot tips or apical meristems has an application particularly in the production of pathogen-free plants.

Shoot culture is often sufficiently reliable with herbaceous plants to be used commercially.[5] The method involves culturing explants such as shoot tips and buds in tubes or other suitable culture vessels containing a nutrient medium. The cultures are maintained in illuminated growth rooms where they grow readily and their axillary buds extend prematurely to produce new shoots. These are excised at intervals of about 1 month and transferred to fresh medium where further axillary shoot production occurs; shoot culture lines can be multiplied indefinitely by this method of sequential subculture. With some cultures, the new shoots are produced from adventitious buds that arise directly from explants and sometimes the shoot proliferation occurs from both axillary and adventitious buds. The production of plantlets is completed by excising the axillary or adventitious shoots and transferring them singly to another type of medium that promotes adventitious root formation. It is essential that all these *in vitro* operations are carried out under sterile conditions with the use of airflow cabinets to minimize contamination with microorganisms which will quickly kill the culture.

In angiosperm it has been common, even with cultures originating from adult trees, for more than 70% of the shoots to be rooted within 4 to 6 weeks of transfer to rooting medium.

Furthermore, the shoots of some species have been dipped in auxin and rooted directly into compost in humid greenhouses. This high degree of success with rooting allied to the rapid shoot proliferation suggests that mass propagation of a wide range of angiosperm trees should now be feasible. However, it is emphasized that many species remain recalcitrant. Notable examples of this are *Citrus* spp. where all attempts to develop plantlets from shoot tips have failed.[6]

In gymnosperms, rooting *in vitro,* even with shoots originating from seed or seedling explants, has been generally much more difficult than with angiosperms and there has been only a small amount of success. Nevertheless, shoots of *Pseudotsuga pinaster* and several other species have been rooted directly into compost in a greenhouse as was done with some of the angiosperms. However, a further problem with gymnosperms has been that a proportion of the rooted plantlets has been plagiotropic and therefore unsuitable as forest trees. This problem has been especially serious with *P. menziesii.*[7] As a result of difficulties with the establishment of cultures, rooting, and the plagiotropic habit, the range of gymnosperm forest trees that can be mass produced at present appears to be limited. At the AFOCEL laboratories in France, where propagation *in vitro* of gymnosperms appears to be most advanced, the only adult trees that have been propagated up to 100,000 times are *Sequoadendron sempervirens.* However, it is considered that similar propagation is feasible with adult trees of *P. pinaster, S. giganteum, Taxodium distichum, Cryptomeria japonica,* and *Cunninghamia lanceolata.*[7,8]

B. MULTIPLICATION BY ADVENTITIOUS SHOOTS

Adventitious shoots are also used as suitable explant material for *in vitro* propagation of plants. These shoots are stems or leaf structures that arise naturally on plant tissues which are located in sites other than at the normal leaf axil regions. The structures include bulbs, corms, tubers, rhizomes, stems, etc. It may be mentioned here that almost all of these organs can also be used as cuttings in the conventional propagation context.

Recent developments in tissue culture techniques have further made it easy to propagate the plants through adventitious shoots. Leaves of *Begonia* and some other ornamental plants produce shoots on their leaves or scale leaves, respectively. Similar types of adventitious shoot development can be induced by manipulating the levels of growth regulators in media and by the preconditions of parent plants and explants in culture. However, extensive callusing should be avoided. This type of method has been very successful in micropropagation of several plants.

Bulbs and corms grow from meristems at the bases of leaves and scales where they join the basal plate. Almost all genera can regenerate shoots from such meristem regions. It has been possible to form multiple shoots from scale and leaf base explants in medium containing both cytokinin and auxin. Buds, bulbils, or cormels under these conditions form adventitious shoots around the swollen base tissues. Sections of young flower stems within a bulb or the ensheathing leaves of a corm respond favorably to auxin or auxin plus cytokinin-supplemented media. Levels of fidelity in propagated material of these species are generally quite high and this is probably due to the fact that adventitious shoots arise from only one or two layers of cells or in some cases only single epidermal cells. Micropropagation has been very useful in bulbous species. In these plants annual rates of multiplication through conventional propagation techniques are generally below 5, whereas by micropropagation, levels of shoot multiplication are achieved which are $\times$ 10 to $\times$ 10^3. When comparing conventional methods with micropropagation methods of bulb production, micropropagation results in overall multiplication levels of $\times$ 5 to $\times$ 10 or more per annum.[4]

Direct regeneration of adventitious shoots on organ explants can be obtained in a wide variety of plants, but in many of these adequate multiplication may be obtained more readily by the use of axillary shoots which are likely to be genetically more stable. Many plants, however, form axillary shoots for rapid propagation. In most cases the most suitable organ for adventitious regeneration is the conventional propagation through much younger, more meristematic tissue which shows more vigorous regeneration.

Leaf cuttings or one or more shoots arising from the base of each petiole have been used for propagation of *Saintpaulia ionantha.*[9] Two-millimeter-thick sections of petioles have been reported to yield up to 20,000 plantlets from each petiole.[10] *Petunia hybrida*[11] and *Kalachoe blossfeldiana*[12] may similarly be propagated in large numbers from small leaf or stem pieces.

Conventional means such as chipping, scaling, and scooping have been employed to increase multiplication in many bulkons species where natural propagation rates are slow.[9,13] The tissue culture technique enables this type of shoot bulbil production to be greatly increased not only by facilitating more prolific regeneration from smaller explants, but also by making possible the continued recycling of *in vitro* material for continuous production. Although shoots can be obtained *in vitro* from a number of organs, the most regenerative regions are the basal plate.[14] *In vitro* shoots can be induced on scale or leaf piece explants on media containing auxin plus cytokinin.[15,16] Developing *in vitro* shoots from a number of bulbous plants can be split up and used as secondary explants for further adventitious shoot production.[17,18] Husscy[4] developed an effective method of continuous propagation of adventitious shoots applicable to many species that form bulbs or corms. In this method shoots induced on suitable explants *in vitro* are trimmed to approximately 5 mm, which includes 2 to 3 mm of basal plate tissue. In some plants such as *Iris* a mixture of axillary and adventitious shoots arises from between the young leaves and scales. In others such as *Narcissus, Nerine, Hyacinthus,* and *Leucojum* special dominance is very pronounced, and to obtain a good response, the main apex has to be destroyed after trimming. This is achieved by making two vertical cuts at right angle to the level of the apex without penetrating low enough to split the basal plate. The same procedure is repeated on each of the clusters of adventitious shoots that arise from between the bruncate scales.

Just as shoots and roots can be induced on explants in an adventitious manner, so too can somatic embryos. Adventitious embryo formation is distinct from the type of embryo formation which occurs via a callus stage. Adventitious embryos can be formed directly from groups of cells within the original explant, e.g., pollen (microspores) and mesophyll cells of the leaf, or from individual cells within the original explant, e.g., the epidermal cells of leaf bases or of primary embryoids. Embryoids are, in fact, formed *in vitro* from diverse tissues and organs in many species. For instance, the orchid *Malaxis pludosa* produces a large number of embryoids at the tips of its leaves; many species and cultivars of *Citrus* and *Mangifera* (mango) exhibit polyembryony in which nucellar cells lining the micropylar end of the embryo sac give rise to embryoids, which push their way into developing endosperm tissue for nutrition and further growth. The nucellar embryoids which develop are diploid and have often been used as traditional sources of clonal material of these fruit trees. In cultures, direct embryogenesis can be induced on certain explants. For instance, leaf pieces of *Arabusta* coffee trees can be induced to form embryos directly when cultured on a basal MS medium devoid of auxin and containing high levels of cytokinin.[19]

C. MULTIPLICATION THROUGH CALLUS CULTURE

A large population of cells sufficient to produce a desired number of regenerated plants is the first requirement to regenerate a large number of plants. This is accompanied by removing a small portion of the plant (an explant) and placing it in liquid or on solid medium. These media contain various nutrients and hormones sufficient for the cells to divide and reproduce, eventually forming a large collection of cells (suspensions if liquid medium, calli if solid medium). After appropriate time, depending on the species and genotype, this primary suspension callus is large enough to be subdivided into identifiable cell types and transferred to another medium depending on the type of regeneration required.

An *in vitro* callus (an unorganized mass of proliferating cells) may be obtained from almost any type of plant. Callus formation from explants is occasionally spontaneous, but generally requires an auxin in the medium, often in combination with cytokinin. The concentrations of the hormones most effective in producing callus vary with the species and the organ used, but tend to be higher than those needed to induce shoots directly from the explant. In most plants, calli can be detached from the explant, subcultured, and serially propagated on medium containing the hormone levels that were used to induce such calli. On lowering the hormonal levels or appropriately adjusting the auxin/cytokinin ratio, some calli will regenerate shoots or embryos. Mass production of callus followed by shoot regeneration would seem to be the ideal method of large-scale propagation, but some serious drawbacks at present are limiting its use to only a few species.

With repeated subculture which is necessary to obtain a large quantity, the capacity of many calli to regenerate shoots is diminished or even lost. Further tissue and cell cultures in which callusing is produced tend to be of low value as a means of micropropagation through the induction of organogenesis or embryogenesis. This is due to the relatively high incidence of aneuploidy and polyploidy sometimes associated with the tissues and regenerated plants obtained from meristems or meristemoids, i.e., meristematic centers consisting of a spherical mass of small, isodiametric cells with dense cytoplasm and high nucleo-cytoplasmic ratio formed on calli.[20]

This does not mean that micropropagation through a callus is not capable of producing regenerants of uniform type. In certain cases the propagator can visibly distinguish aberrant regenerants at the first step of the multiplication procedure, i.e., when shoots are dissected away from calli. Sometimes aberrant shoots are distinguishable by their vitrified or glassy appearance so that these can be eliminated from further propagation steps. However, such an approach is not sufficient guarantee against the inadvertent multiplication of chimeric shoots or mutants.

Notwithstanding this, *in vitro* propagation via organogenic or embryogenic calli is unavoidable in the case of certain economically important species, particularly in some of the cereals, forage legumes, forest trees, and tropical palms.

The production of many thousands of plantlets from calli either derived from cell suspensions or protoplasts isolated from a single leaf constitutes unique cases of cloning. This aspect has been discussed in detail in Chapter 1 of Volume I. Somaclonal-based variation in the regenerants is a common occurrence[21] and, clearly, callus-based cloning methods require special consideration in the context of the definition of the term cloning and its subsequent use in micropropagation.

Over and above such technicalities, there are also an increasing number of reports in the literature on the establishment of stable regenerative calli either by chance or, in some cases, by deliberate selection of organogenic or embryogenic sectors of callus cultures. This has led to increased regeneration potentials in cultures of some traditionally recalcitrant types. For instance, Heyser and Nabors[22] successfully regenerated plants in a consistent manner from secondary calli of oats (*Avena sativa*) using this approach. Also, calli have been obtained which can regenerate genotypically uniform plants for up to 10 to 14 years. These stable calli have been described for genera such as *Lilium,*[23] *Freesia,*[24] *Chrysanthemum,*[25] tomato,[26] and *Hemerocallus.*[27] Recently, highly regenerative calli have been worked out for several crops including rice, wheat, sorghum, millet, oat, soybean, and pigeon pea.[28] The major callus types in cereals recognized by scientists working at the TCCP laboratory are "embryogenic" (E) and "nonembryogenic" (NE) (Figure 1). Embryogenic callus is transluscent to opaque in appearance and milky white to yellow or yellow-green in color. It appears as compact, nodular region of nonvacuolated cells which produce small somatic embryos. Embryos further develop into whole plants when cultured on appropriate media. In contrast, NE callus is crystalline and transparent and appears as a loosely packed friable

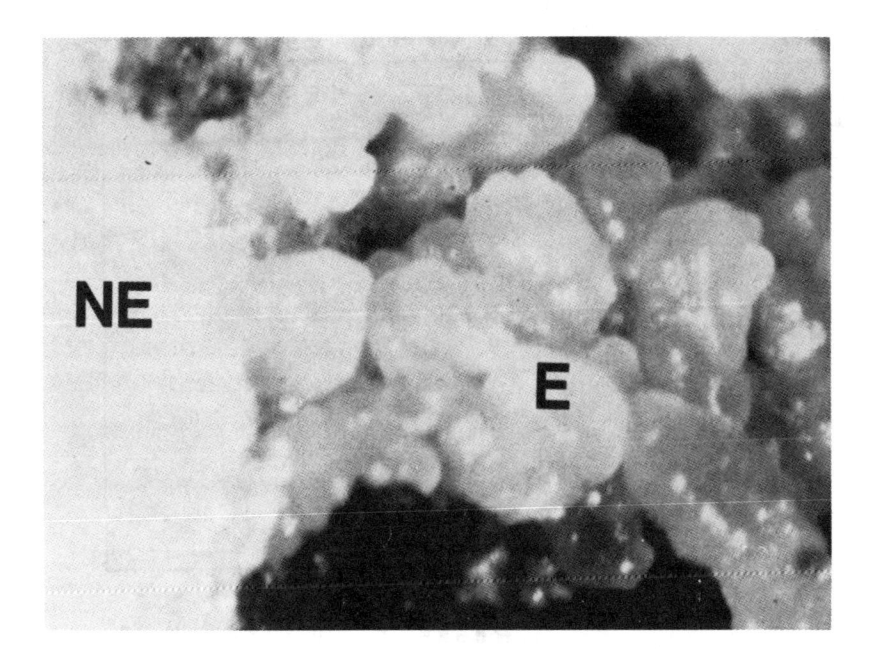

FIGURE 1. Callus culture showing friable nonembryogenic (NE) callus and compact embryogenic (E) callus. (From TCCP [Tissue Culture for Crop Project], Progress Report, Colorado State University, Fort Collins, 1987. With permission.)

region of vacuolated cells. These two callus types have been noted by number of workers.[22,29,30] The E callus has a much higher frequency of regeneration than the NE callus (Figure 2). Heyser and Nabors[29] were able to develop callus in millet (from various plant parts) and long-term totipotent tissue cultures of proso millet (*Panicum miliaceum*) which readily regenerated to normal fertile plants. About 32% of embryogenic calli formed shoots as compared to 2% of nonembryogenic calli cultured on auxin (Figure 3). These types of calli, when obtained, are of great value, since they can be suitably subdivided by random dissection or by placing them in a homogenizer to produce many thousands of propagules in a single operation. They are, therefore, ideal for some semimechanized and large-scale forms of rapid clonal propagation.

This method of plant culture has the potential for even more rapid propagation than shoot culture. As with shoot culture, the physiological status of the initial explant has been of prime importance, and calli from seeds and seedlings have usually had the greatest capacity for differentiation. Roots, shoots, embryoids, and sometimes plantlets have been obtained from just such juvenile material of a limited range of angiosperm and gymnosperm trees.[31] Furthermore, in the case of *Ulmus americana* and *Liquidambar styraciflua,* cell suspensions produced by shaking calli from seedlings have produced embryoids and plantlets.[32,33]

Plants have been regenerated from calli from adult trees of a very limited range of species. Shoots were regenerated from cambial explants of trees of *U. campestris* that were approximately 180 years of age,[34] and there has been similar regeneration in the case of calli from shoots of forest clones of *Populus* spp.[35] With *Prunus,* shoots have been regenerated from several species, but only when the callus was initiated from plants that had been rejuvenated through propagation *in vitro* by shoot culture.[36] However, these callus systems from adult trees are not as effective as the corresponding shoot culture systems for the sustained and rapid production of plants. Notable exceptions to this are several species of *Coffea* which have been multiplied with high frequency through embryoids produced in calli from leaf explants.[37] Highly rejuvenated tissue of the adult tree is nucellar tissue, and with

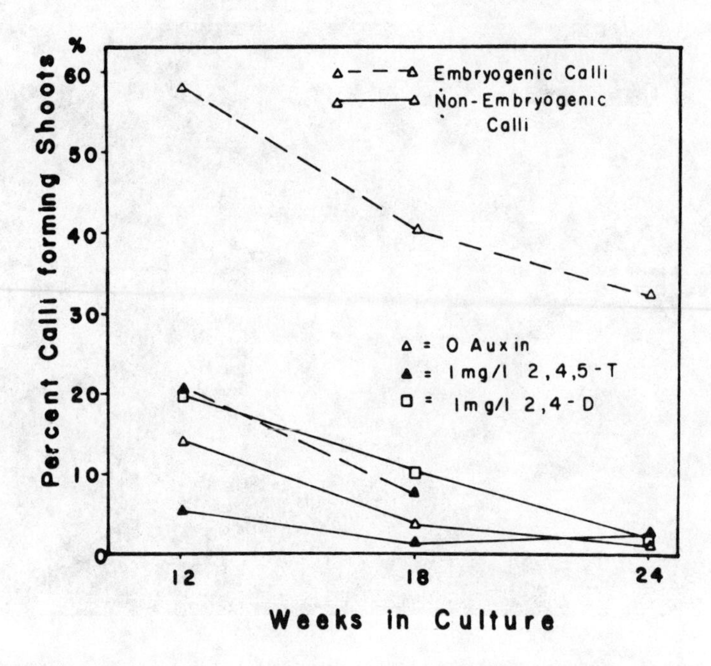

FIGURE 2. Shoot formation by embryogenic and nonembryogenic calli cultured second through fourth passage (12 to 24 weeks after initial explant culture) on 0 or 1 mg/l 2,4,5-T or 2,4-D. Sample sizes were 15 to 210 vials per point for nonembryogenic and 19 to 40 vials per point for embryogenic calli. (From Heyser, J. W. and Nabors, M. W., *Crop Sci.*, 22, 1070, 1982. With permission.)

Citrus spp. embryoid production in callus from such tissue has led to the propagation of proven clones.[6] Embryoids have also been produced from nucellar callus of apple, but at a much lower frequency than with *Citrus*.[38]

Propagation through callus has also been successful and useful where vegetative propagation is not possible through any other method. Oil palm (*Elaeis guineensis*), in which vegetative propagation is not normally possible, has been successfully propagated through callus culture. Initial studies were carried out on seedlings, and shoots were developed in calli from apical shoot tissue following transfer to a series of media containing varying concentrations of sucrose and mineral salts.[39] Embryoids, from which plants developed, were also reported in calli evolved from root and leaf-base tissue of seedlings. Details of the media were not described, but embryogenesis was reported as being more a function of the origin of the initial explant than of subsequent cultural conditions.[40] Subsequently, embryoids were induced in callus evolved from adult palms following incubation in media with cytokinins and auxins for between 5 months and 2 years. However, once the embryogenesis was initiated, it was retained for at least 3 years with about 50 plantlets being produced per gram of callus, which increased in weight between 10- and 30-fold every 2 months.[41] These techniques have been improved to produce embryoids much more rapidly.[42] The information of roots and embryoid-like strucutres has been reported in calli evolved from seedling and adult palms of coconut (*Cocos nucifera*), but plants have not yet been obtained.[43-45]

III. FACTORS AFFECTING MORPHOGENESIS AND PROLIFERATION RATE

The utilization of plant tissue culture for plant regeneration and propagation on a large

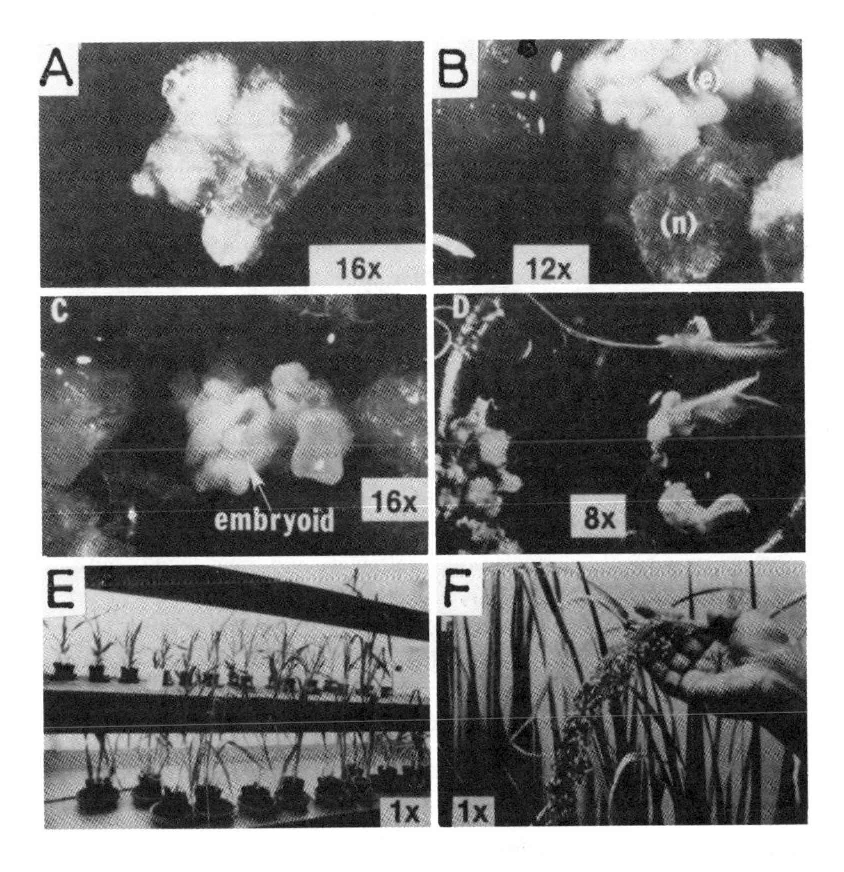

FIGURE 3. Callus initiation, embryogenic callus formation, and embryogenesis in proso millet. (A) Root callus initiated from mature seeds (magnification × 16), (B) embryogenic (e) and nonembryogenic (n) callus (magnification × 12), (C) embroid formation (magnification × 16), (D) complete plant formation by embryogenic calls (magnification × 8), (E) regenerated plants in pots (magnification × 1), and (F) flowering by regenerated plants (magnification × 1). (From Heyser, J. W. and Nabors, M. W., *Crop Sci.*, 22, 1070, 1982. With permission.)

scale requires a quantitative approach which can lead to an optimization of the conditions associated with vegetative plant propagation. The effects of various factors on morphogenesis and proliferation rate are measured in terms of their influence on the relative incidence of organogenesis or embryogenesis, and the number of propagules which can be regenerated per given amount of culture over a given period of time or a single culture generation. Several factors influence the expression of morphogenesis and proliferation rates in micropropagation systems and major factors are described below.

A. NATURE OF THE DONOR PLANT AND EXPLANT

Axillary shoot cultures have usually been initiated from shoot tip or nodal bud explants. Adventitious shoot cultures have also been initiated from such explants and also from portions of internodes, leaves, cotyledons, hypocotyls, and embryos. The choice of explant depends on the species being used and on various methods of shoot proliferation. The reaction of the explant also depends on the season and stage of growth of the parent plant. For instance, bulb scale explants of *Lilium speciosum* regenerated bulbils freely when taken from plants during summer or winter.[46] Flower stem explants of *Tulipa,* on the other hand, gave rise to shoots only during storage; once stem elongation had commenced, the capacity of this organ to regenerate was lost.[47]

In several plants the physiological status of the explant has been of prime importance. Explants from adult forest trees were frequently slow to commence growth *in vitro,* or failed completely, unless selected from rejuvenated tissues. Such tissues may occur naturally, for example, as the vigorous shoots, commonly known as ''suckers'', that arise directly from the roots of various species or as the masses of adventitious buds that are a feature of the ''ligotubers'' of *Eucalyptus* spp. In plants where juvenation does not occur naturally, it has been induced by various treatments such as the grafting of shoots onto seedlings, shoot pruning, the maintenance of high fertilizer levels, vegetative propagation, or spraying with cytokinins.[8]

With more highly cultivated crops such as fruit trees, the necessity of taking explants from rejuvenated tissues has seldom been emphasized, but this is probably because most of the treatments that have been used to rejuvenate forest trees feature in the routine management of cultivated trees. However, even with highly cultivated apple trees, propagation *in vitro* has been achieved most readily when the initial explants have been taken from the shoots of young nursery trees within about 1 month following bud-break, with cold storage being used to maintain a year-round supply of trees at this early stage of development.[48,49] Explants must be free from microorganisms when placed on nutrient media and this has usually been achieved by surface sterilization with solutions of sodium or calcium hypochlorite. Such treatments have failed with some plants presumbly because of penetration of tissues by microorganisms.[8,50] Antibiotics and systemic fungicides have proved useful to sterilize the shoot tips of plants in such situations.

Explants of seeds and seedlings have usually responded readily to culture *in vitro* and rapid proliferation of axillary or adventitious shoots has been achieved with just such juvenile material of species of *Betula,*[51] *Cinchona,*[52] coffee,[53] *Eucalyptus,*[54] *Liquidambar,*[55] *Pinus,*[56,57] *Picea,*[58,59] and *Pseudotsuga,*[60] Explants from seeds have also been quite useful in regeneration of calli in several cereals which are difficult to propagate by any other vegetative method. Initial explants were taken from normal aerial shoots of nursery or field-grown trees. These have included shoots from fruit trees, for example, rootstocks of apple,[49,61] scion cultivars of apple,[62-68] scions of pear,[63] and rootstocks of peach[69] to give the potential for the production of at least 1 million shoots annually.

In the case of adult trees, like gymnosperms the use of explants from rejuvenated tissue has usually been essential for the achievement of rapid shoot proliferation.[8,70] In legumes and cereal tissue culture, mature seeds are the most convenient explant source to obtain regenerable callus.[28] Either the entire seed can be used as in rice or the seed can be soaked overnight and the mature embryo removed. In a few plants like corn and sorghum, developmentally immature embryos, immature leaves, and immature influorescences are the best explants for establishing regenerable cultures. The greatest amount of smooth-shiny E callus in soybean was produced from cotyledons (Table 1).

The most suitable explants for tissue cultures are those in which there is a large proportion of either meristematic tissue present or cells which retain an ability to express totipotency. Successful cultures are rarely obtained from senescing tissues. In tree species, the regeneration potential of tissue culture diminishes with each year of plant maturation, even though in some cases juvenile characteristics are apparently maintained.[60] Tissues excised from the more recently produced parts of a herbaceous plant are also more regenerative than those from older regions. Also, embryogenesis in somatic cells has been generally associated with cultures established from embryo explants rather than mature, nonembryonic tissues. In this respect the season and stage of growth of the parent plant may be critical in determining the behavior of the explant in culture. There are many examples of this. For instance, Litz

TABLE 1
Formation of Smooth-Shiny E Callus from Developmentally Different Cotyledonary Segements on Media Containing 10 mg/l 2,4-D

Immature embryo lengths (mm)	Cultures with embryos (%)	Embryos per culture (no.)
2	—	—
3	6	0.06
4	36	2.04
5	59	2.00
6	39	2.42
7	38	1.23
8	44	2.00
9	25	1.00
10	13	0.13

From TCCP (Tissue Culture for Crops Project), Progress Report, Colorado State University, Fort Collins, 1987. With permission.

and Conover[71] found that optimum conditions for the establishment of papaya tissue cultures occurred during the hot summer months and also during the transitional months of April and November under the Florida climatic regime. Flower stem explants of *Tulipa* give rise to shoots only when excised during the dry storage (dormant) phase; once elongation of stems has commenced following dormancy, the capacity for these explants to regenerate is lost.[47] Similarly, nodal explants of *Dioscorea alata* only produce axillary shoot growth when excised from donor plants growing under a 16-h photoperiod, conditions which induce active vegetative growth. Under 12-h photoperiods, donor plants give explants which show either no growth or only prolific callus development.[72]

Familarity with a plant's natural propagation mechanism is frequently helpful in determining the more suitable explant sources. Thus, sections of leaves are the most suitably employed in cases where plants normally regenerate from leaves while sections of bulb, root, stem, flower, ovary, nucellus, cotyledon, and other structures may have applicability with certain species.

B. COMPOSITION OF CULTURE MEDIA

Only principal basic nutrients that are transported in the vascular system are required for the growth of plant organs, tissues, or cells. Other essential elements are sucrose and certain amino acids, vitamins, and growth factors.[73,74] The exact composition of the medium, however, has to be adjusted according to the requirements of the different groups of plants, and some species and cultivars require additional supplements to sustain adequate growth. The nutrient medium is generally solidified with agar (0.7 to 1.0 w/v), but liquid cultures are preferable for some plants. An alternative method utilizes a filter paper bridge on wick to support the explant above liquid medium with the free ends of the paper immersed in the medium.

Generally, the formation of shoots is promoted by high levels of cytokinin relative to auxin, while the reverse favors the development of roots. This holds for a large number of plants, but it is by no means universal. The levels of hormones required for organogenesis vary from species to species. A number of cytokinin compounds are available, but those mostly used are 6-benzylaminipurine (BAA), 6-furfurylaminopurine, and 6-(γ,γ, dimethlallyamino)-purine. Different auxins used are indole-3-acetic acid (IAA), the naturally oc-

TABLE 2
Hormonal Composition of LS Media for Embryo Initiation, Maintenance, and Plant Regeneration from Various Explants of Soybean cv. Prize

Intiation Media (mg/l)

1 IAA	2 2,4-D
1 IAA + 0.5 BAP	2 2,4-D + 0.5 BAP
1 IAA + 5 BAP	22,4-D + 5 BAP
1 2,4-D	10 2,4-D + 0.132 ABA
1 2,4-D + 0.5 BAP	5 IAA
1 2,4-D + 5 BAP	5 IAA + 0.5 BAP
10 2,4-D	5 IAA + 0.5 BAP
2 IAA	5 2,4-D
2 IAA + 0.5 BAP	5 2,4-D + 0.5 BAP
2 IAA + 5 BAP	5 2,4-D + 1 BAP
	10 2,4-D + 0.264 ABA

Maintenance Media (mg/l)

	+ 0.2 2,4-D	+0	Proline
	+ 0.2 2,4-D	+25 m*M*	Proline
	+ 0.5 2,4-D	+0	Proline
	+ 0.5 2,4-D	+25 m*M*	Proline
Minus	+ 0.2 2,4-D	+0	Proline
NH_4NO_3	+ 0.2 2,4-D	+25 m*M*	Proline
	+ 0.5 2,4-D	+0	Proline
	+ 0.5 2,4-D	+25 m*M*	Proline
Minus NH_4NO_3	+ 0.2 2,4-D	+0	Proline
Plus 800	+ 0.2 2,4-D	+25 m*M*	Proline
NH_4Cl	+ 0.5 2,4-D	+0	Proline
	+ 0.5 2,4-D	+25 m*M*	Proline

Regeneration Media (mg/l)

No growth regulators (controls)
0.104 GA
0.103 GA + 0.132 ABA
0.103 GA + 0.132 ABA + 0.102 IBA
0.110 zeatin
0.110 zeatin + 0.132 ABA
0.110 zeatin + 0.132 ABA + 0.102 IBA
0.110 zeatin + 0.132 ABA + 0.103 GA

From TCCP (Tissue Culture for Crops Project), Progress Report, Colorado State University, Fort Collins, 1987. With permission.

curring auxin stable analogues of IAA such as indole-3-butyric acid (IBA) and 2,4-dichlorophenoxy acetic acid (2,4-D), 2,4,5-trichlorophenoxyacetic acid (2,4,5-T), 4-chlorophenoxyacetic acid (CPA), and 4-amino-3,5,6-trichloropicolinic acid (TCP). These hormones induce rapid callus proliferation. High levels of these substances, especially 2,4-D, strongly suppress organogenesis.

It is necessary to work out different types of variables which include hormones (type and concentration), nitrogen source, and amino acid requirement for raising callus from different explant.[28] In fact, three types of media with different variables are required for raising the callus up to its regeneration (Table 2). Mature embryos have been used as explant

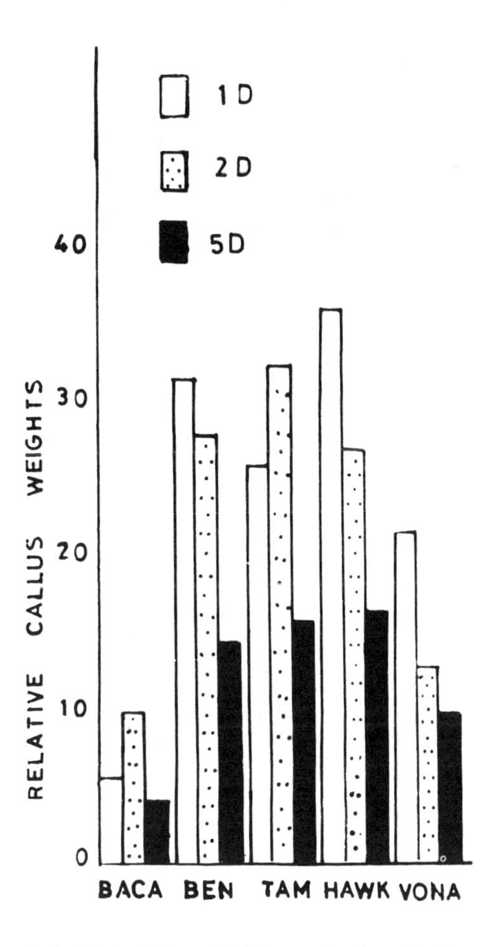

FIGURE 4. Effect of 2,4-D on callus formation from mature embryos of five cultivars of winter wheat. (From TCCP [Tissue Culture for Crop Project], Newsletter No. 7, Colorado State University, Fort Collins, July 1987. With permission.)

sources for callus induction in wheat.[75] LS media supplemented with various concentrations of 2,4-D in the presence or absence of KIN (kinetin) or BAP were used for the initiation of callus formation (at TCCP laboratory[28]). The callus formation was cultivar specific and dependent on the growth hormones in the medium. In general, lower concentrations of 2,4-D (1 mg/l) induced more callus formation (Figure 4), but the cultures were very rooty. Cytokinin did not give significant increases in callus production, except BAP (Figure 5) and KIN (Figure 6) showed beneficial effects on Vona and Howk cultivars of wheat, respectively.

In general, a suitable growth medium is adequate for achieving both stages I and II, with modifications only being necessary for stage III of micropropagation. However, the type and relative proportion of growth-regulating substances present in the initial culture medium will largely determine the regeneration potential of a culture system, the micropropagation of a wide range of crops. Within any given culture system, the level and type of growth regulator mixtures, as well as those of carbon, potassium, phosphate, and nitrogen sources, and the medium pH and buffers used, if any, will affect its overall propagation performance. For instance, the capacity of cell cultures to utilize NH_4^+ as sole nitrogen sources depends on maintaining the medium pH above 5.0.[76] The availability of NH_{4+} as opposed to NH_3^- as sources of nitrogen to a tissue is particularly important regarding the development of somatic embryogenesis in a culture. The leaves of phosphate present in

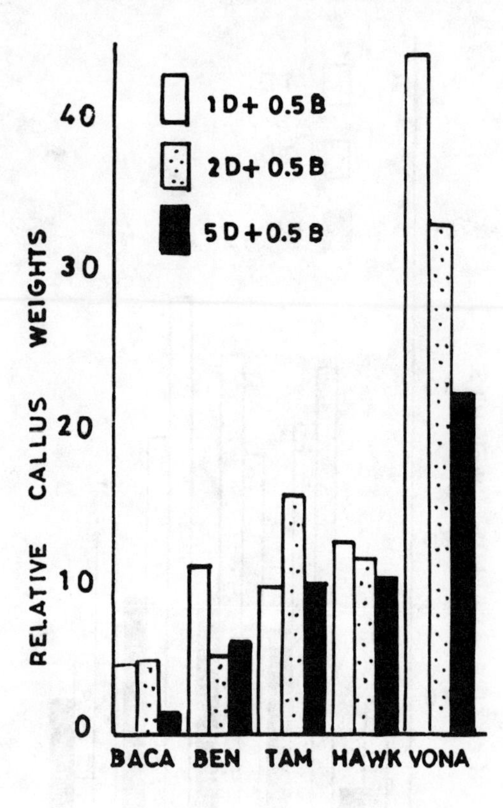

FIGURE 5. Effect of BAP on callus formation from mature embryos of winter wheat. (From TCCP [Tissue Culture for Crop Project], Newsletter No. 7, Colorado State University, Fort Collins, July 1987. With permission.)

media can influence the number and size of shoots produced in shoot tip cultures of tropical foliage plants.[77] The relative levels of potassium influence the number of embryos produced in wild carrot suspension cultures; while embryo productivity is maximal at 10 to 50 m*M* potassium, growth is maximal at 1 m*M*.

Defined media consisting of mineral nutrients, sugar, vitamins, and growth regulators have been used to culture shoots. The macro- and micronutrients have usually been based on the medium of Murashige and Skoog,[78] with sucrose at a level of 2 or 3% (w/v) as the most common carbohydrate source. The presence of a cytokinin, usually BAP at a concentration of about 1 mg/l, has been essential for the proliferation of both axillary and adventitious shoots,[79,80] while addition of an auxin such as IBA and gibberellic acid A_3 has sometimes enhanced shoot elongation. Furthermore, the presence of the phenolic compound phloroglucinol (1,3,5-trihydroxybenzene) at a concentration of about 150 mg/l has enhanced the shoot proliferation of some apple cultivars and also a limited number of other tree species. This compound is a breakdown product of phloridzin, a major phenolic constituent of apple trees, and its action may take place via effects on the metabolism of auxin.[79,80] Activated carbon, at about 2%, often has been an essential medium constituent for the elongation of shoots of gymnosperms. Possibly, this carbon absorbs growth inhibitors exuded by the plant tissues or contains impurities such as monophenylamines which promoted growth.[81,82] The pH of media has been adjusted to 5.2 to 5.8 before the liquid media have been solidified with agar. Media have also been used in liquid with filter paper supports, or the use of shaking has been used to improve the aeration of the cultures.

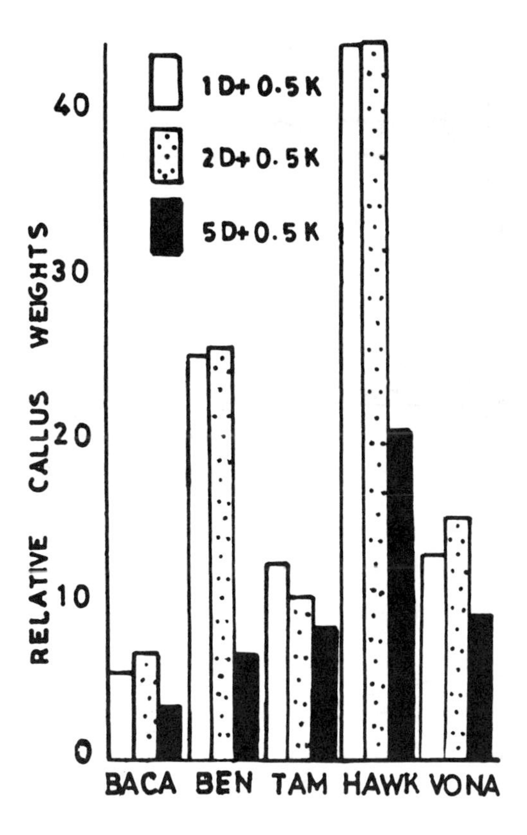

FIGURE 6. Effect of KIN on callus induction from mature embryos of winter wheat. (From TCCP [Tissue Culture for Crop Project], Newsletter No. 7, Colorado State University, Fort Collins, July 1987. With permission.)

C. CULTURE ENVIRONMENT

Certain environmental conditions such as light, temperature, and gas phases influence the growth of plant *in vivo*. These factors also appear to have significant effect on plant materials held in micropropagation. Plant cells, tissues, and organs cultured on media containing a readily available source of energy such as sucrose would be expected to be less dependent on photosynthesis.

Little or no photosynthesis occurs in organs or shoots cultured on media containing sucrose, but some light is necessary for morphogenesis and chlorophyll formation. Fluorescen tubes are generally used as a light, with intensities falling in the range 1000 to 5000 lx. There has been little work on the optimal conditions of light and temperature, but various reports show that these can be important in some species. Optimum light intensity for shoot formation in *Gerbera* and a large number of herbaceous genera is in the region of 100 lx.[83] Intensities as low as 300 lx and as high as 3000 to 10,000 lx are strongly inhibitory.

Optimum daylight is around 16 h for a wide range of plants.[83] In *Allium cepa* more adventitious shoots were formed in 16 d light than in 8 d.[17] However, several studies have indicated that light apparently absorbed by photosynthetic pigments plays an important role in inducing morphogenesis in cultured tissues. For instance, the incidence of *Asparagus* spear production increased in light.[84] Kato[85] showed that bud induction in excised leaf segments of the lily *Helonipsis orientalis* is controlled to a certain extent by a photosynthetic system. This was concluded after the photosynthetic inhibitors DCMU and AT were found to inhibit bud formation in the light, while morphactin, which has a stimulatory effect on

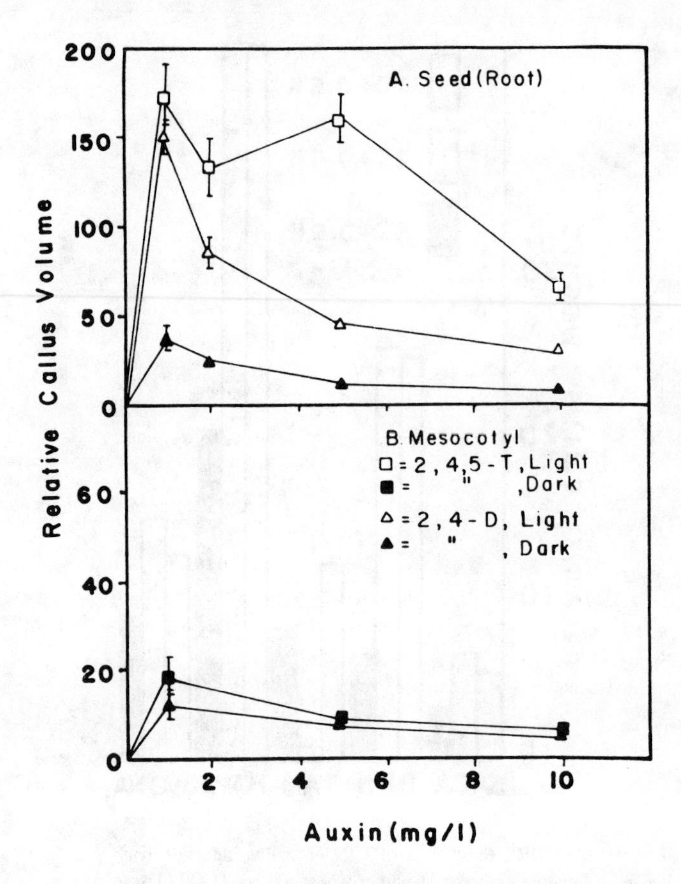

FIGURE 7. Callus initiation by (A) light and dark cultured seed (root callus) and (B) dark grown mesocotyls cultured in dark on various concentration of 2,4-D and 2,4,5-T. Sample size was 15 to 49 vials per point. (From Heyser, J. W. and Nabors, M. W., *Crop Sci.*, 22, 1070, 1982. With permission.)

carbohydrate synthesis, stimulated *in vitro* both the red and blue ends of the visible spectrum. Seibert[86] found that blue light in the region of 467 nm was effective in inducing bud formation in tobacco callus cultures, and Kadkade and Seibert[87] later showed that adventitious shoot formation in lettuce is regulated by phytochrome. As little as 5 min of 660-nm light at 2.5W m^2 each day during the second week of culture led to a doubling in the number of shoots produced. There are also reports that red light stimulates induction of flower buds on thin-layer epidermal sections of tobacco and that far-red light treatments stimulate root production.[88] Several micropropagation systems have now been described in which morphogenesis is induced by appropriate red/far-red light treatments. This alleviates to a certain extent the need for the addition of growth regulators to the culture medium and thus reduces the risk of genetic abnormalities occurring in the micropropagation stages I and II. For most cultured plants, a 660-nm light treatment stimulates shoots, while 740-nm light stimulates roots. The intensity of light required for most micropropagation purposes is around 1000 lx. However, higher light intensities of up to 10,000 lx have been found to be beneficial, particularly during stage III of micropropagation. In millet less callus was initiated by dark cultured seeds than by light grown seeds in the presence of 2,4-D (Figure 7).[29] This illustrates the importance of optimizing cultural conditions for each species under propagation. The effect of photoperiod, as would be expected, often reflects the relative sensitivity of the individual species being propagated. For example, tobacco has an optimum photoperiod of 16 h,[89] while cauliflower has 9 h.[90]

Apart from light, another obvious environmental variable is temperature. Plants in culture are most often grown at 25°C for convenience. However, there are species of plants that have temperature optima for morphogenesis which contrast with this arbitrary regime. For instance, cultures of *Begonia* × *Cheimantha* hybrids produced greatest numbers of shoots on petiole segments at 18°C,[91] while shoot tip cultures of *Asparagus officinalis* grew best at 27°C.[84]

A third factor of the culture environment affecting the performance of tissue cultures is the constitution of the gas phase within the culture vessels. Gases which are metabolically active and could have possible effects on morphogenesis include ethylene, oxygen, carbon dioxide, ethanol, and acetaldehyde. In general, ethylene is antagonistic to morphogenesis and promotes unorganized growth of cells (i.e., callus formation). Of significance here is the fact that the levels of ethylene and other volatiles may be influenced by the types of closures used and the practice of flaming flasks with alcohol or gas burners for sterilization purposes. Beasley and Eaks[92] have found that this practice introduces variable levels of ethylene into flasks. Although most of the gas diffused out of flasks in the first 2 h, in cases where flasks are tightly closed high levels of ethylene can persist for substantial periods. These workers noticed significant differences in the growth and proliferation of excised ovules of cotton when these were grown in sealed incubators as compared to others that were grown in ones that were opened regularly. Clearly, the type of closure used on culture vessels needs to be checked in the wake of these observations, especially since ethanol and acetaldehyde have now also been implicated as potent inhibitors both of organogenesis and embryogenesis.[93] Raised levels of carbon dioxide in the absence of ethylene are known to lead to enhanced greening in some cell culture systems. However, the interaction of all these gases is undoubtedly complex and more detailed studies need to be carried out to ascertain real effects of the gas phase on micropropagation systems. It is one component of the micropropagation environment which should not be overlooked any longer.

D. NATURE OF GENOTYPE

Certain genotypes lend themselves more than others to vegetative propagation. Different genotypes can, therefore, be expected to react differently *in vitro* to a given set of cultural conditions designed to promote proliferation of a given species of plant. Examples of the different *in vitro* responses of quite closely related genotypes are frequent in the literature. For example, tissue culture initiation and plant regeneration were examined by Rines and McCoy[94] for several genotypes from three hexaploid oat species. These were the cultivated oat, *Avena sativa,* and two wild oats, *A. sterilis* and *A. fatua*. Various types of tissue cultures were initiated from immature embryos with so-called "regenerable" cultures being characterized by organized chlorophyllous primordia present in compact, yellowish-white, highly lobed callus. Apart from other factors, such as the preconditioning of donor plants, embryo size, and 2,4-D concentrations used in the culture initiation medium, there was a strong effect of genotype on the frequency with which regenerable-type cultures were produced. Among 23 *A. sativa* cultivars tested, cv. Lodi and two related lines gave the highest frequencies (up to 80%) of regenerable-type cultures. Of 16 *A. sterilis* lines tested, only three produced regenerable-type cultures at frequencies of less than 20%. Seven out of 32 *A. fatua* lines tested produced regenerable-type cultures at frequencies greater than 45%. In all three species these cultures were capable of plant regeneration after more than 12 months in culture (equivalent to nine to ten subcultures). This work has led to the suggestion that since genotype influences culture initiation frequency and culture type, screening of genotypes and selection among segregating populations might prove a fruitful approach in the improvement of micropropagation capabilities in oats and, indeed, in other more recalcitrant cereals. This approach of purposefully breeding or selecting more suitable genotypes which lend themselves to micropropagation techniques was first attempted by Bingham et al.[95]

whereby lines of alfalfa were intentionally bred and selected for their regeneration ability *in vitro*. Similar marked influences of genotype are observed in the types of morphogenesis obtained in particular types of tobacco explants. Tran Thanh Van[96] has described how different species of *Nicotiana* determine the regenerating ability of thin cell layers of tobacco in culture.

IV. APPLICATIONS

Apart from the ability to rapidly multiply plants, micropropagation also offers a means for circumvention of many crop diseases.[97] This is due to the eradication of pathogenic agents either as part of the decontamination phase of explant preparation during stage I, or as the result of culturing small-enough explants which contain few or none of the inefective agents. Further propagation and disease testing of regenerated plants kept under controlled and protected conditions well away from vectors or any potential pathways of reinfection can lead to the maintenance of disease-free plant materials at reduced cost and with less effort. Subsequent bulking by conventional means can then be carried out as and when the need arises. The small size of micropropagules and their ability to proliferate in a soil-free environment facilitate the convenient storage, handling, and rapid dissemination of propagated materials by air transport across international phytosanitary barriers.[98] Linked with these benefits of micropropagation is the suitability of tissue cultures for storage of germplasm either by cryopreservation or by limited culture growth. Using these methods, stocks of germplasm can be maintained in a stable condition for many years. The major application of micropropagation is discussed below.

A. PRODUCTION OF PATHOGEN-FREE PLANTS

This is a specialized area of micropropagation which is based on general aseptic culture procedures and meristem-tip and shoot-tip culture techniques.[99] Diseases of crop plants caused by pathogens such as viruses, viroids, fungi, nematodes, bacteria, and mycoplasma reduce the yield, quality, and vigor of crops. Virus diseases like potato leaf roll virus (PLRV) or potato virus Y (PVY), for example, can cause up to 95% reductions in the tuber yield of potato crops, and potato virus X (PVX) infections can lead to reductions in tuber yields of between 5 and 75%, depending on the virus strains involved, the cultivar of potato infected, the environmental influences exerted, and the types of latent virus infections present. Since many virus diseases and other pathogens like nematodes, bacteria, and mycoplasma are transmitted by vegetative propagation procedures, there is frequently an essential need to eradicate pathogens from certain elite lines of plant material. With nematode infestation and the majority of bacterial infections, this can be achieved most effectively by culturing explants removed from pathogen-free parts of a plant. However, since the majority of viruses infect plants in a systemic manner, this type of approach is not sufficient and more specialized methods are required to achieve the elimination of virus infections. The strategies involved for the production of viral-free plants through *in vitro* propagation are varied and depend upon the type of plant to be evolved.

1. Meristem Culture

The more differentiated vascular tissues occur away from the meristem (towards the older tissues of the stem), the vascular elements of the leaf primordia are still very incipient, and they have not yet made contact with the main strand of the vascular system in the stem. Therefore, virus particles which may be present in the vascular system can reach the meristematic region of the apex only through cell-to-cell movement. This is one of the main reasons that virus concentration decreases acropetally towards the meristem of the apical bud as well as the axillary buds in a virus-infected plant. Whether other factors such as the production of virus inhibitory substances by the meristematic cells or the effect of hormones in the culture medium play a role in the elimination of viruses by meristem culture has not been conclusively proved.

Isolation of the meristem tip, under aseptic conditions, and its culture on an adequate medium lead to the development of plantlets. This development, in principal, follows a pattern similar to that in the entire plant: the cells of the meristem continue to divide and the differentiation of the tissues continues. The nutrition of the excised portion of the plant is supplied by the artificial medium. This technique, called meristem culture, is used to produce pathogen-free plants.

Bacteria and fungi can be easily eliminated since they are not entering the meristem tip. Of course, if the surface sterilization of the bud is insufficient or the tools are not well sterilized, bacteria and fungi will grow in the meristem tissue, since the culture medium for the meristem is also ideal for the growth of most of these microorganisms. However, as long as sterilized conditions are used and portions are excised in which the incipient vascular elements have not yet reached the main vascular strands, bacterial and fungal pathogens will present no problems.

Meristem tips are carefully dissected away from the apical or lateral shoot buds with the aid of a binocular microscope. Provided trimming and dissection procedures are carried out under aseptic conditions, most meristems can be placed directly into culture without sterilization. Meristem tips can be cultured on either liquid or semisolidified nutrient media,[100] supplemented with growth regulators which induce plantlet regeneration. Culture media are generally those based on modified White's media.[101] Growth regulators used vary within the range 0.1 to 2.0 mg l^{-1}. In the cases of potato shoot tips presoaking in 10 mg l^{-1}, GA is beneficial as this treatment breaks dormancy in the tuber sprouts.

After a period of between 50 and 100 d, the few plantlets or adventitious shoots obtained can be further micropropagated and the regenerants tested for the presence of viruses or other pathogens, if appropriate. Disease testing is an essential component of pathogen-free production schemes. Recent improvement in the techniques of PAGE and ELISA allows early and sensitive detection of viroid and virus infections, respectively.

Meristem tip culture, either with or without heat therapy, has been used for the elimination of viruses for several plant species,[102] and the benefits of this are well recognized. Several crops freed of virus yielded 60 to 90% more stem weight than infected stocks. Chrysanthemum, narcissus, carnations, and other flower crops freed of major virus diseases produce both improved size of blooms and increased number of blooms per plant. In other cut flower crops like *Pelargonium,* 20 to 30% more cuttings are produced per plant and rooting capacity is also improved in virus-freed material. Particularly impressive have been the effects of freeing tuber crops of viruses on tuber yield and quality. When potatoes are freed from potato virus M, 10% higher fresh weight yields of tubers and a greater number of marketable tubers per plant are produced.

The combination of meristem tip culture with small-scale and large-scale cloning of plant material has allowed the successful integration of pathogen-tested crop lines into commericial-scale production systems. In the case of some fruit crops, particularly *Citrus* and *Prunus,* advantage has been taken of *in vitro* grafting. Grafted plants are desirable for a number of reasons, including the fact that they possess juvenile characteristics normally associated with seedlings of nucellar or zygotic origin, i.e., they mature and fruit sooner. Besides this, many cultivars have been commericially propagated by grafting for many years and little is known about the disease and pest responses and cold-hardiness.[103] Grafting of virus-free shoot tips directly onto appropriate root stock seedlings *in vitro* has been proved to be an effective procedure for *Citrus,*[104,105] *Prunus,*[106] and apple.[107]

Many economically important trees are infected throughout by viruses as well as being susceptible to attacks by systemic bacteria and fungi. Disease-free trees may be obtained in

TABLE 3
Pathogenic Condition of Mother Plants Prior to Treatment for Disease Elimination

Variety	Procedence	Viral disease	Indexing[a]		
			dsRNA	ELISA	Grafting
Secundina (A)	Palmira	CsXV	+	+	−
Secundina (B)	Palmira	Mosaic	+	+	+
M. Col 33	Palmira	CsXV, Mosaic, FSD	+	+	+
M. Col 113	Quilcace	CsXV, FSD	+	+	+
M. Par 51	Paraguay	CCMV	+	+	

[a] Indexing carried out by Cassava Program virologists.

From Roca, W. M., Szabados, L., and Hussain, A., Biotechnology Reaearch Unit, Annu. Rep., Centro Internacional de Agricultura Tropical Cali, Colombia, 1987.

small numbers by meristem shoot tip culture and heat treatment. Virus-free rootstocks have been propagated *in vitro* in England for commercial distribution overseas. Since this material was maintained under the sterile conditions of culture *in vitro,* it was possible, following the appropriate testing, to guarantee its disease-free status, and this simplified greatly the quarantine procedures involved in the distribution. Furthermore, methods have been described for the storage at 3°C of apple shoot cultures.[108] This would enable disease-free material to be maintained for many months with a minimum of storage area and labor. Such a storage also gives added convenience in relation to the distribution of disease-free material.

Plants regenerated from nucellar callus are usually virus free, since such pathogens are rarely seed transmitted. Viruses have been eliminated by this method from *Citrus* clones that do not produce nucellar. Plants of *Citrus,* whether produced *in vitro* or *in vivo,* exhibit undesirable juvenile characteristics such as the presence of thorns and delayed sexual maturity. For this reason attention has been transferred to meristem shoot tip grafting onto seedlings as a solution to the problems of virus elimination in *Citrus.*[104]

2. Thermotherapy and Chemotherapy

Experiments carried out with different virus-host systems have shown that treatment of plants with elevated temperatures (thermotherapy) leads to a reduction in virus concentration.[100,109] Different reasons have been given to explain this phenomenon. Competition for sites to synthesize nucleic acids and proteins between the fast-dividing host cells and the virus particles may lead to a change in the balance between synthesis and degradation of virus particles. In addition, the nucleic acid of the virus, the carrier of its genetic information, is usually protected from the attack of degrading enzymes consisting of many subunits. At elevated temperatures the linkage between these two subunits becomes weaker and temporary holes may open and permit the attack of nucleases, so leading to the inactivation of virus and a decrease in the virus concentration.

When thermotherapy has been applied to potato tubers, reduction of virus concentrations was achieved except for PLRV. The same is the case with cassava stakes exposed to elevated temperatures.[110] However, thermotherapy applied to the whole plant as well as to the sprouted tubers, followed by meristem culture, has been successfully used for elimination of viruses in potato.[111,112] While this is in use for many vegetatively propagated crops, the detailed conditions have yet to be determined experimentally for each crop and each part to be treated.

The elimination of cassava virus also depends on the size of the shoot tip explant used and on whether thermotherapy has been applied to stakes during sprouting.[110] Four viral diseases are common in cassava plants (Table 3). In a research project carried out to eliminate

TABLE 4
Effect of Thermotherapy and Size of Shoot-Tip Explants on the Elimination of Four Cassava Viral Diseases as Shown by Indexing Tests

		Treatment		Indexing[a] (No. +/total)	
Variety	Disease	Thermo-therapy	Tip size	ELISA	dsRNA
Secundina (A)	CsXV	Yes	Small	0/12	0/7
		Yes	Large	0/1	—
		No	Small	0/2	0/2
		No	Large	5/5	4/5
Secundina (3)	Mosaic	Yes	Small	0/7	0/7
		No	Small	0/1	—
		No	Large	5/8	—
M.Col 33	CsXV, Mosaic, FSD	Yes	Large	0/7	—
		Yes	Large	0/2	—
		No	Small	1/4	0/1
		No	Large	4/5	2/2
M.Col 113	CsXV, FSD	Yes	Small	0/1	—
		No	Small	0/1	—
M.Par 51	CCMV	Yes	Small	0/1	—
		Yes	Large	2/2	—

[a] Indexing carried out by Cassava Program Virologists.

From Roca, W. M., Szabados, L., and Hussain, A., Biotechnology Research Unit, Annu. Rep., Centro Internacional de Agricultura Tropical Cali, Colombia, 1987.

these four viral diseases, two infected plants per variety were desprouted and one was exposed to thermotherapy (40°C day and 35°C night) during 3 weeks; the other plants were kept in the greenhouse. Shoot tip explants of two sizes — "small" with one to two primordia (0.1 to 0.2 mm) and "large" with four to six leaf primordia (0.5 to 0.7 mm) — were excised from both plants and cultured separately in standard cassava meristem culture medium under 28°C (day) and 24°C (night), 12-h photoperiod and 1000 to 2000 lx illumination. After 3 to 4 weeks, each individual plantlet was micropropagated by nodal cuttings to increase the number of daughter plants for use in indexing (Table 4).[101] The initial results showed that all four viral diseases can be eliminated, even in cases where two or three are infecting the same plant. Roca et al.[101] intensified the work on thermotherapy and cassava meristem tip culture in order to clean up field germplasm bank accessions from frog skin disease, to recover plants free from insect pest damage, and as preventive measure to minimize risk in the exchange of materials between CIAT and various countries. A total of 517 clones were processed through this technique (Table 5).

Heat treatment does not eliminate most viruses, but it reduces virus multiplication and translocation in the plant.[101] For instance, caribbean mosaic symptoms do not show up in new cassava leaves grown at 35°C, but the disease symptoms only decrease from severe to mild when the cassava crop was grown from stem cuttings from stakes sprouted at 40°C (day) and 35°C (night) for 30 d.[113] Treating PVX-infected potato plants with high temperature decreased virus infectivity. Dilution point tests of the treated plants suggested a decrease in virus concentration due to heat treatment.[114] As expected, elimination of PVX to shoot tip culture was much higher from heat-treated than from nontreated plants (Table 6). In practice thermotherapy permits excising slightly larger explants, which may be devoid of viruses.[115] Higher temperature also tends to increase the infectivity of the potato spindle tuber viroid

TABLE 5
Cassava Clones Processed through Thermotherapy and Meristem Culture in 1986

Cause for cleaning	No. clones
Frog skin disease: CIAT field collection	165
Damage by pests and weather: CIAT field collection	38
Preventive cleaning	
CIAT elite clones for distribution	27
CIAT nonelite clones for distribution	11
Clones introduced from Thailand	7
Clones introduced from Malaysia	64
Clones introduced from Indonesia	47
CIAT clones for *in vitro* storage	158
Total	517

From Roca, W. M., Szabados, L., and Hussain, A., Biotechnology Research Unit, Annu. Rep., Centro Internacional de Agricultura Tropical Cali, Colombia, 1987.

TABLE 6
PVX Elimination through Thermotherapy of Infected Plants Followed by Shoot Tip Culture

	Relative infectivity of PVX in *G. globosa* leaves inoculated with leaf sap[a] of			
	Clone 800244		Clone 720057	
Potato leaf sap dilution	Heat treated[b]	Control	Heat treated[b]	Control
1/10	+	+++	+	+++
1/100	−	++	+	+++
1/1000	−	+	−	++
1/10000	−	+	−	+
PVX elimination (%) by shoot tip culture[c]	90	5	40	2

[a] No. local lesions per half *G. globosa* leaf inoculated at end of heat treatment.
[b] 36°C (day) and 30°C (night), for 15 d.
[c] 0.4—0.5 mm in size (unpublished data from work done at Centro Internacional de la Papa, Lima, 1977).

From Roca, W. M., *Proc. Biotechnology in International Agriculture Research,* IRRI, Manila, 1984, 3. With permission.

(PSTV) in potatoes;[114] low temperature (8°C) decreases viroid concentrations in the tissues. Cultures of apical meristems from low temperature-treated plants resulted in 30% viroid elimination.[116] Thus, using low instead of high temperature may help to eliminate viroids through *in vitro* techniques.

As an alternative to thermotherapy, chemotherapy has recently shown promising results in both potato and cassava. A nucleoside analogue, virazole, known for its broad spectrum effect against animal DNA and RNA viruses has shown good results when applied to the potato plant as a spray or in hydroponic culture, followed by meristem culture.[101] Preliminary results of culturing meristem tips in the presence of 100 ppm virazole in the medium were promising.[117] In the case of cassava, promising results have also been obtained, even at lower virazole concentrations.

Thermotherapy is routinely used at CIP to eliminate viruses in potatoes. Optimal eradication rates have been obtained when the plant was decapitated before thermotherapy. A temperature regime of 36°C for 8 h under continuous light of high intensity (10,000 lx) improves elimination rates.[118] If possible, plants are treated over 4 weeks. Throughout this time, axillary buds grow rapidly. Meristems are isolated from apical buds as well as axillary buds, and cultivated *in vitro*. After about 6 to 8 weeks, plantlets regenerate from the meristems.

The routine procedure for cassava at CIAT is very similar to that for potato at CIP.[101] Potted stakes with dormant buds are exposed to a thermotherapy of 40°C in the day and 35°C at night, with a 12-h photoperiod. These conditions prevail for 3 to 4 weeks during which time sprouting occurs vigorously; then meristem tips are excised and cultured. Plantlets develop after 6 to 8 weeks. Developing plantlets in potato as well as cassava are propagated *in vitro*. In cassava the propagated plantlets are transferred to pots and then samples are taken for virus testing.

Yield and vigor of local cultivars of cassava may decline because of pathogen accumulation. This decline has been observed in potatoes and other crops infected with viruses.[110,119-121] Producing clean planting material through *in vitro* techniques can significantly raise crop yield and quality. Cassava caribbean mosaic disease, endemic to northern Colombia, has severely affected cv. Secundina. Clean planting material was produced through thermotherapy followed by shoot tip culture. Yield in fresh weight and starch content of roots increased by 70%.[113] After 3 years of planting, yields of *in vitro* processed Secundina have remained stable, whereas yields of the hybrid abruptly decreased in the third year.[110]

In vitro propagation greatly increased the yield of roots and planting material of cassava cultivars freed from virus. Yield and quality continued to increase in the second year, but were lower than in the first (Table 7). In *in vitro* plants, total HCN content in roots of Llanera was 95 ppm and free HCN was 14 ppm; HCN level in CMC-40 roots remained stable. In addition, the point of the stem at which the first branching, i.e., flowering, occurred was strikingly taller in the *in vitro*-derived plants than in the control, and the magnitude of height increased in the first and second year and was larger in Llanera than in CMC-40. The part of the stem below the first branching provided planting material (stakes) of the highest quality.

Using the thermotherapy procedure described above for potato, PSTV cannot be eliminated. This viroid consists of a one-stranded RNA, which is ring shaped and twisted in the form of a supercoil. In this form it is very resistant to attacks by nucleases. It has been shown that elevated temperature favors the multiplication of the viroid.[122] Therefore, only a first test for PSTV is carried out at the end of the thermotherapy period. To eradicate PSTV, a method has been developed based on the observation that in plants grown at low temperature, the concentration of the viroid is low.[123] Therefore, plants were grown at 8°C for 4 months, after which apical domes were excised and cultured. From the plantlets regenerated, 30% were free of PSTV. A clear relationship between explant size and eradication success was observed with improved eradication at small apical dome size.[123] This method is not suitable as a routine technique, since it is very time consuming and costly. However, it may be useful in specific cases in which a valuable clone is thoroughly infected and clean material cannot be obtained.

As in other host-virus systems,[120,121] the success of virus elimination is correlated to the size of the excised portion, both in cassava and potato. The smaller the excised portion, the higher the probability for virus eradication. This seems reasonable in light of the earlier discussion on meristem culture. On the other hand, at least in the case of potato, very small portions are difficult to cultivate or complex media are necessary, leading to callus formation before plantlet development. For cassava at CIAT as well as for potato at CIP, a portion containing the apical dome and one or two leaf primordia are usually excised. The efficiency

TABLE 7
Yield Increases from *In Vitro* Propagation of Cassava

Cultivar	Parameter	Yield increase (%) over control[a] 1st year	2nd year
Llanera	Root yield		
	Fresh wt	53	25
	Dry wt	22	10
	Planting material[b]	47	46
	Plant height	25	0
	Height 1st branching	76	26
CMC-40	Root yield		
	Fresh wt	43	29
	Dry wt	29	2
	Planting material[b]	18	23
	Plant height	6	5
	Height 1st branching	19	14

[a] Conventionally propagated plants, without *in vitro* culture.
[b] Number of commercial size stakes. Percentages are averages of 30—40 plants/treatment-year.

From Roca, W. M., *Proc. Biotechnology in International Agricultural Research,* IRRI, Manila, 1984, 3. With permission.

of virus elimination by meristem culture in potato has been reported to be dependent on the virus itself. The viruses which are restricted to the vascular bundle such as PLRV are usually easy to eradicate. There is a gradual increase in the difficulty to eradicate viruses: PLRV > potato virus A (PVA) > PVY > potato virus M (PVM) > PVX > potato virus S (PVS).[120,121] Applying the above routine thermotherapy procedure, an almost 100% elimination rate has been obtained, except for the PSTV. Similar use of the thermotherapy system described for cassava yielded 100% healthy plants from material infested simultaneously with frost skin and mosaic diseases.[110]

3. Factors in Disease Elimination

Now several *in vitro* procedures are available or can be readily developed to grow plants to maturity from shoot tip explants (0.4 to 0.8 mm), comprising the apical meristem (0.1 to 0.2 mm) and at least one primordial leaf.[3,100,102] All plants regenerated *in vitro* are not necesarily pathogen free, and if the aim is to produce pathogen-free plants the methods should include (1) identifying the pathogens, (2) applying technique for eliminating disease, (3) testing regenerated plants for freedom from pathogens, and (4) propagating healthy plants under conditions that prevent reinfection.

The success of meristem tip culture can depend on many factors. One of the most important is the relative distribution of viruses in the growing tips of donor plants. Some viruses are less tenacious thant others, depending on their relative rates of replication in actively growing tissues in which there is high meristematic activity. Some viruses are present in the very tip of a growing shoot. For example, PVX infections could not be totally

TABLE 8
Individual and Double Virus Eradication Rates from Potato Cultivars through Potato Explant Thermotherapy and Shoot Tip Culture[a]

Subspecies	PVX	PVS	PVX + PVY	PVX + PVS
andigenum				
Cultivars (no.)	6	2	3	3
Elimination (%)	60	50	85	50
tuberosum				
Cultivars (no.)	11	2	5	5
Elimination (%)	66	60	75	54
Total				
Cultivars (no.)	17	4	8	7
$\bar{x}$ Elimination (%)	63	55	80	52

[a] 36°C (day) and 30°C (night), applied to potted cuttings for 15 d; shoot tip size used for culture: 0.4—0.5 mm; virus indexing with indicator plants and serology (unpublished data from work at Centro Internacional de la Papa, Lima, 1978). PVX = potato virus X; PVS = potato virus S; PVY = potato virus Y.

From Roca, W. M., *Proc. Biotechnology in International Agricultural Research,* IRRI, Manila, 1984, 3. With permission.

eradicated from potato plants raised from meristems as small as 0.12 mm in length and which had only a single leaf primordium attached. An estimated 52% of the plantlets obtained from such material still contained PVX infections. Other potato viruses are less tenacious. In a group of slightly larger meristems 0.2 mm in length and containing a single leaf primordium, the regenerants were 100% of PLRV and 70 to 80% free of PVA and PVY. Of these not more than 10% were free of PVX. Therefore, the type of virus infection will determine the size of meristem which must be taken before the complete eradication of virus.

Because virus movement is considerably slower through the symplasm than through the vascular system, the vascular differentiation occurs away from shoot apical meristem and virus concentration decreases acropetally. In the vegetative point, virus titer may be nil. In addition, mitosis in the meristem cells competes with virus multiplication, as occurs in rapidly dividing callus. The regeneration of healthy plants through the culture of shoot tips, despite the presence of virus particles in the apical meristem cells,[102] suggests that *in vitro* conditions help to inactivate the virus. Cell injury carried by excising the shoot tips may induce metabolic alterations that lead to virus degradation.[119] Eliminating disease depends on several interrelated factors, such as the type of virus to be eliminated, the size of the explant for culture, and physical or chemical treatments applied to the infected plants or to the culture.

Viruses vary in how difficult they are to eliminate; those restricted to the phloem can be eradicated by shoot tip culture with little difficulty. The potato viruses according to their increasing difficulty of eradication are as follows: PLRV, PVA, PVY, PVM, PVX, PVS, and PSTV. The cassava virus diseases according to increasing difficulty of eradication are frog skin, common mosaic, caribbean mosaic, and latent virus. The cassava African mosaic has been eradicated without much difficulty. Infection with more than one virus can also be eradicated by shoot tip culture. In several naturally infected *Andigenum* and *tuberosum* potato cultivars, elimination rates of PVX plus PVY were much higher than those of PVX plus PVS, and PVX alone was somewhat easier to eliminate than PVS alone (Table 8). Similarly, 100% cleaning was possible when cassava stakes infected both with frog skin plus latent virus and frog skin plus bacterial blight were sprouted at high temperature before their shoot tips were excised and cultured.[110]

TABLE 9
Influence of Explant Size Used for Culture and Sprouting Temperature of Stakes on Frog Skin Elimination from Cassava

Treatment		
Size of shoot tip[a] (mm)	Sprouting temp[b] (°C)	Frog skin disease-free plants[c] (%)
0.4	26	79
	40/35	100
0.8	26	16
	40/35	85
1.2	26	0
	40/35	65

[a] Height of shoot tip.
[b] Day/night temperature, applied to stakes prior to shoot tip culture, for 15 d.
[c] Av of six plants/treatment. Disease indexing: grafting and electrophoresis.

From Roca, W. M., *Proc. Biotechnology in International Agricultural Research,* IRRI, Manila, 1984, 3. With permission.

The way virus spreads in a plant, efficiency of virus elimination inversely relates to the size of shoot tip explants. However, *in vitro* growth of a very small explant (0.1 to 0.2 mm) is quite slow and often leads to callus formation. Because of this the standard practice has been to culture shoot tips measuring 0.4 to 0.5 mm. Explant size greatly influenced the efficiency of frog skin disease eradication in cassava; furthermore, sprouting of the infected stakes at high temperature enchanced the rate of cleaning up to 100% when 0.4-mm shoot tips were cultured (Table 9).[114]

B. GERMPLASM CONSERVATION AND EXCHANGES

Due to their freedom from microorganisms, small space requirements, and high genotype stability, meristem cultures are suited for germplasm storage and exchange.[124] This potential has been investigated in several plants including potato, cassava, herbage grasses, coffee, sugarcane, asparagus, apple, bananas, strawberry, and grapes.[124]

Plant breeding programs rely heavily on locally adapted ancient plant varieties and their wild relatives as sources of germplasm. Until fairly recently, these varieties had been preserved either in primitive agricultural systems or in their natural habitats. Within the last 20 years it has been recognized that this supply is becoming supplanted by highly bred modern varieties that have been produced by plant breeders working in association with internationally coordinated programs of agricultural improvement.[125] With increased pressures for land use, some of the natural germplasm is being eroded away to such an extent that many authorities fear that unless some moves are made soon, potentially valuable germplasm could be completely lost and future breeding programs could suffer as a consequence.[126] Even in view of the limitations of genetic stability and regenerative ability of plant tissue culture, micropropagation has great values as a potential system of germplasm storage, either as sources of material for cryopreservation or as materials for maintaining for protracted peroids under growth-limiting conditions.

These techniques have no obvious value in preserving valuable tissue culture materials used in plant biotechnology such as protoplasts, pollen embryos, callus and cell suspension cultures, and somatic embryos.[127,128] For the germplasm storage of crop species, however, it would seem that shoot tips, meristems, and embryos are the best types of material for cryopreserving, since, after thawing, high levels of survivability and regeneration are obtained from pea and strawberry meristems following various periods of storage in liquid nitrogen (−196°C). Also, these materials have the best chance of producing genetically stable regenerants, since little callus development occurs during their proliferation, although there have been cases of an increased tendency for meristems to form calli after thawing.[129] There are clearly many more technical details to perfect with cryopreservation and shoot tips and meristems are proving to be promising materials for long-term storage of valuable germplasm.

Potato shoot tip cultures can survive passage lengths of as long as 1 year[130] either by adopting a regime of alternating low temperatures (12°C day, 10°C night) or by maintaining temperature at 10°C together with increasing sucrose levels in media (8% w/v). Incorporation of growth retardants or osmotic agents like mannitol in media can also have beneficial effect on reducing growth rates and increasing subculture intervals.

These details of storage conditions have to be worked out for each plant species. Roca et al.[101] grew cassava variety CIAT-Palmira for two consecutive cycles and evaluated them for their stability in relation to check plants propagated by stem cuttings (stakes). Six morphological, five agronomic, and one biochemical character were evaluated in the second cycle. They found that the majority of characters remained without change during storage (Tables 10 and 11).

In vitro techniques are on the increase for germplasm storage and exchange. The approach basically consists of material from one or several collection sites to a nearby experiment station for planting and isolation followed by meristem tip culture. Subsequently, these cultures are subjected to thermotherapy, *in vitro* micropropagation, disease indexing, and eventually moved to field. However, there are several areas where the approach described above would not be convenient or even feasible due to isolation and lack of laboratory facilities. Roca et al.[101] developed a simple technique to allow collecting cassava germplasm *in vitro* at the site of collection, and viability in such material is maintained for sufficient time until the culture reaches the actual storage institution. They selected actively growing vegetative buds from plants in the CIAT-Palmira germplasm bank and trimmed off the material with an eye on an acrylic plegable small bench in the field. Explants (0.1 to 1.5 cm) were inoculated into semisolid medium containing antimicrobial agents. Test tubes (14 × 100 cm) containing culture media and explant were caped and held under observation for 2 weeks. Treating the explants prior to inoculation with surface sterilants (70% ethanol) for 15 min followed by 0.1% calcium hypochlorite (for 5 min) allowed only up to 30% explant survival and contamination persisted. However, when the explants, after surface sterilization, were inoculated in media containing antibiotics, contamination was greatly reduced. The incorporation of fungicide (benlate) to the medium was highly phytotoxic, but the antibiotic, rifampycin, gave highest survival rates, followed by rifamycin and then by trimethoprim (Table 12).[101]

Growth limitation is also proving to be a reliable and cost-effective method of maintaining stable germplasm. Using low temperatures and media containing osmotic or hormonal inhibitors under oxygen limitation, shoot tip cultures can be maintained satisfactorily without transfer for periods of up to 1 year.[128]

C. SMALL-SCALE CLONING, SEED PRODUCTION, AND MASS PROPAGATION

In breeding work, micropropagation is particularly useful for the maintenance and multiplication of modest numbers of special genotypes or potential new cultivars including any products of genetic engineering involving *in vitro* procedures. In these cases, propagation by means of axillary or apical shoot development is desirable because genetic conservativeness must be absolute. However, since these procedures involve handling individual shoots, the work is labor intensive and therefore generally quite costly.

TABLE 10
Phenotypic Stability of Cassava Plants Retrieved from *In Vitro* Storage and Plants Propagated by Various Techniques: 10-Month-Old Plants (cv. M.Col 1505)[a]

	Plant height (m)	Roots per plant (no.)	Root F. W. per plant (lb)	Harvest index
In vitro storage	2.0 B-A	6.1 A	4.1	0.5
Somatic embryogenesis	1.7 B	5.6 A-B	2.5	0.5
Meristem culture	2.2 B-A	6.2 A	4.8	0.5
Rapid propagation	1.8 B	2.9 B	1.6	0.4
Stakes	2.5 A	8.0 A	4.3	0.3

[a] Values with same letter are significantly not different at 95% confidence level.

From Roca, W. M., Szabados, L., and Hussain, A., Biotechnology Research Unit, Annu. Rep., Centro Internacional de Agricultura Tropical Cali, Colombia, 1987.

TABLE 11
Phenotypic Stability of Cassava Plants Retrieved from *In Vitro* Storage and Plants Propagated by Various Techniques: 6-Month-Old Plants (cv. M.Col 1505)

	% plant for each character				
Morphological characters	*In vitro* storage[a]	Somatic embryogenesis	Meristem culture	Rapid propagation[b]	Stakes
Petiole length					
Short	3	35	3	2	3
Medium	70	40	8	30	10
Large	27	25	89	68	87
Pubescence					
Scarce	6	70	6	5	2
Modcrate	94	30	94	95	98
Color apical leaves					
Light green	82	52	98	98	92
Purple green	18	48	2	2	8
Color mature leaves					
Light green	18	22	2	2	8
Dark green	82	78	98	98	98
Color stem					
Silver green	98	75	98	98	98
Gray green	2	25	2	2	2

[a] Stored *in vitro* for 4 years.
[b] Two-node cutting technique.

From Roca, W. M., Szabados, L., and Hussain, A., Biotechnology Research Unit, Annu. Rep., Centro Internacional de Agricultura Tropical Cali, Columbia, 1987.

For seed production from clone's parents, a major limiting factor is the high degree of genetic conservation that is required. This restricts the type of micropropagation used for axillary bud multiplication which is usually at levels of less than × 10^3. There are only a few applications currently being made under this category. They include the production of F_1 hybrid seed lines in crops like cauliflower, where individual parent clones can be bulked

TABLE 12
Effects of Antibiotics and a Fungicide Incorporated into the Culture Medium on Bacterial and Fungal Contamination of Cassava Shoot Apex Explants (1.0—1.5 mm in size) and the Final Survival of Explants

Antibiotic or fungicide to medium	Contamination		Survival (%)
	Bacterial	Fungal	
No surface sterilization	4/10	6/10	0
No antibiotic nor fungicide			
Surface sterilization	1/10	2/10	30
No antibiotic nor fungicide			
Rifamycin (30 mg/l)	0/10	1/0	90
Rifamycin (30 mg/l)	0/10	1/10	80
Trimethoprim (30 mg/l)	2/10	1/10	60
Rifamycin + Trimethoprim	0/0	0/10	50
Benlate (30 mg/l)	2/10	0/10	0

From Roca, W. M., Szabados, L., and Hussain, A., Biotechnology Research Unit, Annu. Rep., Centro Internacional de Agricultura Tropical Cali, Columbia, 1987.

for the production of more uniform seed,[131] the production of male sterile lines of onion by micropropagation to provide an alternative to difficult backcrossing methods, and the production of asparagus for producing high quality supermale and female homozygous lines, from which desirable all-male hybrids can be produced. The latter material is of high quality and produces higher yields of spears than pistillate plants.[97]

The most reliable types of micropropagation are achieved through axillary shoot proliferation. However, a much larger number of plants can be produced from a given amount of explant material within a short period of time when adventitious shoots or embryos are induced in callus or cell suspension cultures. If a large proportion of the cultured cell population is made up of totipotent cells in cell clusters, extremely high numbers (10^5 to 10^6) of plants can be produced from a few 100-ml cultures within a single-culture generation. This type of mass propagation can best be achieved using liquid culture media for state II of micropropagation.

Plant regeneration through somatic embryogenesis is well documented and usually requires a two-step procedure. In many species during the induction period proembryos and embryos are formed on the explant in the presence of auxin.[132] An efficient and reproducible plant regeneration system, initiated in somatic tissues, has been devised for cassava (*Maninot esculenta*).[132]

Somatic embryogenesis has been induced from shoot tips and immature leaves of *in vitro* shoot cultures of 15 cassava genotypes (Figures 8 to 10). Somatic embryos developed directly on the explants when cultured on a medium containing 4 to 16 mg/l 2,4-D (Figure 11). Secondary embryogenesis has been induced by subculture on solid or liquid induction medium. Long-term cultures were established and maintained for up to 18 months by repeated subculture of the proliferating embryos (Figures 12 and 13). Plantlets were developed from primary and secondary embryos in the presence of 0.1 mg/l BAP, 1 mg/l GA_3, and 0.01 mg/l 2,4-D. Regenerated plants were transferred to the field and were grown to maturity (Figures 14 and 15).

Mass propagation of vegetables like celery and carrot which readily form somatic embryos in liquid cultures has been achieved under laboratory as well as field conditions. There

FIGURE 8. Somatic embryos of cassava developing on immature leaf explant. (Magnification × 20.) Arrow shows somatic embryo's initials. (From Szabados, L., Hoyos, R., and Roca, W. M., *Plant Cell Rep.*, 6, 248, 1987. With permission.)

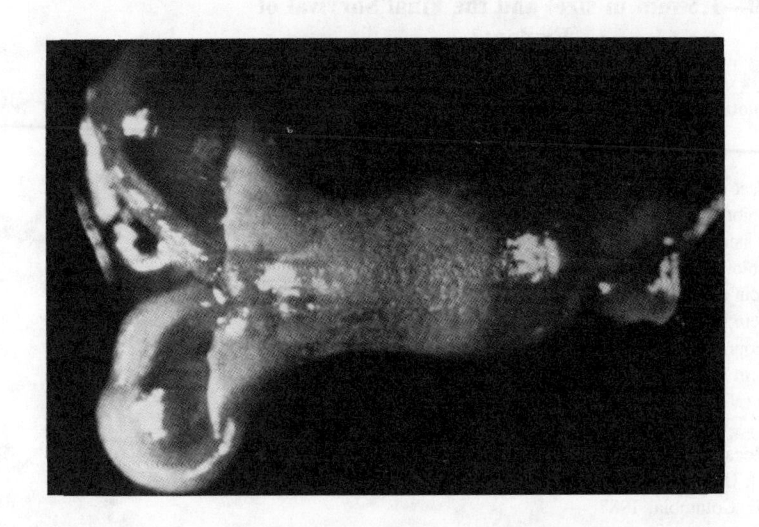

FIGURE 9. Cassava somatic embryo (Magnification × 80.) (From Szabados, L., Hoyos, R., and Roca, W. M., *Plant Cell Rep.*, 6, 248, 1987. With permission.)

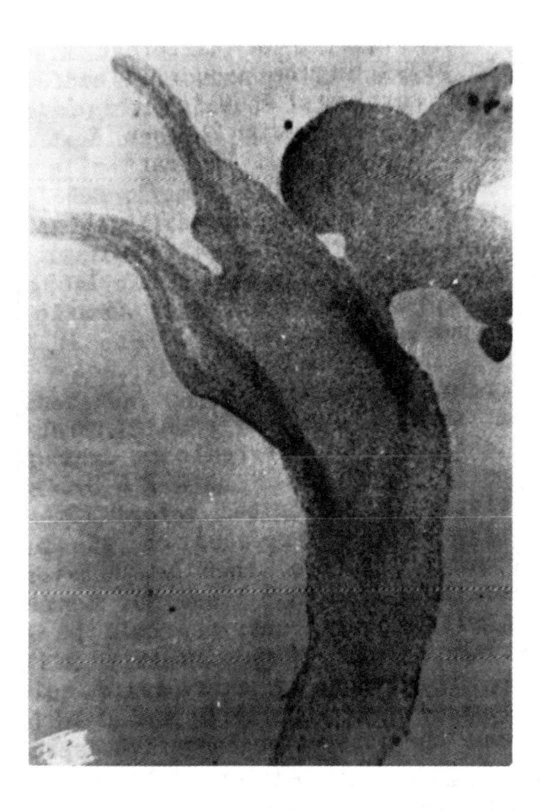

FIGURE 10. Longitudinal section of cassava somatic embryo with initials of secondary embryos. (Magnification × 100.) (From Szabados, L., Hoyos, R., and Roca, W. M., *Plant Cell Rep.*, 6, 248, 1987. With permission.)

are, however, not many species which can be induced to form somatic embryos under appropriate conditions on semisolid media and which also respond well under the liquid culture conditions. Therefore, the concept of producing large quantities of plants via somatic embryogenesis on a large scale is not yet practically feasible for those crops which would benefit most, such as cereals, legumes, fruits, etc.[28] The moderate propagation rates achievable for some plantation crops by adventitious embryo and shoot formation have led to the ability to clonally propagate crops which would otherwise only be propagated by seed.

For some of the monocotyledonous tropical palm crops such as coconut, oil, and date palm, only moderate propagation rates have been achieved. Embryogenic cultures have been obtained and are being used to produce "clonal" plant material for subsequent assessment in field trials.[132] Several large-scale clonal propagation programs are under way in which elite specimens of these plantation crops are being micropropagated. Unilever Ltd. has been undertaking the planting of cloned oil palms in Malaysia since 1977 and these plantings have now been extended to 12,000 palms from 30 clones. Assessments have yet to be completed on the range of variability in the yields of these trees.[103]

In addition to the exploitation of somatic embryogenesis as a means of producing large numbers of cloned plants, there are several other areas of potentially useful mass propagation methods using micropropagation. These include the induction of minitubers and other dormant propagules in multiple shoot cultures. For instance, axillary shoot cultures of potato in the presence of suitable levels of cytokinin and GA form small tubers (or minitubers) in quite large numbers.[4] This makes the transportation of the material easier and in some cases might lend itself to mechanization in the field establishment phase of propagation of seed stocks. Similar minitubers may be produced by tropical tuber crops like *Dioscorea alata*

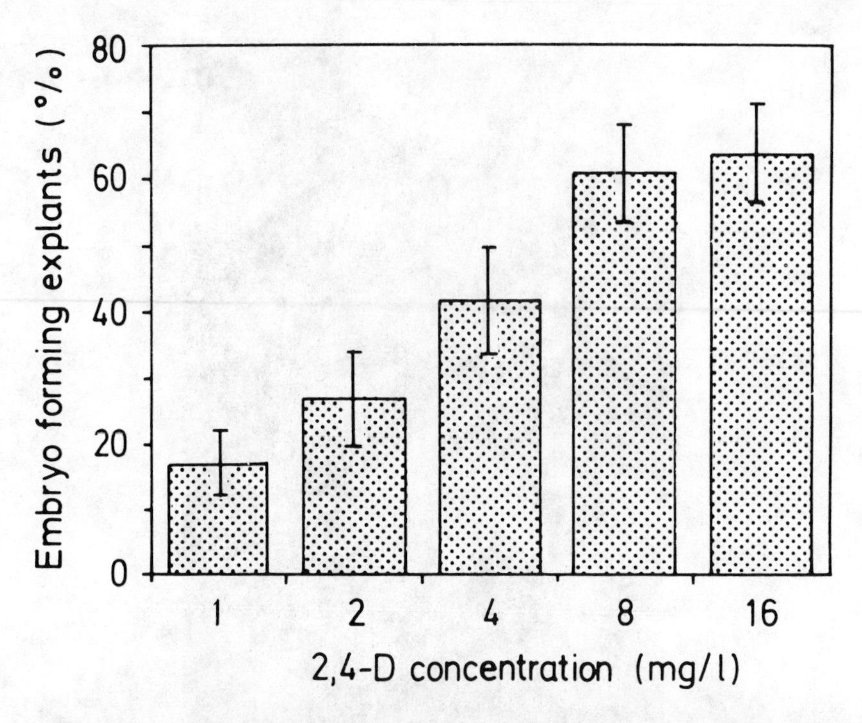

FIGURE 11. The effect of 2,4-D concentration on the formation of cassava somatic embryos on cultured immature leaf explants of the cultivar M Col 1505. (From Szabados, L., Hoyos, R., and Roca, W. M., *Plant Cell Rep.*, 6, 248, 1987. With permission.)

FIGURE 12. Proliferating cassava secondary embryos of a 15-month-old culture (magnification × 20). Insert shows a secondary embryo with new embryo initials (magnification × 50). (From Szabados, L., Hoyos, R., and Roca, W. M., *Plant Cell Rep.*, 6, 248, 1987. With permission.)

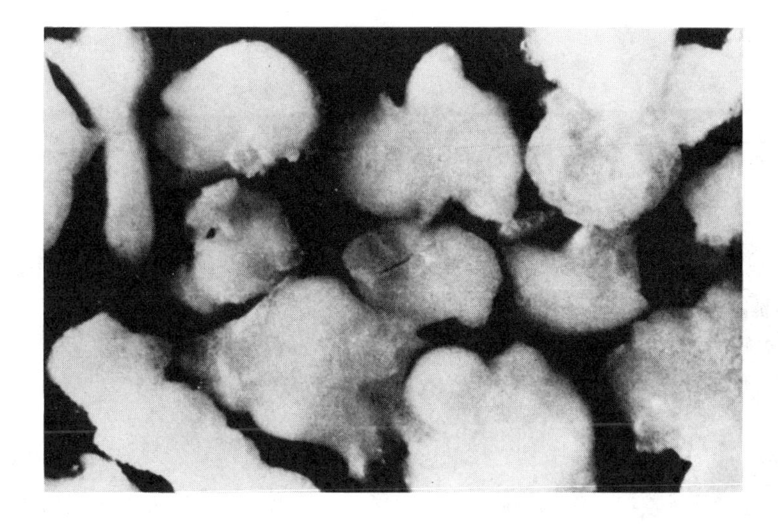

FIGURE 13. Clumps of cassava secondary somatic embryos grown in liquid medium. (Magnification × 15.) (From Szabados, L., Hoyos, R., and Roca, W. M., *Plant Cell Rep.*, 6, 248, 1987. With permission.)

FIGURE 14. Germinating cassava somatic embryos. (From Szabados, L., Hoyos, R., and Roca, W. M., *Plant Cell Rep.*, 6, 248, 1987. With permission.)

when grown in culture and could form the basis of very efficient multiplication and tuber bulking programs, since they would be planted directly into the field and their use would avoid some of the rooting problems associated with field planting of small tuber sets. Methods for *in vitro* propagation of okra (*Ambelmoschus esculentus*) have already been established.[133]

Also of potential at some stage in the future is the use of embryoids produced either through somatic embryogenesis, pollen embryogenesis, or secondary embryogenesis for sowing directly in the field by fluid drilling. Encapsulation of somatic or haploid embryoids

FIGURE 15. Regenerated cassava plants in the field. (From Szabados, L., Hoyos, R., and Roca, W. M., *Plant Cell Rep.*, 6, 248, 1987. With permission.)

at a suitable stage of development (globular, torpedo, or cotyledon stages) protected by gels or other suitable mattrices, which prevent premature desiccation of embryoids and which also contain growth regulatory compounds to control development of the encapsulated embryoids, is being sought. Suitable encapsulation matrices include those in which gels are mixed with peat. It is likely that satisfactory encapsulation of mass-produced embryoid materials for direct drilling will become a feasible method of large-scale clonal planting or as a method for field planting of haploid materials to use directly in field selection trials carried out under specialized cropping or environmental stress conditions.

Ancora[134] has worked extensively on the utilization of an *in vitro* method for micropropagation of globe artichoke. The globe artichoke can be successfully propagated *in vitro* through meristem culture. Thus, it is reasonable to assume that, in future, micropropagation could substitute the conventional methods of propagation, provided that cost is compatible. If this is the case, it could mean a revolution in cultivation practices. If it is confirmed that micropropagated plants are able to produce satisfactory yields in spring, after they have been planted in late August or early September, artichoke cultivation could become annual instead of polyannual. In this case the field could be utilized in summer for another cultivation. Annual cultivation should also have positive implications with regard to the sanitary conditions of artichoke cultivation, which could lead to reduction of chemical treatment actually necessary to control the attacks of pathogens. It is also possible that some expensive cultural operations (the elimination for offshoots after the new growth) would become unnecessary. In fact, it has been observed that in the first period after transfer to the field a micropropagated plant develops only one main shoot and in the second year produces a number of offshoots. Micropropagation allows for an easier clonal selection and, hence, more homogenous cultivation, and is also thought to contribute to mechanical harvesting and, therefore, a reduction in production costs. Finally, a rapid *in vitro* propagation can be of great help for a better evaluation and diffusion of selected lines obtained by breeding.

Potential applications of *in vitro* propagation are well documented, especially for *Citrus*,[6] coffee,[37] and forest trees[33] and include the mass propagation, transport, and storage of disease-free material and breeding programs. Although much of tree propagation *in vitro* remains at the laboratory stages of development, for most of the plants in a few cases the commercialization is likely to occur.

Multiple shoots were induced from the nodal segment of 5-year-old trees of *Eucalyptus grandis*[135] on solid medium. Numerous shoot buds differentiated from leaf and axillary buds were developed into plants in *Morus indica.*[136] Clonal propagation of *Stevia rebandiana* has been established by culturing stem tips with few leaf primordia on agar medium supplemented with high concentration of KIN.[137] Multiple shoots were induced from nodal segments of mature *Eucalyptus* spp. and regenerated into plantlets which.were successfully transferred to pots and fields.[138]

Propagation *in vitro* of tree crops has been regarded as commerically viable only for the initial rapid multiplication of new varieties or in other situations where mother tree stock is limited.[50] However, with the increasing success in these areas, nurserymen are becoming more aware of possible financial benefits of the use of *in vitro* methods for the routine production of plants.[139]

Fruit tree rootstocks are now being mass produced by axillary shoot culture in commerical laboratories in Europe, Canada, and the U.S. Experience of such commerical production is probably greatest in Italy where production from a single laboratory has risen from 38,000 trees in 1977 to more than 2 million trees in both 1980 and 1981. These trees were of high quality, most being rootstocks of peach that were needed to satisfy local demand. The transfer of plantlets to soil accounted for 40 to 80% of the production cost, but, nevertheless, the enterprise has been sufficiently profitable to support substantial research in increasing the efficiency of the transfer to soil and also the propagation of fruit tree section cultivars.[140]

Experience in other commerical laboratories has not been so favorable. Successful application of *in vitro* propagation is not possible with all crops. There have been technical problems such as inconsistent rooting with some apple rootstocks and shoot cultures becoming slow growing with tightly rolled translucent leaves. This latter condition has been described as "vitrified"[141] and has been alleviated in the case of shoot cultures of *Cynara scolymus* by raising the concentration of agar in the medium.[142] Such technical problems, together with the high cost of establishment in soil and the low price presently commanded by some of the root stocks, have resulted in propagation *in vitro* being commercially unattractive in some circumstances. Nevertheless, the general experience with the initial ventures into commercial production appears to have been sufficiently favorable to encourage further investment in several crops.[139] *In vitro* propagation of fruit trees has been applied on a somewhat smaller scale to breeding programs through the rapid production of promising new lines for field trial,[143] and has also proved useful for rapid year-round production of uniform trees for research projects that would otherwise depend on single batches of trees produced annually by conventional methods. Propagation of apple trees consists of budding or grafting the scion cultivar onto rootstocks which are themselves raised by stooling or layering. This process takes 3 years and demands expensive nursery facilities and skills. Propagation of self-rooted trees would be expected to be much more rapid, which could be of major economic advantage, especially in relation to modern high-density orchards. However, most scion cultivars have proved very difficult to root from cuttings and it is only recently that some have been propagated from hardwood cuttings under carefully controlled conditions.[144] In addition, dwarf cultivars have the greatest potential for success as self-rooted trees, but have very slow-growing shoots and could not be expected to produce enough shoot cuttings for rapid propagation. By contrast, a wide range of apple scion cultivars have now been rapidly propagated *in vitro* in various laboratories thoughout the world. Furthermore, at East Malling Research Station in England, the methods have been applied by the fruit tree breeders to their new semidwarf cultivar and also to dwarf types derived from cv. Wijick Mcintosh (a dwarf mutant of cv. McIntosh). In the field, these Wijick types produce single leading shoots and few laterals, whereas *in vitro* they produce axillary shoots as prolifically as the vigorous types of cv. Bramely or cv. Golden Delicious.[143]

Samples of trees of five apple scion cultivars which had been propagated *in vitro* were planted in the orchard at East Malling in 1978 and 1979 for observation and compared with conventional grafted trees. This is a feasibility study of establishing an orchard with self-rooted trees, involving assessment of performance in the absence of control over size and flowering that is presently provided by the appropriate choice of rootstock. All the self-rooted trees have continued to grow vigorously and came into cropping in 1980 or 1981. The rejuvenation *in vitro* that has been referred to in relation to the propagation of such trees did not result in delayed flowering as would be expected if the plants had all the characteristics of seedlings.[145] Planting of self-rooted apple trees has also been in progress at the U.S. Department of Agriculture (USDA) research center, Beltsville, MD since 1979, and the results are similar to those at East Malling.[140] Self-rooted trees, through propagation *in vitro,* could have similar application to the wide range of fruit and other tree crops that are traditionally propagated by grafting onto rootstocks. Thus, self-rooted trees of cherry, plum, and pear have also been produced for orchard trial at East Malling.[50,145,146]

Forest trees have been propagated traditionally by seeds. Propagation *in vitro* from seeds and seedlings has some application, mainly to increase output from types that are low in seed production.[5] However, the most important application is in the mass production of elite adult trees and this is expected to produce dramatic improvements in forest productivity.[33] Such mass production may be feasible with at least some species in the near future. Already more than 2000 trees have been propagated *in vitro* from selected adult gymnosperms at the AFOCEL institute in France for assessment in nursery and forest trials. It is envisaged that such trees will soon be used to establish nurseries of high-quality rejuvenated trees for subsequent forest planting.[7,147]

As in the case of forest trees, rapid propagation of elite material is the most important application and this is already in progress on a limited scale. The first clonal palms from the Unilever Laboratory were planted in the field in Malaysia in 1977 and these plantings have now been extended to 12,000 palms from 30 clones. At the present early development stage of the propagation methods, the cost of this clonal material is about five times that of seedlings. However, it is estimated that this extra cost will be recovered in 5 years through the 30% higher yield which is expected from the best clones and that thereafter there will be an additional annual profit of about 30%.[148]

V. CONCLUSIONS AND FUTURE PROSPECTS

The effectiveness of *in vitro* methods in improving vegetative propagation coupled with the advantages of built-in disease protection makes them an attractive practical alternative to conventional techniques with many types of plant. Shoots may be multiplied rapidly and considerable savings made in the time and space required by conventional procedures. Disease-tested material can be bulked in large quantities without the costly precautions that are otherwise needed to prevent reinfection during propagation. Once a disease-free material is established in culture, it will remain safe from reinfection until planted out. With some crops, *in vitro* method provides, or could provide, the only practicable means of producing clonal material. These obvious advantages, however, may very easily offset one or more of the problems and disadvantages that are inherent in the tissue culture method. Many workers have become over enthusiastic about the possibilities of micropropagation without considering its limitations.

The maintenance cost of tissue laboratories is many times higher than the running costs. The cost of labor constitutes the largest single item of expenditure. Also, highly trained personnel with adequate knowledge of microbiological techniques are required to work in such laboratories. All plants cannot be cultured by supplying the appropriate metabolites suitable for *in vitro* conditions, and in some species shoot multiplication or regeneration is difficult or impossible. Many plants produce an excessive amount of phenolic substances whose oxidation products darken the tissues and medium and strongly inhibit growth. Addition of charcoal has been reported to improve the growth, but activated charcoal also binds to hormones and other metabolites,[149,150] and this lack of selectivity limits its use in tissue culture media.[151]

An additional problem with cultured plants is the development of swollen, distorted leaves which become irreversibly translucent and eventually necrotic, a condition that may lead to the death of the shoot cluster. This has been referred to as "watersoaking" by Hussey[152] and described as vitrification by Debergh et al.[153] The phenomenon occurs in shoots cultured on medium containing cytokinin and can generally be prevented by making sure that the concentrations are no higher than necessary.

The occurrence of contamination by slowly growing and often microscopic invisible microorganisms, particularly bacteria, is a serious problem. It is important that the original cultures be examined microscopically and tested with microbe-detecting media so that any infected cultures can be rejected. Generally, such cultures become infected to *Bacillus subtilus*. In addition, bacteria are carried within the tissue of the plants which are difficult to detect and remove.[154]

An occasional and often serious problem encounted during large-scale micropropagation of some plants is the persistence of certain types of slow-growing saprophytic or pathogenic bacteria that survive the initial decontamination procedures. Such contaminants, e.g., *Pseudomonas* spp., *Erwinia* spp., and *Bacillus* spp., can persist for many culture generations without being noticed. These conditions can cause chlorosis in the propagated plantlets.[154] *Pseudomonas* infections severely limited proliferation rates in tissue cultures of papaya initiated from field-grown plants.[71] In extreme cases, effects of these types of infections may be reduced by the use of appropriate levels of antibiotics incorporated in culture media. An associated problem is the inadvertant propagation of plants which are infected with latent or symptomless types of viruses or mycoplasmas which may markedly reduce the vigor and proliferation rates of cultures.

A further problem more frequently associated with micropropagation of tree species is the accumulation of inhibitory substances in the growth medium during initiation of cultures. Explants of these species often produce excessive amounts of phenolic substances, the oxidation products of which often strongly inhibit growth. Where the problem is confined to the reaction of the initial explant, it may be prevented by dissecting tissue for culture under the surface of liquids or by incorporating ascorbic acid or citric acid in culture media. In the case of teak tissue cultures, polyvinyl polypyrrolidone has proved an effective amendment to culture media.[155] By contrast, some phenolics have been used to great advantage. Phloroglucinol incorporated in media supporting apple shoot tip culture of M-7 rootstocks produced a two- to threefold increase in shoot proliferation and rooting. Various other beneficial effects of this and other phenolics on rejuvenation *in vitro* have now been recorded on apple and some other fruit crops.[103] Activated charcoal, when added at levels of 1 to 2% (w/v) in media, can sometimes be beneficial by adsorbing inhibitory concentrations of growth regulatory substances such as ABA, which might be present in the original explant or produced by proliferating tissues. It is not uncommon in callus cultures to find that the regenerative capacity of some cultures declines over prolonged subculture. There may be several reasons for this, one of which may be the development of habituation in some of the cell population. As a result of epigenetic changes brought on by the cultural process, sectors of the culture may become noncompetent, while others retain their competence for morphogenesis. Experience with culture of embryogenic calli of grasses and cereals has shown that early recognition and physical separation of embryogenic sectors are critical if stable embryogenic tissue cultures are to be developed and then maintained. Results from TCCP experiments have revealed that in wheat, rice, and millet the regeneration of plants depends on the ability to maintain E callus in culture. In the case of wheat, a regeneration has been obtained after maintaining callus for 20 months in culture.[28]

Each crop species has unique growth requirements and, therefore, specific establishment techniques must be refined for individual crop. Plant propagated *in vitro* often cannot be transferred readily to an open soil environment. A weaning stage has usually been used in which the plants have been transplanted from *in vitro* conditions into humid conditions in greenhouses, and the humidity has then been reduced gradually over 3 to 4 weeks to that of the open soil environment. During this stage, great care is necessary with trees, particularly with the water regime, as large losses can occur. Antitranspirants have been used to reduce the vulnerability of apple trees at this stage with promising results. Some fruit tree plantlets established readily in soil, but did not grow subsequently unless first chilled or treated with gibberellins.[156] Plantlets of hardy crops with well-developed root systems can be planted directly in pots of a soil or soilless mixture. Best results are obtained by placing these plants in shaded areas and keeping them moist for a period of 5 to 10 d before moving them into direct sunlight. Fertilization is generally begun after the plants are moved into full sunlight.

With callus culture, there have been numerous reports of cytological irregularities leading to the regenerated plants being mutant or polyploid; this has been the finding with plants from calli of *Populus* spp.[35] Such instability appears to be characteristic of calli with many large parenchymatous cells growing under the influence of growth substances, especially 2,4-D or NAA. A callus from the nucellus of a *Citrus* sp. was composed almost entirely of small meristematic cells and the regenerated plants were genetically uniform. Furthermore, this callus became habituated and did not require growth regulators, and this could have also contributed to the genetic stability.[6] Calli of oil palm have appeared to be genetically stable since the plants from the regenerated embryoids have been uniform within clones. The anatomical structure of these calli has not been described.[148]

These assessments of the variability of plants propagated *in vitro* indicate that genetic variation has not been a major problem with at least some of the shoot culture and callus systems in use. Nevertheless, it is highly desirable to assess this variability at the propagation stage before expensive field planting has been completed.

Recent progress has been excellent and already the success of *in vitro* propagation with adult trees and palms is bringing important practical benefit; the future propagation *in vitro* is an invaluable aid to the cropping of trees appears to be assured. Thus far, shoot culture methods are outstanding in their applicability to the widest range of tree species. Nevertheless, there remains such scope for improving these methods on the commerical scale and also for applying them to subjects that presently remain recalcitrant. Thus, there is a particular need to improve knowledge of juvenility and maturation and also of the control of shoot proliferation, adventitious root formation, and the reactions of plantlets to transfer to soil. Some of the recalcitrant types may yield to the application of methods for inducing rejuvenation and also to modification of existing culture methods such as the changes in the inorganic and cytokinin components of the culture medium that led to the successful propagation of *Liquidambar styraciflua*.[33]

Application of *in vitro* propagation to the production of desirable new plants could be of great values, since breeding by normal sexual hydridization is very labor intensive and can take many years. Furthermore, there are many commercially important crops that are sterile and improved variants cannot be obtained by conventional breeding. In the near future, it is conceivable that some of these limitations will be overcome by the production of mutants and polyploid types through regeneration from callus and single cells that are genetically unstable or have been subjected to mutagenic agents.[145] In the longer term, application of pollen culture and protoplast culture techniques would be expected to transform possibilities for the breeding of trees as for other crop species. Already there have been some significant achievements in this respect.

The application of *in vitro* propagation in the production of pathogen-free plants has been described in detail. At present the elimination of viruses appears to be the major contribution of *in vitro* propagation. Already significant achievement has been made in crops like potato and cassava. However, it is not always possible to remove the viruses completely and in such situations the use of thermotherapy and chemotherapy has been suggested.

REFERENCES

1. **Shull, G. H.,** 'Phenotype' and 'clone', *Science,* 35, 182, 1912.
2. **Stout, A. B.,** The nomenclature of cultivated plants, *Am. J. Bot.,* 27, 339, 1940.
3. **Murashige, T.,** Plant propagation through tissue cultures, *Ann. Rev. Plant Physiol.,* 25, 135, 1974.
4. **Hussey, G.,** In vitro propagation of *Narcissus, Ann. Bot.,* 49, 707, 1983.
5. **Holdgate, D. P.,** Propagation of ornamentals by tissue culture, in *Applied and Fundamental Aspects of Plant Cell, Tissue and Organ Culture,* Reinert, J. and Bajaj, Y. P. S., Eds., Springer-Verlag, Berlin, 1977, 18.
6. **Button, J.,** International exchange of disease-free citrus clones by means of tissue culture, *Outlook Agric.,* 8, 155, 1977.
7. **Boulay, M.,** La micropropagation des arbres forestries, *C. Seances Acad. Agric. Fr.,* 66(8), 697, 1980.
8. **Franclet, A.,** Rejeunissement des arbres adultes en vue de leur propagation vegetative, in *Annales de Recherches Sylvicoles,* AFOCEL, Etudest et Recherches, No. 12, 6/79, Micropropagation d'Arbres Forestiers, 1979, 3.
9. **Hartmann, H. T. and Kester, D. E.,** *Plant Propagation, Principles and Practices,* Prentice-Hall, Englewood Cliffs, NJ, 1975.
10. **Bilkey, P. C., McCown, B. H., and Hildebrandt, A. C.,** Micropropagation of African Violet from petiole cross sections, *HortScience,* 7, 289, 1978.
11. **Daykin, M., Langhans, R. W., and Earle, E. D.,** Tissue culture of the double petunia, *HortScience,* 11, 35, 1976.
12. **Smith, R. H. and Nightingale, A. E.,** In vitro propagation of Kalanchee, *HortScience,* 14, 20, 1979.
13. **Alkema, H. Y. and van Leeuwn, C. J. M.,** Propagation of a number of miscellaneous bulb crops by means of the twin scale method, *Bloembollencultuur,* 88, 32, 1977.
14. **Hussey, G.,** Propagation of some members of the Liliaceae, Iridacear and Amaryllidaceae by tissue culture, in *Petalid Mono-cotyledons,* Brickell, C. D., Cutler, D. F., and Gregory, M., Eds., Academic Press, New York, 1980, 33.
15. **Pierik, R. L. M. and Steegmans, H. H. M.,** Effect of auxins, cytokinins, gibberellins, abscissic acid and ethephon on regeneration and growth of bulbets on excised bulb scale segments of hyacinth, *Physiol. Plant.,* 34, 14, 1975.
16. **Seabrook, J. E. A., Cumming, B. G., and Dionne, L. A.,** The in vitro induction of adventitious shoot and root apices on *Narcissus* (daffodil and narcissus) cultivar tissue, *Can J. Bot.,* 54, 814, 1976.
17. **Hussey, G. and Falavigna, A.,** Origin and production of in vitro adventitious shoot in the onion, *Allium cepa* L, *J. Exp. Bot.,* 31, 1675, 1980.
18. **Hussey, G.,** In vitro propagation, in *Tissue Culture for Plant Pathologists,* Ingram, D. S. and Helgeson, J. P., Eds., Blackwell Scientific, London, 1980, 51.
19. **Dublin, P.,** Direct somatic embryogenesis on fragments of *Arabusta* coffee tree leaves, *Cafe Cacao The (Paris),* 25, 237, 1981.
20. **Torrey, J. G.,** The initiation of organized development in plants, *Adv. Morphog.,* 5, 39, 1966.
21. **Barbier, M. and Dulieu, H.,** Early occurrence of genetic variants in protoplast cultures, *Plant Sci. Lett.,* 29, 201, 1983.
22. **Heyser, J. W. and Nabors, M. W.,** Long term plant regeneration, somatic embryogenesis and green spot formation in secondary oat *(Avena sativa)* callus, *Z. Pflanzenphysiol.,* 107, 153, 1982.
23. **Sheridan, W. F.,** Tissue culture of the monocot *Lilium, Planta,* 82, 189, 1968.
24. **Davies, D. R.,** Speeding up the commerical propagation of freesias, *Grower,* 77, 711, 1972.
25. **Earle, E. D. and Langhans, R. W.,** Propagation of *Chrysanthemum* in vitro. II. Production, growth and flowering of plantlets from tissue cultures, *J. Am. Soc. Hortic. Sci.,* 99, 352, 1974.
26. **De Langhe, E. and De Bruijne, E.,** Continuous propagation of tomato plants by means of callus cultures, *Sci. Hortic.,* 4, 221, 1976.
27. **Krikorian, A. D., Staicu, S. A., and Kann, R. P.,** Karyotype analysis of a daylily clone reared from aseptically cultured tissues, *Ann. Bot.,* 47, 121, 1981.
28. TCCP (Tissue Culture for Crop Project), Progress Report, Colorado State University, Fort Collins, 1987.
29. **Heyser, J. W. and Nabors, M. W.,** The regeneration of proso millet from embryogenic calli derived from various plant parts, *Crop Sci.,* 22, 1070, 1982.

30. **Heyser, J. W., Dykes, T. A., DeMott, K. J., and Nabors, M. W.,** High frequency, long term regeneration of rice from callus culture, *Plant. Sci. Lett.*, 29, 175, 1983.
31. **Abbott, A. J.,** Practice and promise of micropropagation of woody species, *Acta Hortic.*, 79, 113, 1978.
32. **Durzan, D. J. and Lopushanski, S. M.,** Propagation of American Elm via cell suspension cultures, *Can. J. For. Res.*, 5, 273, 1975.
33. **Sommer, H. E. and Brown, C. L.,** Embryogenesis in tissue cultures of sweetgum, *For. Sci.*, 26, 257, 1980.
34. **Jacquiot, C.,** Plant tissue and excised organ cultures and their significance in forest research, *J. Inst. Wood Sci.*, 16, 22, 1966.
35. **Lester, D. T. and Berbee, K. G.,** Within-clone variation among black polar trees derived from callus culture, *For. Sci.*, 23, 122, 1977.
36. **Druart, P. L.,** Plantlet regeneration from root callus of different *Prunus* species, *Sci. Hortic.*, 12, 339, 1980.
37. **Sondahl, M. R. and Sharp, W. R.,** Research in *Coffea* spp. and application of tissue culture methods, in *Plant Cell and Tissue Culture,* Sharp, W. R., Larsen, P. O., Paddock, E. F., and Rahavan, V., Eds., Ohio State University Press, Columbus, 1977, 527.
38. **Eichholtz, D. A. and Robitaille, H. A.,** Asexual apple embryos in vitro, *Compact Fruit Tree,* 13, 142, 1980.
39. **Rabechault, H., Ahee, J., and Guenin, G.,** Recherches pour la culture in vitro des embryos de palmier a huile *(Elaeis guineensis* jacq.), *Oleagineux,* 27, 1972.
40. **Jones, L. H.,** Propagation of clonal oil palms by tissue culture, *Oil Palm News,* 17, 1, 1974.
41. **Rabechault, H. and Martin, J.,** Multiplication vegetative du palmer a huile (*Elaeis guineensis* Jacq.) a l' aide de culture de tissues Foliaires, *C. R. Seances Acad. Paris Ser. D.*, 283, 1735, 1976.
42. **Ahee, J., Arthuis, P., Cas, G., Duval, Y., Guenin, G., Hanower, J., Lievoux, D., Loiret, C., Malaurie, B., Pannekier, C., Raillur, D., Varechon, C., and Zuckermann, L.,** La multiplication vegetative in vitro a huile par embryogenese somatique, *Oleagineux,* 36, 113, 1981.
43. **Eeuwens, C. J.,** Mineral requirement of coconut tissues, *Physiol. Plant.*, 36, 23, 1976.
44. **Guzman, E. V., Rosario, A. G., and Ubalde, E.,** Proliferation, growth and organogenesis in coconut embryo and tissue culture, *Philipp. J. Coconut Stud.*, 3, 1, 1978.
45. **Fulford, R. M., Passey, A. J., and Butler, M.,** Vegetative propagation of coconuts by tissue culture, Report of East Malling Research Station for 1979, 1980, 184.
46. **Robb, S. H.,** The culture of excised tissue from bulb scales of *Lilium speciosum, J. Exp. Bot.*, 8, 348, 1957.
47. **Wright, N. A. and Alderson, P. G.,** The growth of tulip tissues in vitro, *Acta Hortic.*, 109, 263, 1980.
48. **Jones, O. P.,** Effects of cytokinins in xylem sap from apple trees on apple shoot growth, *J. Hortic. Sci.*, 48, 181, 1973.
49. **Jones, O. P., Hopgood, M. E., and O'Farrell, D.,** Propagation in vitro of M26 apple rootstocks, *J. Hortic. Sci.*, 52, 235, 1977.
50. **Jones, O. P., Pontikis, C. A., and Hopgood, M. E.,** Propagation in vitro of five apple scion cultivars, *J. Hortic. Sci.*, 54, 155, 1979.
51. **McCown, B. and Amos, R.,** Initial trials of the commercial micropropagation of birch selections, *Int. Plant Propagators Soc. Combined Proc.*, 29, 387, 1979.
52. **Hunter, C. S.,** In vitro culture of *Cinchona ledgeriana* L, *J. Hortic. Sci.*, 54, 111, 1979.
53. **Dublin, P.,** Induction be bourgeons neoformes at embryogenese somatique. Deux voies de multiplication in vitro des careiers cultives, *Cafe Cacao The (Paris),* 24, 121, 1980.
54. **Durand, R. and Boudet, A. M.,** Le bouturage in vitro de l'eucalyptus, in *Annales de Recherches Sylvicoles,* AFOCEL, Etude et Recherches, No. 12, 6/79, Micropropagation d'Arbres Forestiers, 1979, 57.
55. **Sommer, H. E. and Brown, C. L.,** Application of tissue culture to forest tree improvement, in *Plant Cell and Tissue Culture,* Sharp, W. R., Larsen, P. O., Paddock, E. F., and Raghvan, V., Eds., Ohio State Univerisity Press, Columbus, 1977, 461.
56. **Sommer, H. E. and Brown, C. L.,** Plantlet formation in pine tissue cultures, *Am. J. Bot.*, 61(Suppl.), 11, 1974.
57. **Rancillac, M.,** Mise au point d'une methode de multiplicat vegetative in vitro du Pin maritime *(Pinus pinaster),* in *Annales de Recherches Sylvicoles,* AFOCEL, Etudes et Recherches, No. 12, 6/79, Micropropagation d'Arbres Forestiers, 1979, 41.
58. **Campbell, R. A. and Durzan, D. J.,** Induction of multiple buds and needles in tissue cultures of *Picea glauca, Can. J. Bot.*, 53, 1652, 1976.
59. **Von Arnold, S. and Eriksson, T.,** Induction of adventitious buds on embryos of Norway spruce grown in vitro, *Physiol. Plant.*, 44, 283, 1978.
60. **Cheng, T. Y.,** Adventitious bud formation in cultures of Douglas fir *(Pseudotsuga menziesii), Plant Sci. Lett.*, 5, 97, 1975.

61. **Cheng, T. Y.,** Micropropagation of clonal fruit tree rootsocks, *Compact Fruit Tree,* 12, 127, 1979.
62. **Abbott, A. J. and Whiteley, E.,** Culture of Malus tissues in vitro. I. Multiplication of apple plants from isolated apple apices, *Sci. Hortic.,* 4, 183, 1976.
63. **Lane, D. W.,** Regeneration of apple plants from shoot meristem tips, *Plant Sci. Lett.,* 13, 281, 1978.
64. **Jones, O. P. and Hopgood, M. E.,** The successful propagation in vitro of two rootstocks of *Prunus:* the plum rootstock *pixy (P. insititia)* and the cherry rootstock F 12/1 *(P. avium), J. Hortic. Sci.,* 54, 63, 1979.
65. **Zimmerman, R. H. and Broome, O. C.,** Apple cultivar micropropagation, in Proc. Conf. on Nursery Production of Fruit Plants through Tissue Culture. Applications and Feasibility, U.S. Department of Agriculture, Beltsville, MD, 1980, 54.
66. **Sriskandarajah, S. and Mullins, M. G.,** Micropropagation of Granny Smith apple and factors affecting root formation in vitro, *J. Hortic. Sci.,* 56, 71, 1981.
67. **Quoirin, M. and Lepoivre, P.,** Etude de milieux adaptes aux cultures in vitro de Prunnus, *Acta Hortic.,* 78, 437, 1977.
68. **Jones, O. P.,** Propagation in vitro of apple trees and other woody fruit plants, methods and applications, *Sci. Hortic.,* 30, 44, 1979.
69. **Zuccherelli, G.,** Metodologie nella multiplicazion industriale. In vitro dei portainnesti clonali del pesio. Pescomardorlo GF 677, Susino GF 43, Damasco 1869, S. Givliano 655/2, in *Technique di Culture in Vitro per la Propagazione su Vasta Scala Della Specie Ortoflorafrutticole,* Bellini, E., Ed., Consigilio Nazion delle Ricerche, Pistoia, 1979, 147.
70. **David, A., David, H., Faye, M., and Isemukali, K.,** Culture in vitro et micropropagation du pin maritime, in *Annales de Recherches Sylvicoles,* AFOCEL, Etudes et Recherches, No. 12, 6/79, Micropropagation d Arbes Forestiers, 1979, 33.
71. **Litz, R. E. and Conover, R. A.,** Effect of sex type, season and other factors on in vitro establishment and culture of *Carica papava* L. explants, *J. Am. Soc. Hortic. Sci.,* 106, 792, 1981.
72. **Mantell, S. H., Haque, S. Q., and Whitehall, A. P.,** Clonal multiplication of *Dioscorea alata* L. and *D. yotundara* Poir. yams by tissue culture, *J. Hortic. Sci.,* 53, 95, 1978.
73. **Gamborg, O. L., Murashige, T., Thorpe, T. A., and Vasil, I. K.,** Plant tissue culture media, *In Vitro,* 12, 473, 1976.
74. **Huang, L. and Murashige, T.,** Plant tissue culture media: major constituents; their preparation and some applications, *Tissue Cult. Assoc. Man.,* 3, 539, 1977.
75. TCCP (Tissue Culture for Crop Project), Newletter No. 7, Colorado State University, Fort Collins, July 1987.
76. **Dougall, D. K. and Verma, D. C.,** Growth and embryo formation in wild-carrot suspension cultures with ammonium ion as a sole nitrogen source, *In Vitro,* 14, 180, 1978.
77. **Miller, L. R. and Murashige, T.,** Tissue culture propagation of tropical foliage plants, *In Vitro,* 12, 197, 1976.
78. **Murashige, T. and Skoog, T.,** A revised medium for rapid growth and bioassays with tobacco tissue cultures, *Physiol. Plant.,* 15, 473, 1962.
79. **Jones, O. P. and Hatfield, S. G. S.,** Root initiation in apple shoot cultured in vitro with auxins and phenolic compounds, *J. Hortic. Sci.,* 51, 495, 1976.
80. **Jones, O. P.,** Effect of phloridzin and Phloroglucinol on apple shoot, *Nature,* 262, 392, 1976.
81. **Carbanne, F., Martin-Tanquy, J., and Martin, C.,** Phenolamines associees a l' induction florale et al' etat reproducteur, *Physiol. Veg.,* 15, 429, 1977.
82. **Boulay, M.,** Multiplication et clonage rapide du Sequoia sempervirens par la culture in vitro, in *Annales de Recherches Sylvicoles,* AFOCEL, Etudes et Recherches, No. 12, 6/79, Micropropagation d' arbers Forestiers, 1979, 49.
83. **Murashige, T.,** Manipulation of organ culture in plant tissue cultures, *Bot. Bull. Acad. Sin.,* 18, 1, 1977.
84. **Hasegawa, P. M., Murashige, T., and Takatori, F. N.,** Propagation of *Asparagus* through shoot apex culture. II. Light and temperature requirements, transplantability of plants and cytological characteristics, *J. Am. Soc. Hortic. Sci.,* 98, 143, 1973.
85. **Kato, A.,** The involvement of photosynthesis in inducing bud formation on excised leaf segments of *Helionopsis orientalis* (Liliaceae), *Plant Cell Physiol.,* 19, 791, 1978.
86. **Seibert, M.,** The effects of wavelengths and intensity of growth and shoot initiation in tobacco callus, *In Vitro,* 80, 435, 1973.
87. **Kadkade, P. G. and Seibert, M.,** Phytochrome regulated organogenesis in lettuce tissue culture, *Nature,* 170, 49, 1977.
88. **Tran Thanh Van, K.,** Regulation of morphogenesis, in *Plant Tissue Culture and Its Biotechnological Application,* Barz, W., Reinhard, E., and Zenk, M. H., Eds., Springer-Verlag, Berlin, 1977, 367.
89. **Murashige, T. and Nakano, R.,** Chromosome complement as a determinant of the morphogenic potential to tobacco cells, *Ann. J. Bot.,* 54, 963, 1967.
90. **Margara, J.,** Etude des fecteurs de la reoformation de bourgeous en culture in vitro chez la choufleur *(Brassica oleraceae* L., var. *Botrytis), Ann. Physiol. Veg.,* 11, 95, 1969.

91. **Fonnesbech, M.,** Temperature effects on shoot and root development from *Begonia* × *chemimantha* petiole segments grown in vitro, *Physiol. Plant,* 32, 282, 1974.
92. **Beasley, C. A. and Eaks, I. L.,** Ethylene from alcohol lamps and natural gas burners, *In Vitro,* 13, 263, 1979.
93. **Thomas, D. S. and Murashige, T.,** Volatile emissions of plant tissue cultures. I. Identification of the major components, *In Vitro,* 15, 654, 1979.
94. **Rines, H. W. and McCoy, T. J.,** Tissue culture initiation and plant regeneration in hexaploid species of oats, *Crop. Sci.,* 21, 837, 1981.
95. **Bingham, E. T., Hurley, L. V., Kaatz, D. M., and Saunders, J. W.,** Breeding alfalfa which regenerates from callus tissue in culture, *Crop Sci.,* 15, 719, 1975.
96. **Tran Thanh Van, K.,** Control of morphogenesis by inherent and exogenously applied factors in thin cell layers, in *Perspectives in Plant Cell and Tissue,* Vasil, I. K. Ed., Academic Press, New York, 1980, 175.
97. **Harney, P. M.,** Tissue culture propagation of some herbaceous horticultural plants, in *Application of Plant Cell and Tissue Culture to Agriculture & Industry,* Tomes, D. T., Ellis, B. E., Horney, P. M., Kasha, K. J., and Peterson, R. L., Eds., University of Guelph, Guelph, Canada, 1982, 187.
98. **Roca, W. M., Bryon, J. E., and Roca, M. R.,** Tissue culture for international transfer of potato genetic resources, *Am. Potato J.,* 55, 691, 1979.
99. **Ingram, D. S. and Helgeson, J. P.,** *Tissue Culture Methods for Plant Pathologists,* Blackwell Scientific, Oxford, 1980.
100. **Quak, F.,** Meristem culture and virus free plants, in *Applied and Fundamental Aspects of Plant Cell, Tissue and Organ Culture,* Reinert, J. and Bajaj, Y. P. S., Eds., Springer-Verlag, Berlin, 1977, 598.
101. **Roca, W. M., Szabados, L., and Hussain, A.,** Biotechnology Research Unit, Annu. Rep. Centro Internacional de Agricultura Tropical Cali, Colombia, 1987.
102. **Roca, W. M.,** Clonaje de cottcelulas vegetales *in vitro:* application en agriculttura, *Rev. Colcien.,* 4, 14, 1986.
103. **Jones, O. P.,** In vitro propagation of tree crops, in *Plant Biotechenology,* Mantell, S. H. and Smith, H., Eds., SEB Seminar 103, Cambridge University Press, Cambridge, 1983, 139.
104. **Navarro, L., Reistacher, C. N., and Murashige, T.,** Improvement of shoot-tip grafting in vitro for virus-free citrus, *J. Am. Soc. Hortic. Sci.,* 100, 471, 1975.
105. **Yentsey, C. O.,** A method for virus-free propagation of citrus-shoot tip grafting, *Citrus Ind.,* 59, 39, 1978.
106. **Martinez, J., Hugard, H., and Jonard, R.,** The different grafting of shoot tops realized in vitro between peach (*Prunus persica* Batsch), apricot (*Prunus armeniaca* L.) and myrobolan (*Prunus cesarifera* Ehrh.), *C. R. Acad. Sci. Ser. D.,* 288, 759, 1979.
107. **Huth, W.,** Culture of apple plants from apical meristems, *Gartenbauwissenschaft,* 43, 163, 1978.
108. **Lundergan, C. A. and Janick, J.,** Low temperature storage of in vitro apple shoots, *HortScience,* 14, 514, 1979.
109. **Kassanis, B.,** Heat inactivation of leaf-roll virus in potato tuber, *Ann. Appl. Biol.,* 37, 339, 1950.
110. CIAT (Centre International de Agricultura Tropical), Cassava tissue culture, Genetic Resources Unit, Cali, Colombia, 1983.
111. **Stace-Smith, R. and Mellor, F. C.,** Eradication of potato virus X and S by thermotherapy and axillary bud culture, *Phytopathology,* 58, 199, 1968.
112. **Pennazie, S.,** Potato therapy; meristem tip culture combined with thermotherapy, *Riv. Ortoflorofruttic. Ital.,* 5, 446, 1971.
113. CIAT (Centro Internacional de Agricultura Tropical), Resources Unit Annual Report, Colombia, 1982.
114. **Roca, W. M.,** In vitro clonal propagation to eliminate crop diseases, in *Biotechnology International Research,* IRRI, Manila, 1985, 3.
115. **Walkey, D. G. A.,** In vitro methods for virus elimination, in *Frontiers of Plant Tissue Culture,* Thorpe, T. A., Ed., University of Calgary, Calgary, 1978, 245.
116. **Lizarraga, R. E., Salazar, L. E., Roca, W. M., and Schilde-Rentschler, L.,** Elimination of potato spindle tuber viroid by low temperature and meristem culture, *Phytopathology,* 70, 754, 1980.
117. **Schilde-Rentschler, L. and Roca, W. M.,** Virus, elimination in potato and cassava, in Global Workshop on Root and Tuber Crop Propagation Proc. Workshop, CIAT, Cali, Colombia, 1986, 89.
118. **Roca, W. M., Espinoza, N. O., Roca, M. R., and Bryan, J. E.,** A tissue culture method for the rapid propagation of potatoes, *Am. Potato J.,* 55, 691, 1978.
119. **Mellor, F. C. and Stace-Smith, R.,** Virus-free potatoes by tissue culture, in *Applied and Fundamental Aspect of Plant Cell, Tissue and Organ Culture,* Reinert, J. and Bajaj, Y. P. S., Eds., Springer-Verlag, Berlin, 1977, 616.
120. **Accatine, P.,** Papa corahilla libre de virus mediante cultive de meristemas, *Agric. Tec. (Santiago),* 26, 34, 1966.
121. **Kassanis, B.,** The use of tissue cultures to produce virus-free clones from infected potato varieties, *Ann. Appl. Biol.,* 45, 422, 1957.

122. **Sanger, H. L. and Ramm, K.,** Radioactive labelling of viroid RNA, in Modification of the Information Content of Plant Cells, Proc. 2nd John Innes Symp., Markham, R., Ed., Norwich, England, 1975, 230.
123. **Lizarraga, R. E., Salazar, L. E., and Schile-Rentschler, L.,** Effect of meristem size on eradication of potato spindle tuber viroid, in Proc. Int. Congr. on Research for the Potato in the Year 2000, Centre Internacional de la Pap (CIP), Lima, Peru, 1982.
124. **Roca, W. M., Rodriguez, J., Beltran, J., Roa, J., and Mafla, G.,** Tissue culture for the conservation and international exchange of germplasm, in Proc. 5th Int. Congr. on Plant Tissue and Cell Culture, Tokyo, Japan, 1982.
125. **Henshaw, G. G. and O'Hara, J. F.,** In vitro approaches to the conservation and utilization of global plant genetic resources, in *Plant Biotechnology,* Mantell, S. H. and Smith, H., Eds., Cambridge University Press, Cambridge, 1983, 219.
126. **Frankel, O. H. and Hawkes, J. G.,** *Crop Genetic Resources for Today and Tomorrow,* Cambridge University Press, Cambridge, 1975.
127. **Withers, L. A.,** The development of cryopreservation technique for plant cell, tissue and organ cultures, in *Plant Tissue Culture,* Fujiwara, A., Ed., Japanese Association of Plant Tissue Culture, Tokyo, 1982, 793.
128. **Wither, L. A.,** Germplasm storage in plant biotechnology, in *Plant Biotechnology,* Mantell, S. H. and Smith, H., Eds., Cambridge University Press, Cambridge, 1983, 187.
129. **Grout, B. W. W., Westcott, R. J., and Henshaw, G. G.,** Survival of shoot meristems of tomato seedlings frozen in liquid nitrogen, *Cryobiology,* 15, 478, 1978.
130. **Westcott, R. J.,** Tissue culture storage of potato germplasm. Minimal growth storage, *Potato Res.,* 24, 331, 1981.
131. **Crisp, P. and Walkey, D. G. A.,** The use of aseptic meristem culture in cauliflower breeding, *Euphytica,* 23, 305, 1974.
132. **Szabados, L., Hoyos, R., and Roca, W. M.,** In vitro somatic embryogenesis and plant regeneration of cassava, *Plant Cell Rep.,* 6, 248, 1987.
133. **Mangat, B. S. and Roy, M. K.,** Tissue culture and plant regeneration of Okra *(Abelmoschus escutentus), Plant Sci.,* 47, 57, 1986.
134. **Ancora, G.,** Globe artichoke (*Cynara scolymus* L.) in *Biotechnology in Agriculture and Forestry,* Vol. 2, Bajaj, Y. P. S., Ed., Springer-Verlag, Berlin, 1986, 471.
135. **Lakshmi Sita, G. and Shobha Rani, B.,** In vitro propagation of *Eucalyptus grandis* L. by tissue culture, *Plant Cell Rep.,* 4, 63, 1985.
136. **Mhatre, M., Bapat, V. A., and Rao, P. S.,** Regeneration of plants from culture of leaves and axillary buds in mulberry *(Morus indica* L.), *Plant Cell Rep.,* 4, 78, 1985.
137. **Tamura, Y., Nakamura, S., Fukui, H., and Tabata, M.,** Clonal propagation of *Stevia rebaudiana* Bertoni by stem-up culture, *Plant Cell Rep.,* 3, 183, 1984.
138. **Gupta, P. K., Mehta, U. J., and Muscarenhas, A. F.,** A tissue culture method for clonal propagation of mature trees of *Eucalyptus torelliana* and *Eucalyptus camaedulensis, Plant Cell Rep.,* 2, 296, 1983.
139. **Anon.,** Raisers, neglet micropropagation, *Grower (London),* 96(23), 2, 1981.
140. **Zimmerman, R. H.,** The laboratory of micropropagation at Cesana, Italy, *Int. Plant Propagators Soc. Combined Proc.,* 29, 398, 1979.
141. **Natavel, L. M.,** L'utilisation des culture in vitro pour la multipication de quelques especes legumieres et fruitieres, *C. Seances Acad. Agric. Fr.,* 66(8), 681, 1980.
142. **Debergh, P. C. and Maene, L. J.,** A scheme for commercial propagation of ornamental plants by tissue culture, *Sci. Hortic.,* 14, 335, 1981.
143. **Longbottom, H., Tobutt, K. R., and Jones, C. P.,** Propagation in vitro of fruit trees, Report of East Malling Research Station for 1979, 1980, 187.
144. **Child, R. D. and Hughes, R.,** Factors influencing rooting in hardwood cuttings of apple cultivars, *Acta Hortic.,* 79, 43, 1978.
145. **Gayner, J. A., Hopgood, M. E., Jones, O. P., and Watkins, R.,** Propagation in vitro of fruit Plants, Report of East Malling Research Station for 1980, 1981, 145.
146. **Cehl, V. H.,** Own-root varieties, Report of East Malling Research Station for 1978, 1979, 60.
147. **Franclet, A., David, A., David, H., and Bouley, M.,** Premier mise en evidence morphologie d'un rejeunissement de meristemes primaires caulinaires de Pin maritime age (*Pinus pinastev* Sol.), *C. R. Acad. Sci. Paris Ser. D.* 290, 927, 1980.
148. **Corley, R. H. V., Wong, C. Y., and Wooi, K. C.,** Early results from the first oil palm clone trials, in Oil Palm in Agriculture in the Eighties, PORIM Conf., Kuala Lumpur, June 1 to 27, 1981.
149. **Constantin, M. J., Henke, R. R., and Mansur, M. A.,** Effect of activated charcoal on callus growth and shoot organogenesis in tobacco, *In Vitro,* 13, 293, 1977.
150. **Weatherhead, M. A., Burden, J., and Henshaw, G. G.,** Some effects of activated charcoal as an additive to plant tissue culture media, *Z. Pflanzenphysiol.,* 89, 141, 1978.

151. **Hussey, G. and Hepher, A.,** Clonal propagation of sugar beet plant and the formation of polyploids by tissue culture, *Ann. Bot.*, 42, 477, 1978.
152. **Hussey, G.,** In vitro propagation of the onion *Allium cepa* by axillary and adventitious shoot proliferation, *Sci. Hortic.*, 9, 227, 1978.
153. **Debergh, P., Harbaoui, Y., and Lemeur, R.,** Mass propagation of globe artichoke *(Cynara scolymus):* evaluation of different hypotheses to overcome vitrification with special reference to water potential, *Physiol. Plant.*, 53, 181, 1981.
154. **Knauss, J. F. and Miller, J. W.,** A contaminant, *Erwinia carotovora* affecting commerical plant tissue cultures, *In Vitro,* 14, 754, 1978.
155. **Gupta, P. K., Nadgir, A. L., Mascarenhas, A. F., and Jagannathan, V.,** Tissue culture of forest trees, clonal multiplication of *Tectona grandis* L. (Teak) by tissue culture, *Plant Sci. Lett.*, 17, 259, 1980.
156. **Speigel-Roy, P. and Kochba, J.,** Mutation breeding in citrus, in *Induced Mutations in Vegetatively Propagated Plants,* International Atomic Energy Agency, Vienna, 1973, 91.

Chapter 3

PROTOPLASTS IN CROP IMPROVEMENT

I. INTRODUCTION

An isolated protoplast is a plant cell in which the outer wall has been mechanically or enzymatically removed. By appropriate selection of the enzyme mixture it is possible to liberate billions of protoplasts in a few hours. The successful agricultural application of biotechnology in crop improvement depends, to a considerable extent, on plant regeneration from protoplasts. A number of enzymes have become available in the last few years which have wide-ranging wall degradation properties and varying degrees of purity. This has resulted in the isolation of protoplasts from several plants.

Protoplasts provide a most reliable source of individual cells and are the only current means of obtaining somatic hybrid plants. Protoplasts will also be the material of choice for the integration of alien cloned genes into target plants. Thus, the availability of a protoplasts-to-plant system in a given crop is the key consideration for application of cell manipulation in that crop. The progress in this field of endeavor during the last 15 years is impressive. In this chapter the current status of protoplast culture, protoplast fusion techniques available for the improvement of crops, and utilization of protoplasts for the integration of useful genes into plants have been described.

II. METHODOLOGY

A. PROTOPLAST ISOLATION AND REGENERATION

The initial observation by Cocking[1] that plant protoplasts could be released from root tip cells using a fungal cellulase in 0.6 *M* sucrose has sprung the revolutionary and potentially most exciting (yet technically specialized) branch of plant cell culture. Protoplasts have now become the most useful materials for plant cell manipulations. The removal of cell walls, while simultaneously conserving the cytoplasmic and nuclear constituents of the cells necessary for subsequent cell wall deposition and cell division, leaves the plasmalemma membrane fully exposed. This facilitated experiments designed to investigate and manipulate the physical and chemical properties of the membranes and to follow the effect of the endocytosis of foreign particles (like DNA and RNA molecules, intact virus particles, microbes, and organelles) on daughter cells and subsequently regenerated plants.

1. Isolation

Protoplast isolation can be carried out by means of many protocols designed for specific types of materials. A good aseptic approach and aptitude for fine manipulations are always essential. Important components of the isolation procedure for plant protoplasts are the removal of the cell walls without causing irreversible damage to the released protoplasts[3] and the maintenance of a suitable osmotic environment to stabilize the protoplasts. The technique of protoplast isolation has been approached through several different ways. The mechanical method is dependent on preliminary plasmolysis of cell within tissues and the subsequent dissection of the tissue and deplasmolysis to release the preformed protoplasts. However, this method generally produces relatively low yields of viable protoplasts and their behavior in subsequent culture is also affected by the presence of substances released from damaged cells. These techniques have been improved upon by the use of more gentle, less injurious enzymic methods for releasing protoplasts. By 1969, several potent and partially purified cell wall-degrading enzymes were available such as one extracted from *Trichoderma*

TABLE 1
Some General Types of Protoplast Commercial Enzymes Used for Isolation of Protoplasts

Enzyme	Commonly used concentration (%)	Organism or source	Commercial source
Cellulases	1.0—2.5	Basidomycetes	Kinki Yakult Biochemical, Nishinomiya, Japan; Kyowa Hakko Kogyo Co., Japan; Plenum Scientific, Hackensack, NJ
Cellulase RS	1.0	*Trichoderma reesi*	Yakult Honsha Co., Japan; Calibiochem, San Diego, CA
Cellulysin	1—3	*T. reesi*	Calibiochem, San Diego, CA; Yakult Honsha Co., Japan
Hemicellulase	0.2—0.5	*Aspergillus niger*	Sigma Chemical, St. Louis, MO
Hemicellulase HP-150 rhozyme	1—2	*Aspergillus*	Rohm and Hass Co., Philadelphia; Corning Glass, Corning, NY
Meicelase	1—4	*T. reesi*	Meiji Seika Kaisha Ltd., Tokyo
Pectic acid acetyl transferase	0.1—1.0	—	Hoechst, Germany
Pectinase	0.5—1.0	*A. niger*	Sigma Chemical, St. Louis, MO
Pectinase macerase (Macerozyme)	0.5—1.0	*Rhizopus* sp.	Yakult Honsha Co., Japan; Kiniki Yakult Biochemical, Nishinomiya, Japan; Calbiochem, San Diego, CA
Pectinol Ac	0.15	*A. niger*	Corning Glass, Corning, NY
Pectolyase Y-23	0.1	*A. niger*	Kikkoman Shoya Co. Ltd., Japan

viride, a mixture of cellulase and pectinase (Driselase) of basidiomycete origin, and a macerozyme obtained from *Rhizopus* spp. which is rich in pectinase (Table 1). Although partial purification and further desalting of some of these enzymes are sometimes necessary to avoid deleterious effects on viability of certain protoplasts, most of the commercially available enzymes like Pectolyase Y-23, Onozuka R-10, Meicelase, Rhozyme, and Mecerozyme R-10 are desalted and are of sufficiently high quality to give rapid release and high yields of protoplasts from most sources.

Mesophyll cells of leaves and cultured cell suspensions are the two most commonly used materials in experimental work. Their relative usefulness is determined to a large degrcc by the stringency required for genetic uniformity and for morphogenic potential (by either organogenesis or embryogenesis) in the isolated protoplasts. Occasionally, other tissues provide better alternatives, especially with regard to the latter requirement. For example, seedling cotyledons have proved excellent material for isolated totipotent protoplasts in *Brassica oleracea, Datura innoxia,* and *Medicago sativa,*[4] roots of young seedlings have proved convenient sources of totipotent protoplasts of *Brassica* spp.,[5] and flower petals have been efficient sources of protoplasts of ornamental *Nicotiana* spp.[6]

Leaf mesophyll tissues of different genotypes vary in their requirements for protoplast release. In some species such as *Calystegia, Arachis, Asparagus,* and *Ipomoea,* it is possible to obtain a preparation of free cells by mechanical means as a first step to protoplast isolation. Due to the physical structure of the leaves, shearing action in a glass homogenizer is all that is needed to disrupt the tissue into separate cells. After filtering through muslin to remove debris, protoplasts are released using cellulase treatments. Bilkey and Cocking[7] have also used nonenzymatic methods to release particular types of callus-derived protoplasts from tissues cultured on 2,4-dichlorophenoxy acetic acid (2,4-D)-supplemented media. The cells produced in the presence of this auxin have particularly thin cell walls such that by simply teasing cells, large numbers of protoplasts can be released. In other cases, cells are first separated from each other using pectinase and then transferred to a second medium containing cellulase to degrade cell walls for protoplast release. This two-step procedure is not necessary in most cases and present-day methods employ a simultaneous degradation of pectin and cellulose cell wall constituents in a single step. One eventuality which has to be considered in the case of the one-step approach, however, is the possibility that spontaneous fusion of protoplasts caused by fusion of neighboring cells still linked by plasmodesmatal strands may occur. Another factor which should not be overlooked is the fact that the sequential method most often leads to the production of protoplasts predominantly from the palisade layer, whereas the mixed enzyme method produces protoplasts from palisade, spongy, and upper epidermal tissues. This means that protoplasts are often derived from leaf cells with divergent physiological and genetic characteristics, which may have an important bearing on their subsequent behavior in culture and also the genotypes regenerated. Enzyme mixtures used vary with species and materials being employed as sources of protoplasts (Table 2). Examples of various enzyme mixtures in different osmotica used in protoplast osmoticum mixtures at a stable temperature (20°C) overnight for up to 16 h are adequate for protoplast release in most species. Before isolation, leaf tissues of suitable physiological age and condition (preferably obtained from young plants grown under optimal, yet definable and reproducible, environmental conditions in growth cabinets) are prepared. After brief immersion in 70% ethanol, leaves are surface sterilized in a weak (2.5%) sodium hypochlorite solution for 15 to 30 min. After several washes in sterile, distilled water, large amounts of leaf material are dried between layers of sterile tissue paper. Prior to enzyme treatment, the lower epidermis of leaves is removed with fine forceps so that enzymes can penetrate into intercellular spaces. As an alternative to peeling (which is not practicable on small leaves of seedlings or for some types of material obtained from plantlets propagated *in vitro* or on cotyledon material), tissue is "feathered" into 1- to 2-mm strips and enzyme penetration is achieved under vacuum.

Protoplasts obtained from suspension and callus cultures can be isolated without so much preparation, although as Uchimiya and Murashige[46] observed, the stage of growth of cell cultures is an extremely important factor when high yields of protoplasts are required. These workers found that in tobacco batch suspensions, cells at the 4- to 5-d stage of culture in a 14-d subculturing schedule are the most suitable for protoplast isolation and that enzyme mixtures in which Macerozyme R-10 were included at 0.2% w/v gave optimal (30%) release of protoplasts. Protoplasts can also be isolated, albeit at low levels, from immature pollen using lyophilized preparations of Helicase.

A method has been developed for the reuse of cell wall-digesting enzymes to isolate protoplasts from actively growing suspension culture of plant cells by Saxena and King.[47] They suggested that protoplasts could be satisfactorily prepared as many as three times using the same enzyme mixture without any loss in yield or viability of the isolated protoplast.

a. Factors Affecting Protoplast Isolation

i. Leaf Mesophyll Protoplasts

It is possible to isolate protoplasts from virtually any plant tissue, but most frequently mesophyll cells have been used for the isolation of protoplasts from different species. The isolation of protoplasts depends on the concentration of the enzyme osmoticum, and incubation period. Another factor which is important in determining the regeneration of cultured protoplasts is the genotype of the plant itself. Different cultivars of the same species have been reported to show different responses for regeneration of protoplasts. Out of 14 cultivars of *Lycospersicon esculentum* only a few could be regenerated.[48] There are also some individual differences within the cultivars of the same plant for protoplast isolation and regeneration.[49,50]

Various physical and chemical factors have marked influence on the isolation and growth of protoplasts (Table 3). Successful isolation of the protoplasts from mesophyll of *N. tabacum*

TABLE 2
Enzyme Mixtures for the Isolation of Protoplast from Different Plants

Species	Source	Enzyme mixture (w/v)	pH	Osmoticum	Ref.
Beta vulgaris	Cell suspension	1% Driselase 2% Pectinase 4% Cellulase	6.0	0.8 *M* mannitol	8
Brassica juncea	Leaf mesophyll	1% Pectinase 1% Cellulase	5.8	0.6 *M* mannitol	9
B. napus	Hypocotyl	2% Rhozyme HP-150 4% Meicelase 0.3% Macerozyme R-10	5.7	13% mannitol	10
B. rapa	Suspension culture	4% Cellulase 2% Driselase	5.6	0.25 *M* mannitol	11
Citrus aurantium	Callus	0.3% Pectinase 0.2% Cellulase 0.1% Driselase	5.7	0.14 *M* sucrose and 0.56 *M* mannitol	12
Crepis capillaris	Suspension culture	4% Cellulase 2% Driselase	5.6	0.25 *M* mannitol	11
Elaeis guineensis	Suspension culture	1% Driselase	5.6	0.3 *M* glucose	13
Glycine tabacina	Suspension culture	1.6% Cellulysin 0.8% Pectolyase Y-23 0.2% Macerase	5.6	—	14
Hadysarum coronarium	Leaf mesophyll	2% Cellulase 2% Rhozyme 1% Macerozyme	5.5	0.3 *M* mannitol	15
Hyoscymus muticus	Suspension culture	4% Cellulase 2% Driselase	5.6	0.25 *M* mannitol	11
Ipomoea batatas	Callus	3% Cellulase 0.5% Macerozyme 0.3% Pectolyase Y-23	5.5	0.7 *M* mannitol	16
Linum usitatissum	Roots and cotyledons	2% Rhozyme HP-150 4% Meicelase 0.3% Macerozyme	5.6	13% mannitol	17
Lithospermum erythrorhizon	Cell Suspension	0.5% Macerozyme 1.0% Driselase R-10 2.5% Cellulase	—	—	18
Lycopersicon esculentum	Cotyledon	1% Driselase 1% Macerozyme 1% HUP Cellulase	5.7	0.7 *M* sorbitol	19
	Suspension culture	4% Cellulase 2% Driselase	5.6	0.2 *M* mannitol	11
Macleava spp.	Leaf mesophyll	0.2% Pectine-acid transeliminase 0.5% Cellulase	5.6	0.5 *M* mannitol	20
Malus domestica	Leaf mesophyll	0.5% Macerozyme R-10 2% Cellulase Onozura R-10	7.0	9% mannitol	21
Medicago sativa	Cell suspension	0.1% Cellulase 1% Driselase 1% Rhozyme HP-150	5.8	7.2% mannitol	22
Medicago spp.	Leaf suspension	2% Rhozyme 4% Meicelase 0.3% Macerozyme	5.6	13% mannitol	
Nicotiana alata cv. Nicky Red	Flower petal	0.5% Cellulase 0.5% Macerase 0.25% Driselase	5.7	0.38 *M* glucose	6

TABLE 2 (continued)
Enzyme Mixtures for the Isolation of Protoplast from Different Plants

Species	Source	Enzyme mixture (w/v)	pH	Osmoticum	Ref.
N. plumbaginifolia	Haploid suspension culture	2.4% Cellulase R-10 2% Rhozyme HP-150 0.3% Macerozyme R-10	5.8	0.4 *M* mannitol	24
	Callus	0.2% Driselase 0.5% Cellulase	5.6	0.4 *M* sucrose	25
N. sylvestris	Leaf mesophyll	0.1% Cellulase 0.02% Macerozyme	5.8	8% mannitol	26
N. tabacum	Cell suspension	1.0% Cellulase R-10 0.5% Pectinase 0.5% Hemicellulase	5.8	0.35 *M* glucose	27
	Cell suspension	2% Driselase 2% Cellulase 0.5% Macerozyme	—	Seawater	28
	Leaf mesophyll	0.5% Macerozyme R-10	—	0.6 *M* mannitol	29
	Leaf mesophyll	2% Cellulase 0.5% Macerozyme	—	Seawater	28
	Leaf mesophyll	2.0% Cellulase 0.5% Macerozyme	5.2	0.7 *M* mannitol	30
	Epidermis	4.0% Meicelase 0.4% Macerozyme	5.8	0.7 *M* mannitol	31
N. tabacum cv. White Burley	Pollen tetrad	0.75% Helicase	5.6	0.3 *M* sucrose	32
N. tabacum var. Samsun	Stem pith	0.2% Pectine acid tran-seliminase	8.0	0.6 *M* mannitol	33
N. tabacum var. Xanthi	Leaf mesophyll	4% Meicelase 0.4% Macerozyme	5.8	0.71 *M* mannitol	3
Petunia hybrida	Suspension culture	4% Cellulase 2% Driselase	5.6	0.25 *M* mannitol	11
Phaseolus aureus	Roots	2% Rhozyme 4% Meicelase 0.3% Macerozyme	5.6	13% mannitol	5
Populus	Leaf mesophyll	0.5% Cellulase 0.1% Macerase	—	0.55 *M* mannitol	34, 35
Physalis alkenkengi	Shoot buds	1% Cellulase Onozuka R-10 0.2% Macerozyme R-10	5.8	0.6 *M* mannitol	36
P. ixocarpa	Shoot buds	1% Cellulase Onozuka R-10 0.2% Macerozyme R-10	5.8	0.6 *M* mannitol	36
P. minima	Shoot buds	1% Cellulase Onozuka R-10 0.2% Macerozyme R-10	5.8	0.6 *M* mannitol	36
P. peraviana	Shoot buds	1% Cellulase Onozuka R-10 0.2% Macerozyme R-10	5.8	0.6 *M* mannitol	36
P. pruinosa	Shoot buds	1% Cellulase Onozuka R-10 0.2% Macerozyme R-10	5.8	0.6 *M* mannitol	36
Solanum tuberosum cv. Bintje	Axenic shoots	1.5% Cellulase 0.3% Macerozyme	5.5	0.6 *M* mannitol	37
S. melongena	Leaf mesophyll	0.25% Cellulase R-10 0.05% Hemicellulase 0.25% Pectinase	5.6	0.3 *M* mannitol	38
	Leaf mesophyll	0.25% Cellulase 0.2% Driselase 0.1% Macerase	5.6	0.3 *M* mannitol	39

TABLE 2 (continued)
Enzyme Mixtures for the Isolation of Protoplast from Different Plants

Species	Source	Enzyme mixture (w/v)	pH	Osmoticum	Ref.
S. tuberosum	Cell suspension	1% Cellulysin 0.25% Macerase 0.1% Pectolyase	5.6	0.73 *M* mannitol	40
	Leaf mesophyll	1% Cellulysin 0.1% Macerase	5.6	0.4 *M* mannitol	41
Trifolium repens	Leaf epidermis	2% Cellulase R-10 1% Macerozyme R-10 2% Rhozyme HP-150	5.7	15% sucrose	42
Trigonella spp.	Cell suspensions	2% Rhozyme 4% Meicelase 0.3% Macerozyme	5.6	13% mannitol	43
T. corniculata	Leaf mesophyll	2% Rhozyme 4% Meicelase 0.3% Macerozyme	5.6	9% mannitol	44
Viga aconitifolia	Callus	Cellulase Macerozyme	5.8	12% mannitol	45

TABLE 3
Effect of Various Factors on Isolation and Growth of Protoplasts

Species/crop	Factor	Effect	Ref.
Avena sativa	Polyamines	Enhance thymidine incorporation and protoplast stabilization	51
	Pretreatment of leaves with senescence retardants	Cycloheximide, kinetin, L-lysine, L-arginine, putrescine, and cadavarine showed spontaneous lysis of protoplasts and increase in protein synthesis	51
Beta vulgaris	Addition of casein hydrolysate and yeast extract	Stimulation of protoplast division	8
Brassica sp.	Salts	KNO_3 (2000 mg/l) as sole macronutrient with glucose as osmoticum/carbon source eliminated brown particle production	52
B. rapa	Combination of agarose plating and bead culture	Improved plating efficiency of protoplasts	11
Cereals	pH	Protoplasts isolated at low pH (4.6) were less viable than those isolated at pH 5.4	53
Citrus anrantium	Macerozyme	Higher plating efficiency was obtained when Macerozyme rather than Pectinase was included	54—56
Crepis capillaris	Combination of agarose plating	Improved plating efficiency of protoplast	11
Elaeis guineensis	Cell suspensions nurse culture	Increased survival rate and callus formation	13
Gycine max	Osmotic stress	Inhibition of thymidine incorporation into protoplast	57

TABLE 3 (continued)
Effect of Various Factors on Isolation and Growth of Protoplasts

Species/crop	Factor	Effect	Ref.
Haplopappus gracilus	Donor cell suspension growth medium	Increased auxin concentration and decreased sucrose or glucose concentrations, addition of L-cysteine or L-methionine, and reduction of NH_4-NO_3 increased viability and yield of protoplasts, respectively	58
Hedysarum coronarium	Explant source	Only cotyledon protoplast exhibited appreciable plating efficiency	15
Hyoscyamus muticus	Combination of agarose plating and bead culture	Improved plating efficiency of protoplast	11
Ipomoea batatas	Temperature	No division at 22—24°C; optimal 28—30°C	59
I. batatas	Light	Accumulation of anthocyamin	16
Lilium speciosum × *henryi*	Sucrose concentration in cell culture	Callus growth for 3 weeks on sucrose-free medium yielded more stable protoplasts	60
Linum usitatissimum	Explant source	Plant regeneration only from protoplasts derived from root and cotyledons	61
Lithospermum erythrorhizon	Replacement of sucrose with glucose and addition of coconut milk to culture medium	Increased frequency of colony formation	18
Lycopersicon esculentum	Fertilization of the donor plants	Increase in protoplast release by feeding with ammonium, decrease by supplementation with calcium	62
	Salts	Reduced macronutrient concentrations were necessary for sustained divisions	63
	Maintenance of callus culture for a long time	No regeneration into complete plants	19, 64, 65
	Combination of agarose plating and bead culture	Improved plating efficiency of protoplast	11
L. peruvianum	Osmoticum	Replacement of mannitol with glucose enhances plating efficiency	63
	Temperature above 42°C	Severe injury to protoplasts	40
Medicago sativa	Light	Only protoplasts in darkness formed colonies	66
	Leaf age	Protoplasts from youngest, barely expanded leaves divided more quickly than protoplasts from older leaves	49
	Cell suspension	Agarose treatment improved plating efficiency	22
Nicotiana tabacum	Plating density	Below a minimal density the plated protoplast degenerated	67
	Water	Water containing impurities decreased variability of cultures	68
	Carbon source	Less sugar required for protoplasts than for cells; sucrose is essential carbon source; glucose and cellobiose also promoted cell division	69

TABLE 3 (continued)
Effect of Various Factors on Isolation and Growth of Protoplasts

Species/crop	Factor	Effect	Ref.
	Growth regulators	2,4-D, IAA not necessary for cell division; NAA had a promotive effect; auxin not required for cell wall regeneration; cytokinin not necessary but kinetin is better than BA	69
N. tabacum cv. Xanthi	Isopropyl *N*-phenyl carbonate (IPC)	A high proportion of the plants regenerated in the presence of IPC were sterile	67
N. tabacum	Calcium treatment	During winter months, supplementary calcium application to plants was necessary to enhance protoplast stability; spring/summer calcium feedings had no beneficial effects	70—72
	Combination of agarose plating and bead culture	Improved plating efficiency of protoplasts	11
N. rustica	Auxin concentration in cell culture medium	Cell suspension grown with 2,4-D, as auxin gave higher yields of good quality protoplasts than when NAA was used as auxin	71
N. sylvestris	Addition of organic acids, and high cell densities	Better yield and regeneration of protoplasts	70
N. plumbagnifolia	Haploid cell suspension	Protoplasts isolated from cells after 4 years regenerated into plants	24
Oryza sativa	Ammonium	NH_4NO_3 inhibited growth of protoplast calli, whereas NH_4SO_3 was stimulatory	73
	Nurse cell culture	Improved yield of protoplasts and high regeneration	74
Petunia hybrida	Combination of agarose plating and bead culture	Improved plating efficiency of protoplasts	11
Pisum sativum	pH, osmoticum	pH 6.2, more dividing protoplasts than pH 5.5; L-glutamine (2—5 m*M*) improved protoplast survival; optimum osmoticum was 0.4 *M* for mannitol or sorbitol, 0.3 *M* for glucose	75
Populus sp.	Elimination of ammonium, agar, exudate buildup	Better development of protoplasts	34, 35
Rosa sp.	Cell density	Optimum 1—2 × 10^4 protoplasts per milliliter	76
	Growth phase of cell culture	Accelerated subculture schedule prior to protoplast isolating increased protoplast viability	76
Solanum tuberosum	Leaf (mesophyll) photoperiod for donor plant	Short photoperiods (dim light) gave consistently high protoplast yield	77
	Temperature	Higher temperature and higher light intensity resulted in lower yield of protoplasts	78

TABLE 3 (continued)
Effect of Various Factors on Isolation and Growth of Protoplasts

Species/crop	Factor	Effect	Ref.
	Low temperature pretreatment	Leaves preincubated at 4°C in salt solution yielded more stable protoplasts	77
	Activated charcoal	Culture ability of protoplasts increased	41
	Temperature above 42°C	Severe injury to protoplasts	40
S. melongena	Light intensity and time	Plants were regenerated only at 1500 lx after 7—10 d	39
Spinacia oleracea	Growth phase	Cells only in rapidly dividing stage grew	79
Stylosanthes guianensis	Growth phase	Logarithmic growth phase was optimal for protoplast isolation and culture	80
Trigonella corniculata	MS medium with 2.0 mg/l NAA and 0.5 mg/l BAP	Protoplast formed somatic embroids at high frequency	44
Trifolium repens	2.0 mg/l 2,4-D and 1.0 mg/l zeatin	Best callus growth	42
Triticum aestivum	Temperature	Culturing temperature optima 22°C	80
Triticum sp.	Carbon source/osmotium	Glucose is the best source, budding occurs if osmotic strength is too low	73
Vicia hajastana	Density	Protoplasts minimal media grew at 5 × 10^3 ml; with supplemented medium protoplasts could grow as low as 1—2 ml	81
	Enriched medium	Protoplasts cultured at low densities required amino acids, TCA organic acids, nucleic acid basis, sugars, and sugar alcohol for growth	81

plants at 22°C required 15 h daylight (10,000 to 20,000 lx) with weelky feedings of high nitrogen fertilizer.[68] In addition, the age of the plants is also crucial, with plants older or younger than 40 to 60 d being unsatisfactory for protoplast isolation. Cells taken from plants which had flowered were responsible for poor protoplast release. Maintenance of young plants of alfalfa at 21°C, 12,000 lx (12 h), and of fertilizers applied daily before using the donor plant for protoplasting was reported to be suitable for protoplast isolation.[49] Protoplasts obtained from older alfalfa leaves were less viable than those isolated from the youngest leaves. Plants also required high concentrations of enzymes. In addition, it was found that protoplast viability was increased considerably in leaflets (with the lower epidermis removed) precultured for 36 to 48 h on cell culture medium prior to protoplast isolation. The preculturing helped to reduce the environmental shock to the donor tissue that otherwise would have occurred during protoplast isolation. Plants of *Solanum tuberosum* were raised from tubers and required environmentally controlled growth rooms under high light intensity (15,000 lx, 12 h) for successful protoplasting.[77] These plants were further grown for 4 to 10 d at lower intensity (7000 lx, 6 h) before using them for protoplast isolation. Raising temperature to 24°C, 70 to 75% relative humidity, and daily application of fertilizer were also cited as important factors to maximize plant growth.

Protoplast isolation from shoots grown *in vitro* and acclimated to culture conditions has been found to be much easier.[82,83] Donor material is already sterile and therefore sterilization procedures which could damage leaf tissue are not necessary.

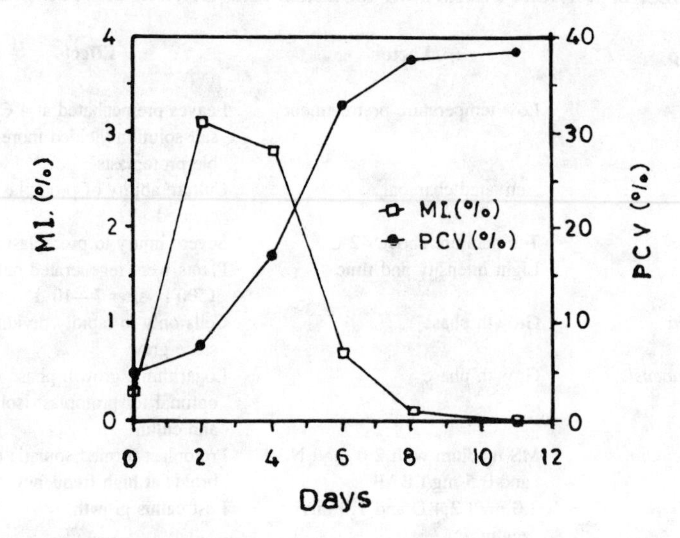

FIGURE 1. Growth of *Stylosanthes guianensis* CIA 136 cell suspension for protoplast isolation and fusion experiments. M1: mitotic index = no. mitotic cells/no. total cells × 100; PCV: packed cell volume = volume of sedimented cells/total culture volume × 100; days: cultures period. (From Roca, W. M., Szabados, L., and Hussain, A., Annual Report, Biotechnology Research Unit, CIAT, Colombia, 1987. With permission.)

ii. Cell Culture Protoplasts

Protoplasts from cell suspension and callus cultures can also be isolated without much preparation. The success of protoplasting is dependent on the stage of growth of the culture. Cells taken from late logarithmic growth phase are best suited for protoplast isolation (Figure 1). Substantial reduction in protoplast yield has been reported if cultures were used more than 3 d after subculturing in *Haplopappus gracilis*.[58] Increased auxin concentrations and reduced sucrose concentrations 1 d before protoplast isolation increased the yield of protoplast. It was suggested that protoplast release was optimal in cells cultured on nitrogen sources (such as ammonium sulfate medium) for alfalfa, flax, rice, soybean, tobacco, and wheat.[84] The growth of cell cultures was optimal in potassium nitrate medium and poorest in ammonium sulfate medium.[85] Glutamine medium was good for soybean, while arginine medium resulted in high protoplast release from cultures of rice.

iii. Enzymes

Several enzymes for isolation of protoplasts are available (Table 1). Enzymes or enzyme mixtures used vary with species and material being employed as a source of protoplasts (Table 2). These enzymes singly or in mixtures degrade cell components, primarily cellulose, hemicellulose, and pectin. Cellulase RS, a highly purified product, is the strongest commercial cellulase and is useful for digesting the thick cell walls of cultured cells. However, Meicelase is a weaker enzyme preparation that will digest fragile, mesophyll cells without excess damage. Most frequently, a cellulase is combined with a weak pectinase to achieve single-step isolation of protoplasts. Macerase, a weak enzyme, is commonly used along with pectinase; however, Pectolyase Y-23 has been suggested as an alternative for difficult species. Pectolyase, which has high pectin lyase and polygalacturonase activities, has been reported to release tobacco leaf protoplasts in 25 min when combined with cellulysin.[30]

Commercial preparations invariably contain toxic substances like nucleases and impurities which may damage protoplasts. Several laboratories, therefore, desalt enzyme prior to use. However, high purity cellulases have been reported less efficient than partially purified preparations in isolated protoplasts, indicating that in addition to cellulase, some additional undefined activity is necessary for protoplast isolation.

iv. Osmoticum

Removal of the cell wall during protoplasting takes away the outer support of the cell, and the osmotic strength of the isolation medium must therefore be balanced. Without this balance the protoplasts can plasmolyse and burst. Further, slight plasmolysis of the protoplasts is necessary during isolation, as excess osmotic pressure has been reported to impair metabolism[86] and cell wall regeneration. The regulation of the osmotic pressure of the medium is generally done by the addition of sugars or sugar alcohols. Mannitol and sorbitol are most frequently used, with mannitol preferred for isolation of leaf mesophyll protoplasts. Glucose is often used as osmoticum for cultured cells.[87] The osmoticum for a protoplast culture medium may be metabolically active (e.g., glucose or sucrose) or inert (e.g., mannitol). Active substances are metabolized, thereby gradually lowering the osmotic strength of the medium.

Different osmotic strengths, depending on the species from where protoplasts are obtained, have been used (Table 2). For mesophyll protoplasts, the environmental conditions under which donor material is grown influence the osmotic concentration needed for successful isolation of protoplasts.[77]

v. Miscellaneous Isolation Conditions

Various other conditions such as dark or low light intensity, pH, etc. have also been reported to affect the protoplast isolation.[85] Gentle shaking is used for cell culture protoplasts.[81] Appropriate temperature during isolation of protoplasts and stationary incubation is also needed for fragile mesophyll protoplasts.[88] The temperature during isolation of protoplast has been varied over a wide range. For example, Vasil and Vasil[89] recommended 14°C for corn protoplasts, while 27°C was optimal for tomato protoplasts.[48] The minimum requirements for a protoplast isolation medium are a salt mixture, an osmoticum, and cell wall-degrading enzymes. By manipulating the above-mentioned conditions, viable protoplasts can generally be isolated. Dissolving the enzyme directly into the protoplast culture medium[81] avoids the shock of transfer of protoplasts from isolation solution to the culture medium.

Dai et al.[90] reported improved procedure for the isolation and regeneration of potato protoplasts. One of the most important steps in the isolation and culture of the protoplasts was maintaining the osmotic potential of the enzyme solution at or near the isotonic level of the enzyme tissue during the digestion of cell wall. This culture procedure was reported to be less complex and more expedient than the multiple media changes reported by Haberlach et al.[91] Protoplasts of several spring and winter varieties of *B. napus* were isolated from hypocotyl tissue and were found to divide at a high frequency without browning in modified Shepard's medium.[92] This high efficiency of proliferation was sustained through plant regeneration with all varieties.

Until recently, tomato protoplasts were regarded as recalcitrant in tissue culture because of the lack of efficient and reproducible culture methods. Tan et al.[93] described several conditions including preconditioning of the donor plants and a two-step regeneration procedure which resulted in a rapid and reproducible method for regeneration of tomato cultivars. The procedures were reported to be better as compared to those reported previously by several workers.[90,94]

2. Protoplast Purification

Enzymatic isolation of protoplasts from leaf or cultured cells is followed by purification

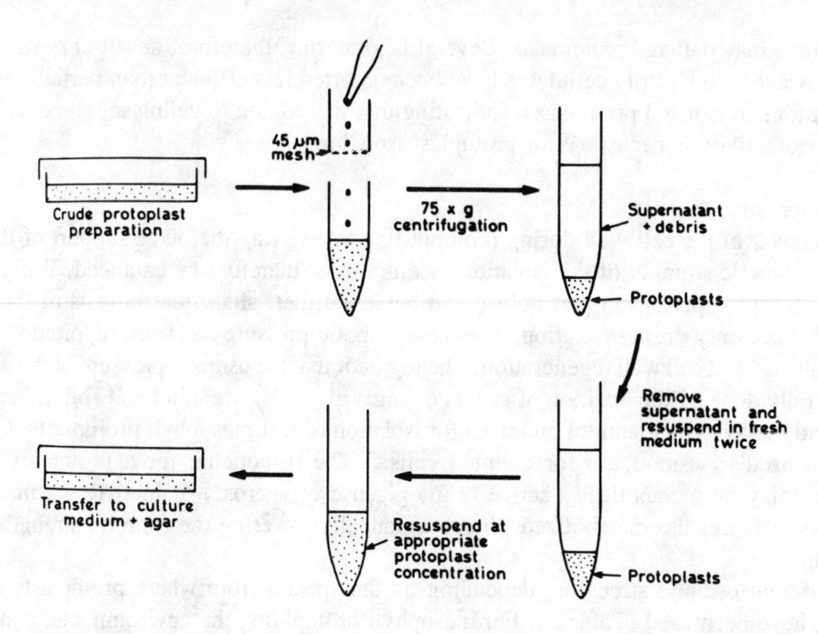

FIGURE 2. Diagrammatic scheme for purification of crude protoplast preparation. (From Dodds, J. H., *Plant Genetic Engineering*, Dodds, J. H., Ed., Cambridge University Press, London, 1985, 5. With permission.)

of the protoplast. This is mainly done by filteration-centrifugation and floatation. During filteration-centrifugation the crude protoplast preparation, consisting of protoplast, enzyme, and debris, is passed through a filter, usually 40 to 100 µm, which retains undigested cell clumps and large debris. The remaining mixture of protoplast, small debris, and enzymes is then centrifuged to precipitate protoplasts while debris continues to float (Figure 2). The enzyme is removed and fresh medium is added before centrifugation. The protoplasts are washed in this way to remove the enzyme and then are placed in protoplast culture medium. This method is very effective for purifying protoplasts because the same osmoticum is used throughout the purification (Figure 3).

The filteration-centrifugation step is often too rough for delicate protoplasts. Therefore, an alternative (flotation-purification method) to the filteration-centrifugation method has been suggested.[27] Protplasts have a lower density than organelles or cell wall fragments. A concentrated solution of sucrose or sorbitol is mixed with the crude protoplast prparation by using the appropriate speed of centrifugation, pure protoplast float, and debris sediment. Concentrations of sucrose used for floatation have varied from 0.3[77] to 0.6 *M*.[95] Likewise, centrifugation speeds have ranged from 40 to 80, to 350 × *g*.[96] This method may be preferable for protoplasts which will not withstand passage through a filter; however, the protoplasts must be able to tolerate centrifugation in concentrated sucrose solution (culture medium and growth regulators).

Protoplast culture media have a profound effect on protoplasting (Table 3). B5[27] and MS[97] cell culture media are the bases of two most common protoplast culture media, 8p[81] and NT,[30] respectively. The retention of same osmoticum and high calcium used in the protoplast isolation medium is preferred. Sucrose, glucose, mannitol, and sorbitol are most commonly used as osmoticum. Many components of these media have been varied to achieve optimum protoplast growth. Carbon source, in particular, has been varied extensively. Uchimiya and Murashige[69] reported that tobacco protoplasts grew equally well on sucrose, cellobiose, or glucose. However, sucrose is preferred for brome grass,[98] while sucrose and glucose are mixed (2:1) for tomato protoplasts.[48]

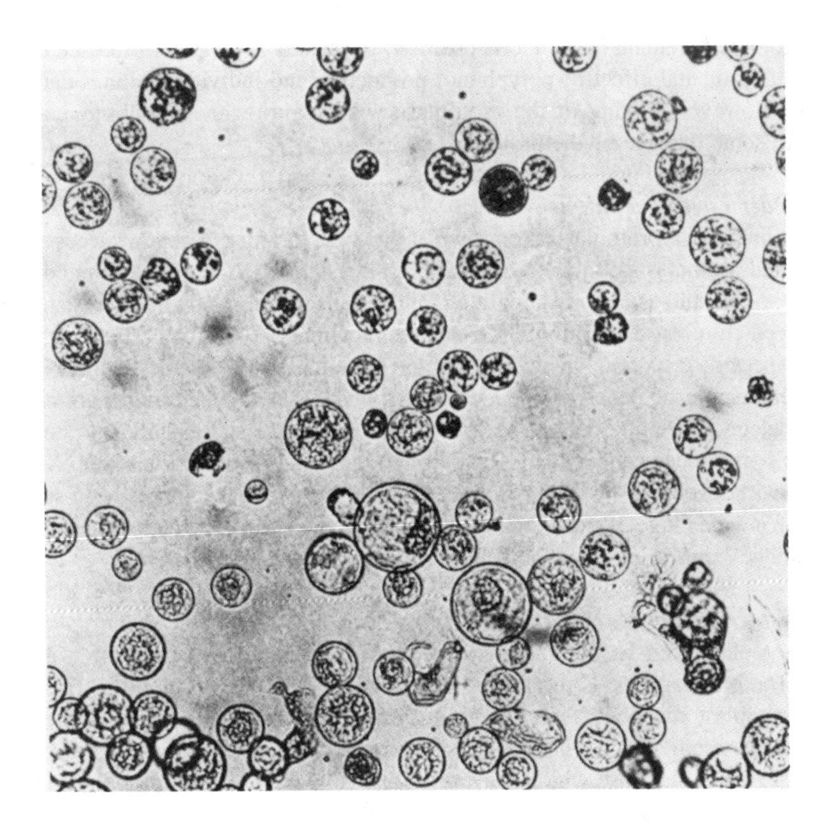

FIGURE 3. Isolated potato mesophyll protoplasts. (From Dodds, J. H., *Plant Genetic Engineering*, Dodds, J. H., Ed., Cambridge University Press, London, 1985, 5. With permission.)

In protoplast culture media, growth regulators have been varied extensively. 2,4-D is the most commonly used auxin. However, Uchimiya and Murashige[69] observed a higher plating efficiency when naphthylene-1 acetic acid (NAA) was used instead of 2,4-D or indole-3 acetic acid (IAA). In general, each species has different growth-regulator requirements. There are also conflicting reports on the requirement for a cytokinin in protoplast culture media. For instance, in tobacco, cytokinin is apparently not necessary to initiate cell divisions, but von Arnold and Eriksson[99] reported 2,4-D and 2-isopentyl adenine (2ip) together for cell division in pea mesophyll protoplasts. In addition, pH of the medium, though studied in less detail, also appears to vary in protoplast culture media.[100,101]

3. Culture of Protoplasts

Several options for culturing the purified protoplast are available depending upon the source and condition of the protoplasts. The simplest method to culture protoplast is liquid droplet culture.[98] Protoplasts are placed with a Pasteur pipette in 0.1- to 0.2-ml droplets into sterile plastic petri dishes and the plates are then sealed with parafilm to avoid desiccation and contamination. Though this method has the advantage that fresh medium can be easily added in known concentrations, the protoplasts aggregate to the center of each drop. This could prove to be harmful if protoplasts produce large amounts of phenolics or other compounds which harm neighboring protoplasts.

Another commonly used technique is the agar culture method. Protoplasts are mixed with agar (0.6%) either before[102] or after cell walls have been regenerated. Small volumes (1 to 2 ml) of agar mixture are then poured into plates and sealed with parafilm. In this method, protoplasts remain fixed in one position, physically separated from each other. This lessens the detrimental effect of polyphenol production and individual clones can be easily monitored. However, mixing of the protoplasts with warm agar and subsequent agitation may lead to some damage of the protoplasts.

a. The Feeder Layer Technique

A large number of protoplasts are grown at one time in order to obtain successful growth of the isolated protoplasts. Below a minimal cell density which is often around 10^4/ml, the cells do not divide and usually disintegrate rapidly. This phenomenon interferes significantly with plant cell cultures, especially in those cases in which only a small number of surviving cells are expected, e.g., mutagenized cells and protoplasts of somatic fusion products. A similar phenomenon of density dependence was observed with cultured mammalian cells before its detection in plant cells. The former was overcome successfully by the use of an underlay of X-irradiated cells.[103] Fifteen years ago, a similar technique based on X-irradiated protoplasts was developed by Galun and associates.[104-106] Since then it has been successfully employed in a wide range of experiments. This technique is commonly known as the "feeder layer technique" and it involves the following steps.

i. Protoplast Irradiation

The protoplasts can be X-irradiated by either a roentgen apparatus or a gamma-ray source.[107] The optimal dose is the lowest dose which "completely" inhibits (i.e., by over 99.99%) protoplast division. This dose may differ from one species to another, but the source of tissue from which the protoplasts were derived has a far greater effect on the optimal dose. Thus, the dose which prevents division of *N. tabacum* mesophyll protoplasts is about 5 krad and is similar to the dose required for other Solanaceae mesophyll protoplasts.[107] In contrast, the dose required to prevent mitosis in cell suspension-derived *N. tabacum* protoplasts is 20 krad[107] and is similar to the dose required for other cell suspension or callus-derived protoplasts, e.g., *Citrus* protoplasts.[105,108] The irradiated protoplasts should be washed once or twice. This probably removes toxic-free radicals. For simplicity, the protoplasts can be irradiated while still in the enzyme solution, just before collection, and thereafter washed as usual.

ii. Plating of Irradiated Protoplasts

The first step involves the mixing of irradiated protoplasts in agar medium together with the unirradiated protoplasts. The unirradiated protoplasts are referred to as the target protoplasts. In the second step the irradiated protoplasts are mixed in agar medium and plated in petri plates as an underlay; upon solidification, the target protoplasts are plated on top as an overlay in a softer agar medium. In the third step irradiated protoplasts are plated in liquid medium to which the target protoplasts are added.

In the first report of the use of a feeder layer,[104] it was noted that the closer the contact between irradiated and target protoplasts, the better the recovery of the latter. Therefore, steps 1 and 3 have a certain advantage over step 2. On the other hand, in step 2 the feeder plates can be prepared and stored in a cool place ahead of time (up to at least 1 week before use). Step 3 is the procedure of choice if the target protoplasts should preferably be cultured in liquid medium. In that case the contact is maximal and the irradiated protoplasts can be prepared ahead of time.[107]

The plating density of the irradiated protoplasts is crucial as shown by Raveh et al.[104] If their density is too high, they have an inhibitory effect on the target protoplasts; if their density is too low, they do not have a feeding effect. Their optimal concentration is equivalent to the lowest concentration in which the same protoplasts would divide satisfactorily when not irradiated.

Irradiated feeder layers were very useful in the recovery of small numbers of somatic fusion calli.[109,110] They were also used successfully in promoting divisions in otherwise poorly dividing protoplast systems. Thus, protoplast division and colony formation in *S. chacoense* and *S. tuberosum* were substantially improved by plating over a *Nicotiana* feeder layer.[111] This method is also routinely used in *Citrus* protoplast work in which feeder layers of one species efficiently promote protoplast division of other species.[107] It could be used for mutant isolation, especially when a feeder layer is prepared from a resistant species and the target protoplasts of another species are screened for the same resistance.

4. Cell Wall Regeneration, Cell Division, Organogenesis, and Whole Plant Regeneration

Isolated protoplasts begin to develop a new cell wall within a few hours.[2,31,96,112] This wall formation can be detected by ultraviolet microscopy and the use of calcaflor stain. It takes only a few days for the normal wall formation. Under optimum conditions, the cells begin to divide and eventually give rise to callus colonies. It is from these callus colonies that new intact plants are regenerated.

The ability to induce organ formation on callus clumps by altering the composition of the culture medium is one of the classic studies in plant tissue culture. By manipulating the culture medium it is possible to induce shoot formation or protoplast-derived callus.[27,106] The regenerated shoots can then be excised and rooted and transferred to sterile soil, thus returning to the normal whole plant state (Figure 4). The stage of rooting and transfer to soil is extremely delicate; the plant is moving from a highly protected sterile environment to the nonsterile and more harsh soil environment. After root induction, the small plantlets should be moved to sterile pots and maintained in a growth room with controlled humidity. The *in vitro* plantlet has a very underdeveloped cuticle and needs the humidity control to prevent water-loss wilting. The humidity should be gradually reduced to slowly harden the plants until they are ready to be transferred to pots and later to the field.

B. PROTOPLAST FUSIONS AND REGENERATION OF SOMATIC HYBRIDS

Earlier methods of transfer of genes into cultivated crops were mainly limited to interspecific hybrids. Several plant varieties have been produced and released. Important among them are tomato, tobacco, barley, potato, wheat, etc. *S. dismissum* has been used to transfer disease resistance into cultivated potatoes.[113] Now a variety of tobacco is available which has resistance to diseases. This variety has been derived from three different wild *Nicotiana* species.[114] Despite the enormous benefits which have been obtained through interspecific hybridization in crop improvement, its limitations led to the discovery of new technologies to create genetic variations and to transfer genes. Protoplast fusion is one of such techniques which has offered a unique opportunity to introduce variation and to transfer genes between different varieties or species where interspecific hybridization fails. In this chapter a brief account of the experimental and methodological procedures leading to the formation of somatic hybrids, their regeneration into complete plants, and subsequent application of protoplast fusion in crop improvement is discussed.

1. Fusion and Isolation

Isolation, culture, and fusion of protoplasts and the techniques used for this purpose are now well established.[27] For the production of somatic hybrids, the fusion between leaf-derived protoplasts of one species with cell culture-derived protoplasts of the second species has become very common. Polyethylene glycol (PEG) is used quite frequently to induce protoplast fusion and its use was first reported by Kao and Michayluk.[87] However, PEG, when added at 20 to 30%, causes agglutination of protoplasts. A high frequency of protoplast fusion is obtained when PEG diluted at high pH in the presence of calcium is added. This

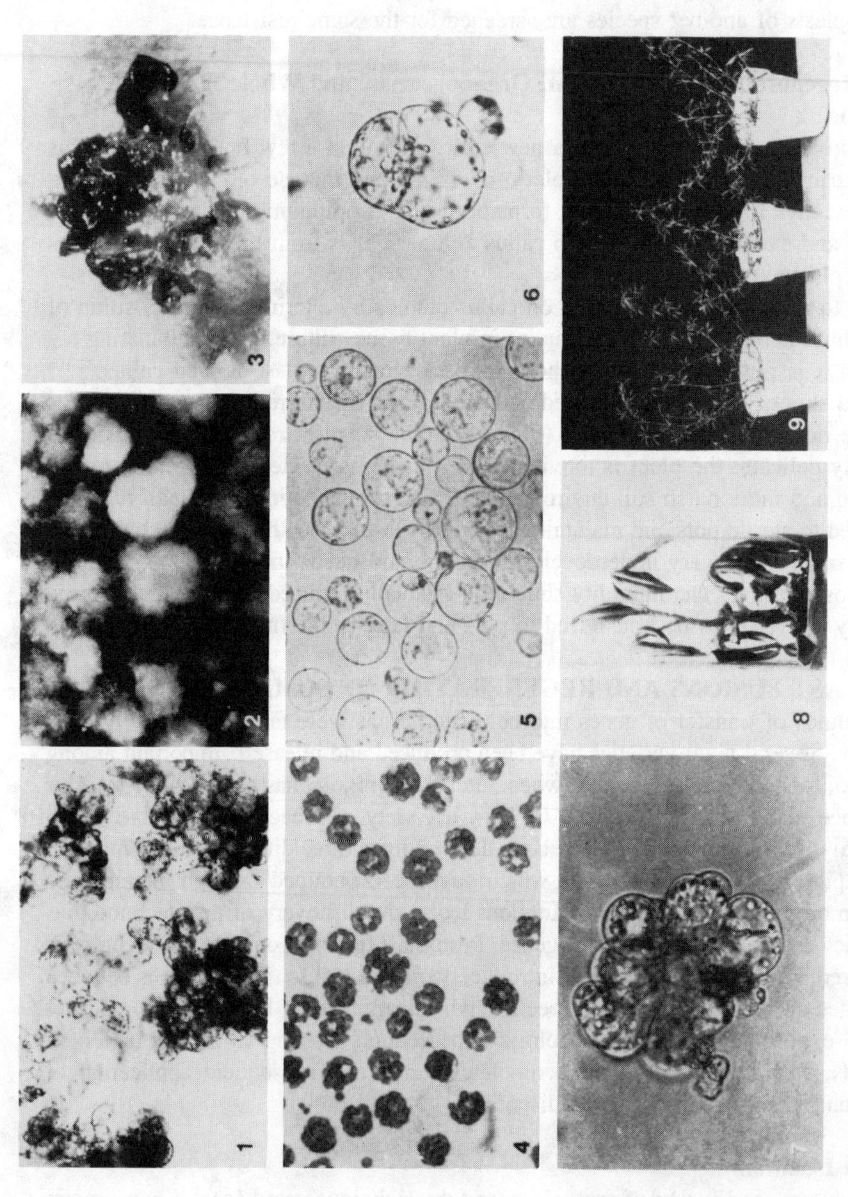

FIGURE 4. (1) Cell suspension culture of *Stylosanthes guianensis* var. 2243 (magnification × 200); (2) cell suspension culture of var. 136 with compact, proembryo-like colonies (magnification × 100); (3) regenerating colonies from cell suspensions of var. 136 (magnification × 65); (4) protoplasts from leaf mesophyll of var. 2243 (magnification × 400); (5) protoplasts from cell suspension of var. 2243 (magnification × 400); (6) first cell division in an isolated mesophyll, protoplasts of var. 2243 (magnification × 540); (7) colony formation from mesophyll protoplasts of var. 2243 (magnification × 540); (8) shoot formation in a protoplast-derived callus of var. 136; (9) potted plants regenerated from mesophyll protoplasts of var. 2243. (From Szabados, L. and Roca, W. M., *Plant Cell Rep.*, 5, 174, 1986. With permission.)

has even resulted in 100% fusion in certain cases.[115] The mechanisms of action of PEG are quite complex. It has been suggested that PEG acts as a molecular bridge, dissociating the plasmalemma, resulting in the intercellular connections. Immediately after fusion, protoplasts that are cultured in appropriate medium regenerate cell walls and undergo mitosis. As the fusion is not 100% and it is a random process, it results in the formation of a mixed population of parental cells, homokaryotic fusion products, and heterokaryotic fusion products of hybrids. The next step, therefore, is the separation of hybrid cells and regenerated hybrids.

a. Complementation Methods

Several methods are used for this purpose. The most common method of identification is based on the fact that hybrid cells display genetic complementation for recessive mutations and physiological complementation for *in vitro* growth requirements. The complementation method for the successful separation and isolation of auxin autotrophic somatic hybrids was used for the first time by Carlson et al.[116] following fusion of two *Nicotiana* species, each with an auxin requirement for cell growth on auxin-free culture medium.

The most frequently used method, the albino gene complementation method (to isolate somatic hybrids following fusion of two distinct homozygous recessive albino mutants of *N. tabacum*), was used by Melchers and Labib.[117] In this method a population of protoplasts isolated from a genetically recessive albino is fused with either a population of protoplasts isolated from a second nonallelic albino mutant or with a population of normal green mesophyll protoplasts. Douglas et al.[71] isolated green interspecific hybrid shoots following fusion of chlorotic *N. rustica* protoplasts with *N. tabacum* protoplasts. Similarly, Schieder[83] fused protoplasts of two diploid homozygous albino mutants of *D. innoxia* A1/5a and A7/1S, that had been induced by X-ray treatment.

It is not necessary to use two albino mutants to recover somatic hybrid shoots in many cases. A single recessive albino mutation is enough in the isolation of somatic hybrid plants when used in combination with a morphological trait, a biochemical mutant, or a growth response. For example, a culture medium can be selected which favors both regeneration of the albino species and prohibits regeneration of the green species. As the green species cannot regenerate, all green calli or shoots that are recovered represent putative somatic hybrids. Interspecific somatic hybrids of *Datura, Daucus, Nicotiana,* and *Petunia,* as well as several intergeneric somatic hybrids, have been recovered by using this combination of genetic and physiological complementation of albino mutant. Morphological markers have also been used in combination with an albino mutation to distinguish putative somatic hybrid plants derived from wild-type parental protoplasts. Dudits et al.[118,119] fused albino *D. carota* protoplasts with wild-type *D. capillifolius.* As both *D. capillifolius* and the hybrid protoplasts were capable of regeneration, isolation of green shoots was insufficient to distinguish these two lines. However, the morphology of the leaves in the hybrid plants more closely resembled *D. carota* leaves, which were used to distingush between these two lines.

Modification of these selection procedures to recover several interspecific somatic hybrids in the genus *Nicotiana* has been used by Evans et al.[120] By utilizing a semidominant albino mutation (Su/Su, *N. tabacum*) in one parent, each population of protoplasts can be uniquely identified when shoots are regenerated. The mesophyll protoplasts of wild *Nicotiana* species produce only dark-green shoots,[121] while albino protoplasts produce only albino shoots when regenerated.[27] The protoplast fusion products that contain a mixture of green and albino genetic information have been visually distinguished from the parental regenerates as light-green shoots.[122]

The induction of dominant resistant and recessive auxotrophic mutants induced *in vitro* have also been used to isolate somatic hybrids. Amino acid analog resistance mutants[123] and two nitrate reductase-deficient mutants[124] have also been proposed for the recovery of mature hybrid plants. On the other hand, some variants have been sucessfully used to recover somatic hybrid plants. Maliga et al.[125] used a kanamycin-resistant variant of *N. sylvestris*, KR 103, isolated from cultured cells, as a genetic marker to recover fusion products between *N. sylvestris* and *N. knightiana*. The streptomycin-resistant (SRI) mutation is encoded in cytoplasmic DNA and those somatic hybrids that contained *N. tabacum* chloroplast DNA expressed streptomycin resistance. The SRI, mutant of *N. tabacum*, isolated from cultured cells, was used to recover (1) intraspecific hybrids with *N. tabacum*,[126] *N. sylvestris*,[127] and *N. knightiana*.[128] Isolation of cyclohexamide-resistant plants of carrot from cultured cells made it possible to identify somatic hybrids between resistant and albino lines of *D. carota*. The somatic hybrid plants were both cycloheximide resistant and green. Similarly, Kameya et al.[129] used cell line C 123 of *D. carota* that simultaneously expressed 5-methyltryptophan (5 MT) and azetidine-2-carboxylate (A 2C) resistance to identify interspecific hybrids between *D. carota* and *D. capillifolius*. Selection for hybrid cells was based on the resistance to 5 MT. The callus reinitiated from somatic hybrid plants expressed complete resistance to 2 AC, but only intermediate resistance to 5 MT.

The complementation method has not been used on a wider scale in higher plants. However, this method has been successfully used to isolate somatic hybrids in *Sphaerocarpus*[130] and *Physcomitrella*.[131] Thus, there is every possibility of this method becoming applicable to higher plants. The major limitation is, however, the paucity of higher plant auxotrophs. In the first such report, Glimelius et al.[124] fused the two different types of nitrate reductase-deficient mutants of *N. tabacum* (nia and cnx). Neither mutant line could be grown with nitrate as sole nitrogen source, while hybrids could regenerate shoots in the nitrate media.

The genetic basis of protoplast regeneration has not been studied in detail. However, there are indications from fusion experiments that the ability to regenerate plants from protoplast behaves as a dominant trait in most of the cell hybrids. It is interesting to note that even in the absence of regeneration of any of the parental lines, the hybrid line is capable of regeneration in most of the hybridization experiments. This has led to the development of culture media that permit growth of hybrid cells while inhibiting the growth of at least on parental cell line. For example, Maliga et al.[125] fused KR 103, the kanamycin-resistant line of *N. sylvestris* that is incapable of plant regeneration, with *N. knightiana* protoplasts, also incapable of regeneration, and were able to recover interspecific somatic hybrid plants. Thus, in some cases, hybrid cells have been produced that are capable of regeneration while neither parent line can be regenerated.

Another method to recover somatic hybrids is metabolic complementation. In this method parental cells are first treated with an irreversible biochemical inhibitor, such as iodoacetate or diethylpyrocarbonate. After the treatment and fusion, only hybrid cells are capable of regeneration. Pretreatment with iodoacetate has been used to recover somatic hybrids between *N. sylvestris* and *N. tabacum*[127] and *N. plumbaginifolia* and *N. tabacum*.[132] In these cases, though the hybrids were able to regenerate, neither of the parent lines treated with iodoacetate was capable of regeneration.

b. Visual Selection

This method has been used to isolate several hybrids from the parental lines. Visual identification of hybrid cells is, however, the most tedious method. This generally involves mechanical isolation of hybrid cells which have some visual differences from the parental cell line. The growth pattern of the hybrid callus is often more vigorous than parental callus. Schieder[133-135] has suggested that all interspecific *Datura* hybrids have much better callus growth than either parental line. Similarly, *N. glauca* + *N. langsdorffii* somatic hybrids could be preselected based on superior growth. For example, fusion products between green chloroplasts and colorless protoplasts that contain distinct starch granules due to growth on sucrose-supplemented medium can be easily distinguished immediately after fusion is over.[85] Soon after PEG treatment, the fusion products contain starch granules in one half of the cell and chloroplasts in the other half. Thereafter, chloroplast diffuses out evenly in the cell and during first cell division the chloroplasts are clumped around the nuclear material in many hybrids. The chloroplasts become colorless, thereafter making it difficult to distinguish between parental cell lines and hybrids. Similarly, petal + leaf fusion products and petal + cell culture fusion products can also be readily distinguished.[136] The petal pigment, usually vacuolar, is originally separated within the fused cell, but eventually becomes evenly distributed through the fused cell. In some cases, this new mixture of protoplast contents produces cells with unique coloration.[6] Shortly after transfer to protoplast culture medium, the color of the flower petals (usually vacuolar) diffuses, and as with leaf + cell culture fusion products it can be used as a cell marker for a few days after fusion.

Meadows and Potrykus[137] described a method for a staining nuclei with Hoechst 33258 in unfixed protplasts. This use of vital fluorescent probes with plant protoplasts may be helpful for the identification, selection, and subsequent culture of hybrid fusion products in an automatic cell-sorting system.

Using visual markers and mechanical separation, Kao and associates were able to isolate and monitor the growth of several individual heterokaryocytes.[88,138] Using this method, they were able to isolate and then monitor the growth of intergeneric and interfamilial cell hybrids. Kao and Michayluk[81] developed a protoplast medium for successful culture of these isolated hybrids even when they were present at low densities. A modification of the microisolation method in which individual heterokaryocytes were cultured in very small volumes, thereby resulting in an effective density of 2.4×10^3 cells per milliliter, was suggested by Gleba and Hoffmann.[139] This method resulted in a very high plating efficiency and was used to recover complete plants. A combination of this microisolation method with nurse culture techniques was used to develop a still better procedure to recover somatic hybrid plants.[128] Menczel et al.[128] placed a single microisolated heterokaryocyte in a droplet containing albino protoplasts capable of rapid growth. The albino protoplasts supported the growth of the single isolated heterokaryocyte which was subsequently identified as a regenerated green shoot among the albino shoots.

These methods, though difficult and tedious, have the ability to monitor the cell and plant development of individual clones, which appears to make them more popular. Cell suspension protoplasts of *Stylosanthes guianensis* (CIAT 136) were fused with leaf mesophyll protoplasts of *S. capitata* (CIAT 1019) or with leaf mesophyll protoplasts of *S. macrocephala* (CIAT 2286).[80] Protoplasts were fused with a standard PEG containing high Ca^{+2} and high pH treatment. Fused protoplasts were easy to distinguish due to the presence of visible cytoplasmic markers characteristic in each of the parental line protoplasts: leaf mesophyll protoplasts have green chloroplasts, while cell suspension protoplasts have more dense cytoplasm and usually starch grains (Figure 5). O'Connell and Hanson[65] reported a selection and screening method for the identification of a number of somatic hybrid callus clones following fusion of *L. esculentum* protoplasts and *L. pennellii* suspension culture protoplasts. Visual selection for callus morphology combined with high fusion frequency and irradiation of one parental protoplast type resulted in selection of a callus clone population containing a high proportion of somatic hybrids. Somatic hybrid plants were regenerated following calcium-high pH fusion of the unidirectional sexually imcompatible cross of *Petunia parodii* wild-type leaf mesophyll protoplasts with protoplasts from a cytoplasmic-determined chlorophyll deficient mutant of *P. inflata*.[140] Genetic complementation to chlorophyll synthesis and sustained growth in selective media were used to visually identify the hybrid callus.

c. Microisolation and Fluorescent Cell Sorting

Several physical methods to separate somatic hybrid protoplasts from mixtures of hybrid and parental protoplasts are in use nowadays. Despite the wide range of methods available, the simplest and more reliable method for physical isolation of heterokaryocytes is micro-

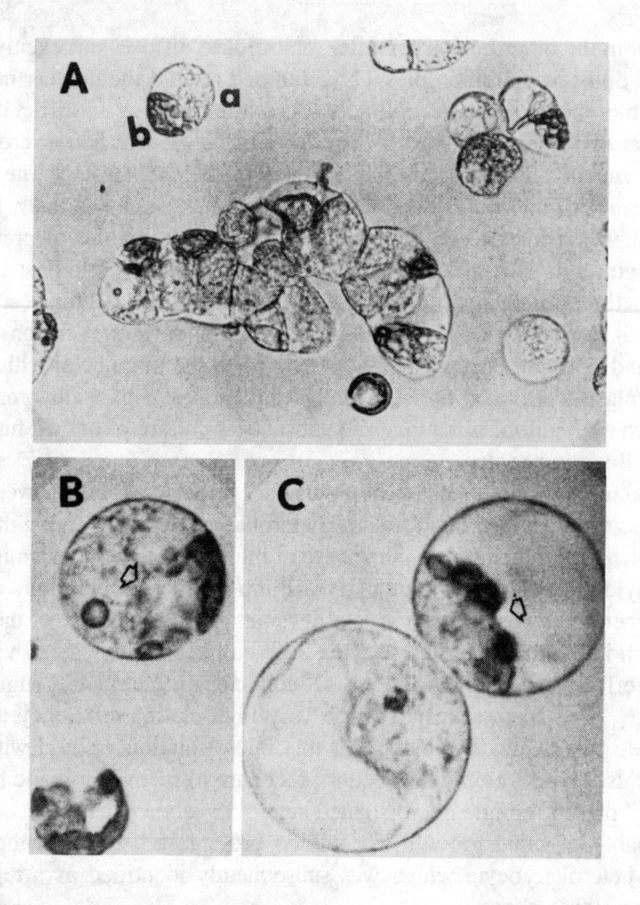

FIGURE 5. Early events in protoplast fusion between *S. guianensis* and *S. capitata*. (A) Heterokaryon of colorless protoplasts from cell suspension cultures of *S. guianensis* (a) induced to fuse with chlorophyllic chloroplasts of *S. capitata* (b). Note that both single above and multiple below fusions can occur. (B) A hybrid protoplast just after fusion. Note borderline between parental protoplasts still apparent (arrow). (C) Hybrid protoplast 1 day after fusion. Note unfused protoplasts (arrow). (From Roca, W. M., Szabados, L., and Hussain, A., Annual Report Biotechnology Research Unit, CIAT, 1987. With permission.)

isolation. Galbraith and Mauch[141] have suggested the use of cell sorter to separate protoplasts differentially stained with fluorescent dyes, but the high cost of the cell sorter precludes its widespread application. Harms and Potrykus[142] used isoosmotic density gradient to separate fusion products. By centrifuging a protoplast mixture for 2 to 4 min at 50 to 100 × *g* in KMC (potassium-magnesium-calcium) sucrose density gradients, they were able to enrich for heterokaryocytes. However, this method is not 100% effective. A modified form of this technique was successfully applied by Patnaik and associates,[143] who labeled cell suspension protoplasts with fluorescein isothiocyanate and fused them with chloroplast containing mesophyll protoplasts. When observed by the fluorescence microscopy technique, the chloroplasts fluoresce red because of the chlorophyll, and the isothiocyanate cells exhibit a highly characteristic apple-green cytoplasmic fluorescence. Patnaik and associates used micromanipulative methods to separate out green fluorescing cytoplasm containing red fluorescing cytoplasm.[143]

A fluorescence flow cytometry and cell sorting system originally designed for separation of cancer cells and noncancer cells is now becoming popular to isolate somatic hybrids.[144] The technical operation and construction of the machine used for this purpose are extremely complex. The two parent protoplast populations are labeled with different fluorescent tags, one red and one green; the fusion products will contain both fluorescent markers at the same time. When a mixture of parental protoplasts and fusion products is introduced into the machine, it prepares droplets of an appropriate size so as to allow only a single protoplast or fusion product per droplet. The droplet is then electronically scanned to determine its fluorescence properties. It will contain parental type 1 and type 2 fusion hybrid. The droplet is then released between two electrical plates that can have a positive or negative charge or be neutral. The charge on the plate will deflect the droplets, allowing separation of three cell types. It follows, therefore, that after running the mixture through the machine, there will be three tubes, one of which will contain fusion products for further analysis. This technique is still very much in the experimental stage and does not suggest great potential for the future. In addition, this method is not so efficient and powerful selection methods, utilizing mutants induced *in vitro,* may be necessary to isolate interspecific and intergeneric hybrids between more distantly related species. Several attempts to develop more efficient techniques are being made and several scientists have used different methods to identify and isolate hybrid cells.[124-129,145-158]

d. Identification by Electrophoresis

In situations where no visible or genetic markers are available at cell level, possible hybrid cell colonies are selected on the basis of their electrophoretic isoenzyme pattern. Esterase patterns of *S. guianensis, S. capitata,* and *S. macrocephala* differed sharply even at the callus level.[80] Screening through a sufficient number of regenerated colonies, these differences permit the selection of putative hybrids. In a fusion experiment of *S. guianensis* × *S. capitata,* 6 possible hybrid colonies were selected from 78 hybrids and protoplasts screened electrophoretically (Figure 6). Plant regeneration was induced in these colonies and 64 plantlets were obtained.

By biochemical screening through electrophoresis, it is possible to identify hybrid colonies and regenerate plants whose hybridity can be confirmed later. This is a real advantage in the somatic hybrid production of crop plants where the low protoplast culture efficiency does not permit single cell culture (necessary for mechanical isolation of hybrid cells by micromanipulator) and no selective markers are available at cell level.

e. Use of Genetically Transformed Protoplast Cells

Despite numerous selection systems for somatic hybrids, it is evident that selection is still the bottleneck in the production of many desirable fusion hybrids. The situation was greatly improved by the production of the nitrate reductase-deficient (NR^-) and streptomycin-resistant double mutant of *N. tabacum.*[159] Such a double mutant combining a recessive mutation in the nuclear genome (NR^-) and a dominant mutation in the chloroplast genome (SR^+) can be used as a universal hybridizer with any wild-type cells.[160] Selection against unfused protoplasts of both types is possible by addition of streptomycin to the culture medium without a reduced nitrogen source. The establishment of a comparable general selection system, but involving only the nuclear genomes, became feasible with the possibility of applying the method of direct gene transfer[161,162] on NR^- mutants using resistance against kanamycin sulfate (K^+) as a selectable marker. Kanamycin resistance has been shown to be a dominant nuclear marker,[161-163] which can now be easily introduced into the genome of dicots as well as monocots. Protoplasts from plant material thus produced can be used for hybrid formation with any wild-type protoplasts. Unfused transformed mutant protoplasts (NR^-K^+) can be eliminated by culture on a medium without a reduced nitrogen source, whereas selection against unfused wild-type cells is provided by the addition of kanamycin sulfate to the culture medium. The application of this system became feasible after the reports of Brunold et al.[164] They presented evidence of somatic hybrids selected with this procedure

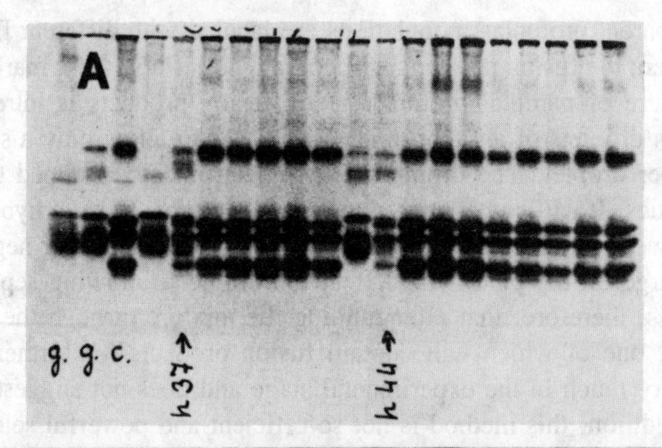

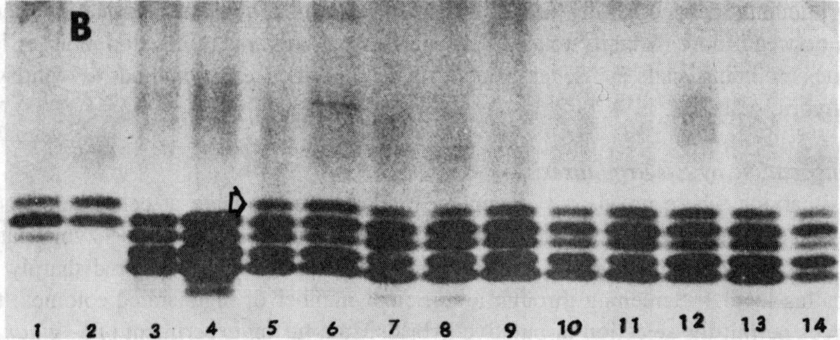

FIGURE 6. Selection of possible somatic hybrids by electrophoretic isoenzyme analysis of fusion colonies. (A) Two possible hybrid colonies (h 37 and h 44) out of 16 colonies analyzed. Parental colonies: g (*S. guianensis* CIAT 136) and c (*S. capitata* Ciat 1019). (B) Patterns of parental callus and regenerated plantlet of *S. guianensis* CIAT 136 (slots 1 and 2), parental callus and regenerated plantlet of *S. capitata* CIAT 1019 (slots 3 and 4), the patterns of fusions colonies selected as in (A) (slots 5, 7, 9, 11, and 13), and the patterns of plantlets regenerated from these colonies (slots 5 to 14 show an additional distinct band corresponding to parental line *S. guianensis* [arrow] and *S. capitata* parental line). (From Roca, W. M., Szabados, L., and Hussain, A., Annual Report, Biotechnology, Research Unit, CIAT, 1987. With permission.)

using a nonmorphogenic *N. tabacum* cell line resistant to kanamycin and lacking nitrate reductase (NR^-K^+) and *N. tabacum,* which carried resistance to streptomycin as an additional selectable marker ($NR^+K^-SR^+$).

III. APPLICATIONS

In order to harvest the fruits of protoplast technology, it is becoming increasingly clear that first a successful protoplast-to-plant system must be developed for a particular plant under consideration. Once available for a particular crop, desired genes can be transferred through direct uptake of DNA or by several other methods now available. Such cells can then be regenerated into complete plants. Transformed plants can subsequently be tested in the field for the inheritance of transformed character from the parents of progeny. This will finally lead to the formation of a new plant variety and its subsequent incorporation into the breeding programs (Figure 7) or genera through conventional breeding techniques because the hybrid plants are generally sexually incompatible. Protoplast fusion can be used to overcome this problem, particularly in the transfer of desired genes from the wild-type plant species to cultivated plant species. As we will see later, such fusions have been successful in producing compatible sexual hybrid plants.

Transferring subcellular organelles, such as nuclei, on chloroplasts or mitochondria

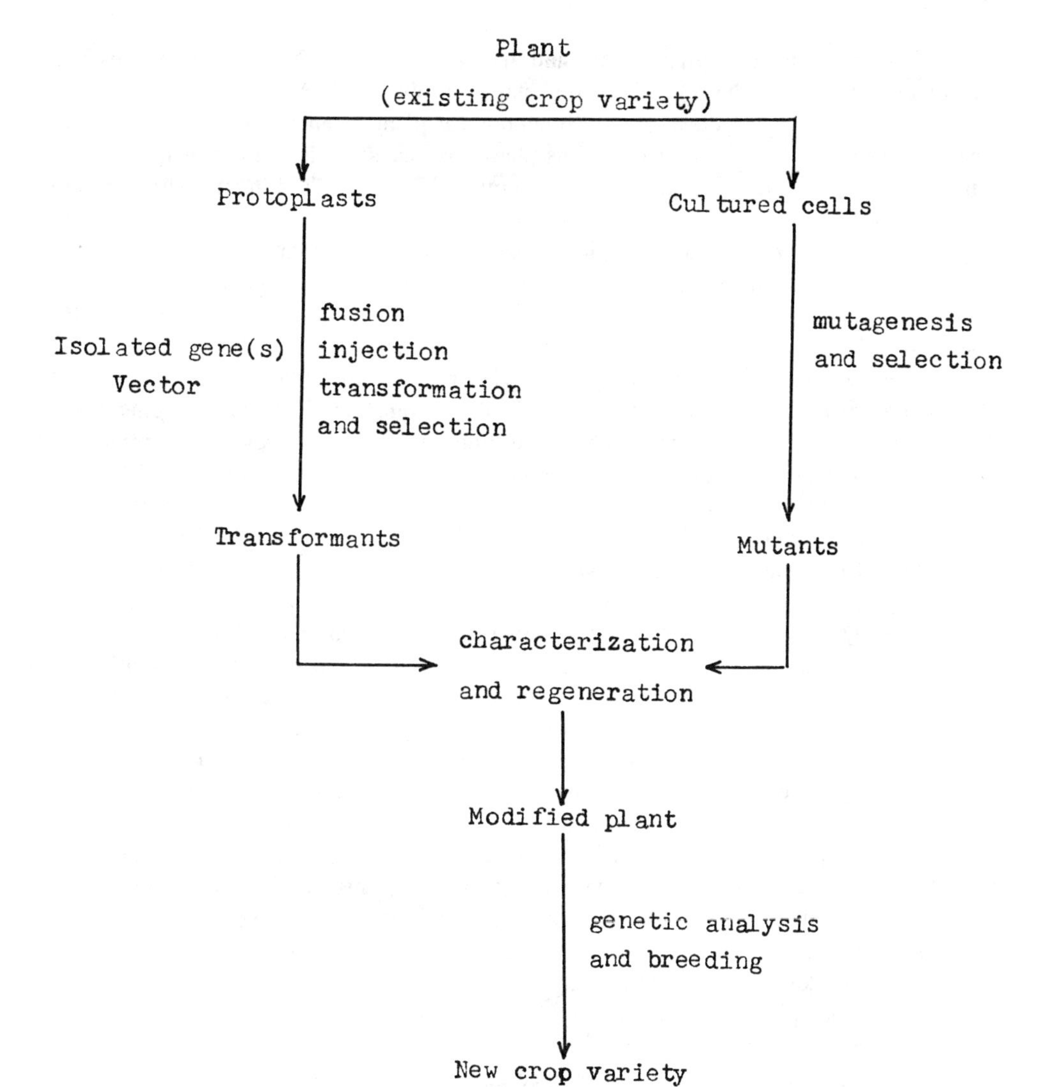

FIGURE 7. Steps to create new variety of plants by using protoplast technology.

from one plant species to another has also been accomplished by protoplast technology. There has also been substantial progress in the transfer of organelle-based characters to somatic hybrids from wild-type plants to cultivated species. Several somatic hybrids will be in the market in the coming 5 years. The progress made in this direction during these years is also discussed.

A. REGENERATION OF PLANTS FROM PROTOPLASTS

Implementation of novel approaches to plant cell genetics and crop improvements requires an efficient protoplast system, i.e., reproducible techniques for the regeneration of functional plants from isolated protoplasts. Isolated protoplast cells also have obvious advantage over calli and cell suspension. Much effort was therefore invested to furnish methods which will result in mass production of protoplasts.

Remarkable progress has been made in this field during the last 20 years.[165-171] The first[172] and subsequent protoplasts-to-plant systems were developed in model plants such as *Nicotiana* spp., *Petunia* spp., *Hyoscyamus muticus,* and in carrot.[173] Subsequently, such systems became available in gramineae[174] and in orchard crops.[105] Galun and associates have done extensive work on the regeneration of plants in tobacco.[145-158]

It is also necessary to develop rapid methods for plant regeneration from protoplasts, because the extended duration of the callus phase may cause undesirable genetic variation in the regenerated plants.[175] Rapid plant regeneration is also required in analysis of genetic modifications and genetics of transformed or fused protoplasts. Firoozabady[176] described a simple and efficient method for the rapid regeneration of plants from tobacco *(N. tabacum)* mesophyll protoplasts in which the callus stage is minimized and a large number of diploid plants are regenerated from colonies derived from individual protoplasts. Six to 7 d after protoplast isolation, colonies are placed on double filter-feeder plates that consist of a strong regeneration medium (SMpi) containing 7.5 mg/l of 6-(γ,γ-dimethylallyamino)-purine (2iP) and 0.1 mg/l of *p*-chlorophenoxyacetic acid (PCPA). Complete plants were regenerated in about 5 weeks after transfer to a rooting medium (Figures 8 and 9). However, upon remaining on shoot regeneration medium, 50 to 75% shoots were regenerated from single colonies derived from individual protoplasts. The early exposure of colonies to the regeneration medium initiated rapid shoot induction in microcalli at a high frequency. Previously, the culture period for tobacco protoplasts was shortened by Raveh and Galun[106] and Evans et al.[85] to regenerate shoots in 7 and 6 to 7 weeks, respectively. In the procedure developed by Firoozabady,[176] the culture period was further reduced and large numbers of plants were regenerated. In addition, the author also suggested a reduction in somaclonal variation.

The genus *Brassica* includes a wide range of crop species with great economic value worldwide. Plants have been regenerated from different genotypes of *Brassica* including *B. napus, B. oleracea,*[177] and *B. junacea.*[161] Protoplasts isolated from stem tissue of *B. alboglabra* Bailey divided rapidly and formed microcalli on medium supplemented with 2,4-D (0.2 mg/l) and 6-benzyladenine (BA) (0.2 mg/l).[178] On shoot regeneration medium, the presence of 0.25 to 1 mg/l BA was most effective in inducing shoot and plant regeneration. Protoplasts of *B. oleracea* were able to undergo embryogenesis and developed into intact plants.[179] Protoplasts of *B. nigra* isolated from stem peels of recemes have been successfully regenerated into plants for the first time by Chuong et al.[180] Of the calli obtained from protoplasts, 10% developed into plantlets.

Low frequency plant regeneration from haploid protoplasts isolated from stem embryos of *B. napus* has been obtained.[181,182] Li and Kohlenbach[73] reported that 27% calli produced from haploid mesophyll protoplasts developed embryos. The development of regeneration procedures for stem protoplasts from these haploids is a prerequisite for their subsequent use in fusion experiments to combine cytoplasmic traits in *B. napus*. Chuong et al.[180] described the isolation and culture of protoplasts from *B. napus* haploid plants.

Solanum melongena (egg plant) is an important nontuberous *Solanum* species and its protoplasts have been sucessfully regenerated into complete plants.[38,39] However, this species is susceptible to several microbial pathogens. Thus, there are attempts to transfer disease-resistant genes from a wild species *S. torvum* through somatic hybridization. *S. torvum* is resistant to the fungal soil-borne disease verticillum wilt, fusarium wilt, bacterial wilt, and to root knot nematode to which *S. melongena* is susceptible. As a first step in this direction a protocol has been designed to isolate and regenerate plants from *S. torvum* by Guri et al.,[183] paving the way of somatic hybridization experiments to transfer disease resistance genes to cultivated *Solanum* spp.

S. etuberosum (possesses frost resistance and potato leaf roll virus resistance) is another member of the family Solanaceae which can prove useful to transfer genes for frost resistance and potato leaf roll virus resistance to *S. tuberosum* through somatic cell hybridization. A protoplast-to-plant system has been developed for *S. etuberosum* recently.[184] Comprehensive reports on the regeneration of potato protoplasts are beginning to appear.[185-188] Dunwell and Sunderland[189] were the first to succeed in regenerating potato plants from protoplasts. Ha-

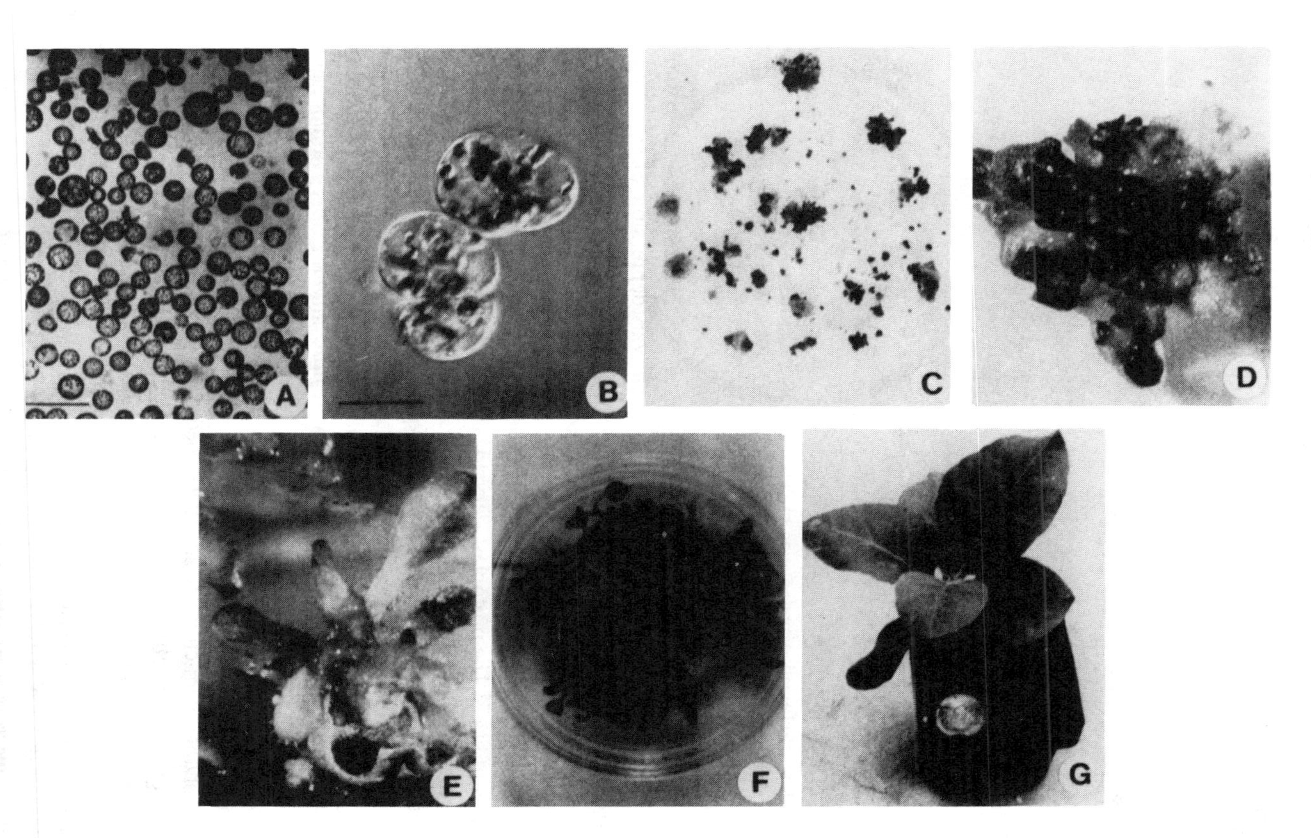

FIGURE 8. Plant regeneration from tobacco mesophyll protoplasts by the rapid method. (A) Freshly isolated protoplasts (bar = 100 μm); (B) first division after 2 to 3 d in culture (bar = 30μm); (C) globular embroids, meristematic mantles, and shoot tips developed on SMPi medium (XI); (D) meristematic mantles and shoot tips at higher magnification (magnification × 10); (E) shoots (magnification × 10); (F) shoot multiplication and micropropagation on SMPi medium (XI); (G) the regenerated plant after transfer to soil (magnification × 0.2). (From Firoozabady, E., *Plant Sci.*, 46, 127, 1986. With permission.)

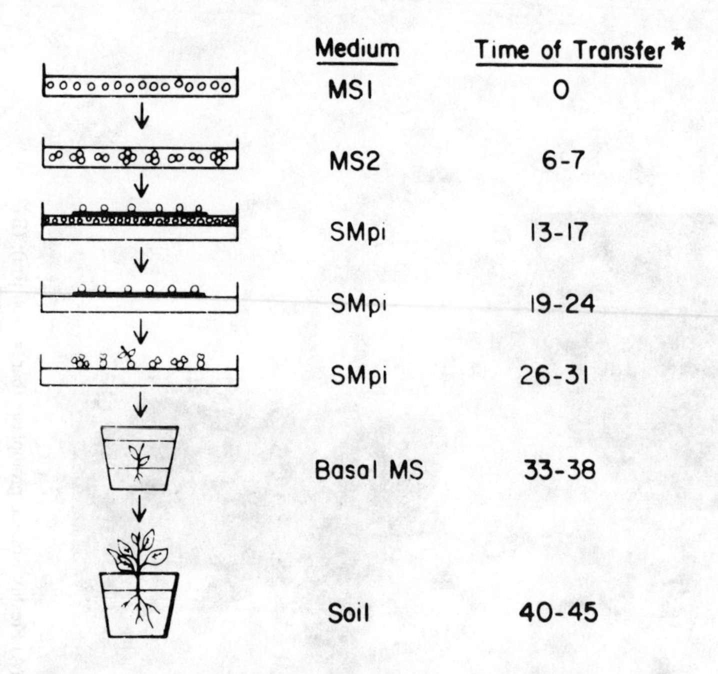

FIGURE 9. Schematic representation of the culture procedure for rapid plant regeneration. (From Firoozabady, E., *Plant Sci.*, 46, 127, 1986. With permission.)

berlach et al.[91] could regenerate protoplasts from *S. brevidens, S. demissum, S. etuberosum,* and *S. pennelli.* Protoplast-derived colonies of *S. melongena* var. depression were successfully regenerated into fertile plants in potting compost.[38] Guri and Izhar[39] discovered a medium for relatively high plating efficiency of *S. melongena* which regenerated into plants only within 6 weeks.

A high yield of viable protoplasts was obtained from callus cultures derived from shoot spices of *Vigna aconitifolia.*[45] The protoplast-dervied callus formed multiple shoots. Koblitz and Koblitz[19] described a method for isolation and culture of protoplasts from cotyledon mesophyll cells of tomato (*Lycopersicon esculentum* cv. Nadja$^+$). Regenerated plantlets grew into normal plants which produced mature fruits. However, these authors were not able to regenerate plants from *L. esculentum* cv. Lukullus whereas they were successful in regenerating plants from protoplasts of *L. esculentum* var. *flammatum, L. peruvianum,* and *L. pimpinellifolium.*[64]

Mesophyll protoplasts of *Trigonella corniculata* derived from callus regenerated into plants in the medium lacking growth hormones.[4,44] Numerous culture media combinations failed to regenerate shoots from callus tissue developed from protoplasts of *Physalis* spp.[36] High yields of protoplasts were obtained by enzymatic digestion of leaf tissue derived from axenic shoot cultures of *P. minima.*[190] Isolated protoplasts regenerated into plants which were established in soil. High yields of viable protoplasts were also obtained by enzymatic treatment from cotyledonary leaves of various greenhouse-grown cultivars of *Phaseolus vulgaris.*[191] The protoplasts, however, could not be regenerated into whole plants.

Though plantlets were formed from protoplasts of *Helianthus annus,* in other species only callus was obtained.[192,193] In *Gossypium hirsutum,* it has still not been possible to regenerate complete plants from protoplasts, although regeneration up to callus stage has been obtained.[194] Cultured protoplasts from cell suspensions of *Pelarogonium aridium, P.* × *hortorum,* and *P. peltatum* divided to form callus.[195] On agar-solidified regenerative medium such protoplast-derived calli underwent plant regeneration at frequencies approaching 100% for *P. aridium* and 10% for *P.* × *hortorum.* Under similar conditions shoot primordia arose in 5% of *P. peltatum* calli, but these never developed into plants. However, following a liquid shake culture regime, whole plants were induced in 20% of *P. peltatum* and the regeneration ability was improved to 60% in *P.* × *hortorum.*

Protoplast isolated from young unexpanded leaves of wild lettuce species *(Lactuca saligna)* was successfully regenerated into complete plants.[196] This is significant because this wild species is a potential source of novel resistance to downy mildew, powdery mildew, turnip mosaic virus, cucumber mosaic virus, and cabbage looper. Once efficient techniques are developed for protoplast regeneration, this species can be used to transfer genes to *L. sativa* through somatic hybridization.

In recent years there have been reports of plant regeneration from protoplast-derived callus in a number of forage legumes including *Medicago sativa,*[49,66] *M. glutinosa,*[23] *M. arborea,*[165] *Trifolium repens,*[42] *Lotus corniculatus,*[42] *Onobrychis viciifolia,*[42] and *Hedysarum coronarium.*[15] Successful regeneration from protoplast has been reported for several legumes including *Vigna aconitifolia*[45,197,198] and *Psophocarpus tetragonolobus.*[199] However, protoplasts could only be cultured to callus stage in *V. uniquiculata, V. sublobata,* and *V. mungo.*[200]

An important aspect of somatic hybridization is the availability of a selection system capable of recovering a limited number of hybrid colonies from a larger number of colonies derived from parental protoplasts and homokaryons. Mutants resistant to streptomycin have proved very successful in designing selection schemes aimed at somatic hybrid recovery in the genus *Nicotiana.*[126-128,159,201] Hamill et al.[202] described recovery of streptomycin-resistant variants from protoplast-derived cell colonies of *O. viciifolia,* which may be useful in improving the nutritional value of forage crops by somatic hybridization because of its high protein content and leaf tannin-conferring bloat-safe characteristics.

Cereals generally pose a number of problems and it is difficult to regenerate plants from protoplasts. More basic and fundamental work regarding the cultural requirements of cereals will lead to breakthrough in this field. Reports on plant regeneration from protoplast of gramineae are very few. Plantlets have been regenerated from protoplasts in grass species[89,203] and recently plant regeneration from rice protoplasts has been described.[204] In both reports regenerating cell lines derived from either mature embryo calli or immature embryo calli were selected, but the frequency of regeneration was very low. Kyozuka et al.[74] developed a novel nurse culture method for effective plant regeneration from the protoplast of rice *(Oryza sativa).* Culture methods have now been developed for plant regeneration in rice. The nurse culture methods use the agarose bead-type culture in combination with actively growing nurse cells that are either in the liquid part of the culture or inside a culture plate insert, placed in the center of the dish. Protoplasts isolated either from the primary seed calli or suspension cultures of various callus origins divided and formed colonies with a frequency of 10%, depending upon the protoplast source and genotype. The presence of nurse cells was essential for protoplast division.

This high-frequency (up to 50%) plant regeneration system from rice protoplasts by novel nurse culture methods[74] was further used to investigate the performance of such plants in the field.[205] Plants showed more stems, with more panicles, fewer spikelets per panicle, slightly lower fertility, and similar lighter kernal weight. Grain yield of protoplast-derived plants was more than that of respective controls. Though several variants were observed, a few plants showed no chromosomal anomalies. About 80% of the plants were normal. These findings are encouragning for the possibility of breeding rice using protoplasts.

Scientists at Colorado State University have recently begun work on protoplast isolation, culture, and plant regeneration and genetic manipulation of cereal crops.[206] The goal of the project is to develop a system of protoplast regeneration that will be applicable to a wide variety of monocots with minimal variation in procedures. Exhaustive research is being carried out on enzyme combinations, concentrations, source, and purity. In addition, extensive studies are being conducted on temperature and time of incubation, agitation speed, lighting conditions, osmotica, purification techniques, culture nutrients, hormones, culture density, culture conditions, and optimum time to add fresh medium. Work on isolation and purification of protoplasts from rice, proso millet, corn, and wheat has already been completed. Isolations and purifications have also been successfully completed on a wide variety of dicots and less agronomically important monocots to test the efficacy of the procedures developed. Protoplasts are generally isolated from embryogenic callus cultures or suspension cultures of these cells.

Success has been obtained in regeneration of the whole plant from protoplasts of an herbaceous plant, like pioneer elm *(Ulmus)*.[207] Only plantlets could be regenerated from protoplasts of Chinese medicinal herb *Rehmannia glutinesa*.[208] Improvement of woody perennials, protoplasts from totipotent somatic cells, is in some respects equivalent to the egg in genetic potential. The regeneration of complete plants and manipulating the protoplasts have now become possible even in woody plants. Galun and associates reported initially that embroids and plants can be derived from protoplasts isolated from nucellus calli of orange *(Citrus sinensis)*.[12,105] Regeneration of plants from protoplasts in *Citrus* is of interest, because this plant belongs to a group of woody plants having a conspicuous and long juvenile phase differing from other existing protoplast systems in several physiological aspects. In addition, sexual reproduction and hybridization in *Citrus* are hampered by heterozygosity, sterility, and nucellar embryony;[55,209] reproduction by protoplast culture should be useful in this genus. Studies by Vardi et al.[12] established the regeneration of complete plants from protoplasts derived from nucellus calli of sour orange *(C. aurantinum)*, mandarin *(C. reticulata)*, grapefruit *(C. paradisi)*, and lemon *(C. limon)*. Vardi et al.[210] were also able to develop a system from protoplast-to-tree in *Microcitrus* based on protoplasts derived from a sustained embryogenic callus. A related genus *Microcitrus* is also important because it is a wild species which can be used to transfer several characters to *Citrus*. Plant of Trovita orange *(C. sinensis)* have been successfully regenerated from protoplasts.[211-213] In addition, the protoclones developed were all uniform without any variation.[214]

Russell and McCown[34,35] began their research on protoplast technology in *Populus* with the poplar hybrid NC-5339 *(P. alba* × *P. grandidentata)*. Protoplasts isolated from leaves were cultured and complete plants regenerated. The successful culture of protoplasts of the poplar hybrid NC-5339 relied on the avoidance of inhibitory factors during the culture period to facilitate protoplast development. A floating disk technique was therefore developed in which the protoplasts were grown in a monolayer together with a polyester screen disk at the surface of a liquid medium. Though the protoplasts developed into a complete plant, the developing poplar protoplasts progressed through five characteristic stages during a 6- to 8-week period before cytokinesis. This slow protoplast development limits the usefulness of the poplar protoplast system. A refined method for rapid development of protoplast was thus developed.[215] The leaf protoplasts of hybrid poplar were cultured in contact with screen disks floated in liquid medium. Protoplast culture was influenced by the growth medium, purification procedure, plating density, coconut water, and absence of casein hydrolysate supplement. The protoplast regeneration was very rapid in this medium and plants developed quite fast.

Though the protoplasts from oil palm *(Elaeis guineansis)* in nurse medium formed callus, complete plants could not be regenerated.[13] Protoplast isolation from apple leaf tissue of *in vitro* cultured shoots was possible if very young leaves from buds were used.[50] Digestible leaves were found in apple shoots of the Akero 14 d after transferring the shoots. Protoplast yield was considerably improved if the apple shoots were cultured in medium containing 0.5 m*M* L-methionine. Plants were regenerated from protoplast-derived calli of five tested cultivars with a frequency of 17 to 50%, and were isolated from cell suspension-derived somatic embryos of *Coffea canephora.*[216] After repeated subculture in the medium with 0.5 mg/l, each of kinetin, 2,4-D, and other growth regulators resulted in the globular embryos which upon subculture grew to plantlets.

For woody perennials protoplast technology is not much beyond the preparative stage, with plant regeneration from protoplasts being restricted to a few species. Ochatt et al.[217] were able to regenerate plantlets from protoplasts of sweet cherry (*Prunus avium* L). These results, coupled with recent developments for several fruit species, suggest that wood fruit crops will, in the near future, be amenable to techniques of genetic manipulation based on protoplasts. Protoplasts of another woody plant olive *(Olea europaea)* have been isolated and regenerated up to callus stage.[218]

Protoplast-to-plant systems are now available for several economic crops[219-227] and the list will expand substantially in the near future covering several crop species (Table 4).

B. PROTOPLAST TRANSFORMATIONS BY DIRECT, CHEMICALLY MEDIATED, AND MICROINJECTION METHODS

The genetic manipulation and modification of plants present some special challenges. Most molecular genetics until recently have been done with simple unicellular organisms, and to a lesser extent with laboratory animals. The application of molecular genetics through protoplasts is relatively more recent and, consequently, at an early stage of technical development.[251]

The protoplast system is quite useful for the transfer of foreign genes. DNA can be offered to protoplast cells in four different ways:[252] (1) pure DNA, (2) calcium, phosphate-DNA coprecipitate, (3) DNA encapsulated in liposomes, and (4) by direct injection of the DNA via microinjection. DNA uptake can be enhanced by a heat shock treatment, by a high pH treatment, and/or by the addition of PEG, poly-L-ornithine (PLO), or polyvinyl alcohol (PVA).

A striking breakthrough came with direct DNA transfer into protoplasts, resulting in the recovery of transformed cells.[253] Most recently, transformation of bacterial plasmids has been reported in several monocotyledonous[254-256] and dicotyledonous species.[161,257] Uchimiya et al.[258] have been able to directly transfer foreign genes into tobacco protoplasts. Protoplasts of *Vicia rosea* were transformed by spheroplasts of *Agrobacterium tumefaciens* harboring nopaline type-T plasmids.[251]

The chemicals PEG and PVA, together with Ca^{2+} and high pH, induce protoplasts to take up naked DNA. This technique has resulted in stable integration of foreign DNA in regenerated *N. tabacum* plants. The integrated gene was shown to be inherited in a Mendelian fashion in a subsequent generation.[161] The transformation frequencies obtained by using chemically mediated DNA uptake have been low when compared to those following *Agrobacterium* cocultivation. Thus, the technique of DNA uptake would be most useful for graminaceous monocotyldedons or other species which are not susceptible to or difficult to transform with *Agrobacterium.* Protoplasts of *Lolium multiflorum* (Italian rye grass) and *Triticum monococcum* (diploid wheat) have been transformed using chemically mediated DNA uptake.[254,255,259]

Several workers have used direct DNA uptake to transform dicotyledonous protoplasts by Ti plasmids[253] and *Escherichia coli*-based vectors.[161,162] The transformation frequency in these experiments was low, but could be increased significantly using a procedure incorporating PEG treatment, heat shock, and electroporation.[260] Recently, Chen et al.[261] transformed sugarcane (*Saccharum* hybrid) by kanamycin resistance gene at a frequency of approximately 8 in 10^7 (as compared to 1 in 10^4 to 10^6 in the previous reports) following PEG-induced uptake of Sma1 linearized pABD1 plasmid. Transient expression of chimaric genes (neomycin phosphotransferase fused with different promotors) in suspension culture-derived protoplasts of maize, barley, rice, and rye has also been obtained.[262] The introduction

TABLE 4
Some Common Plants from Which Calli, Plantlets, and Whole Plants Have Been Regenerated from Protoplasts

Plant species	Source of protoplasts	Final stage of differentiation	Ref.
Atropa officinalis	Cell culture	Plants	228
Beta vulgaris	Cell suspension	Callus	8
Brassica junacea	Mesophyll	Plantlets	9
B. napus	Mesophyll	Plants	5, 73, 181, 229
	Hypocotyl	Plantlets	10
	Haploid stem peel	Plants	180
B. nigra	Stem peels	Plants	180
B. oleracea	Mesophyll	Plants	179
	Mesophyll	Plants	4, 5
	Stem tissue	Plants	178
Citrus aurantium	Callus	Plants	12
C. limon	Callus	Plants	12
C. reticulata	Callus	Plants	12
C. paradisi	Callus	Plants	12
C. sinensis	Callus	Plantlets	105
	Nucellar callus	Plant	211, 213
Coffea canephora	Suspension culture	Plants	216
Cyamopsis tetragonoloba	Mesophyll	Plantlets	230
Digitalis lanta	Mesophyll	Plants	73
Daucus carota	Roots	Embroids	129
	Cell culture	Plants	173
Datura innoxia	Mesophyll	Plants	83
Elaeis guineensis	Cell suspension	Callus	13
Eruca sativa	Mesophyll	Plants	231
Glycine canescens	Embryo	Plantlets	232
G.max	Cotyledons	Callus	233
G. soja	Cell suspension	Callus	14
G. tabacina	Cell suspension	Callus	14
Hedysarum coronarium	Cotyledons	Plants	15
Gossypium hirsutum	Cell suspension	Callus	194
Howorthia magnifica	Callus	Plants	234
Helianthus annus	Hypocotyl and cotyledons	Plantlets	192, 193
H. praecox	Hypocotyl and cotyledons	Callus	192
H. scaberimus	Hypocotyl and cotyledons	Callus	192
H. rigidus	Hypocotyl and cotyledons	Callus	192
Ipomoea batatas	Stems tissue	Plants	235
	Mesophyll	Callus	59
Lactuca sativa	Mesophyll	Plants	226
Linum alpinum	Cotyledons and hypocotyls	Callus	17
L. altaicum	Cotyledons and hypocotyls	Callus	17
L. grandiflorum	Cotyledons and hypocotyls	Callus	17
L. lewisii	Cotyledons and hypocotyls	Plantlets	17
L. narbonense	Cotyledons and hypocotyls	Callus	17
L. usitatissimum	Root and cotylendons	Plants	61
Lithospermum erythrorhizon	Cell suspension	Callus	18
Lycopersicon esculentum cv. Nadja	Cotyledon mesophyll	Plants	19, 48, 63
L. esculentum cv. Lukullus	Callus culture	Plantlets	64
L. esculentum	Leaf mesophyll	Plants	236
L. peruvianum	Cell suspension	Plants	40, 237
	Callus cultures	Callus	64
L. pimpinellifolium var. ribesoides	Callus culture	Callus	64
Malus domestica	Leaf mesophyll	Callus	21

TABLE 4 (continued)
Some Common Plants from Which Calli, Plantlets, and Whole Plants Have Been Regenerated from Protoplasts

Plant species	Source of protoplasts	Final stage of differentiation	Ref.
Medicago spp.	Mesophyll	Plants	23, 49, 238
M. arborea	Mesophyll	Plants	165
M. sativa	Cell suspension	Callus	22
Microcitrus	Callus	Plants	210
Nicotiana plumbaginifola	Haploid cell suspension	Plants	24
N. sylvestris	Mesophyll	Plants	239
N. tabacum	Mesophyll	Plants	172
	Epidermis	Plants	31
Onobrychis viciifolia	Leaf mesophyll	Callus	42
Oryza sativa	Callus	Plants	74
	Callus	Plants	240
	Anther callus	Plants	241
Olea europaca	Leaves and cotyledons	Callus	218
Petunia spp.	Mesophyll	Plants	242
P. alkenkengi	Shoot buds	Callus	36
P. hybrida	Mesophyll	Plants	96
P. ixocarpa	Shoot buds	Callus	36
P. peruviana	Shoot buds	Callus	36
P. pruinosa	Shoot buds	Callus	36
P. parodii	Mesophyll	Plants	243—246
Physalis minima	Mesophyll	Plants	190
P. minima	Shoot buds	Callus	36
Phaseolus vulgaris	Cotyledonary leaves	Callus	191
Picea glauca	Cell suspension	Callus	247
Populus	Shoot culture	Plants	34, 35
Prunus avium	Cell suspension or leaf mesophyll	Plantlets	217
Pyrus malus	Mesophyll	Plants	50
Rehmannia glutinosa	Mesophyll	Plantlets	208
Senecio vulgaris	Mesophyll	Plants	248
Spinacia oleracea	Cell suspension	Callus	79
Solanum brevidens	Mesophyll	Plantlets	91
S. desmissum	Mesophyll	Plantlets	91
S. dulcamara	Shoot tip	Plants	219
S. etuberosum	Mesophyll	Plantlets	91
	Mesophyll	Plants	184
S. melongena	Mesophyll	Plants	38
	Mesophyll	Plants	39
S. pennelli	Mesophyll	Plants	91
S. torvum	Mesophyll	Plants	183
S. tuberosum	Mesophyll	Plants	249
	Cell suspension	Plants	40
	Mesophyll	Plants	41
S. xanthocarpum	Mesophyll	Plantlets	250
Trigonella corniculata	Mesophyll	Plants	5
	Mesophyll	Plants	4, 44
T. foenumgraecum	Mesophyll	Plants	197
Trifolium repens	Mesophyll	Callus	42
Vigna aconitifolia	Callus culture	Plantlets	45
V. aconitifolia	Hypocotyl	Plants	198
V. mungo	Hypocotyl	Callus	200
V. sublobata	Hypocotyl	Callus	200
V. unguiculata	Hypocotyl	Callus	200

and expression of foreign genes were performed in both dividing and nondividing protoplasts by applying the PEG transformation method.

Direct transfer of genes has also been reported in *L. multiflorum* by Potrykus et al.[256] using DNA transformation of protoplasts from a nonmorphogenic cell culture. A construction consisting of expression signals from gene VI of CaMV joined to the aminogylcoside phosphotransferase gene from transposon Tn5 conferred resistance to antibiotic G-418 to cell colonies arising from transformed protoplasts. Similarly, Lorz et al.[254] also reported gene transfer to cereal cells mediated by protoplast transformation with naked DNA. Protoplasts isolated from cultured cells of *T. monococcum* were incubated in the presence of PEG with circular and linear plasmid DNA. Transformed cells were selected in a medium containing kanamycin. The transformation of foreign genes by direct method into cereals is a significant achievement. At present efficient protoplast culture and regeneration of cereal are possible only to the callus stage, and, therefore, predominantly nonmorphogenic lines can be used to study cereal transformation. Intensive efforts are, however, needed to improve protoplast culture and plant regeneration conditions for all cereal crops.

DNA sequences have been introduced into a variety of plant species.[263-267] Genetic transformation of *N. tabacum* protoplast was achieved by incubation of protoplast with plasmid DNA calcium phosphate coprecipitate followed by the fusion of the protoplasts in the presence of PVA and subsequent exposure to high pH.[165] A derivative of the plasmid pBR322 containing a chimeric gene, consisting of nopaline synthase promotor, the coding region of the aminoglycoside phosphotransferase gene Tn5, and the polyadenylation signal region of the octapine synthase gene, was used for these transformation experiments. The chimeric gene conferred resistance of transformed cells to kanamycin. A similar approach for Ti-plasmid-independent gene transfer using the procedure of Krens et al.[253] has also been carried out successfully by Paszkowski et al.[161] These authors used hybrid genes containing the CaMV gene promotor. Subsequent studies revealed that this transformed DNA is incorporated into the nuclear genome in a stable Mendelian fashion.[255,256] Conceptually, there is no host range problem to be expected from this method, which introduces only the structural gene with expression signals without the use of bacterial intermediates.

Microinjection, the mechanical introduction of DNA into protoplasts or cells, is another technique being developed for direct delivery of DNA. In general, micropipettes are made with the aid of a mechanical puller. Micromanipulators are used to guide the pipette through which the injection solution is delivered into the cell by the aid of pressure from a syringe. This method was initially applied to systems consisting of cells of sufficient size, e.g., oocytes and mammalian tissues. The technique may be more broadly applicable than the previously described methods for DNA transformation. Not only can DNA be infected into plant protoplasts, but the potential exists for infecting DNA directly into pollen grains, microspores, suspension cells, embryos, or ovules.[268,269] In addition, this method has been proposed for the transfer of cell organelles and for the manipulation of isolated chromosomes.[270] The difficulty of this technique lies in the physical coordination of both an instrument to inject the DNA, as well as a means to immobilize the cell to be injected.

Immobilization of protoplasts using methods such as agarose embedding, agar embedding,[271] polylysine-treated glass surfaces,[272] and suction holding pipettes[270] is currently being investigated. Once injection has been achieved, the injected cell must be properly cultured to ensure its continued growth and development. Intranuclear injections of tobacco protoplasts, using the holding pipette and hanging drop system, have resulted in the integration of foreign DNA in 14% of the samples.[270] More recently, methods for immobilizing plant protoplasts for microinjection have been published.[251,271-273] Linsefors and Brodelius[274] immobilized protoplast of *Daucus carota* and *Cathraranthus rosens* by entrapment in gel-forming polysaccharides. However, efficient transformation of plant cell has yet to be demonstrated by microinjection. Several factors may contribute to this lack of success, but perhaps one important factor has been the inability to easily target specific cellular compartments for injection.

C. ELECTROPORATION OF PROTOPLASTS AND TRANSFER OF FOREIGN GENES

Isolated genes have been introduced into eukaryotic cells to define the requirements for their expression *in vivo* and to develop genetic recombinants.[275] In plants infection with *A. tumefaciens* has been established as the most effective method of transforming dicot species (this aspect has been described in detail in Chapter 5). However, it is difficult to transfer genes even through *Agrobacterium* to monocotyledonous plants like cereals and legumes which form the major food crops. Monocot transformation has been successful using PEG-facilitated DNA uptake,[254-256] brome mosaic virus,[276] and electroporation.[254-256] Electroporation, subjecting cells to brief pulses of electric current to permeable biomembranes, was developed for use in animal cell hybridization and transformation. Recently, scientists have used this technique to transfer a wide variety of genes in plants.

In electroporation two types of electric pulses have been used; one is a square pulse which is given by a constant power, and the second is a pulse in which a peak of voltage is given by a capacitor which declines exponentially. Using an exponentially decaying pulse, Fromm et al.[277] introduced bacterial CAT (chloramphenicol transferase) gene into tobacco, carrot, and maize protoplasts. Okada et al.[278] demonstrated that the same type of pulse caused infection of tobacco protoplasts with tobacco mosaic virus (TMV) particles or their RNA. The square pulse was used by Nishiguchi et al.[279] and they reported optimum square pulse conditions of electroporation for introduction of TMV particles into tobacco mesophyll protoplasts. A longer electric pulse was necessary to induce TMV particle entry.

Electroporation has several uses in transient gene expression studies. Nonintegrated DNA is free from position effects which have been observed in stable transformants and require the pooling of many independent events to eliminate promotor strength or other regulatory factors involved in the control of gene expression. In addition, the functions of gene construction can be evaluated rapidly using electroporated DNA. Recently, Hauptmann et al.[267] evaluated electroporation as a universal method for delivering DNA to protoplasts of dicots and monocots as determined by monitoring transient expression of several chimeric gene constructs. Expression was obtained in the dicot species of *D. carota, Glycine max,* and *Petunia hybrida* and the monocot species of *T. monococcum, Pennisetum purpureum, Panicum maximum, Saccharum officinarum,* and a double cross trispecific hybrid between *Pennisetum purpureum, P. americanum,* and *P. squadmulatum.* The kinetic of delivery of DNA to monocots was similar to those of dicot protoplasts, except that the expression was 10 to 100 times lower in the monocots. The optimal compromise between DNA uptake and expression vs. cell survival was determined first for *D. carota* and was found to be applicable to other species.

Conditions were established for induction of both TMV and CMV RNAs into tobacco mesophyll protoplasts by electroporation.[279] Optimal injection was achieved with several direct current pulses. Changing the position of the protoplasts within the chamber between electric pulses was essential for acheivement of high rates of injection.

It has been demonstrated that transient expression of a bacterial gene (with a plant viral promotor sequence) is possible in both nonregenerable corn and carrot protoplast system.[277] Analysis of callus derived from electroporated corn protoplasts has revealed stable integration of the intact engineered gene.[277] Enhanced transformation frequencies have been reported when protoplasts were exposed to a series of treatments beginning with heat shock followed by low concentrations of PEG and electroporation.[236,260]

Chimeric genes containing the chloramphenicol acetyltransferase (CAT) coding sequence were introduced into protoplasts of suspension culture of tobacco cells using improved conditions of electroporation by Okada et al.[251] They found that the absence of the nuclear membrane in mitotic cells favors delivery to the nucleus of the exogenous DNA introduced into the cytoplasm. Nishiguchi et al.[279] developed an electroporation procedure for the transformation of carrot protoplasts with Ti-plasmid DNA from *A. tumefaciens*. Protoplast regeneration, somatic embryogenesis, and plantlet regeneration were not affected by electroporation conditions selected for DNA uptake. Pullock et al.,[280] however, were able to transfer mesophyll protoplasts of *Petunia hybrida* and *N. plumbaginifolia* with *A. tumefaciens*. Protoplasts of *B. napus* have been transformed by the electroporation technique with the plasmid pABD1 which contained a selectable gene conferring resistance of kanamycin.[281] Plants raised were fertile and expressed resistance to kanamycin in the sexual progeny.

High level expression of foreign coding regions introduced into a number of plant cell types has been effectively obtained using the 35S promotor region from cauliflower mosaic virus.[282] Using a PEG-electroporation technique Lin et al.[233] were successful in introducing foreign genes into soybean protoplasts which regenerated into calli. CAT activity was detected in protoplasts 6 h after introduction of a 35S-CAT-nopaline synthase 3′ chimeric gene. The highest activity was detected in 3-d-old electroporated protoplast cultures indicating transient expression of the introduced genes.

Boston et al.[283] have been successful in constructing chimeric genes containing the coding sequences of a bacterial CAT gene under the control of two different promotors, the mannopine synthase promotor from the Ti plasmid, pTiA6 of *A. tumefaciens* and the promotor from a maize zein gene.[284] These plasmids were introduced into carrot protoplast which was made permeable by electroporation.[283] Both these promotors were used effectively for direct synthesis of CAT in electroporated carrot protoplasts.

Ou-Lee et al.[275] found that electric pulse-mediated DNA uptake and expression can be applied to several graminae species and suggested that ease and convenience of this system will alow rapid evaluation of potential vectors for monocot transformation and for delineating important regions of plant promotors. The bacterial CAT gene was expressed in protoplasts of three important graminaceous plant species after introduction of the gene by electroporation.[275] Gene transfer occurred when high-voltage electric pulses were applied either directly or indirectly (without a node contact) to a solution containing plasmid DNA and protoplasts of rice, wheat, or sorghum. The indirect method was more rapid, resulted in higher protoplast viability, and was subject to contamination than the direct contact method. Gene expression of approximately equal magnitude resulted when the CAT gene was fused to either the 35S promotor of cauliflower mosaic virus or the copia long terminal repeat of *Drosophila*. It is interesting that *Drosophila* promotor (copia) not only directed synthesis of functional CAT enzyme in plant cells, but was effective as one of the strongest promotors known to function in plants.

The development of a method for transfer of genes through protoplasts is of much significance when coupled with the recent information that protoplasts in these graminaceous monocotyledonous plants can be induced to regenerate into whole plants. Ou-Lee et al.[275] were successful particularly with rice protoplasts. Rice is an important crop plant and it is possible to regenerate cell lines and calli from rice protoplasts.[74,285] The gene transformation also provides the opportunity to study the stable integration and long-term expression of genes introduced into rice. Successful plant regeneration from protoplast-derived callus of rice has been achieved[74] and it may now be possible to study, at the whole plant level, the stable inheritance and expression of genes including those of agronomic interest, introduced into rice protoplasts.

Fromm et al.[277] have shown that with appropriate modification the electroporation technique is also applicable to most protoplasts, and they have achieved the stable transformation of maize tissue cultures by this method.[264] This technique has also been successful in transformation of genes in soybean.[232] Successful application of standard genetic engineering procedures to soybean has been limited by the lack of an efficient transformation system and the ability to regenerate transformed tissues. Oncogenic transformation of soybean by virulent *Agrobacterium* strains and the axenic culture of excised tumors on hormone-free media have been reported, but disarmed Ti vectors have not yet been used successfully. Thus, electroporation may prove quite handy in transfomation of genes in such plants where an *Agrobacterium*-mediated transformation system fails. Christou et al.[232] genetically engineered soybean protoplasts of several commercially important cultivars using chimeric genes coding for resistance to the aminoglycoside antibiotics (kanamycin and G418). Electroporation of protoplasts with chimeric APH II gene and subsequent selection on media supplemented with kanamycin resulted in the recovery of calli resistant to the antibiotic. Enzyme assays for APH II activity and Southern blot hybridization confirmed the expression of the foreign DNA and its stable integration into the soybean genome. Only root formation could be induced in these transformed calli and roots maintained expression of the APH II gene.

The drawback of this technique, as with chemically mediated DNA uptake, is its present limitation to plant protoplasts (requiring a protoplast regeneration system to produce transformed plant). Investigators are now attempting to apply electroporation techniques to a wide range of plant tissues such as partially digested cells,[273,286] with the hope of broadening the applicability of this technique.

D. SELECTION OF MUTANTS

As regeneration of complete plant has become possible in several crops, work has already begun to modify protoplasts by several techniques. Regeneration of such protoplasts into complete plant can lead to the formation of plants with desired characters. The resistant lines have already been obtained by direct selection of protoplasts for tolerance to toxic chemicals.[287] Several mutant lines which include mutants for methionine sulfoximine,[116] valine,[239] and aminoethyl cystine[70] have been selected in *Nicotiana* spp.

Mutagenized *Datura innoxia* protoplasts which resulted in plants with altered pigment patterns were obtained by Schieder[83] and Krumbiegel.[288] Mutagenized tobacco protoplasts were plated in the presence of isopropyl *N*-phenyl carbonate (IPC) in an attempt to isolate IPC-resistant cells.[289] However, a high proportion of the plants regenerated in the presence of IPC were sterile. Only one mutant plant produced protoplasts which were more resistant to IPC than the parental variety.

A procedure was developed by Aviv and Galun[289] for selection of nutritional mutants of tobacco protoplasts. The procedure was based on selective killing of dividing cells and the use of X-irradiated protoplasts as feeder layer. Though several authors have claimed to obtain mutant cell lines, until recently—only in a very few cases—study has been conducted to assess the relationship between ploidy levels and irradiation doses, and cases in which mutants, selected at the cell level, produced a mutant plant were rarely documented.[288,290] One such reported mutant was a valine-resistant *N. tabacum* plant isolated by Bourgin.[239] In addition, mesophyll protoplasts of higher plants, which were capable of dividing and giving rise to mature flowering plants, offer a good system for mutant isolation.[291-293] The availability of androgenic haploid plants increases the probability of obtaining a recessive mutant.[292] Indeed, mutant *Hyoscyamus*[76] and *Nicotiana* plants[132] were isolated using haploid mesophyll protoplasts. Vunsh et al.[294] were also able to obtain protoplasts derived from haploid and diploid *N. sylvestris* and *N. tabacum*. Such protoplasts were exposed to ultraviolet light and valine-resistant cell lines were selected. The ratio between number of verified valine-resistant cell lines and the initial number of ultraviolet-exposed protoplasts enabled the estimation of the following order of mutation frequency: haploid *N. sylvestris* > haploid *N. tabacum* > diploid *N. sylvestris*. These plants which retained the valine resistance and transmitted it to their sexual progeny were derived from the resistant cell lines. Their selection of mutants also emphasized the need for multiple stage selection to obtain metabolic mutants in angiosperm.[239,295,296]

Only a relatively small number of cell lines isolated in cultured plant cells have been derived from protoplast culture. The number of species in which the protoplast system can conveniently be used is also small. If available, however, the protoplast system is the best chance to recover auxotrophic mutations and to regenerate genetically unaltered plants. Interpretation and scoring of data are also simple in this single cell system. Protoplast-derived mutants in the immediate future might make an impact through their utilization in developing transformation and gene cloning methods in plants. However, application of protoplast cultures to produce improved plants through mutant selection also depends on the successful regeneration of protoplasts in crops.

E. PROTOPLASTS AS A TOOL FOR QUANTITATIVE ASSAY OF PHYTOTOXIC COMPOUNDS

Plant protoplasts and cell suspensions may provide suitable tools for sensitive bioassay of phytotoxins. Several reviews have summarized the importance and current status of phytotoxin studies with plant cell cultures and protoplasts.[297-299] Behnke and Lonnendonker[300] found that potato cell suspensions were killed when placed in the presence of a 50-fold diluted culture filterate of *Phytophthora infestans* as compared to a 2.5-fold diluted filterate required to kill callus tissue or to cause phytotoxic symptoms on potato leaves. Bioassay of *Helminthosporium maydis* toxin on maize plants having T cytoplasm indicated that leaf protoplasts were 10 to 100 times more sensitive than intact roots or leaves of such plants.[301]

P. infestans causes trunk gummosis, collar and root rot, damping-off, and brown rot in *Citrus* spp. Culture filterates of this fungus were reported to contain high and low molecular weight nonspecific phytotoxic components.[302] The low molecular weight component was partially purified and characterized as hydrophilic and acidic compound capable of inhibiting root elongation of germinated lemon seeds, inhibiting rootlet development, and causing wilt symptoms in lemon seedlings.[302] The possibility of obtaining disease-resistant plants by *in vitro* screening with nonhost-specific phytotoxins has been demonstrated by applying culture filterates of *Alternaria solani* to potato protoplasts[249,303] or by treating potato callus tissues with *P. infestans.*[304] Selection for resistance to Southern corn leaf blight caused by fungus *H. maydis* race T was obtained by culturing corn cells on agar containing the T-toxin, which is a host-specific toxin produced by the fungus, and plants from the T-toxin-resistant cells were regenerated.[298] Since regeneration of *Citrus* plants from protoplasts is possible,[56,105,108] it might be feasible to use citrus protoplasts or calli for the development of plants resistant to *P. citrophthora* phytotoxins. Breiman and Galun,[302] however, worked out the conditions for isolating resistant protoplasts, but were not able to raise plants resistant to *P. citrophthora.*

F. REGENERATION OF COMPLETE PLANTS WITH DESIRED CHARACTERS FROM SOMATIC HYBRIDS

1. Interspecific Hybrids

The discovery of PEG[87] as a fusogen has given a great impetus to somatic hybridization studies, as this chemical is easy to manipulate, causes a high frequency of fusion, and seems to have no deleterious effects on the fusion product. There are a limited number of reports on the successful regeneration of somatic hybrids, though success has been obtained in several cases which include *Nicotiana langsdoriffi* + *N. glauca,*[116,117,305] *Petunia hybrida* + *P. parodii,*[243] *Dacus carota* + *D. capillifolius,*[306] *Datura innoxia* + *D. stramonium,*[136] and *N. sylvestris* + *N. knightiana.*[145,146] Extending this work to new plants is a matter of refining the technique and needs an imaginative mind and a skilled hand. Undoubtedly, there are problems; for example, incompatibility at certain levels needs to be explored. Chromosomal mosaics, deletions, and eliminations are common in cells after such fusions, though some of these variations are useful in introducing and increasing genetic diversity.

Several interspecific and intergeneric somatic hybrids regenerated through protoplast fusions are now available (Table 5) and many more will be included in the list in the coming few years. Several authors have completed the details of somatic hybrid plants.[227,367] Galun and associates have contributed extensively and their recent work now reveals much regarding the nature and utilization of somatic hybrids for crop improvement.[314,315,368-373] They have used *Nicotiana* most extensively in their somatic hybridization experiments. Some more additions of *Nicotiana* somatic hybrids include *N. tabacum* + *N. sylvestris,*[85] *N. tabacum* + *N. otophora,*[85] *N. tabacum* + *N. glutinosa,*[319] *N. tabacum* + *N. bigelovii,*[374,375] *N. sylvestris* + *N. begelovii,*[333] and *N. tabacum* + *N. rustica.*[28] *N. tabacum* + *N. repanda*[316] and *N. tabacum* + *Salpiglossis sinuata*[316] hybrids are between sexually incompatible species and the hybrid plants contained far less than the amphiploid number of chromosomes. The *N. tabacum* + *N. repanda* hybrids reached maturity, had greatly reduced pollen viability, and have not yet been successfully backcrossed.

Leaf mesophyll protoplasts of nitrate reductase-deficient streptomycin-resistant mutant of *N. tabacum* were fused with cell suspension protoplasts of wild-type *Petunia hybrida.*[160] Somatic hybrid cell colonies were selected for streptomycin resistance and nitrate reductase deficiency. Six lines showed nuclear somatic hybrids, possessing the chloroplast of *N. tabacum* at an early stage of development. However, after 6 to 12 months in culture, genomic incompatibility was observed resulting in the loss of most of the tobacco nuclear genome, and the regenerated plants possessed the chloroplast of *N. tabacum* in a predominantly *P. hybrida* nuclear background.

Terbutryn-resistant plastids of *N. plumbaginifolia* TBR mutant were introduced into *N. tabacum* plants by protoplast fusion following X-irradiation of TBR protoplasts.[376] Cybrid plants were resistant to high levels of atrazine. The cybrids were, however, associated with reduced vigor of the plants.

Other hybrids including *Solanum tuberosum* + *S. chacoense, S. tuberosum* + *S. nigrum,*[336] and *Brassica oleracea* + *B. compestris* are also very common. While all these hybrids are between sexually compatible species, they demonstrate that interspecific somatic hybridization is not restricted to *Nicotiana*. Somatic hybrid plants of genus *Lycopersicon* can also be produced by fusion of protoplasts of two species, one of which cannot be regenerated from protoplasts.[310] This procedure enables the introduction of certain desirable traits of wild-type members of the genus into cultivated tomato species which do not regenerate from protoplasts.

The utilization of protoplast fusions for the improvement of *Solanum* species has been stimulated by advances in the protoplast cultural capability. Kowalczyk et al.[377] have demonstrated that plants can be regenerated from *S. viarum* and *S. dulcamara* leaf mesophyll protoplasts. *S. viarum* has been suggested as the main alternative source of steroid drug precursors to the medicinal yam. It is an annually grown prickly undershrub which has a wide distribution throughout the Indian subcontinent. The steroidal glycoalkaloid, solasodine, is contained in its mature fruits. Improvements required are a spineless strain to facilitate harvesting, an increase in the solasodine content of the berries, and resistance to disease caused by the fungus *Fusarium oxysporium*. Protoplast technology offers the opportunity to improve the agronomic characters of *S. viarum* through protoplast cloning and somatic fusion with other alkaloid-producing species such as *S. dulcamara*. The combination of the superior characters, through somatic hybridization between these two species, could result in an increase in total plant solasodine yield, a reduction in spines, and a greater environmental tolerance.

Compared with other solanaceae fusions, such experiments with potatoes are rare. Melchers[378] was the first to regenerate somatic hybrids involving potato. He combined a potato dihaploid with a chlorophyll-deficient mutant of *S. lycopersicum*. All hybrids were

TABLE 5
Interspecific and Intergeneric Hybrid Plants Produced via Somatic Hybridization

Interspecific Somatic Hybrid Plants or Fusion Products

Parents	Ref.
Brassica oleracea + *B. compestris*	307
B. oleracea + *B. compestris*	177
Corchorus ditorius + *C. capsularis*	308
Datura innoxia + *D. candida*	135
D. innoxia + *D. sanguinea*	134
D. innoxia + *D. stramonium*	134
Daucus carota + *D. capillifolius*	129, 306, 309
Lycopersicon esculentum + *L. pennelli*	65
L. peruvianum + *L. pennellii*	237
Lycopersicon (cultivar) + *Lycopersicon* (wild)	310
Medicago sativa + *M. falcata*	311
Nicotiana glauca + *N. langsdorffi*	116, 305
N. knightiana + *N. umbratica*	312
N. knightiana + *N. sylvestris*	126
N. paniculata + *N. tabacum*	313
N. sylvestris + *N. bigelovii*	314, 315
N. tabacum + *N. alata*	316
N. tabacum + *N. debneyi*	317, 318
N. tabacum + *N. glauca*	122, 319
N. tabacum + *N. glutinosa*	316, 319
N. tabacum + *N. knightiana*	128, 145
N. tabacum + *N. nesophila*	320, 321
N. tabacum + *N. paniculata*	322
N. tabacum + *N. plumbaginifolia*	128, 323
N. tabacum + *N. repanda*	324
N. tabacum + *N. rustica*	71, 325—327
N. tabacum + *N. stocktonii*	320
N. tabacum + *N. sylvestris*	109, 121, 127, 155, 322, 328, 329
N. tabacum + *N. undulata*	314, 315
Petunia hybrida + *P. axillaris*	330
P. hybrida + *P. parodii*	241, 245, 331—334
P. parodii + *P. inflata*	243
Solanum melongena + *S. sisymbriifolium*	335
S. nigrum + *S. tuberosum*	336
S. tuberosum + *S. brevidens*	337, 338
S. tuberosum + *S. chacoense*	339
S. tuberosum + *S. lycopersicon*	329
S. tuberosum + *S. phureja*	15, 23, 340
S. tuberosum + *S. stenotonum*	15, 23
Vicia hajastana + *V. narbonensis*	341
V. hajastana + *V. villosa*	138

Intergeneric Somatic Hybrid Plants or Fusion Products

Parents	Ref.
Arabidopsis thaliana + *Brassica campestris*	139, 317, 342
Atropa belladonna + *Nicotiana chinensis*	343
A. belladonna + *N. tabacum*	344
Citrus sinensis + *Poncirus trifoliata*	345
Datura innoxia + *Atropa belladonna*	346, 347
Daucus carota + *Aegopodium podagraria*	119
D. carota + *Hordeum vulgare*	306
D. carota + *N. tabacum*	344, 348
D. carota + *Petunia hybrida*	344, 349

TABLE 5 (continued)
Interspecific and Intergeneric Hybrid Plants Produced via Somatic Hybridization

Intergeneric Somatic Hybrid Plants or Fusion Products (continued)

Parents	Ref.
Hyoscyamus muticus + *N. tabacum*	350
Glycine max + *Brassica napus*	351
G. max + *Caragana arborescens*	352
G. max + *Colchicum autumnale*	353
G. max + *H. vulgare*	138
G. max + *Medicago sativa*	352
G. max + *Melilotus* sp.	352
G. max + *N. glauca*	354
G. max + *N. rustica*	352
G. max + *N. tabacum*	352, 355, 356
G. max + *Pisum sativum*	138, 352, 353
G. max + *Vicia hajastana*	138, 357
G. max + *Zea mays*	138
Lycopersicon peruvianum + *Petunia hybrida*	358
N. tabacum + *Atropa belladonna*	359
N. tabacum + *Hyoscyamus muticus*	360
N. tabacum + *P. hybrida*	82, 160, 361
N. tabacum + *Salpiglossis sinuata*	324
P. hybrida + *A. belladonna*	228, 344
P. hybrida + *Parthenocissus tricuspidata*	242
P. hybrida + *Solanum nigrum*	362
P. hybrida + *Vicia faba*	247, 363
Saccharum officinarum + *Pennisetum americanum*	364
Solanum lycopersicoides + *Lycopersicon esculentum*	365
S. tuberosum + *L. esculentum*	328
Sorghum bicolor + *Z. mays*	366
Vicia hajastana + *Pisum sativum*	138

grossly aberrant, formed no tubers, and were sterile. Binding et al.[336] were able to fuse dihaploid potato genotypes with *S. nigrum,* which was resistant to herbicide atrazine. Both the species were sexually incompatible and the resistance maternally inherited. Plants regenerated from such hybrids showed several traits of both the parents, including resistance to atrazine. Butenko and Kuchko[339] mixed protoplasts of the potato cv. Priekulski Rannii with wild species *S. chacoense* (2n = 24) in a selective medium in which only heterofusions could grow. They obtained 0.5 to 1.0% heterofusions. Two of these plantlets were free from chlorophyll. One hybrid grew into a normal plant with 2n = 60, nearly indistinguishable from sexual hybrid, but it showed a distinct heterosis. The somatic hybrids had electrophoretic peroxidase different from those of sexual hybrids. Fusion of an albino mutant of potato cv. Russet Burbank and the nontuber-bearing species *S. brevidens* resulted in vigorous green plants with morphological traits intermediate between both parents.[338] Pollen did not function and only parthenocarpic fruits were produced.

Austin et al.[379] succeeded in raising normal-looking fusion plants of potato leaf roll virus (PLRV)-resistant *S. brevidens* and *S. phureja/S. stenotomum.* Leaf mesophyll protoplasts of *S. pinnatisectum* (2n = 24) (γ-irradiated and consequently unable to divide) were fused with untreated protoplasts of genomic chlorophyll-deficient mutant (2n = 24) containing the germplasm of *S. tuberosum* and *S. phureja.*[340] Two types of plants differing in their pigmentation characteristics were selected. The regenerants of one group were identified as true somatic hybrids. The regenerants of the second group were corrected for the gene controlling chlorophyll deficiency, but contained species-specific isoenzyme of the potato cultivar only. Phenotypic and flowering characteristics of hybrid plants generated by protoplast fusion between a tetraploid *S. tuberosum* line and diploid *S. brevidens* (formed hexaploid hybrids) were examined by Austin et al.[337] These somatic hybrids showed a wide range of variation under field conditions. Flowering was reduced in protoclones, the majority of the hybrids flowered with viable pollen and set tubers. Protoplasts of 6-azauracil-resistant cell lines of *S. melongena* L. were fused with protoplasts of *S. sisymbiifolium* Lam. to create somatic hybrids between these sexually incompatible species.[335] Colonies selected in medium containing 1 m*M* 6-azauracil and placed in medium containing zeatin were able to regenerate into hybrid plants.

Wild *Solanum* species have been important as sources of useful characteristics in the development of potato cultivars, but some potentially useful species are sexually incompatible. Somatic cell hybridization is one way by which genes of sexually incompatible plants may be combined directly with *S. tuberosum*.[336-338] Fusion of protoplasts from wild species with those of *S. tuberosum* can result in the transfer of disease and pest resistance in potato. Austin et al.[337] reported the incorporation of resistance to PLRV from *S. brevidens* into tetraploid somatic hybrids. A diploid *S. tuberosum* line (selected from a tetraploid *S. tuberosum,* late blight differential lines) have also been fused with those of another accession of *S. brevidens.* Several useful characteristics were incorporated in the hybrids. Further studies by Helgeson et al.[380] among hexaploid somatic hybrids between *S. brevidens* (parent resistant to PLRV) and *S. tuberosum* (late blight differential conferring resistance to *Phytophthora* race O) clearly demonstrated the expression of late blight resistance gene in all of the hybrids. In additon, most of the hybrids were also resistant to PLRV, indicating that the *S. brevidens* genes for PLRV resistance were present and expressed in the somatic hybrids.

Practical incorporation of PLRV resistance from *S. brevidens* into *S. tuberosum* is the sexual transfer of PLRV resistance from somatic hybrids to conventional breeding lines. Two sets of somatic hybrids between *S. brevidens* (2x) and *S. tuberosum* (2x and 4x) were evaluated for male fertility, meiotic regularity, and female fertility by Ehlenfedt and Helgeson.[381] Somatic hybrids were tetraploid from 2x + 2x fusions and hexaploid from 2x + 4x fusions. The hybrid plants were fertile and found to retain important characters from both the parents, and PLRV resistance from *S. brevidens* and late blight resistance from *S. tuberosum* were expressed in the individual hybrids. Transmission of these genes through meiosis and their expression in the sexual progeny have not been studied by these workers. Once this is done, subsequent incorporation of these genes into useful breeding lines will complete the demonstration that protoplast fusion can offer an effective means for breeding the germplasm base of potatoes.

Heterosis for seed yield has been demonstrated for F_1 hybrids of canola. Canola is defined as varieties of *B. napus* and *B. compestris* possessing less than 5% erucic acid in the seed oil and less than 30 μmol glucosinolates per gram of oil-free seed meal. Commercial production of hybrid seed would be facilitated by incorporating cytoplasmic male sterility (CMS) into parental stocks tolerant to the triazine herbicides. Triazine tolerance and CMS are controlled genetically by extranuclear DNA (cytoplasmic triazine tolerance [CTT] encoded in DNA and CMS in the mitochondrial DNA). For commercial triazine-tolerant single-cross hybrid canola production, both cytoplasmic traits must be present in the female parent.[382] Although CTT canola varieties and CMS canola lines exist, it is not possible to combine these traits by conventional approaches, since they are transmitted only through the female parent. However, protoplast fusion techniques can allow biparental transmission of these traits into asexually synthesized cytoplasmic hybrids or cybrids. Pelletier et al.[383] regenerated *B. napus* cybrids which possessed the chloroplast of CTT *B. compestris* and mitochondria from *Raphanus sativus* carrying the "ogu" (ogura) CMS trait. Yarrow et al.[384]

successfully combined the CTT trait with the "nap" CMS trait in the nuclear genomic background of the spring planted canola variety *(B. napus)* Regent following protoplast fusion. Whole plant cybrid regenerants were morphologically normal and produced seed on pollination, demonstrating their potential for incorporation into a breeding program. Major success of cybrid recovery was attributed to selection of fusion products by manual micro-manipulation and culture in the *Nicotiana tabacum* nurse system. Achievements such as this have extensive potential applications in canola as well as promising implications for other crops.

In the genus *Brassica* atrazine resistance gene is also encoded on chloroplast DNA. Atrazine resistance gene has been transferred into the cultivated oil seed crops *B. napus* and *B. compestris* from an atrazine-resistant biotype of *B. compestris.*[382] An atrazine-resistant, male-fertile *B. napus* plant was synthesized by fusion of protoplasts from diploid species *B. oleracea* var. *italica* carrying the ogura male-sterile cytoplasm derived from *R. sativus* and *B. compestris.*[117] Leaf protoplasts from *B. oleracea* var. *italica* carrying the ogura male-sterile cytoplasm derived from *R. stativus* were fused with hypocotyl protoplasts of atrazine-resistant *B. compestris.* Although the hybrid plant was female sterile, it was successfully used to pollinate *B. napus.* Thus a somatic hybrid plant that differed in morphology from both the parents has been regenerated in the medium containing 50 μM atrazine. Allelix scientists have developed canola hybrids with herbicide tolerance and cytoplasmic male sterility via cell fusion techniques.[385] The cybrids showed increased yields and better weed control. Thus, with the development of effectvie canola hybrids, the possibility of their introduction into breeding programs for commercialization may turn hopes into reality.

Producing herbicide-resistant plants can have definite benefits, especially in crop rotation. For instance, corn is naturally resistant to triazine herbicides, whereas soybean is not. Occasionally, soybeans do not grow well in a field the year after triazine-sprayed corn is grown there. In this case, one solution would be to introduce triazine resistance into soybeans. This particular resistance is due to a modified protein in the chloroplast membrane. Therefore, resistance between dissimilar species could be transferred by protoplast fusion or liposome-mediated chloroplast transfer.

A new vegetable called "Senposai" (1000 treasure vegetable), which is being test marketed near Tokyo, is expected to hit the world market this year.[386] Developed by the Tokita Seed Company and the Kirin Brewery Company in Japan, Senposai is a cross between an ordinary cabbage and komatsuna, a type of Chinese cabbage. The breeders have used somatic cell hybridization to produce this new vegetable, which incorporates the cabbage's resistance to high temperature.

Medicago, a forage legume, is cultivated in temperate and subtropical climates. Somatic hybridization provides a means of overcoming sexual incompatibility and of transforming characters from *M. coerulea* to *M. sativa* (alfalfa, lucerne) and to *M. glutinosa.* Protoplasts of *M. sativa* have been developed for use in genetic manipulations, with successful plant regeneration being obtained from mesophyll protoplasts.[49,328] Realistic assessment of gene transfers by somatic hybridization can now be undertaken, since conditions for protoplast isolation, culture, and somatic embryogenesis have been established for *M. coerulea* and *M. glutinosa.*[23]

Protoplast fusion has a great deal of promise for crop improvement; however, it is very difficult to produce new varieties by using these methods. There are several limitations which have to be overcome to make this method successful. Though it is not difficult to fuse the protoplast of any two species, it is also difficult to regenerate efficiently all the somatic hybrids. Though an enormous amount of work is going on in this field, the production of somatic hybrid plants has been mostly limited to the solanaceae. Recent discoveries have revealed the production of the regeneration of protoplasts from species other than solanaceae, widening the scope of this technique for its utilization in crop improvement.

2. Intergeneric Hybrids

In the early years of protoplast fusion studies, imagination and speculation were allowed to take over from logical scientific theory and numerous, rather optimistic, ideas were put forward. It was proposed that hybrids might be made between root crops (carrots and turnips) and top-vegetable crops (cauliflower and cabbage), the idea being to produce a plant with a cabbage top that after harvesting would be dug up to give a second root-vegetable harvest. To date, none of these ambitious ideas has led to success.

Though some success has been obtained in the production of intergeneric somatic hybrids, most of them are sterile[27] and, therefore, cannot be used for the development of new varieties. In most of the cases reported to date, somatic hybrids containing a mixture of genes from two species must be backcrossed to cultivated crop to develop new varieties. It is necessary to achieve intergenomic recombination between two species to transfer useful genes from a wild species into a cultivated crop. Though meiotic chromosome pairing and recombination occur frequently between varieties of a species or closely related species, recombination is minimal in divergent hybrids. Depression of yield would most likely result from transfer of whole chromosome between divergent species.

There is growing evidence that protoplast fusions between distantly related species or genera can be exploited for partial gene transfer. However, most of these combinations appear to be morphologically abnormal and could not be used in a breeding program. This method, therefore, has been used to transfer small amounts of genetic information from *Aegopodium* into carrot,[119] celery into carrot,[387] and tobacco into *Petunia*.[73] However, viability could be improved by backfusing *N. glauca* and *Glycine max* (asymmetric hybrids) to *N. glauca*.[354] These hybrids were more stable than the original hybrid cell lines. Similarly, Dudits[388] was able to improve viability by backfusing *Aegopodium* + carrot asymmetric hybrids to carrot.

Several difficulties are currently being encountered in the regeneration of intergeneric hybrids.[389-393] Wherever such hybrids have been regenerated into whole plants several unusual characters appear with a few anticipated characters. The point raised here can be easily explained from the intergeneric hybrids produced between potato and tomato.

Potato (*S. tuberosum* L.) ($2n = 4X = 48$) and tomato (L. esculentum Mill) ($2n = 2X = 24$) are members of the Solanaceae family but are not sexually compatible. In 1978 Melchers et al.[329] provided evidence for a somatic hybrid plant from fusions between protoplasts of a cultured dihaploid potato line and leaf cells of a chlorophyll-deficient tomato. Hybrid plants displayed morphological features of both parents, and analysis of the RUDP-case large subunit revealed that three plants carried the chloroplastic genome of tomato, whereas a fourth had that of potato. Those plants with a tomato plastome were termed "Tomoffeln" or "topatoes", while the one possessing the plastome of potato were designated "Karmaten" or "pomatoes".[170] Subsequently, additional somatic hybrids were recovered: four topatoes and five pomatoes. None possessed the chromosome number of a true amphitetraploid ($2n = 4X = 48$), and it was not determined whether this was a consequence of chromosome segregation or of the use of mixoploid potato cells as protoplast donors. Some hybrid plants formed "tuber-like stolons" (but no tubers), and none set fertile flowers or fruits.

Shepard et al.[394] have been successful in producing four somatic hybrid plants from fusions between chlorophyll-deficient protoplasts of a variegating protoclone (protoplast-derived clone) of the potato cv. Rutgers and Nova cultivars of tomato. The potato protoclone (774) has been previously described[208] and has a normal complement of 48 chromosomes. One somatic hybrid was identified from regenerated populations of the 774 potatoes crossed with Rutgers, and three resulted from fusions of 774 and Nova. The selection scheme developed for somatic hybrid colonies was based on the following observations. First, tomato mesophyll protoplasts divided in very low efficiency when cultured in the light at 24°C, whereas these conditions are optimal for potato mesophyll protoplasts. Second, small protoplast-derived calli (p-calli) of potato did not grow when abscisic acid (ABA) was included in culture media at a concentration exceeding 0.5 mg/l. The growth rate of tomato p-calli, in contrast, is either unaffected or slightly stimulated at the same ABA levels. Tomato p-calli did not undergo shoot morphogenesis under conditions that were inductive for potato. When green adventitious shoots developed into small plantlets, the final screening characters, which appear when tomato shoots are regenerated from leaf disk callus, were employed: these characters were the formation of lobes and serrations in leaflets and reddish-purple pigmentation in stems.

General morphological characters were consistent for all somatic hybrid plants.[394] The basic plant growth habit was that of a potato-like vine. Terminal and lateral leaflets were deep green in color and displayed serrations and lobing; at 28°C, plants grew vigorously and anthocyanin (red) pigmentation accumulated in stems and on the underside of leaves; vegetative cuttings accumulated anthocyanins throughout and eventually died. White tubers (2 to 11 cm long) were produced that turned reddish purple if exposed to light during development. Floral characters were identical to those of parental Russel Burbank potato except for the 774-Rutgers hybrid whose petals were light yellow. Sterile fruit up to 2.5 cm in diameter having a yellow color at maturity and liberating a tomato-like odor developed on both Rutgers and Nova hybrids. All hybrids were sensitive to root-invading microorganisms and required initial establishment in sterilized vermiculite. Moreover, under routine greenhouse conditions, somatic hybrid plants were susceptible to the powdery mildew fungus, whereas neither the potato nor the tomato parents were susceptible. These characters are found only in the somatic hybrids reported by Shepard et al.[394] and were not reported for those previously described by Melchers,[170] nor have they been observed in potato protoclonal population. Gel electrophoresis of restricted mtDNA and cpDNA from 774 Rutgers and 774-Nova-1 revealed the extranuclear DNAs only of potato, suggesting that the plants were true hybrids. If the plants had displayed mtDNA and cpDNA of both parents, they could have been chimeras composed of a mixture of potato and tomato cells rather than hybrids. Analysis of the small RUDPcase subunit from 774 Rutgers, 774-Nova-1, and 774-Nova-3 plants by isoelectric focusing in polyacrylamide gels established the presence of small subunit polypeptides of both tomato and potato.[394] The 774-Nova-2 plant was not completely examined. Profiles of several isozymes (peroxidase, malate dehydrogenase, esterase, 6-phosphoglucomutase, and polyphenol oxidase) were analyzed from leaf tissue of 774-Rutgers and compared with those of parental tomato and potato and with a random population of 15 regenerated protoclones of Russet Burbank potato. For each enzyme, hybrid extracts shared specific bands with both potato and tomato. However, within the protoclonal population, individuals that shared some tomato-specific bands for each enzyme were identified. No protoclone displayed all of the unique bands of the hybrid.

In another pairing between incompatible members of the same family *Datura innoxia* and *Atropa belladona,* only calli with fleshy leaves were regenerated from synkaryon that retained all chromosomes of both the parents.[346,347] This developmental block was termed "somatic incompability", unlike *Arabidobrassica* cell lines that had lost one or more *Arabidobrassica* chromosomes.[347] The genus *Petunia* also contains compatible and incompatible species. In early studies with sexually compatible individuals, somatic hybrid plants generally had the predicted amphidiploid chromosome number (2n = 4X = 32)[243] Later, hybrid plants were obtained from fusion between *P. parodii* and *P. parviflora* which were not sexually compatible.[246] None were amphidiploid, but one cell line provided approximately 50 plants with a constant chromosome number of 31. The plants set pollen with a fertility quotient of 36% compared with 98 to 99% fertility for parent pollen. In more distant interspecific combinations, there has been complete chromosomal elimination of one parent. However, in a few examples, some genetic expression from the donor was retained despite total chromosomal loss. Hybrid cell lines of *P. hybrida* fused with *Parthenocissus* and cells lacking chromosomes of the latter species expressed peroxidase isozyme patterns of both parents for at least 1 year.[242] Dudits et al.[119] fused protoplasts of an albino nuclear mutant of carrot *(Daucus carota)* with those of chlorophyll-containing *Aegopodium podagraria.* Green plants from three callus lines had only carrot chromosomes, but molecular hybridization suggested integration of small *A. podagraria* chromosome segments. It is thus evident that except in very distant pairings, the extent or direction of chromosome segregation in interspecific hybrid cell lines is largely unpredictable, and for some combinations virtually any chromosome mix is possible.

Significant improvement in *Brassica* has resulted from protoplast fusions. Schenck and Robbelen[52] have shown that resynthesis of *B. napus* from its ancestral diploids *(B. oleracea* and *B. compestris)* is possible via somatic protoplast fusion. Pelletier et al.[383] have extended these studies to a more comprehensive investigation of intergeneric cytoplasmic hybridization in the Cruciferae by protoplast fusion. The agronomic importance of this work is considerable because the plants produced will probably be useful for *Brassica* hybrid seed production. *B. napus* plants were regenerated after fusion between protoplasts bearing cytoplasms of different genera. One type of hybrid produced had *B. napus* chloroplasts and cytoplasmic male sterility trait from *R. sativus,* and another type had chloroplasts of a triazine-resistant *B. compestris* and cytoplasmic male sterility trait from *R. sativus* with nucleus of *B. napus.*

Rapeseed oil contains appreciable quantities of fatty acids, with chain length greater than the usual 18 carbon atoms. Significant amounts of polyunsaturated acids are also present including significant amounts of linoleic and α-linolenic acids, but not γ-linolenic acid,[395] and it may be possible to modify the nature of these fatty acids using protoplast fusions. One breeding objective, which is also applicable to a range of other species including rapeseed, is to convert linolenic acid to γ-linolenic acid by fusion of protoplasts of *B. napus* with those of *Oenothera biennis* (evening primrose). The evening primrose is a rich source of polyunsaturated fatty acid and linolenic acid and is unique in plants in possessing the gene for the production of the enzyme delta-6-desaturase, which catalyzes the conversion of linoleic acid to γ-linolenic acid. By somatic hybridization, including perhaps the irradiation of evening primrose, it might thereby be possible to obtain rapeseed plants with some linolenic acid synthetic capability. Moreover, this breeding objective involving a single gene transfer would be a very suitable challenge for the assessment of transformation in crop improvement. Increased supplies of linolenic acid for human consumption are required, because it seems likely that a functional deficiency of γ-linolenic acid can cause degeneration of arteries.

Leaf mesophyll protoplasts of *L. esculentum* were fused with suspension culture-derived protoplasts of *S. lycopersicoides* and intergeneric somatic hybrid plants were regenerated.[365] Leaf mesophyll protoplasts of a nitrate reduction-deficient streptomycin-resistant mutant of *N. tabacum* were fused with cell suspension protoplasts of wild-type *Petunia hybrida.*[160] All six lines selected for streptomycin resistance and nitrate reductase deficiency represented nuclear somatic hybrids possessing the chloroplast of *N. tabacum* at an early stage of development. However, 6 to 12 months later genomic incompatibility was observed resulting in the loss of most of the nuclear genomes. Mesophyll protoplasts of plastome chlorophyll-deficient, streptomycin-resistant *N. tabacum* were fused with those of wild-type *Atropa belladonna* using the PEG, high pH, high calcium ion, and dimethyl sulfoxide method.[359] In all 41 cell lines were selected and grown as green colonies.

One of the main reasons for the interest in the area of somatic hybridization is that it may have use for crossing plants that have a conventional incompatibility system. It is evident that some somatic crosses also exhibit a form of cellular incompatibility in that it is impossible for the two nuclei to exist intact in the same cytoplasmic environment. Often there is selective elimination of some of the chromosomes and in extreme cases there is a complete elimination of one nucleus.

As protoplast fusion results in summation of chromosome number, in asexually propagated plants such as potato or sugarcane, variation in chromosome number can be tolerated, but most sexually propagated crops, particularly fruits and vegetables, do not tolerate variation of chromosome number. To overcome this difficulty androgenesis has been proposed to manipulate ploidy level. It has also been suggested that haploid plants derived from the anther culture[117] or anthers of somatic hybrids should be used to recover the original ploidy.[133,134] However, it is difficult to extend anther culture methodology to distantly related interspecies hybrids.

Hybrid vigor is well known in sexual hybridization and it has been suggested[396] that somatic hybridization may produce an even greater vigor in hybrids. Critical evaluation of this suggestion is required, since it could result in enhanced yields in many crops, including alfalfa. An attractive feature for the improvement of certain species is that somatic hybridization enables the whole genome to be combined.

There are several unexplored potential applications of protoplast fusion for crop improvement in addition to their applications described above. Several somaclonal variants in protoplast-derived plants have been observed; the impact of this phenomenon on crop improvement has not yet been fully assessed. There is every possibility that somaclonal variation also appears in somatic hybrids.[321]

Protoplast fusion will thus continue to evolve into a role that will be useful for crop improvement as the limitations are successfully overcome. The rapid development in this field with the unique gene combinations developed through protoplast fusion in recent years ensures that several new plant varieties with useful combinations will be obtained in the years to come.

3. Organelle Transfer by Protoplast Fusions and Cytoplasmic Genetic Traits

Proving the hybrid nature of regenerated plants requires a demonstration of genetic contributions from both the parents. Morphological characters have often provided suggestive infomation, but the range of variability observed in plants raised from nonfused protoplasts[249] weakens the value of intermediate morphology as a sole criterion. Results are the most convincing when expression from both parents is in the form of identifiable biochemical markers that are encoded in plastid, mitochondrial, or nuclear DNA. Although relatively few biochemical markers have been analyzed in somatic hybrid plants, considerable differences in expression (or repression) do occur among individual hybrid lines from the same two parental species and even between different plants derived from a single hybrid cell line. Hence, interspecific protoplast fusions do not necessarily yield populations of somatic hybrid plants that manifest a uniform phenotype or that equally express designated molecular markers, even when all possess the predicted amphiploid chromosome number. Both nuclear and extranuclear gene expression may contribute to such differences because fusion produces hybrid cells that at least initially contain mixed organelle as mixed nuclear chromosome populations. The nuclear and extranuclear expressions have been studied by RUDPcase protein in somatic hybrid plants, as this protein acts as a marker for both nuclear and plastid genomes. Ribulose-1,5-bisphosphate carboxylase (RUDPcase) constitutes a major percentage of total protein in green plant tissues. The enzyme is composed of a chloroplast DNA-encoded large subunit and a nuclear DNA-encoded small subunit that exhibits Mendelian inheritance. Both subunits are composed of several discrete polypeptide chains. The RUDPcase large subunits from one parental species or the other, but not both, have regularly been observed in somatic hybrid plants of *Nicotiana*[397,398] and in potato-tomato hybrids.[170] Rarely is plastid segregation unidirectional, unless there is a genetic lesion in one plastid type or the application of selective pressure. Rather, the consensus is that after protoplast fusion, chloroplasts undergo a random sorting which results in the survival of a single plastid type per cell.[399]

Although intolerance of chloroplast mixtures is a consistent feature of individual cells, multiple plastid types do survive within the tissues of a regenerated plant.[354] Iwai et al.[326] reported only the large RUDPcase subunit of *N. tabacum* in a *N. tabacum* + *N. rustica* somatic hybrid plant, but later they found that in a population of nine androgenetic plants regenerated from anthers of the hybrid, two contained only the large subunit of *N. rustica*. Hence, plastids from both parents must have existed in the original plant.[326]

Since chloroplast segregation predictably follows protoplast fusion, transfer of plastid-determined characters would be aided by techniques favoring the survival of the preferred plastid genome. Potential examples include resistance characters that are encoded in plastid DNA. Medgyesy et al.,[147] for example, used streptomycin to select colony populations after fusions between mitotically inactivated (with iodoacetate) protoplasts of a streptomycin-resistant *N. tabacum* line and those of *N. sylvestris*. Both cybrid (cytoplasmic hybrid) and nuclear hybrid plants that expressed streptomycin resistance were obtained. Other plastid markers with *in vitro* selective potential include resistance to tentoxin (liberated by the fungus *Alternaria tenuis*)[110] and resistance to triazine herbicides.[389]

Differential elimination of parental protoplasts can be achieved by utilizing the mannitol sensitivity of *N. sylvestris* protoplast population[400,401] and the X-ray-induced inhibition of cell division of *N. tabacum*.[104] Zelcer et al.[109] were able to successfully transfer cytoplasmic male sterility by fusion between protoplasts of normal *N. sylvestris* and X-ray-irradiated protoplasts of male sterile *N. tabacum*. X-ray-irradiated protoplasts from plants of CMS cultivar of *N. tabcum* were fused with protoplasts of *N. sylvestris* plants. The selection of fusion products was based on the suppression of cell division by X-irradiation of CMS parent protoplasts and use of mannitol medium which is unfavorable to the protoplast of *N. sylvestris*. Two types of plants were found to regenerate from fusion products. Type A had shoot, leaf, and parianth morphologies of *N. sylvestris*, but their anther resembled the anthers of CMS *N. tabacum* parent and were male sterile. Type B plants grew as rosettes and bore flowers with white corollas similar to *N. sylvestris*. These results led to the conclusion that type A plants bore genomes of *N. sylvestris*, while their plastome was partially or entirely of the CMS parent *(N. tabacum)*. Subsequently, a general procedure to obtain such fusion products and to secure the respective cybrid plants was developed by Galun and co-workers.[109,368,400-403] This procedure led to the unidirectional transfer of organelles, either intra-specifically between cultivar and mutants or interspecifically between cultivars and their wild relatives. They called this technique ''donor-recipient'' technique. By this technique, the nuclear of donor protoplasts is arrested with X-rays or gamma-rays, before fusion with recipient protoplasts. With suitable selection procedures, only colonies from heterofusions are retained. These colonies should result from fused protoplasts containing functional nuclei from the recipient only. The inhibition is removed when the recipient protoplasts are fused with irradiated donor protoplasts.[132] The donor-recipient technique with or without modification is a much more efficient tool for organelle transfer than regular somatic hybridization, by which the progeny is composed of hybrid nuclei which are tetraploid. Such nuclei are often troublesome when the hybrids are wanted for crop improvement and breeding purposes. With the development of this technique significant achievement has been made in the transfer of organelle-mediated characters and in the coming years we may see much more information accumulating on this aspect.

Chloroplasts and mitochondria harbor in their respective genomes (i.e., the plastomes and the chondriome) a multitude of genes; several of these were shown to have breeding value and additional important breeding traits are assumed to be coded by the organelle genomes, since these traits are cytoplasmically inherited in crops whose organelles are transmitted maternally.[370,371,404,405] Several functions are known which are under the control of either plastome or chondriom. In additon to being autonomous for rRNAs and tRNAs, the plastome encodes for several polypeptide subunits of enzymes involved in photosynthesis, such as the large subunit of ribulose-1, 5-bisphosphate carboxylase/oxygenase, subunits of the ATP synthetase complex, and other polypeptides of the photosystems. Probably each of these proteins has subunits coded by the nuclear genome, translated on cytoplasmic ribosomes, transferred into the chloroplasts, and finally complexed with plastome-coded polypeptides. Thus, a concerted synthesis and subunit complementation seem to be required.[404] There is good genetic evidence that traits such as streptomycin,[291] lincomycin,[406] and tentoxin[407] resistance, as well as resistance to the herbicide atrazine,[408] are controlled by plastome. The genetic information encoded in the mitochondrial genome (the chondriome) of angiosperms is only beginning to appear,[405] but there is fair evidence from several crop plants that resistance to specific toxins and alloplasmic (i.e., cytoplasmic) male sterility are chondriom controlled. Several enzymes in the mitochondrion seem to have subunits coded by the chondriome. Thus, as in the chloroplast, a concerted synthesis of nuclear and chondriome-coded polypeptides seems to be required for the normal functioning of mitrochondria.

In almost all crop plants chloroplasts are transmitted maternally and probably so are mitochondria.[354] Thus, organelle-controlled characters are rather difficult to handle by conventional breeding. Moreover, since chloroplasts and mitochondria are transmitted together, sexual hybridization cannot cause transfer of only one of these organelles from a donor source into a target cultivar. The aspect on somatic hybridization in relation to organelles of the hybrid progeny has been reviewed in detail by Galun.[370]

The fate of mitochondrial genomes in synkaryons and ultimately somatic hybrid plants is now becoming clear. Belliard et al.[397] regenerated hybrid plants from fusions between sexually compatible *N. tabacum* and a CMS *N. debneyi*. Their results suggested retention of the male sterility character, possibly residing in mitochondrial DNA (mtDNA) in some hybrid plants, along with either coexistence of multiple mitochondrial types or recombination of mtDNA.[71] Aviv and Galun[110] regenerated six classes of somatic hybrid plants from fusions between *N. sylvestris* and X-irradiated *N. tabacum* protoplasts. Of these, four were cybrid classes containing *N. sylvestris* nuclear genomes, and either (or both) of the chloroplast (tentoxin resistance) and cytoplasmic male fertility characters of *N. tabacum*. The degree of male fertility restoration was independent of plastid origin and, hence, was possibly correlated with mitochondrial composition. It is significant that in this instance male fertility was restored rather than eliminated through somatic fusions. Further evidence that a heteroplasmic (mixed cytoplasm) state for male sterility may be maintained for a considerable period comes from experiments with *Petunia* species. Izhar et al.[331] analyzed male-sterile somatic hybrids with the nuclear genome of *P. axillaris* and the cytoplasm of a male sterile *P. hybrida* line. Experiments were sufficient to reject mosaicism. When chloroplasts from two sources are introduced into one heteroplastomic fusion cell and the cell is cultured to result in a hybrid (or cybrid) plant, sorting of chloroplasts follows resulting in fusion plants with homoplastic cells identical in either parental cpDNA.[149] However, despite a sorting out of mitochondrial characters, the resulting ''cybrids'' mtDNA are similar but rarely identical to those of either fusion partners. There are cases in which chloroplasts and mitochondrion-controlled characters (male sterility) sort out independently.[109,110,155] However, either the donor chloroplast or its mitochondria may not be compatible with the recipient nuclei, thus resulting in transfer from the donor of only one of the two kinds of organelles.

The homoplastic constitution of hybrid plants, presumably caused by rapid sorting out of choloroplasts as reported in many of the cases, may be less common than previously assumed, since cybrids containing a mixed chloroplst constitution could be easily established[149,150] and evidence for heteroplastic constitution is now becoming available. The restriction pattern of mitochondrial DNA in protoplast fusion-derived plants has been analyzed by several workers,[372,398,400,402,408-413] and the information at hand clearly indicates that the chondriomes in hybrid and cybrid plants are rarely identical to those in either of the fusion partners. The fate of both chloroplasts and mitochondria-controlled traits in fusion-derived plants has been investigated in several studies.[109,110,155,414] Most of these plants were hybrids rather than cybrids. Thus, the compatibility between a given nuclear genome and alien plastomes and chondriomes could not be revealed. This difficulty was solved by the development of the donor-recipient technique by Zelcer et al.[109] and subsequent development of iodoacetate treatment of the recipient. In order to follow organelle transfer and sorting out under conditions which will be either selective or nonselective for chloroplast transfer, Aviv et al.[403] used albino *(N. tabacum* VBW), or normally pigmented *(N. sylvestris)* plants as protoplast-recipients and *N. alata, N. bigelovii,* and *N. undulata* were used as donors. A rather simple procedure was developed by Galun and associates based on Southern blot hybridization and radioactive cpDNA probes and electrophoresis of total plant DNA digested with specific endonucleases.[149] In this procedure only small amounts of tissue (200 mg or less) are required and ten or more samples can be handled simultaneously. Organelle composition of the cybrid plants was investigated by the analysis of chloroplast DNA restriction pattern, tentoxin sensitivity, chloroplast pigmentation, mitochondrial DNA restriction pattern, and alloplasmic male sterility. The incidence of chloroplast and mitochondrial transfer was strongly facilitated when iodoacetate-treated protoplasts of *N. tabacum* VBW (albino) rather than *N. sylvestris* (normal chloroplasts) protoplasts served as recipient. Mitochondrial restriction patterns in cybrids commonly differed from those of either fusion partner. All fusion combination resulted in at least some male-sterile cybrids, having either donor or recipient plastomes, thus indicating that interspecific donor-recipient protoplast fusion is an efficient means to produce plants with alloplasmic male sterility. Galun and associates[149,415] were also able to develop a refined cybrid selection process by using plastid-dependent chlorophyll deficiency as a convenient and sensitive marker for direct selection. Along with this system these authors also employed chloroplast DNA restriction patterns and plastid-dependent tentoxin sensitivity to enable direct assaying for plastid recombinant types.

Although cytoplasmic exchange can be achieved by breeding, the protoplast fusion method is not only much less time consuming than repeated backcrossing, it also furnishes the only means to unidirectional in either, rather than both, organelle types. Aviv et al.,[403] using the donor-recipient technique, were able to produce cybrids by using two different species of *Nicotiana: N. sylvestris* as recipient and *N. rustica* as organelle donor. Some of the regenerated plants from such cybrids with either type of plastome produced sterile pollen, but none showed anther malformation typical to alloplasmic male sterility. Chondriom identification by mitochondrial DNA restriction analysis of cybrid plants revealed only restriction patterns which were either similar or identical to those of *N. sylvestris,* while no cybrids with *N. rustica* restriction patterns were detected. It was also noticed that every heterofusion led to mtDNA rearrangements, but the degree of rearrangement was variable from very pronounced to minor, and probably to those difficult to detect.

Plastome-encoded traits in callus culture have been studied in the genus *Nicotiana*. Thus, cell lines resistant to lincomycin[406] and streptomycin[291] have been obtained. A high rate of plastome-encoded mutations was induced in *Nicotiana* by exposing seeds to N-nitroso-M-methylurea (NMU) by Fluhr et al.[151] They subjected such seeds to nutritional and *in vitro* selection procedures for systematic isolation of plastome-dependent antibiotic-resistant plants. Multiple flowering lines resistant to streptomycin, spectinomycin, lincomycin, or chloramphenicol were obtained. Sexual hybridization, cybrid formation following protoplast fusion, and *in organello* protein synthesis were used to rigorously assign the mutations to chloroplast genome.

Aviv et al.[315] also found rhodamine-6G (R6G), a lipophilic dye (which degrades mammalian mitochondria), to arrest the division of *Nicotiana* protoplasts. When albino recipient-protoplasts were treated with R6G and fused with X-irradiated (green) donor-protoplasts, only green cybrid plants were obtained. The mtDNA of cybrids was similar to parental mtDNA.

Alloplasmic plants, i.e., plants in which the original cytoplasm is replaced by the cytoplasm of another species, were traditionally produced by an interspecific cross followed by recurrent backcrosses to the pollen parent.[416] The production of alloplasmic plants often leads to CMS resulting in seed parent lines, which are useful in the production of commercial F_1 hybrid seeds. Recently, alloplasmic-like lines were recovered from various protoplast fusion experiments. Most of the reported cases involved experiments with *Nicotiana*.[109,323,397,402] The major differences between sexual crosses-derived alloplasmic lines and alloplasmic-like lines obtained via protoplast-fusion are that in the former all the cytoplasmic components (both chloroplasts and mitochondria) are contributed solely by the alien (original female parent) species. While in the latter, the first event is a coexistence of the two cytoplasmic components in the same fusion product cell, followed by partial or total sorting. Therefore, in alloplasmic-like lines the complete replacement by the alien (donor) cytoplasm is expected to be rare; in most cases novel chloroplasts/mitochondria combinations will be produced, e.g., recipient chloroplasts with donor mitochondria; recipient mitochondria with alien chloroplasts. Moreover, the initial cytoplasmic coexistence may lead to recombination of organellar DNAs. While there is very limited evidence for chloroplast DNA (cpDNA) recombinations,[417] novel mitochondria DNA (mtDNA) restriction patterns, which suggested mtDNA recombinations, are frequently observed.[397,392,408,418,419]

As no correlation was found between chloroplast-type CMS, it is currently assumed that nuclear mitochondrial interaction is controlling male sterility/fertility as found in *Nicotiana*.[155,397,420] There are also indications that male fertility is restored by certain nuclear (restorer) genes. In fact, Aviv et al.[155] have shown that fertility can be restored in *Nicotiana* by transfer of cytoplasm from a fertile donor into an alloplasmic CMS recipient. Galun and associates reported for the first time a case of male fertility restoration by an apparent recombination of mtDNAs from two different CMS *Nicotiana* cybrid lines.[314] Using a donor-recipient technique, they constructed several alloplasmic-like lines of *Nicotiana* in which the original cytoplasms (or part of them) of either *N. tabacum* or *N. sylvestris* were replaced, respectively, either by *N. undulata* or by *N. bigelovii* cytoplasms. These hybridizations resulted into two kinds of CMS cybrid plants: *N. tabacum* with *N. undulata*-like cytoplasm and *N. sylvestris* with *N. bigelovii*-like cytoplasm. Fusion of protoplasts, derived from the above CMS types, led to the development of 21 calli, out of which only one regenerated cybrid produced plants with fertile pollen, short filaments, and slightly tapered anthers. Self-pollination in this cybrid plant resulted in the second-generation progeny with almost normal filaments and anthers. Selfing produced a third generation with numerous plants having normal stamens and fertile pollens. Mitochondrial DNA (mtDNA) analysis of second and third generation revealed a novel pattern which differed from each of the parental CMS cybrids and also from the mtDNA or normal, male-fertile *Nicotiana* species. Thus, mtDNA recombination between different types of CMS can restore male fertility.[314]

Aviv and Galun[374] further undertook investigations to study the stability of recombinant mtDNA in the sexual progenies of cybrids and the analysis of the sibling plants of such progenies. Generally, the mtDNAs of male-fertile, second-generation plants were similar to the mtDNAs of the original recipients, while mtDNAs of the male-sterile second-generation plants were similar to the mtDNAs of the donor *(N. begelovii)*. The analysis of mtDNAs from the third-generation plants indicated stabilization of the chondriomes; no variations were detected between the mtDNAs of plants derived from a given second-generation plant.

4. Nuclear Genetic Traits and Their Transfer through Somatic Hybrids

During early studies, there was considerable hope for new amphidiploid plants from fusion between protoplasts of sexually incompatible individuals. However, as had already been established in animal cell research, somatic combinations between distantly related or unrelated genomes were regularly followed by the elimination of chromosomes from cell lines. The objectives of interspecific fusions have thus shifted away from synthesis of novel amphidiploid plants toward the introduction of small genetic elements from alien species into ones of practical interest. In the few cases in which sexual progeny derived from somatic hybrids has been described, attention has been directed to a small number of genetic loci or to a morphological character of interest.

Chlorophyll mutants, used to select somatic hybrids, have been monitored most frequently in subsequent sexual generations. Melchers[328] analyzed progeny obtained following self-fertilization of diploid intraspecific *N. tabacum* somatic hybrids produced by fusion of two different haploid albino mutants. Segregation ratios were obtained that were statistically consistent with the ratios from comparable diploid sexual hybrids. Anthers of intraspecific tetraploid *D. innoxia* somatic hybrids produced by fusion of two different albino mutants were cultured to examine segregation ratios in androgenetic diploid plants.[133,134] The segregation ratios of green to each albino mutant following anther culture were as expected for autotetraploid somatic hybrid plants that are duplex for two loci, although no double albino mutants (presumably lethal) were recovered. Power et al.[246] observed segragation in R_1 progeny of the albino mutant used to recover somatic hybrids between *P. parodii* and *P. parviflora*. Similarly, Schieder and Vasil[367] reported segregation of albino seedlings in self-fertilized progeny of *Datura innoxia* + *D. stramonium* or *D. discolor* somatic hybrids, and Evans et al.[85,320] reported segregation of light-green and albino seedlings in backcrossed and self-fertilized progeny of *N. nesophila* + *N. tabacum* and *N. sylvestris* + *N. tabacum* somatic hybrids. Segregation ratios have been distorted (non-Mendelian) in hybrids between distantly related species, but this distortion is also observed in distant sexual hybrids.

Morphological traits, though often not ascribed to a single gene, have also been monitored in the progeny of several somatic hybrids. Both *P. hybrida* + *P. parodii* somatic hybrids and amphiploid sexual hybrids were self-fertilized and both segregated in the same manner for flower color variations.[243,421] Schieder[133] has monitored flower color in several *Datura* somatic hybrids. Evans et al.[122] monitored morphological characters in three generations of progeny derived from *N. glauca* + *N. tabacum* somatic hybrids. Adopting the terminology of Chaleff,[422] somatic hybrids that have been regenerated from fused protoplasts are called R/R_o plants, while the progeny of self-fertilized somatic hybrids are R_1 plants. Subsequent self-fertilized progeny have increasing R numbers as subscripts. As previously reported,[122] the somatic hybrids are intermediate between *N. tabacum* and *N. glauca* for all five morphological characters. Morphological characters of somatic hybrids and R_1 plants are comparable to F_1 and F_2 *N. tabacum* × *N. glauca* sexual hybrids.[120] For all traits except corolla length, R_2 plants were intermediate between the two backcross populations. As expected, the backcrosses to *N. glauca* more closely resembled *N. glauca,* while backcrosses to *N. tabacum* more closely resembled *N. tabacum.* Both R and R_1 plants contained the expected chromosome number ($2n = 72$), while chromosome elimination was observed in the next generation. Only two of eight backcrosses to *N. glauca* have the expected chromosome number, as all 24 other plants in this generation lost from 2 to 14 chromosomes. No mosaic plants were detected. No single morphological character in the backcrosses could be directly correlated with the number of chromosomes lost. On the other hand, for most characters, the R_2 plants seemed to resemble the backcrosses to *N. tabacum* more than the backcrosses to *N. glauca.* In this generation, several morphological abnormalities were detected including a plant with very narrow leaves. This R_1 × *N. tabacum* (same generation as R_2) plant had 55 chromosomes. Another R_1 × *N. tabacum* plant contained 56 chromosomes and had several abnormal flowers with deformed corollas. In these flowers anthers were fused to the petals, and no pollen was formed. Pollen viability was extremely variable in the plants of this generation. Several of the backcrosses to *N. glauca* had no viable pollen, while no backcrosses to *N. tabacum* and only one R_2 plant had 0% pollen viability. In general, pollen viability was not correlated with number of chromosomes lost. Isozyme for asparate aminotransferase and alanyl aminopeptidase was monitored for each plant. All R and R_1 plants had all the *N. tabacum* and *N. glauca* bands for these enzymes. Nearly all plants in the next (R_2) generation also retained all *N. glauca* and *N. tabacum* bands for both enzymes. Only one $R_1 \times$ *N. tabacum* plant was identified that lost the *N. glauca* bands for aspartate aminotransferase. However, even this plant retained the *N. glauca* band for alanyl aminopeptidase.

The potato-tomato somatic hybrid plants were cytologically examined at meiosis and mitosis.[394] Observation of root tip cells of the 774 Rutgers hybrid shortly after initial transplanting consistently showed 72 chromosomes, the predicted number of a true amphiploid. Over the next 12 months, numerous vegetative cuttings were made from the hybrid, and root tip cells were analyzed for somatic chromosome number. The process was then repeated for each of the 774-Nova hybrids. Results from these experiments indicated that root tip cells of Nova and Rutgers somatic hybrid cuttings displayed chromosome numbers ranging from 62 to 72 depending on the cutting. The most frequently encountered chromosome number for the Rutgers hybrid was 70, with greater variability observed for the Nova hybrids. These data indicate a degree of mitotic instability and some chromosome segregation in vegetative cuttings, but not wholesale chromosome elimination. Phenotypic variations in the form of abnormal leaflets and color deviations were occasionally observed, particularly among cuttings of Nova-2 hybrid, but they could not be correlated with a specific change in chromosome number. Each of the hybrid plants flowered profusely but produced no viable pollen. Since parental Russet Burbank potato expresses the same deficiency, it is uncertain whether some measure of fertility would be possible with another potato parent. In meiosis, there was clear evidence of chromosome elimination for 774-Rutgers and 774-Nova-1. The remaining two Nova hybrids were not analyzed.

One objective of the study conducted by Shepard et al.[394] was to know whether karyotype stability can be achieved in populations of vegetative cuttings from hybrid plants. To this end, subpopulations from cuttings displaying chromosome numbers up to 72 were made in order to establish whether any lines will stabilize. Moreover, protoplasts have been cultured from somatic hybrid plants of 774-Rutgers to determine whether karyotype stability will prevail in the second somatic (S_2) generation. Shepard et al.[394] were able to raise 50 plants from these protoplasts; one S_2 protoclone displayed the leaf morphology of potato, with no evidence of lobing or anthocyanin pigmentation. Electrophoretic analysis showed the RUDPcase small subunit of both tomato and potato. The predominant chromosome number of root tip cells was 64; hence, although the protoclone had assumed a more potato-like phenotype, it had probably not lost all tomato chromosomes. There are also S_2 protoclones that display a more tomato-like morphology than is seen in their hybrid parent, including more intense red pigmentation, more pointed terminal leaflets, and more extensive leaf serration.

In the last 10 years progress in the development of protoplast fusion, culture, regeneration, and hybrid-selection system has resulted in the formation of several interspecific somatic hybrids. However, if these hybrids are to be useful in breeding programs, they must have some fertility. With the exception of a few cases, interspecific somatic hybrids are sterile. An additional difficulty is their amphidiploid nature which requires an extensive backcrossing to introduce desirable genes into the target crop species. The formation of asymmetric somatic hybrids carrying the complete genome of the recipient species plus a few chromosomes or chromosome segments from a donor species may provide an approach to solve some of these problems. Several methods are available for forming asymmetric hybrids: egg transformation, uptake of isolated chromosomes by protoplasts, and the fusion of protoplast combinations in which most of the chromosomes of one species are spontaneously eliminated.[119] Chromosome elimination can also be induced in somatic hybrids by irradiation of one of the protoplast types prior to fusion. This method has been applied in several laboratories to transfer organelle-encoded traits to form asymmetric nuclear hybrids.

X-ray irradiation prior to fusion does indeed lead to the loss of DNA from the irradiated partner, but relatively large amounts of nuclear DNA from the irradiated partner are still retained. Somatic fusion hybrids between tobacco *N. tabacum* and henbane *(Hyoscyamus muticus)* with prior X-irradiation of one partner revealed that irradiation significantly reduced the amount of chromosomal DNA of the irradiated fusion partner in the somatic hybrid.[423] Irradiation with doses which completely inhibited protoplast division did not prevent the transfer of a substantial amount of DNA into cybrids with the partial genome from the irradiated partner.

X-ray irradiation is now in common use to promote the selective transfer of organelle and nuclear traits in the production of cybrids.[403] Carrot plants regenerated from fusion hybrids between X-ray-irradiated parsley *(Petroselinum hortense)* and untreated carrot *(Daucus carota)* protoplast contained normal complement of carrot chromosome together with an extra chromosome.[387] Fusion of tobacco *(N. sylvestris)* protoplasts with X-ray-irradiated *Datura* or *Physalis* protoplasts led to the correction of nitrate reductase deficiency of the tobacco cell line.[424] Plants regenerated from fusions between X-ray-irradiated *Physalis* and untreated nuclear albino mutant of *Datura* complement.[425] Recently, Bates et al.[426] described the formation of asymmetric somatic hybrids between two species of *Nicotiana*. They fused γ-irradiated mesophyll protoplasts of a kanamycin-resistant (Km^r) nopaline-positive (NoP^+) line of *N. plumbaginifolia* with wild-type protoplasts of *N. tabacum*. Although irradiation induced the loss of *N. plumbaginifolia* chromosomes from the hybrids, selection on kanamycin allowed recovery of plants retaining the chromosome encoding Km^{γ}/NoP^+. Due to the male sterility encountered in the hybrids it was not possible to stabilize the chromosomes that had been transferred from *N. plumbaginifolia* to *N. tabacum*. If the problem of male sterility can be solved, the donor-recipient fusion technique can provide a one-step bridge for the interspecific transfer of nuclear-encoded traits. This approach has also permitted the production of genetically different asymmetric and symmetric hybrids between protoplasts of *S. pinnatisectum* and *S. tuberosum* and *S. phureja*.[340]

IV. STRATEGIES FOR THE TRANSFER OF NEW TRAITS AND DEVELOPMENT OF A NEW VARIETY

A. TRANSFER OF ORGANELLE AND ORGANELLE-CONTROLLED TRAITS

Organelle transmission in most higher plants is uniparental and chloroplasts and mitochondria are transmitted together, maternally, irrespective of whether the plants are self- or cross-pollinated. Transfer of heritable traits by conventional genetic manipulation, as commonly practiced for nuclear genes, is therefore not applicable to organelle genes. Somatic cell hybridization may prove to be an efficient and successful method for the transfer of organelle traits. Though there is no direct evidence in support of this, future work in this direction may provide certain examples of gene transformation through organelles into the plants. Galun and Aviv[427] proposed a scheme to transfer chloroplasts and chondriome (mitochondrial genome)-controlled traits from specific donor plants to recipients. They have discussed in detail procedures for organelle transfer or organelle-controlled trait isolation, culture, and fusion of protoplasts leading to plant regeneration from cultured protoplasts. The scheme as proposed by Galun and Aviv[427] is described here briefly.

1. The Donor-Recipient Protoplast Fusion

Galun and Aviv[427] proposed that for the construction of a plant having in its cells a given nuclear genome but chloroplast and/or mitochondria from another plant, a donor-recipient fusion should be performed. The cybrid plants, derived from such a fusion, will contain recipient protoplast in which the original functional nucleus is retained, but either one or both of the cytoplasmic organelles (i.e., chloroplasts and mitochondria) are exchanged by organelles from a donor plant. It is difficult to carry out unidirectional transfer of organelles, because efficient techniques for specific inactivation of chloroplasts and/or mitochondria are still to be developed. The only available alternative is to suppress the division capability of the nucleus in protoplasts from the donor plant by exposing it to γ- or X-rays. Such a donor plant is fused with recipient protoplast. The fusion product formed in this way contains only one functional nucleus (of the recipient), but a mixture of chloroplasts and mitochondria from both the donor and the recipient. Aviv and Galun[427] suggested that subsequent cell division and plant regeneration through sorting out of organelles leads to the formation of four types of plants: (1) donor's chloroplasts, donor's mitochondria; (2) donor's chlorplasts, recipient's mitochondria; (3) recipient's chloroplasts, donor's mitochondria; and (4) recipient's chloroplasts, recipient's mitochondria. All these plants shall have the recipient-nuclear genome. Galun and Aviv were able to obtain these four types of combinations with certain exceptions. The cybrid plant rarely contained the exact chondriomes of either the donor or the recipient plants.[427] Commonly, the mitochondrial DNA of the cybrids, as expressed by their restriction pattern, had similarity to either of the fusion partners. This suggests that rearrangement or recombination occurs in the chondriome's genophores following the fusion between donor and recipient protoplasts. In addition, in some fusion combinations sorting out of chloroplasts was not complete and the heteroplastomic state was maintained in the sexual progeny. For practical purposes, the lack of pure parental-type chondriomes in the cybrids constitutes an advantage because there is a sorting out of chondriome-controlled traits (e.g., alloplasmic male sterility), hence, the actual aim of the organelle transfer is achieved.

Because the donor-recipient protoplast fusion technique enables the transfer of chlorplasts and/or chondriome-controlled traits from a given donor to a recipient plant, it may be applied to transfer chloroplasts having specific genetic features such as resistance to antibiotics (e.g., lincomycin, streptomycin, spectinomycin), resistance to herbicides (e.g., atrazine), resistance to fungal toxin (e.g., tentoxin), or pigmentation deficiency. This technique may be used to transfer cytoplasmic (better termed alloplasmic) male sterility as well as to test whether or not a pigmentation deficiency is plastome controlled.

Galun and Aviv[427] explained in their scheme the steps required to achieve these aims with respect to *Nicotiana,* because this is the plant which has been worked out extensively in relation to protoplast fusion and regeneration.

a. Nature of Donor and Recipient

Depending on the type of trait to be transferred through this technique a suitable type of donor and recipient plant must be selected. When a specific plastome or plastome-controlled trait is to be transferred into a recipient plant, a donor plant harboring the respective chloroplasts should serve as a source for donor protoplasts. The same approach is applicable for chondriome-controlled traits. In addition, the donor plants should preferably differ from recipient plants in readily recognizable morphological, nuclear-coded traits such as leaf shape, flower shape, and color. This is essential to carry out future selection of the cybrids. Donor's protoplasts with normal-pigmented chloroplasts may be fused with recipient protoplasts having plastome-controlled pigmentation deficiency.[402] Chloroplast transfer is also facilitated following fusion of donor protoplasts containing antibiotic resistance with recipient protoplasts containing sensitive chloroplasts.

Nuclear-controlled traits, such as nuclear-coded chlorophyll deficiency, inability to reduce nitrate, resistance to amino acid analogues and other toxic compounds, as well as suppression of cell division (or protoplasts) in certain culture media, should be applied successfully to establish positive selection for cell colonies resulting from heterofused protoplasts. Galun and Aviv[427] suggested that protoplasts with the capability of dividing and ultimately resulting in functional plants should serve as recipients, which is not essential for donor protoplasts. Utilization of protoplasts from two different sources, such as mesophyll protoplasts as donors and cell suspension protoplasts as recipients, was found to be advantageous because such heterofusions can be readily observed and isolated by micromanipulation.

For specific organelle transfer experiments the recipient protoplasts should harbor a plastome-controlled pigmentation deficiency or another type of chloroplast mutation. Such mutants can be readily obtained in *Nicotiana* by seed treatment with NMU followed by germination in nonselective or selective nutrient medium.[97] A similar method to obtain variegated tomato was also reported. The modification of techniques for mutagenesis can also lead to the formation of plastome mutants induced by NMU in other species.

b. Iodoacetate Perfusion Treatment

It will be advantageous to obtain the elimination of unfused and homofused recipient protoplasts. This can be achieved by perfusion treatment of protoplasts by iodoacetate. This method is based on the finding that in mammalian cells the ratio of hybrid cells can be increased if each of the fusion partners is pretreated with a (different) chemical causing metabolic lesions; complementation in the fusion product releases the metabolic inhibition, but nonfused (and autofused) cells will not recover from the lesions. Maliga and co-workers used iodoacetate in *Nicotiana* to cause metabolic lesion in one of the fusion partner's protoplasts.[132,147] A virtually complete killing of unfused *Nicotiana* protoplasts is obtained by a 30-min exposure to 0.3 to 0.4 m*M* iodoacetate. In practice the protoplasts are released from the maceration enzyme mix, washed once, exposed to iodoacetate, and then washed twice with the same washing solution.

c. Arrest of Nuclear Division in Donor Protoplasts

Nuclear division in plant protoplasts can be arrested by X-ray radiation. The doses causing virtually total division arrest vary considerably among types of protoplasts. Hence, while "diploid" tobacco protoplasts require 5 krad (50 J kg^{-1}), about half of this dose is sufficient to arrest cell division in haploid tobacco protoplasts; protoplasts derived from calli suspension cultures require much higher radiation doses, e.g., about 20 to 50 krad for *Citrus* protoplasts derived from embryogenic calli.[427] When higher radiation doses are required it is more practical to use a cobalt radiation source than an X-ray-emitting Roentgen machine. Menczel et al.[128] reported a γ-ray dose dependence for the efficiency of chloroplast transfer. Ionizing radiation also has side effects; however, even exposure of *Nicotiana* protoplasts to high radiation doses (e.g., 100 krad of γ-rays) did not produce any organelle mutations.[427]

2. Selection, Isolation, and Regeneration

In fusion combinations where no positive selection of heterofusion products is available, as well as in other specific cases, it may be useful to stain the donor and/or the recipient protoplasts before, with fluorescence dyes. This method has already been described in detail. The fusion of protoplast in the donor-recipient technique is carried out by employing a regular procedure of somatic hybridization. Once the fusion is over, the product can be isolated at an early stage or at a later stage of plating in appropriate culture medium. The latter selection can take place right after protoplast plating, at the stage of small calli, or during differentiation of the calli to shoots and roots. Selection at the plating stage is exemplified by fusion between *N. sylvestris* and X-irradiated protoplasts of other *Nicotiana* species. The selection is based on the sensitivity of *N. sylvestris* protoplasts to mannitol in the culture medium. After fusion, the protoplasts are suspended in liquid medium containing 0.6 *M* mannitol, maintained overnight in the refrigerator, and then transferred to a culture room (25°C).[427] About 10 d after plating the suspension is gradually diluted with MS medium and the illumination is increased. Control plates containing nonfused *N. sylvestris* protoplasts should be included in each experiment to ensure that no division occurs in these plates, indicating that the selection against these protoplasts is satisfactory. Selection at the stage of small calli is exemplified by the fusion between X-irradiated protoplasts of *N. tabacum* line 92 (*N. tabacum* nuclei, *N. undulata* chloroplasts) having streptomycin resistance and CMS, and protoplasts from an albino *N. tabacum* (VBW) line. Three to 4 weeks after fusion, the small calli are transferred individually from the liquid medium to plates with solidified (agar) MS medium containing streptomycin (1 mg ml^{-1}). Calli with cells containing streptomycin-resistant chloroplasts will start greening. The latter are then transferred to MS containing 0.8 mg ml^{-1} indoleacetic acid and 2 mg ml^{-1} kinitin to induce shoot regeneration. Finally, shoots are plated in Nitsch agar devoid of growth regulators and the rooted shoots are potted in peat moss. The regeneration of cybrid plants, derived from the donor-recipient fusion technique, is performed as detailed for protoplast-derived plants and for the regeneration of somatic hybrid plants.

3. Identification of Organelle-Controlled Traits

a. Plastome Identification

Somatic hybrid, or cybrid plants, resulting from fusion between protoplasts harboring different plastome compositions are not necessarily homoplastomic. Thus, the analysis of cybrid plants derived from the donor-recipient fusion method may reveal two types of chlorplasts in the same plant, in the same leaf, or even in the same cell. Pigmentation is simple and provides an obvious indication of plastome composition. For example, if the recipient plant is albino and the donor normal green, all the green plants resulting from the respective donor-recipient fusion are probably cybrids with donor's chloroplast. Nevertheless, it is desirable to augment the "observed" transfer of chloroplasts with additional identifications. Fusion between *N. bigelovii* (donor) and *N. tabacum* (recipient) protoplasts may serve as an example.[427] Rather than normal pigmented *N. tabacum,* and albino line, VBW, serves as recipient. Green cybrids with *N. tabacum* morophology are suspected to contain *N. bigelovii* plastome. To verify this assumption the cybrids are tested with respect to tentoxin reactivity (*N. tabacum* and *N. bigelovii* chloroplasts are resistant and sensitive, respectively, to this fungal toxin). In addition, the chloroplast DNA of the cybrids can be analyzed. Variegated cybrids, derived from this fusion, are suspected to have *N. tabacum* plastomes (in albino leaf areas) and *N. bigelovii* (in green areas). This assumption should be verified by cpDNA analysis of the respective leaf areas.

The fungal toxin tentoxin is a cyclic tetrapeptide (cyclo-L leucyl-*N*-methyl-(Z)-dehydrophenylalanylglycyl-*N*-methyl-L-alanyl) produced by certain races of *Alternaria.* It causes chlorosis in certain angiosperm species but does not affect others. Tentoxin is, therefore, a useful chloroplast marker in cybrids derived from fusion between protoplasts which harbor resistant and sensitive chloroplasts, respectively. The tentoxin test can be performed with either the leaves of the cybrid itself or the first generation of seeds. The latter are obtained by self-pollination or cross-pollinating the cybrid with a compatible pollinator. Procedures to identify chloroplast composition in plants resulting from somatic hybridization and cybridization by isoelectric focusing of RUBPcase were developed for *Nicotiana* species, and can readily be modified to fit other genera.[428] Uchimiya[319] developed a microscale procedure for 300 to 500 mg of leaves. The advantage of the isoelectric focusing analysis is that it will reveal the migration of the small subunits of the RUBPcase which are nuclear coded. Hence, the analysis will also identify the nuclear genome of the tested cybrid tissue. The base sequences of chloroplast DNA (cpDNA) from different plant species, belonging to the same genus, are usually sufficiently variable to render restriction endonuclease analysis useful for characterizing the plastome compositions of cybrids derived from the donor-recipient protoplast fusion technique. This analysis is performed either by the large scale or by the small scale method. The techniques for isolating plant cpDNA and subsequent restriction endonuclease analysis by gel electrophoresis and ethidium bromide staining from cybrid plants do not differ from the usual methods employed to characterize cpDNA.[323] The small-scale method is based on total DNA extraction and requires only 100 to 250 mg leaf tissue, thus allowing nondestructive cpDNA characterization of cybrid seedlings or of specific sectors of the same leaf blade. The analysis is based on differential Southern hybridization of blots containing fragmented total DNA with cloned cpDNA fragments. The detailed procedure was described by Galun and Aviv.[427]

b. Chondriome Analysis

The strategy for chondriome analysis in hybrid and cybrid plants, derived from protoplast fusion, is basically the same as presented above for plastome analysis. The methods have been worked out by Galun and Aviv.[427] The chondriome can be identified either by their expression or by mitochondrial DNA (mtDNA) analysis. The expression can be manifested in either biochemical (e.g., ATPase) or in morphological traits (e.g., male sterility).[322] Mitochondrial expression could be used to differentiate between two chondriomes in an indirect manner: resistance or sensitivity to specific compounds (e.g., antibiotics) provided that the fusion partners' mitochondria differ in their reactivity to such compounds. Unfortunately, such differences have not yet been reported in angiosperm mitochondria. Mitochondrial enzymes are either coded by nuclear genes and transferred into the mitochondria or they are composed of polypeptide subunits of which one or more are chondriome coded (e.g., a subunit of ATPase in maize).[322] Functional holoenzyme is thus a result of concerted interaction between chondriome and nuclear genome translation products. The known "expressions" of plant mitochondria are therefore results of specific interactions between a given nuclear genome and the chondriome harbored in the same cell, rather than an autonomous manifestation of the chondriome. The most prominent expression of chondriome/nuclear genome interaction is a morphological one: malformation of floral members, commonly coined CMS but better termed alloplasmic male sterility. While its phenotypic expression varies from morphologically intact anthers containing nonfertile pollen to complete lack of stamens, most and possibly all reported cases of induced alloplasmic male sterility have a common denominator: an aberrant nuclear "cytoplasmic" interaction. Accumulated evidence indicates that the "cytoplasm" in this interaction is the chondriome. CMS is clearly expressed in mature plants. It can therefore be conveniently used in analyzing the chondriomes in the cybrid progeny derived from protoplast fusion. The base sequence of mtDNA, among different species of the same genus and even among different subspecies, is sufficiently variable to serve as a convenient marker for chondriomes in hybrids, and cybrids obtained by somatic hybridization. Moreover, unlike CMS, the analysis of mtDNA can be applied to unorganized cells.

Galun and Aviv[371] reviewed the studies on somatic hybridization in which progeny plants with CMS were reported. In most of these studies one of the fusion partners was derived from a CMS plant and the other partner from a normal fertile plant. More recently, it was shown that even fusions between recipient and donor protoplasts—which were both derived from fertile plants—can result in CMS cybrids. Hence, the donor-recipient protoplast fusion is useful to transfer CMS (from donor to recipient) as well as to "create" CMS. Some alloplasmic male sterilities are clearly expressed by typical floral member malformations, e.g., complete lack of stamens, stigmatoid anthers, petaloid stamens, split corolla. These male sterilities can be determined at anthesis and even in the floral buds. In other cases (e.g., cybrids between *N. sylvestris* and *N. rustica*) the floral members of CMS plants are normal and pollen is produced, but the latter does not germinate. In the latter cases, the CMS of protoplast fusion progeny should be determined by pollen germination.

The analysis of mtDNA in protoplast fusion progenies is based on the restriction patterns obtained by gel electrophoresis of endonuclease-digested mtDNA.[427] The mtDNA is obtained from purified mitochondria isolated from leaf homogenates or from ruptured cell cultures; thus, the isolation and purification techniques include means to exclude nonmitochondrial DNA (e.g., DNAase treatment of isolated mitochondria). When the mtDNA of individual cybrid plants is to be analyzed the leaves from one plant may not furnish enough plant material for mtDNA analysis. Plant material can be increased by either of two means: (1) the plant is pollinated and the sexual progeny will serve as a mtDNA source; (2) an explant is cultured *in vitro* providing cells that are propagated in suspension until sufficient plant material becomes available. Both means are time consuming and improved procedures allowing reliable mtDNA analysis of small plant samples (i.e., of less than 10 g) are desirable. As mtDNA in cybrid plants may exhibit extensive rearrangements relative to the parental mtDNA, Southern hybridization to several mtDNA probes is recommended by Galun and Aviv.[427]

B. DEVELOPMENT OF A NEW VARIETY

Galun and Aviv[429] presented a hypothetical scheme to generate a new plant. The scheme was discussed with reference to *Cucumis melo* (melon). Following are the steps proposed by them.

Step 1a — Protoplast is isolated from a cultivar which will serve as recipient. Simultaneously, protoplast is isolated from a *Cucumis* plant source of the same species with chloroplast containing herbicide resistance genes and will serve as chloroplast donor. Recipient protoplasts are treated as described earlier, fused, and several plantlets are raised.

Step 1b — In this step, the procedure of step 1a is undertaken but donor protoplasts will be obtained from a wild *Cucumis* species whose choice may be based on previous information, indicating that the combination of its ''cytoplasm'' induces male sterility when combined with *C. melo*.

Step IIa — The cybrids resulting from step 1a will be screened for their chloroplast compositions and those having herbicide-resistant chloroplasts will be retained.

Step IIb — The cybrids resulting from step 1b will be screened at the plantlet stage for mitochondrial compositions. Those having wild *Cucumis* mitochondria will be retained or the plantlets will be kept up to flowering and those showing male sterility will be retained.

Step III — Protoplasts from plants retained from step IIa will be fused with protoplasts of plants retained from step IIb. A donor-recipient fusion procedure will be employed leading to cybrids. Protoplasts of either step IIa or step IIb may serve as donors and, respectively, protoplasts of either step IIb or step IIa may serve as recipients. The fusion products will be cultured up to the plantlet stage.

Step IV — The plantlets obtained in the previous step will be screened for both chloroplast and mitochondria. Those containing herbicide-resistant chloroplasts and mitochondria from the wild *Cucumis* will be retained. Rather than testing for mitochondria the screening may be delayed up to flowering and male sterile plants with the required chloroplasts will be retained.

Step V — Plants retained from step IV will be crossed with pollen from a target *C. melo* cultivar and as many plants as possible will be raised to flowering. The plants should be male sterile.

Step VI — Anthers from step V plants will be cultured to induce androgenesis. The experience with tobacco, tomato, and other plants (that male sterility should be manifested beyond the first microspore mitosis) appears to indicate that male sterility will not interfere with androgenesis. The haploids obtained from anther culture will be cultured to plantlet stage.

Step VII — If diploidization does not occur simultaneously an induced diplodization will be performed.

Step VIII — The diploid plants obtained by the previous step will be screened for horticultural characters such as yield and fruit quality, and those closest to the target cultivar will be retained.

Step IX — The latter plants may be propagated by repeated backcrossing with the target cultivar and will serve as breeding material containing the required organelle composition, in order to establish potential seed-parents for commercial F_1 hybrid melon varieties.

The male-sterile breeding lines established in step VIII will serve in conventional breeding procedures[430] to find the best hybrid combination between these lines and normal breeding lines which will serve as pollen parents.

All these steps may not be required for each and every plant during improvement. They may be amended in several ways. For example, if a plant-to-protoplast system would be developed in a target cultivar, then step V, and possibly even steps VII and VIII, can be eliminated.

A role for somatic hybrids in crop improvement cannot be realized unless somatic hybrids can be integrated into conventional breeding programs. For example, although disease resistance may be expressed in a somatic hybrid, undesirable characters are also expressed in the hybrid. These undesirable characters may affect quality or depress yield of the crop. Elimination of deleterious traits is best achieved through repeated backcrossing of a somatic hybrid to the cultivated crop, while continually selecting for desirable characteristics.

When somatic hybrids are produced between distantly related species, the use of these hybrids in conventional breedings is complicated. For example, if an interspecific hybrid is produced in order to incorporate disease resistance from a wild species into a cultivated crop, it would be most advantageous to eliminate all of the wild species genome except the one or few genes of interest. The factors complicate incorporation of desirable genes from a wild species into a crop. First, if the trait of interest is polygenic, then an intact functional block of genes, rather than a single gene, must be transferred in sexual crosses. Transfer of such a block of genes will also entail incorporation of closely linked genes. Some of these closely linked genes may depress yield or have other undesirable characteristics. Second, incorporation of a single gene trait from a wild species into a cultivated crop will necessitate interspecific recombination between genomes. The frequency of such recombination is reduced in hybrids between distantly related species. Double spots, composed of adjacent green and albino sectors, on heterozygous light-green Su/su plants have been interpreted as reciprocal genetic events and verified in *N. tabacum* as the result of mitotic recombination.[85] Similar single and double spots have been observed on light-green leaves of somatic hybrids of albino *N. tabacum* Su/su combined with *N. tabacum* (su/su), *N. sylvestris, N. otophora, N. glauca, N. nesophila,* and *N. stocktonii.* The presence of double spots on leaves of somatic hybrids of widely divergent species such as *N. nesophila* and *N. tabacum* is encouraging, as evident from mitotic recombination between distantly related genomes. Indeed, one somatic hybrid between *N. tabacum* and *N. sylvestris* has a greatly increased frequency of spot formation relative to *N. tabacum* Su/su.[85] The high spotting frequency has been transmitted to sexual progeny of this somatic hybrid. It should be noted though, that spots associated with the Su locus, particularly the nonreciprocal single spots, may be produced by genetic phenomena other than mitotic recombination.

N. tabacum is the most widely grown commercial nonfood plant in the world and it is susceptible to bacterial wildfire disease and fungal black shank. *N. rustica* carries resistance to both types of disease, but sexual crosses between *N. tabacum* lack fertility. It is also possible to produce amphiploid somatic hybrids between these two species by protoplast fusions. Leaf protoplasts of wild-type *N. tabacum* can be fused with wild-type suspension culture protoplasts and heterokaryons isolated directly and cultured to regenerate whole, fertile somatic hybrid.[143] Fertile somatic hybrids can also be produced by fusion of wild-type *N. tabacum*[201] which is both NR^- and SR^+.[159] In this somatic hybrid the presence of the black ovary wall gene from *N. rustica* can be readily demonstrated. Irradiation is being currently assessed to determine its influence on the degree of gene transfer between these two species and to determine the readiness with which disease resistance genes could be transferred from *N. rustica* without the necessity for backcrossing.

Through protoplast fusion, transfer of nuclear and cytoplasmic information can be easily facilitated between plant species. This method will be particularly useful for transfer of single gene or small blocks of closely linked genes. Thus, characters, like resistance to diseases, can be induced with ease.[85] Novel somatic hybrids may also be valuable as bridges in development of distant gene combinations. For example, Evans et al.[320,321] have produced somatic hybrids of *N. nesophila* with cultivated tobacco, but did not recover hybrids using *N. repanda,* a well-characterized species which is a unique source of disease resistance. However, Evans et al.[321] have been able to sexually hybridize the *N. nesopila* and *N. tabacum* somatic hybrid with pollen of *N. repanda,* to produce a bridge hybrid that contains genetic information of all three species. This new hybrid may be used as a novel source of disease resistance after backcrossing to cultivated tobacco. An advantage of this method will be transfer of polygenic traits which are otherwise difficult to be transferred. However, transfer of unlinked polygenic traits using products of protoplast fusion is probably not likely to occur as early as transfer of single genes.

Another application which is emerging out from somatic hybridization is the selection and evaluation of mutant cell lines. In fact, production of mutants proved to be a better tool in the hands of the scientists in the selection of new somatic hybrids. The SRI streptomycin-resistant mutant of *N. tabacum* and kanamycin-resistant line of *N. sylvestris* have been used to recover complete hybrid plants.[431] Thus, the cnx and nia mutants of *N. tabacum* have been used to select[313] and identify intraspecific hybrids when fused with the SRI mutant,[201] albino plastome mutant, and cytoplasmic CMS[420] protoplasts. Selection was based on the observation that nitrate reductase-deficient lines will not grow on NH_4^+/NO_3 selective medium. Several albino mutants and auxotrophs have been isolated for isoleucine, uracil, and leucine from haploid *N. plumbaginifolia.*[432] All six albinos and the three auxotrophs were in different complementation groups and behaved as recessive genes in somatic hybrids.

Improvement using protoplast fusions is at an earlier stage of development in flax, which is an important source of natural fibers, and its seeds contain oil and significant amounts of protein. Several of the wild species of *Linum* (2n = 18) possess many agronomically valuable genes for disease resistance, for instance, to flax rust and draught resistance. Attempts at sexual crosses between these wild species and flax have failed to produce seeds. Here, again, protoplast fusions, including irradiation of wild species, could be used to bring about the desired gene flow. This work can now proceed, since whole flax plants can be regenerated from flax root and cotyledon protoplasts.[61] At a somewhat comparable stage of development is the use of protoplast fusion for lettuce improvements. Here the main breeding problem is the introduction of more durable resistance to the fugus *(Bremia lactucae)* which causes downy mildew. Horizontal resistance to *B. lactucae* is available in the wild lettuce species *Lactuca perennis.* As yet, it has not been possible to cross lettuce (*L. sativa)* with *L. perennis* sexually, so somatic hybridization, coupled with irradiation procedures, is a logical avenue to explore. Encouragingly, whole plant regeneration has been obtained from protoplasts of lettuce roots, cotyledons, and *in vitro* grown shoots[220] and leaf protoplasts.[256] Studies on rice, *Oryza sativa* improvement by fusion of protoplasts are at an earlier stage of development, and it has now become possible to regenerate plants from rice protoplasts.[240] Breeding objectives such as transfer of CMS factors by protoplast fusion and fusions with protoplasts of wild rice species to enhance disease resistance properties and transfer of disease resistance characteristics can be easily achieved.

Trifolium repens (white clover) is an important forage crop in temperate regions. Sexual incompatibility within the genus *Trifolium* and other forage legumes prevents the introduction of desirable characters by conventional means. One such forage legume is sainfoin

(Onobrychis vicifolia), which combines persistency, draught resistance, and high palatability with valuable proteins and nonbloating properties. Conditions for the isolation, culture, and regeneration of mesophyll protoplasts of white clover and sainfoin have now been established; protoplast fusion for somatic hybridization, coupled with heterokaryon isolation procedures, can now be realistically undertaken for somatic hybridization assessments.[42]

Cocking[333] has described the various facts of somatic hybridization utilized at Nottingham. It has been suggested that protoplast fusion may provide a means of introducing characteristics for leaf tennin production from sainfoin *(O. viciifolia)* and Birdsfoot Trefoil *(Lotus corniculatus),* which act as antibloat factors with the main forage legumes white clover *(T. repens)* and alfalfa *(Medicago sativa)*. Sexual incompatibility prevents the introduction of such desirable characters by conventional means. Irradiation of sainfoin and Birdsfoot Trefoil protoplasts is being assessed to achieve the limited gene transfer required for the selection of white clover and alfalfa partial somatic hybirds. Regeneration of plants from protoplasts of parental forage legumes has already been achieved.[224]

V. CONCLUSIONS AND FUTURE PROSPECTS

Potential application of protoplasts is becoming clear with the progress in research in this field. The first requirement for the manipulation of cells of a particular plant for crop improvement is the development of the protoplast-plant regeneration system. This has already been achieved in several plants like *Nicotiana, Petunia, Hyoscyamus, Brassica,* and *Oryza sativa*. Even protoplasts have been regenerated to complete plants in more difficult species like cereals. Successful regeneration has also been carried out in woody plants like *Citrus, Populus,* apple, and coffee plants. This has opened up new directions in the exploitation of protoplasts for the development of new crop varieties in these genera.

Recent advances in plant tissue culture have resulted in the regeneration of whole plants from genetically transformed cells or of increasing number of species. The application of recombinant DNA technology for the enhancement of crop production requires continuous blending of these procedures with those of conventional breeding programs. Current research focuses on the isolation of useful genes as well as DNA sequences affecting plant qualities such as nutritional enhancement, herbicide resistance, photosynthetic efficiency, stress tolerance, nitrogen fixation, and the production of primary and secondary plant products. It is hoped that the utilization of protoplasts to transfer desired genes into the plant system through microinjection and electroporation will provide the tools necessary to improve crop varieties. However, these techniques for the integration of useful genes into plant genomes are still in the experimental stage, but may become useful within the next decade. Now scientists are in a position to modify protoplasts through several means. The above-mentioned goals can also be achieved by mutating protoplast genes and regenerating such cells into complete plants. Techniques have already been developed for the selection and regeneration of nutritional mutants. Plant protoplasts are also providing suitable tools for sensitive bioassay of phytotoxins.

Somatic hybridization has important potential in the following areas: (1) production of fertile amphidiploid somatic hybrids of sexually incompatible species; (2) production of heterozygous lines within a single species which normally could only be propagated by vegetative means; (3) transfer of limited parts of the genome from one species to another by the formation of heterokaryons in which unidirectional sorting of cytoplasmic elements occurs. This enables hybrids with a mixed nuclear component to be obtained against a common cytoplasmic background. Conversely, the irradiation of one of the partners enables the selective loss of one nuclear genome and the combination of single nuclear genomes against a segregated cytoplasmic background as in cybrid formation; and (4) production of novel interspecific and intergeneric crosses between plants that are difficult or impossible to hybridize conventionally, e.g., fusions between protoplasts of *Lycopersicon esculentum* (tomato) and *Solanum tuberosum* (potato) created the "ptomato" first achieved by Melchers et al.[329] Many others have followed, e.g., fusions between protoplasts of *Datura innoixa* and *Atropa belladonna, Arabidopsis thaliana* and *B. compestris,* and *Petunia parodii* and *P. parviflora*. There are limitations, however, to these types of somatic hybridizations, since plants regenerated from some of these combinations are not always fertile and do not produce viable seed. Experience now indicates that the main opportunities for use of somatic hybridization is likely to be more successful if centered around hybridization within the same genus or between closely related genera.

The transfer of limited parts of a genome between two different cell populations by somatic hybridization is of broad significance to plant breeding, particularly with reference to the transfer of cytoplasmic genomic components. The cytoplasmic genomes are important determinants of yield in crops such as hexaploid triticale. However, while it is relatively simple to bring together two plastid genotypes in a common cytoplasm by protoplast fusion, it has been frequently difficult to keep them together especially through meiosis. The use of restriction enzyme fragment "finger printing" has shown that somatic segregation generally occurs following somatic hybridization, so that just one species type of chloroplast DNA remains in a protoplast hybrid between two sexually compatible *Nicotiana* species. Therefore, opportunities for increasing cytoplasmic variability by protoplast fusions appear to be greater with mitochondria than with chloroplasts. Galun and associates have obtained evidence to indicate that mitochondrial recombination occurs in cytoplasmic hybrids of *N. tabacum* by protoplast fusion, and that the mtDNAs present in the cybrids were different from those in the parents and from the expected mixture of the two. An important phenotype under cytoplasmic control which is of immediate breeding and agronomic interest is CMS. Transfer of CMS from *N. tabacum* to *N. sylvestris* by protoplast fusion was first reported by Zelcer et al.[109]

By irradiating *N. tabacum* protoplasts with X-rays, functional nuclei were eliminated and CMS-controlling elements transferred following somatic fusion and integration and recombination of mtDNA from both parents. CMS phenotype is now proving a most useful phenotypic marker in combination with herbicide and antibiotic resistance marker systems. Experimental results of Galun and associates indicated that in some donor-recipient combinations both chloroplasts and mitochondria could be transferred between species and that a rather early sorting out of the two types of organelles took place. In other cases sorting out of chloroplast was delayed beyond sexual reproduction and there were indications that in certain donor-recipient combinations only one of the two types of organelles (chloroplasts) could be transferred. Galun and associates presented hypothetical schemes for unidirectional transfer of organelles, protoplast manipulations, and androgenesis which can be integrated into conventional breeding programs in order to achieve breeding goals such as herbicide resistance, male sterility, and good fruit quality. An alternative method to X-ray irradiation for elimination of the nuclear component of one of the partners in somatic fusion is by the production of enucleate microplasts. This avoids irradiation treatments which are relatively nonspecific and which could possibly also be causing undetected changes in the cytoplasmic genomes.

Resistance of plants to the herbicide atrazine has been reported to have been transferred from bird's rape *(B. compestris)* to cultivated oilseed rape *(B. napus)* via fractionated protoplast (i.e., subprotoplast) fusions.[382] Herbicide resistance of this type is maternally inherited, since the genes responsible for resistance are located in the chloroplast genome. Since conventional sexual hybrids contain and express only maternally derived cytoplasmic genes, the value of protoplast fusions for transfer of cytoplasmic-inherited genes of both parents is of major significance to crop breeding; opportunities are created to generate completely novel types of cytoplasmic mixtures not previously available through conventional sexual crosses.

When chloroplasts from two sources are introduced into one heteroplastomic fusion cell and the cell is cultured to result in a hybrid or cybrid plant, it is common that sorting out of chloroplasts follows with certain exception. This results in fusion of plants with homoplastomic cells identical in either cpDNA. The fate of mitochondria in somatic hybrids and cybrids, resulting from protoplast fusion, was followed in only a few cases. Contrary to chloroplasts, the restriction patterns of mtDNA of hybrids and cybrids are similar, but not identical, to either of fusion partners. There are indications that the coexistence of different chondriomes in one fusion cell is not a necessary condition for mtDNA rearrangement in *Nicotiana*. Despite a sorting out of mitochondrial characters, the resulting cybrid mtDNAs are similar, but rarely identical, to those of either fusion partner. Generally, agriculture has been very successful in utilizing sexual hybridization and mutation breeding from crop improvement. Often many such genes are involved and only very rarely have these actually been identified. Where possible, direct transfer of such genes into plants using suitable vectors is desirable and protoplasts offer additional opportunities in this respect. Of particular interest is the recent work in *Nicotiana* which has shown that irradiation of pollen provides a valuable tool for plant donor and recipient genotypes.[433,434] Irradiated pollen has been utilized to facilitate limited gene transfer in wheat.[435] The exciting vista is that it may be possible to obtain limited gene transfer even between sexually isolated species by protoplast fusion, if one of the fusion partners is suitably irradiated. With suitable selection procedures, this would be equivalent to having directly transferred such genes using suitable vectors; it would avoid the need to identify the genes required in plant improvement and be readily applicable to polygenic transfers. Concerted efforts are now required to evaluate the extent of limited gene transfer that can be obtained as a consequence of protoplast fusions. The plant breeder will often wish to use such fusions to transfer only a single or a small number of characters from one parent to another.

Until the process of directed transformation with cloned genes reaches a higher level of sophistication, protoplast fusion offers a means for introducing genes from unconventional sources. Somatic hybrid plants such as the pomato are not of immediate value, just as is true of interspecific crosses between most distant sexually compatible species. In the latter, considerable backcrossing is needed to eliminate unwanted portions of the alien genome. Novel somatic hybrid plants are thus only the starting point of a genetic introgression scheme. If the hybrid plants are sexually compatible with either parent, removal of nondesirable material can be straightforward. Where this is not true, protoplast regeneration from interspecific hybrids that undergo continuous chromosome segregation should provide novel genomic mixes. Moreover, since evidence is accumulating, at least in potato, that chromosome translocations are frequent in plants regenerated from mesophyll protoplasts, the coexistence of genomic sets may also allow translocation of, for example, tomato chromosome segments into potato chromosomes. Hence, unidirectional chromosome segregation combined with translocation or substitution could allow recovery of one parental phenotype with minor contributions from the other.

Protoplast isolation, fusion, and subsequent regeneration of parasexual hybrids appear to be one of the most significant developments in the field of tissue culture. This will have far-reaching implications on crop improvement. Although still in the formative stage, this technology has already played an important role in opening new vistas and has awakened the interest of plant physiologists, molecular geneticists, and plant breeders. A new vegetable Senposai developed by somatic fusions by Tokita Seed Company and Kirin Brewery Company in Japan is already in the market. Canola hybrids are at the verge of commercialization. In the future, protoplast manipulation will be one of the most frequently used tools in agricultural research with unlimited potential for genetic engineering and plant improvement.

The economically important cereal and grass crops have generally proved to be notoriously recalcitrant to manipulation *in vitro*. Regeneration of plants from single cells, a prerequisite for cellular and molecular manipulation, has proven to be especially difficult. Consequently, a group of plants has until recently remained outside the mainstream of plant biotechnology. The discovery and exploitation of embryogenic tissue cultures, in which plant regeneration takes place by the formation of embryos from single somatic cells, have led to the development of efficient procedures for plant regeneration in almost all of the important species of grasses, and recovery of mature plants from protoplasts in crops such as maize, rice, and sugarcane.[436] These results, along with the success in somatic hybridization and the demonstration of transient as well as stable expression of introduced genes in grass cells and plants, provide challenging opportunities for the genetic manipulation and improvement of this group of food crops. It is argued, however, that a far better understanding of growth, development, physiology, and molecular biology-genetics of plants — and continuous dialogue and interaction with plant breeders and geneticists — are required for the effective use of protoplasts for crop improvement.

REFERENCES

1. **Cocking, E. C.,** A method for the isolation of plant protoplasts and vacuoles, *Nature (London),* 187, 962, 1960.
2. **Cocking, E. C.,** Virus uptake, cell wall regeneration and virus multiplication in isolated plant protoplasts, *Int. Rev. Cytol.,* 28, 89, 1970.
3. **Evans, P. K. and Cocking, E. C.,** Isolated plant protoplasts, in *Plant Tissue and Cell Culture,* Street, H. E., Ed., Blackwell Scientific, Oxford, 1977, 103.
4. **Lu, D. Y., Pental, D., and Cocking, E. C.,** Plant regeneration from seedling cotyledon protoplasts, *Z. Pflanzenphysiol.,* 107, 59, 1982.
5. **Xu, Z. H., Davey, M. R., and Cocking, E. C.,** Plant regeneration from root protoplasts of *Brassica, Plant Sci. Lett.,* 24, 117, 1982.
6. **Flick, C. E. and Evans, D. A.,** Isolation, culture and plant regeneration from protoplasts isolated from flower petals of ornamental *Nicotiana* species, *Z. Pflanzenphysiol.,* 109, 379, 1983.
7. **Bilkey, P. C. and Cocking, E. C.,** A non-enzymatic method for the isolation of protoplasts from callus of *Saintpaulia ionantha* (African Violet), *Z. Pflanzenphysiol.,* 105, 285, 1982.
8. **Szabados, L. and Gaggero, C.,** Callus formation from protoplasts of a sugarbeet cell suspension culture, *Plant Cell Rep.,* 4, 195, 1985.
9. **Chatterjee, G., Sikadar, S. R., Das, S., and Sen, S. K.,** Regeneration of plantlets from mesophyll protoplasts of *Brassica juncea* (L.) Czern., *Plant Cell Rep.,* 4, 245, 1985.
10. **Chuong, P. V., Pauls, K. P., and Beversdorf, W. D.,** A simple culture method for *Brassica* hypocotyl protoplasts, *Plant Cell Rep.,* 4, 4, 1985.
11. **Shillito, R. D., Paszkowski, J., and Potrykus, I.,** Agarose plating and a bead type culture technique enable and stimulate development of protoplast-derived colonies in a number of plants, *Plant Cell Rep.,* 2, 244, 1983.
12. **Vardi, A., Spiegel-Roy, P., and Galun, E.,** Plant regeneration from Citrus protoplasts: variability in methodological requirement among cultivars and species, *Theor, Appl. Genet.,* 62, 171, 1982.
13. **Bass, A. and Hughes, W.,** Conditions for isolation and regeneration of viable protoplasts of oil palm *(Elaeis guineensis), Plant Cell Rep.,* 3, 169, 1984.
14. **Gamborg, O. L., Daw, B. P., and Stahlhut, R. W.,** Cell division and differentiation in protoplasts from cell cultures of *Glycine* species and leaf tissue of soybean, *Plant Cell Rep.,* 2, 213, 1983.
15. **Arcioni, S., Mariotti, D., and Pezzotti, M.,** *Hedysarum coronarium* L. in vitro conditions for plant regeneration from protoplasts of various explants, *Plant Physiol.,* 121, 141, 1985.
16. **Nishimaki, T. and Nozue, M.,** Isolation and culture of protoplasts from high anthocyanin producing callus of sweet potato, *Plant Cell Rep.,* 4, 248, 1985.
17. **Barakat, M. N. and Cocking, E. C.,** An assessment of the cultural capabilities of protoplasts of some wild species of *Linum, Plant Cell Rep.,* 4, 164, 1985.
18. **Maeda, Y., Fuji, Y., and Yamada, Y.,** Callus formation from protoplasts of cultured *Lithospermum erythrorhizon* cells, *Plant Cell Rep.,* 2, 179, 1983.
19. **Koblitz, H. and Koblitz, D.,** Experiments on tissue culture in the genus *Lycopersicon* Miller mesophyll protoplast regeneration to plants in *Lycopersicon esculentum* cv. 'Nadja'+, *Plant Cell Rep.,* 1, 143, 1982.

20. **Lang, H. and Kohlenbach, H. W.,** Differentiation of alkaloid cells in cultures of *Macleava* mesophyll protoplasts, *Planta Med.*, 46, 78, 1982.
21. **Kouider, M., Hauptmann, R., Widholm, J. M., Skirvin, R. M., and Korban, S. S.,** Callus formation from *Malus* × *domestica* cv. 'Jonathan' protoplasts, *Plant Cell Rep.*, 3, 142, 1984.
22. **Holbrook, L. A., Reich, T. J., Iyer, V. N., Haffner, M., and Miki, B. L.,** Induction of efficient cell division in alfalfa protoplasts, *Plant Cell Rep.*, 4, 229, 1985.
23. **Arcioni, S., Davey, M. R., Dos Santos, A. V. P., and Cocking, E. C.,** Somatic embryogenesis in tissues from mesophyll and cell suspension protoplasts of *Medicago coerulea* and *M. glutinosa, Z. Pflanzenphysiol.*, 106, 105, 1982.
24. **Barfield, D. G., Robinson, S. J., and Shieds, R.,** Plant regeneration from protoplasts of long term haploid suspension cultures of *N. plumbaginifolia, Plant Cell Rep.*, 4, 104, 1985.
25. **Hoffman, F. and Adachi, T.,** *Arabidobrassica:* chromosomal recombination and morphogenesis in asymmetric intergeneric hybrid cells, *Planta,* 153, 586, 1981.
26. **Caboche, M.,** Nutritional requirement of protoplast-derived, haploid tobacco cells grown at low cell densities in liquid medium, *Planta,* 149, 7, 1980.
27. **Gamborg, O. L., Shyluk, J. P., and Shahin, E. A.,** Isolation, fusion and culture of plant protoplasts, in *Plant Tissue Culture: Methods and Applications in Agriculture,* Thorpe, T. A., Ed., Academic Press, New York, 1981, 115.
28. **Nakata, K. and Oshima, H.,** Cytoplasmic chimaericity in the somatic hybrids of tobacco, in *Plant Tissue Culture,* Fujiwara, A., Ed., Japanese Association of Plant Tissue Culture, Tokyo, 1982, 641.
29. **Zelcer, A. and Galun, E.,** Culture of newly isolated tobacco protoplasts: precursor incorporation into protein, RNA and DNA, *Plant Sci. Lett.*, 7, 331, 1976.
30. **Nagata, T. and Takebe, I.,** Plating of isolated tobacco mesophyll protoplasts on agar medium, *Planta,* 99, 12, 1971.
31. **Davey, M. R., Frearsen, E. M., Withers, L. A., and Power, J. B.,** Observations on the morphology, ultrastructure and regeneration of tobacco leaf epidermal protoplasts, *Plant. Sci. Lett.*, 2, 23, 1974.
32. **Bajaj, Y. P. S.,** Potential of protoplast culture work in agriculture, *Euphytica,* 23, 633, 1974.
33. **Harms, C. T., Lorz, H., and Potrykus, I.,** Multiple-drop array (MDA) technique for the large scale testing of culture media variation in hanging microdrop cultures of single cell systems. II. Determination of phytohormone combinations for optimal division response in *Nicotiana tabacum* protoplast culture, *Plant Sci. Lett.*, 14, 237, 1979.
34. **Russell, J. A. and McCown, B. H.,** Culture and regeneration of *Populus* leaf protoplast isolated from non-seedling tissue, *Plant Sci.*, 45, 133, 1986.
35. **Russell, J. A. and McCown, B. H.,** Techniques for enhanced release of leaf protoplasts in *Populus, Plant Cell Rep.*, 5, 284, 1986.
36. **Bapat, V. A. and Schieder, O.,** Protoplast culture of several members of the genus *Physalis, Plant Cell Rep.*, 1, 69, 1981.
37. **Bokelmann, G. S. and Roest, S.,** Plant regeneration from protoplasts of potato. (*Solanum tuberosum* cv. Bintje), *Z. Pflanzenphysiol.*, 109, 259, 1983.
38. **Jia, J. and Potrykus, I.,** Mesophyll protoplasts from *Solanum melongena* var. depressum Barley regenerate fertile plants, *Plant Cell Rep.*, 1, 71, 1981.
39. **Guri, A. and Izhar, S.,** Improved efficiency of plant regeneration from protoplasts of egg plant (*Solanum melongena* L.), *Plant Cell Rep.*, 3, 247, 1984.
40. **Adams, T. L. and Townsend, J. A.,** A new procedure for increasing efficiency of protoplast plating and clone selection, *Plant Cell Rep.*, 2, 165, 1983.
41. **Carlberg, I., Glimelius, K., and Eriksson, T.,** Improved culture ability of potato protoplasts by use of activated charcoal, *Plant Cell Rep.*, 2, 223, 1983.
42. **Ahuja, P. S., Lu, D. Y., Cocking, E. C., and Davey, M. R.,** An assessment of the culture capabilities of *Trifolium repens* L. (White clover) and *Onobrychis vicifolia* Scop. (Sainfoin) mesophyll protoplasts, *Plant Cell Rep.*, 2, 269, 1983.
43. **Dos Santos, A. V. P., Davey, M. R., and Cocking, E. C.,** Cultural studies of protoplasts and leaf callus of *Trigonella corniculata* and *T. foenum-graecum, Z. Pflanzenphysiol.*, 109, 227, 1983.
44. **Lu, D. Y., Davey, M. R., and Cocking, E. C.,** Somatic embryogenesis from mesophyll protoplasts of *Trigonella corniculata* (Leguminosae), *Plant Cell Rep.*, 1, 278, 1982.
45. **Krishnamurthy, K. V., Godbole, D. A., and Mascarenhas, A. F.,** Studies on a draught resistant legume: the moth bean, *Vigna aconitifolia* (Jacq), *Plant Cell Rep.*, 3, 30, 1984.
46. **Uchimiya, H. and Murashige, T.,** Evaluation of parameters in the isolation of viable protoplasts from cultured tobacco cells, *Plant Physiol.*, 54, 936, 1974.
47. **Saxena, P. K. and King, J.,** Reuse of enzyme for isolation of protoplasts, *Plant Cell Rep.*, 4, 319, 1985.
48. **Morgan, A. and Cocking, E. C.,** Plant regeneration from protoplasts of *Lycopersicon esculentum* Mill, *Z. Pflanzenphysiol.*, 106, 97, 1982.

49. **Kao, K. N. and Michayluk, M. R.,** Plant regeneration from mesophyll protoplasts of Alfalfa, *Z. Pflanzenphysiol.*, 96, 135, 1980.
50. **Wallin, A. and Welander, M.,** Improved yield of apple leaf protoplasts from *in vitro* cultured shoots by using very young leaves and adding L-methionine to the shoot medium, *Plant Cell Tissue Organ Cult.*, 5, 69, 1985.
51. **Kaur-Sawhney, R., Flores, H. E., and Galston, A. W.,** Polyamine-induced DNA synthesis and mitosis in oat leaf protoplasts, *Plant Physiol.*, 65, 368, 1980.
52. **Schenck, H. R. and Robbelen, G.,** Somatic hybrids by fusion of protoplasts from *Brassica oleracea* and *B. compestris, Z. Pflanzenzuecht.*, 89, 278, 1982.
53. **Chin, J. C. and Scott, K. J.,** A large scale isolation procedure for cereal mesophyll protoplasts, *Ann. Bot.*, 43, 23, 1979.
54. **Vardi, A.,** Isolation of protoplasts in Citrus, *Proc. Int. Soc., Citric.*, 2, 575, 1977.
55. **Vardi, A. and Speigel-Roy, P.,** Citrus breeding, taxonomy and the species problems, *Proc. Int. Soc., Citric.*, 3, 51, 1978.
56. **Vardi, A.,** Studies and isolation and regeneration of orange protoplasts, in *Production of Natural Compounds by Cell Culture Methods, Proc. Int. Symp. Plant Cell Culture,* Alfermann, A. W. and Reinhard, E., Eds., Gesellsch. Strahl, Umweltforsch., Munich, 1978, 234.
57. **Cress, D. E.,** Osmotic stress inhibits thymidine incorporation into soybean protoplast DNA, *Plant Cell Rep.*, 1, 186, 1982.
58. **Wallin, A., Glimelius, K., and Erisksson, T.,** Pretreatment of cell suspensions as a method to increase the protoplast yield of *Haplopappus gracilis, Physiol. Plant.*, 40, 307, 1977.
59. **Bidney, D. L. and Shepard, J. F.,** Colony development from sweet potato petiole protoplasts and mesophyll cells, *Plant Sci. Lett.*, 18, 335, 1980.
60. **Simmonds, J. A., Simmonds, D. H., and Cummings, B. G.,** Isolation and cultivation of protoplasts from morphogenetic callus cultures of *Lilium, Can. J. Bot.*, 57, 512, 1979.
61. **Barakat, M. N. and Cocking, E. C.,** Plant regeneration from protoplast derived tissue of *Linum usitatissimum* L. (Flax), *Plant Cell Rep.*, 2, 314, 1983.
62. **Cassels, A. C. and Barlass, M. A.,** Method for the isolation of stable mesophyll protoplasts throughout the year under standard conditions, *Physiol. Plant.*, 42, 236, 1978.
63. **Mühlbach, H. P.,** Different regeneration potentials of mesophyll protoplasts from cultivated wild species of tomato, *Planta,* 148, 89, 1980.
64. **Koblitz, H. and Koblitz, D.,** Experiments on tissue culture in the genus *Lycopersicon* Miller: shoot formation from protoplasts of tomato long-term cell cultures, *Plant Cell Rep.*, 1, 147, 1982.
65. **O'Connell, M. A. and Hanson, M. R.,** Somatic hybridization between *Lycopersicon esculentum* and *Lycopersicon pennellii, Theor. Appl. Genet.*, 70, 1, 1985.
66. **Dos-Santos, A. V. P., Outka, D. E., and Cocking, E. C.,** Somatic embryogenesis in tissues derived from leaf protoplasts and leaf explants of *Medicago sativa, Z. Pflanzenphysiol.*, 99, 261, 1980.
67. **Aviv, D. and Galun, E.,** Isolation of tobacco protoplasts in the presence of isopropyl N-phenylcarbamate and their culture and regeneration into plants, *Z. Pflanzenphysiol.*, 83, 267, 1977.
68. **Watts, J. W., Moloyashi, F., and King, J. M.,** Problems associated with the production of stable protoplasts of cells of tobacco mesophyll, *Ann. Bot.*, 38, 667, 1974.
69. **Uchimiya, H. and Murashige, T.,** Influence of nutrient medium on the recovery of dividing cells from tobacco protoplasts, *Plant Physiol.*, 57, 424, 1976.
70. **Negrutiu, I. and Muller, J. F.,** Culture condition of protoplast derived cells of *Nicotiana sylvestris* for mutant selection, *Plant Cell Rep.*, 1, 14, 1981.
71. **Douglas, G. C., Keller, W. A., and Setterfield, G.,** Somatic hybridization between *Nicotiana rustica* and *Nicotiana tabacum*. II. Protoplast fusion and selection and regeneration of hybrid plants, *Can. J. Bot.*, 59, 220, 1981.
72. **Cassels, A. C. and Cocker, F. M.,** Seasonal physiological aspects of the isolation of tobacco protoplasts, *Physiol. Plant.*, 56, 69, 1982.
73. **Li, L. and Kohlenbach, H. W.,** Somatic embryogenesis is quite a direct way in cultures of mesophyll protoplasts of *Brassica napus* L., *Plant Cell Rep.*, 1, 209, 1982.
74. **Kyozuka, J., Hayashi, Y., and Shimamoto, K.,** High frequency plant regeneration from rice protoplasts by novel nurse culture methods, *Mol. Gen. Genet.*, 206, 408, 1987.
75. **Gamborg, O. L., Shyluk, J., and Kartha, K. K.,** Factors affecting the isolation and callus formation in protoplasts from the shoot apices of *Pisum sativum* L, *Plant Sci. Lett.*, 4, 285, 1975.
76. **Strauss, A. and Potrykus, I.,** Callus formation from protoplasts of cell suspension cultures of Rosa 'Paul's Scarlet', *Physiol. Plant.*, 48, 15, 1980.
77. **Shepard, J. F. and Totten, R. E.,** Mesophyll cell protoplasts of potato, *Plant Physiol.*, 60, 313, 1977.
78. **Tavazza, R. and Ancora, G.,** Plant regeneration from mesophyll protoplasts in commercial potato cultivars (Primura, Kennebec, Spunta, Desiree), *Plant Cell Rep.*, 5, 243, 1986.

79. **Nakagawa, H., Tanaka, H., Oba, T., Ogura, N., and Iizuka, M.,** Callus formation from protoplasts of cultured *spinacia oleracea* cells, *Plant Cell Rep.*, 4, 148, 1985.
80. **Roca, W. M., Szabados, L., and Hussain, A.,** Annual Report Biotechnology Research Unit, CIAT, Colombia, 1987, 1.
81. **Kao, K. N. and Michayluk, M. R.,** Nutritional requirements for growth of *Vicia hajastana* cells and protoplasts at a very low population density in liquid media, *Planta,* 126, 105, 1975.
82. **Binding, H.,** Somatic hybridization experiments in Solanaceous species, *Mol. Gen. Genet.*, 144, 171, 1976.
83. **Schieder, O.,** Hybridisation experiments with protoplasts from chlorophyll-deficient mutants of some Solanaceous species, *Planta,* 137, 253, 1977.
84. **Fukunaga, Y. and King, J.,** Effects of different nitrogen sources in culture media on protoplast release from plant cell suspension cultures, *Plant Sci. Lett.*, 11, 241, 1978.
85. **Evans, D. A., Bravo, J. E., Kut, S. A., and Flick, C. E.,** Genetic behavior of somatic hybrids in the genus *Nicotiana: N. otophora, N. tabacum* and *N. sylvestris* + *N. tabacum, Theor. Appl. Genet.*, 65, 93, 1983.
86. **Ruesink, A. W.,** Leucine uptake and incorporation by *Convolvulus* tissue culture cells and protoplasts under severe osmotic stress, *Physiol Plant.*, 44, 48, 1978.
87. **Kao, K. N. and Michayluk, M. R.,** A method for high-frequency intergeneric fusion of plant protoplasts, *Planta,* 115, 355, 1974.
88. **Kao, K. N.,** Chromosomal behaviour in somatic hybrids of soybean — *Nicotiana glauca, Mol. Gen. Genet.*, 150, 225, 1977.
89. **Vasil, V. and Vasil, I. K.,** Isolation and culture of cereal protoplasts. II. Embryogenesis and plantlet formation from protoplasts of *Pennisetum americanum., Theor. Appl. Genet.*, 56, 97, 1980.
90. **Dai, C., Mertz, D., and Lambeth, V.,** Improved procedures for the isolation and culture of potato protoplasts, *Plant Sci.*, 50, 79, 1987.
91. **Haberlach, G. T., Cohen, B. A., Reichert, N. A., Baer, M. A., Towill, L. E., and Helgeson, J. P.,** Isolation, culture and regeneration of protoplasts from several related *Solanum* species, *Plant Sci. Lett.*, 39, 67, 1985.
92. **Barsby, T. L., Yarrow, S. A., and Shepard, J. F.,** A rapid and efficient alternative procedure for the regeneration of plants from hypocotyl protoplasts of *Brassica napus, Plant Cell Rep.*, 5, 101, 1986.
93. **Tan, M. L. C., Rietveld, E. M., Marrewijk, G. A. M., and Kool, A. J.,** Regeneration of leaf mesophyll protoplast of tomato cultivars *(L. esculentum):* factors important for efficient protoplast culture and plant regeneration, *Plant Cell Rep.*, 6, 172, 1987.
94. **Shahin, E. A.,** Totipotency of tomato protoplasts, *Theor. Appl. Genet.*, 69, 235, 1985.
95. **Day, D. A., Jenkins, C. L. D., and Hatch, M. D.,** Isolation and properties of functional mesophyll protoplasts and chloroplasts from *Zea mays, Aust. J. Plant Physiol.*, 8, 21, 1981.
96. **Willison, J. H. M. and Cocking, E. C.,** The production of microfibrils at the surface of isolated tomato fruit protoplasts, *Protoplasma,* 75, 397, 1972.
97. **Murashige, I. and Skoog, F. A.,** Revised medium for rapid growth and bioassays with tobacco tissue cultures, *Physiol. Plant.*, 15, 473, 1962.
98. **Kao, K. N., Gamborg, O. L., Muller, R. A., and Keller, W. A.,** Cell division in cells regenerated from protoplasts of soyabean and *Haplopappus gracilis, Nature (London) New Biol.*, 232, 124, 1971.
99. **von Arnold, S. and Eriksson, T.,** A revised medium for growth of pea mesophyll protoplasts, *Physiol. Plant.*, 39, 257, 1977.
100. **Nagata, T. and Ishii, S.,** A rapid method for isolation of mesophyll protoplasts, *Can J. Bot.*, 57, 1820, 1979.
101. **Durand, J., Potrykus, I., and Donn, G.,** Plantes tissues de protoplastes de *Petunia, Z. Pflanzenphysiol.*, 69, 26, 1973.
102. **Cella, R. and Galun, E.,** Utilization of irradiated carrot cell suspensions as feeder layer for cultured *Nicotiana* cells and protoplasts, *Plant Sci. Lett.*, 19, 243, 1980.
103. **Puck, T. T. and Marcus, P. I.,** A rapid method for viable cell titration and clone production with HeLa cells in tissue culture: the use of X-irradiated cells to supply conditioning factors, *Proc. Natl. Acad. Sci. U.S.A.*, 41, 432, 1955.
104. **Raveh, D., Hubermann, E., and Galun, E.,** *In vitro* culture of tobacco protoplasts: use of feeder techniques to support division of cells plated at low densities, *In Vitro,* 9, 216, 1973.
105. **Vardi, A., Speigel-Roy, P., and Galun, E.,** *Citrus* cell culture: isolation of protoplasts, plating densities, effect of mutagens and regeneration of embryos, *Plant Sci. Lett.*, 4, 231, 1975.
106. **Raveh, D. and Galun, E.,** Rapid regeneration of plants from tobacco protoplasts plated at low densities, *Z. Pflanzenphysiol.*, 76, 76, 1975.
107. **Aviv, D. and Galun, E.,** The feeder layer technique, in *Cell Culture and Somatic Cell Genetics of Plants,* Vasil, I. K., Ed., Academic Press, New York, 1984, 199.

108. **Vardi, A. and Raveh, D.**, Cross feeder experiments between tobacco and orange protoplasts, *Z. Pflanzenphysiol.*, 78, 350, 1976.
109. **Zelcer, A., Aviv, D., and Galun, E.**, Interspecific transfer of cytoplasmic male sterility by fusion between protoplasts of normal *Nicotiana sylvestris* and X-ray irradiated protoplasts of male sterile *N. tabacum, Z. Pflanzenphysiol.*, 90, 397, 1978.
110. **Aviv, D. and Galun, E.**, Restoration of fertility in cytoplasmic male-sterile (CMS) *Nicotiana sylvestris* by fusion with X-irradiated *N. tabacum* protoplasts, *Theor. Appl. Genet.*, 58, 121, 1980.
111. **Kuchko, A. A. and Butenko, R. G.**, The culture of protoplasts isolated from *Solanum tuberosum* and *Solanum chacoense*, in *Use of Tissue Cultures in Plant Breeding*, Novak, F. J., Ed., Czechoslovakian Academy of Science Institute Experimental Botany, Prague, 1977, 441.
112. **Pojnar, E., Willison, J. H. M., and Cocking, E. C.**, Cell wall regeneration by isolated tomato fruit protoplasts, *Protoplasma*, 64, 460, 1967.
113. **Ross, H.**, Wild species and primitive cultivars as ancestors of potato varieties, in *Broadening the Genetic Base of Crops*, Zeven, A. C. and van Harten, A. M., Eds., PUDOC, Wageningen, Netherlands, 1979, 237.
114. **Collins, G. B., Litton, C. C., Legg, P. D., and Smiley, J. H.**, Registration of Kentucky 15 tobacco, *Crop Sci.*, 18, 694, 1978.
115. **Vasil, I. K., Vasil, V., Sutton, W. D., and Gikes, K. L.**, Protoplasts as tools for the genetic modification of plants, in *Proc. 4th Int. Symp. on Yeast and Other Protoplasts*, University of Nottingham, Nottingham, U. K., 1975, 82.
116. **Carlson, P. S., Smith, H. H., and Dearing, R. D.**, Parasexual interspecific hybridization, *Proc. Natl. Acad. Sci. U.S.A.*, 69, 2292, 1972.
117. **Melchers, G. and Labib, G.**, Somatic hybridization of plants by fusion of protoplasts. I. Selection of light resistant hybrids of haploid light sensitive varieties of tobacco, *Mol. Gen. Genet.*, 135, 277, 1974.
118. **Dudits, D., Kao, K. N., Constabel, F., and Gamborg, O. L.**, Fusion of carrot and barley protoplasts and division of heterokaryocytes, *Can. J. Genet. Cytol.*, 18, 263, 1976.
119. **Dudits, D., Hadlaczky, G., Bajszar, G., Koncz, C., Lazar, G., and Horvath, G.**, Plant regeneration from intergeneric cell hybrids, *Plant Sci. Lett.*, 15, 101, 1979.
120. **Evans, D. A., Bravo, J. E., and Gleba, Y. Y.**, Somatic hybridization: fusion methods, recovery of hybrids and genetic analysis, *Int. Rev. Cytol.*, Suppl. 16, 143, 1983.
121. **Evans, D. A.**, Somatic hybridization within the genus *Nicotiana, Plant Physiol.*, 63 (Suppl.), 117, 1979.
122. **Evans, D. A., Wetter, L. R., and Gamborg, O. L.**, Somatic hybrid plants of *Nicotiana glauca* and *Nicotiana tabacum* obtained by protoplast fusion, *Physiol. Plant.*, 48, 225, 1980.
123. **White, D. W. R. and Vasil, I. K.**, Use of amino acid analogue-resistant lines for selection of *Nicotiana sylvestris* somatic hybrids, *Theor. Appl. Genet.*, 55, 107, 1979.
124. **Glimelius, K., Eriksson, T., Grafe, R., and Muller, A. J.**, Somatic hybridization of nitrate-deficient mutants of *Nicotiana tabacum* by protoplast fusion, *Physiol. Plant.*, 44, 273, 1978.
125. **Maliga, P., Lazar, G., Joo, F., Nagy, A. H., and Menczel, L.**, Restoration of morphogenic potential in *Nicotiana* by somatic hybridisation, *Mol Gen. Genet.*, 157, 291, 1977.
126. **Wullems, G. J., Molendij, K. J., and Schilperoort, R. A.**, The expression of tumor markers in intraspecific somatic hybrids of normal and crown gall cells from *Nicotiana tabacum, Theor, Appl. Genet.*, 56, 203, 1980.
127. **Medgyesy, P., Menczel, L., and Maliga, P.**, The use of cytoplasmic streptomycin resistance: chloroplast transfer from *Nicotiana tabacum* into *Nicotiana sylvestris, Theor, Appl. Genet.*, 56, 203, 1980.
128. **Menczel, L., Nagy, F., Kiss, Z. R., and Maliga, P.**, Streptomycin resistant and sensitive somatic hybrids of *Nicotiana tabacum* plus *Nicotiana knightiana:* correlation of resistance to *Nicotiana tabacum* plastids, *Theor. Appl. Genet.*, 59, 191, 1981.
129. **Kameya, T., Horn, M. E., and Widholm, J. M.**, Hybrid shoot formation from fused *Daucus carota* and *D. capillifolius* protoplasts, *Z. Pflanzenphysiol.*, 104, 459, 1981.
130. **Schieder, O.**, Selektion einer somatischen hybride nach fusion von protoplasten auxotropher mutanten von *Sphaerocarpus donnellic, Aust. Z. Pflanzen Physiol.*, 74, 357, 1974.
131. **Ashton, N. W. and Cove, D. J.**, The isolation and preliminary characterisation of auxotrophic and analogue resistant mutants of the moss, *Physcomitrella patens, Mol. Gen. Genet.*, 154, 87, 1977.
132. **Sidorov, V. A., Menczel, L., Nagy, F., and Maliga, R.**, Chloroplast transfer in *Nicotiana* based on metabolic complementations between irradiated and iodoacetate treated protoplasts, *Planta*, 152, 341, 1981.
133. **Schieder, O.**, Genetic evidence for the hybrid nature of somatic hybrids from *Datura innoxia* Mill, *Planta*, 141, 333, 1978.
134. **Schieder, O.**, Somatic hybrids of *Datura innoxia* Mill + *Datura discolor* Bernh. and of *Datura innoxia* Mill + *Datura stramonium* L. var. *tatula* L. I. Selection and characterization, *Mol. Gen. Genet.*, 162, 113, 1978.
135. **Schieder, O.**, Somatic hybrids between a herbaceous and two tree *Datura* species, *Z. Pflanzenphysiol.*, 98, 119, 1980.

136. **Potryrus, J.,** Fusion of differentiated protoplasts, *Phytomorphology,* 22, 91, 1972.
137. **Meadows, M. G. and Potrykus, I.,** Hoechst 33258 as a vital stain for plant cell protoplasts, *Plant Cell Rep.,* 1, 77, 1981.
138. **Kao, K. N., Constabel, F., Michayluk, M. R., and Gamborg, O. L.,** Plant protoplast fusion and growth of intergeneric hybrid cells, *Planta,* 120, 315, 1974.
139. **Gleba, Y. Y. and Hoffmann, F.,** Hybrid cell line *Arabidopsis thaliana* + *Brassica compestris:* no evidence for specific chromosome elimination, *Mol. Gen. Genet.,* 162, 257, 1978.
140. **Schnaberlauch, L. S., Kloc-Bauchan, F., and Sink, K. C.,** Expression of nuclear-cytoplasmic genomic incompatibility in interspecific *Petunia* somatic hybrid plant, *Theor. Appl. Genet.,* 70, 57, 1985.
141. **Galbraith, D. W. and Mauch, T. J.,** Identification of fusion of plant protoplasts. II. Conditions for the reproducible fluorescence labelling of protoplasts derived mesophyll tissue, *Z. Pflanzenphysiol.,* 98, 129, 1980.
142. **Harms, C. R. and Potrykus, I.,** Enrichment for heterokaryocytes by the use of iso-osmotic density gradients after plant protoplast fusion, *Theor. Appl. Genet.,* 53, 49, 1978.
143. **Patnaik, G., Cocking, E. C., Hamill, J., and Pental, D.,** A simple procedure for the manual isolation and identification of plant heterokaryons, *Plant Sci. Lett.,* 24, 105, 1982.
144. **Redenbaugh, K., Ruzing, S., Bartholomew, J., and Bassham, J. A.,** Characterisation and separation of plant protoplasts via flow cytometry and cell sorting, *Z. Pflanzenphysiol.,* 107, 65, 1982.
145. **Maliga, P., Kiss, Z. R., Nagy, A. H., and Lazar, G.,** Genetic instability in somatic hybrids of *Nicotiana tabacum* and *Nicotiana knightiana, Mol. Gen. Genet.,* 163, 145, 1978.
146. **Maliga, P., Lorz, M., Lazar, G., and Nagy, F.,** Cytoplast-protoplast fusion for interspecific chloroplast transfer in *Nicotiana, Mol. Gen. Genet.,* 185, 211, 1982.
147. **Medgyesy, P., Menczel, L., and Maliga, P.,** The use of cytoplasmic streptomycin resistance: chloroplast transfer from *Nicotiana tabacum* into *Nicotiana sylvestris,* and isolation of their somatic hybrids, *Mol. Gen. Genet.,* 179, 693, 1980.
148. **Lazar, G. B., Dudits, D., and Sung, Z. R.,** Expression of cyclohexamide resistance in carrot somatic hybrids and their segregants, *Genetics,* 98, 347, 1981.
149. **Fluhr, R., Aviv, D., Edelman, M., and Galun, E.,** Cybrids containing mixed and sorted out chloroplasts following interspecific somatic fusion in *Nicotiana, Theor. Appl. Genet.,* 65, 289, 1983.
150. **Fluhr, R., Aviv, D., Galun, E., and Edelman, M.,** Generation of heteroplastidic *Nicotiana* cybrids by protoplasts fusion: analysis for plastid recombinant types, *Theor. Appl. Genet.,* 67, 491, 1984.
151. **Fluhr, R., Aviv, D., Galun, E., and Edelman, M.,** Efficient induction and selection of chloroplast-encoded antibiotic resistant mutants in *Nicotiana, Proc. Natl. Acad. Sci. U.S.A.,* 88, 1485, 1985.
152. **Galun, E.,** New studies in plant cell culture, invited lecture at ''The Road Ahead'', Int. Futures Conf., Grahamstown, July 3 to 7, 1978, 1.
153. **Galun, E. and Aviv, D.,** Is somatic hybridization by protoplast fusion applicable to crop improvement?, *IAPTC Newsl.* 25, 2, 1978.
154. **Galun, E. and Aviv, D.,** Plant cell genetics in *Nicotiana* and its implications to crop plants, *Monogr. Genet. Agaria,* 4, 153, 1979.
155. **Aviv, D., Fluhr, R., Edelman, M., and Galun, E.,** Progeny analysis of the interspecific somatic hybrids: *Nicotiana tabacum* (CMS) + *Nicotiana sylvestris* with respect to nuclear and chloroplast markers, *Theor. Appl. Genet.,* 56, 145, 1980.
156. **Zelcer, A. and Galun, E.,** Culture of newly isolated tobacco protoplasts: cell division and precursor incorporation following a transient exposure to coumarin, *Plant Sci. Lett.,* 18, 185, 1980.
157. **Galun, E.,** Plant protoplasts as physiological tools, *Annu. Rev. Plant Physiol.,* 32, 237, 1981.
158. **Galun, E.,** Screening for chloroplast composition in hybrids and cybrids resulting from protoplast fusion in angiosperm, in *Methods in Chloroplast Molecular Biology,* Edelman, M., Hallick, K. G., and Chua, N. H., Eds., Elsevier/North-Holland, Amsterdam, 1982, 139.
159. **Hamill, J. D., Pental, D., Cocking, E. C., and Muller, A. J.,** Production of a nitrate reductase deficient streptomycin resistant mutant of *Nicotiana tabacum* for somatic hybridization studies, *Heredity,* 50, 197, 1983.
160. **Pental, D., Hamill, J. D., Pirrie, A., and Cocking, E. G.,** Somatic hybridizaton of *Nicotiana tabacum* and *Petunia hybrida.* Recovery of plants with *Petunia hybrida* nuclear genome and *N. tabacum* chloroplast genome, *Mol. Gen. Genet.,* 202, 342, 1986.
161. **Paszkowski, J., Shillito, R. D., Soul, M. W., Mandak, V., Hohn, T., Hohn, B., and Potrykus, I.,** Direct gene transfer to plants, *EMBO J.,* 3, 2717, 1984.
162. **Hain, R., Stabel, P., Czernilofsky, A. P., Steinbib, H. H., Herrera-Estrella, L., and Schell, J.,** Uptake, integration, expression and genetic transmission of a selectable chimaeric gene by plant protoplasts, *Mol. Gen. Genet.,* 199, 161, 1985.
163. **Harms, C. T.,** Somatic hybridization by plant protoplast fusion, in *Protoplasts,* Potrykus, I., Harms, C. T., Hinnen, A., Hutter, R., King, P. J., and Shillito, R. D., Eds., Lecture Proceedings Birkhauser, Basel, 1983, 69.

164. **Brunold, C., Kruger-Lebus, S., Saul, M. W., Wegmuller, S., and Potrykus, I.,** Combination of kanamycin resistance and nitrate reductase dificiency as selectable markers in one nuclear genome provides universal somatic hybridizer in plants, *Mol. Gen. Genet.*, 208, 469, 1987.
165. **Mariortti, D., Arcioni, S., and Pezzotti, M.,** Regeneration of *Medicago arborea* L., plants from tissue and protoplast cultures of different organ origin, *Plant Sci. Lett.*, 37, 149, 1984.
166. **Lazzeri, P. A., Heldebrand, D. F., and Collinis, G. B.,** A procedure for plant regeneration from immature cotyledon tissue of soybean, *Plant Mol. Biol. Rep.*, 3, 160, 1985.
167. **Cocking, E. C.,** Plant cell protoplasts, isolation and development, *Annu. Rev. Plant Physiol.*, 23, 29, 1972.
168. **Evans, D. A. and Bravo, J. E.,** Plant protoplast isolation and culture, *Int. Rev. Cytol.*, 16 (Suppl.), 33, 1983.
169. **Gengenbach, B. G., Green, C. E., and Donovan, C. M.,** Inheritance of selected pathotoxin resistance in maize plants regenerated from cell cultures, *Proc. Natl. Acad. Sci. U.S.A.*, 74, 5113, 1977.
170. **Melchers, G.,** The future, *Int. Rev. Cytol.*, 11B (Suppl.), 241, 1980.
171. **Barwale, U. B., Kerns, H. R., and Widholm, J. M.,** Plant regeneration from callus cultures of several soybean genotypes via embryogenesis and organogenesis, *Planta*, 167, 473, 1986.
172. **Takebe, I., Labib, G., and Melchers, G.,** Regeneration of whole plants from isolated mesophyll protoplasts of tobacco, *Naturwissenschaften*, 58, 318, 1971.
173. **Grambow, H. J., Kao, K. N., Miller, K. A., and Gamborg, O. L.,** Cell division and plant development from protoplasts of carrot cell suspension cultures, *Planta*, 103, 48, 1972.
174. **Vasil, I. K.,** Plant cell culture and somatic cell genetics in cereals and grasses, in *Plant Improvement and Somatic Cell Genetics*, Vasil, I. K., Scowcroft, W. R., and Frey, K. J., Eds., Academic Press, New York, 1982, 179.
175. **Keller, W. A., Setterfield, G., Douglas, G., Gleddie, S., and Nakamura, C.,** Production, characterization and utilization of somatic hybrids of higher plants, in *Application of Plant Cell Tissue Culture to Agriculture and Industry*, Tomes, D. T., Ellis, B. E., Harney, P. M., Kasha, K. J., and Paterson, R. L., Eds., University of Guelph, Guelph, Canada, 1982, 81.
176. **Firoozabady, E.,** Rapid plant regeneration from *Nicotiana* mesophyll protoplasts, *Plant Sci.*, 46, 127, 1986.
177. **Robertson, D., Palmer, J. D., Earle, E. D., and Mutschfer, M. A.,** Analysis of organelle genomes in somatic hybrid derived from cytoplasmic male-sterile *Brassica oleracea* and atrazine resistant *B. compestris*, *Theor. Appl. Genet.*, 74, 303, 1987.
178. **Pua, E. C.,** Plant regeneration form stem-derived protoplasts of *Brassica alboglabra* Bailey, *Plant Sci.*, 50, 153, 1987.
179. **Fu, Y. Y., Jia, S. R., and Lin, Y.,** Plant regeneration from mesophyll protoplast culture of cabbage *(Brassica oleracea* var. *captitata)*, *Theor. Appl. Genet.*, 71, 495, 1986.
180. **Chuong, P. V., Pauls, K. P., and Beverdorf, W. D.,** Plant regeneration from *Brassica nigra* (L) Koch. stem protoplasts, *In Vitro*, 23, 449, 1987.
181. **Thomas, E., Hoffman, F., Potrykus, J., Wenzel, G.,** Protoplast regeneration and stem embryogenesis of haploid androgenetic rape, *Mol. Gen. Genet.*, 145, 245, 1976.
182. **Kohlenbach, H. W., Wenzel, G., and Hoffman, F.,** Regeneration of *Brassica napus* plantlets in cultures from isolated protoplasts of haploid stem embryo as compared with leaf protoplasts, *Z. Pflanzenphysiol.*, 105, 131, 1982.
183. **Guri, A., Volokita, M., and Sink, K. C.,** Plant regeneration from leaf protoplasts of *Solanum torvum*, *Plant Cell Rep.*, 6, 302, 1987.
184. **Pellow, J. W. and Towill, L. E.,** Colony formation and plant regeneration from mesophyll protoplasts of *Solanum tuberosum*, *Plant Cell Tissue Organ Cult.*, 7, 11, 1986.
185. **Davey, M. R. and Kumar, A.,** Higher plant protoplasts: retrospect and prospect, *Int. Rev. Cytol.*, 16 (Suppl.), 219, 1983.
186. **Reisch, B.,** Genetic variability in regenerated plants, in *Handbook of Cell Culture*, Evans, D. A., Sharp, W. R., Ammirato, P. V., and Yamada, Y., Eds., Macmillan, New York, 1982, 748.
187. **Scowcroft, W. R.,** Genetic Variability in Tissue Culture, Impact on Germplasm Conservation and Utilization, Ret. Int. Board Plant Genetic Resources, Food and Agriculture Organization, Rome, 1984, 41.
188. **Lorz, H.,** Spontane und induzierte Variabilitat bei in vitro Kulturen, Vortrage f., *Z. Pflanzenzüch.*, Geisenheim, 1985.
189. **Dunwell, J. M. and Sunderland, U.,** Anther culture of *Solanum tuberosum* L., *Euphytica*, 22, 317, 1973.
190. **Gupta, P. P.,** Regeneration of plants from mesophyll protoplasts ground cherry (*Physalis minima* L), *Plant Sci.*, 43, 151, 1986.
191. **Crepy, L., Barros, L. M. G., and Valente, V. R. N.,** Callus production from leaf protoplasts of various cultivars of bean (*Phaseolus vulgaris* L.), *Plant Cell Rep.*, 5, 124, 1986.

192. **Bohorova, N. E., Cocking, E. C., and Power, J. E.,** Isolation, culture and callus regeneration of protoplast of wild and cultivated *Helianthus* species, *Plant Cell Rep.*, 5, 256, 1986.
193. **Lenee, P. and Chupeau, Y.,** Isolation and culture of sunflower protoplasts (*Helianthus annuus* L.): factors influencing the variability of cell colonies derived from protoplasts, *Plant Sci.*, 43, 69, 1986.
194. **Saka, K., Katterman, F. R., and Thomas, J. C.,** Cell segregation and sustained division of protoplasts from cotton (*Gossypium hirsutum* L.), *Plant Cell Rep.*, 6, 470, 1987.
195. **Yarrow, S. A., Wu, S. C., Barsby, T. L., Kemble, R. J., and Shepard, J. F.,** The introduction of CMS mitochondria to triazine tolerant *Brassica napus* L., var. ''Regent'', by micromanipulation of individual heterokaryons, *Plant Cell Rep.*, 5, 415, 1986.
196. **Brown, C., Lucas, J. A., and Power, J. B.,** Plant regeneration from protoplasts of a wild lettuce species (*Lactuca saliga* L.), *Plant Cell Rep.*, 6, 180, 1987.
197. **Shekhawat, N. S. and Galston, A. W.,** Mesophyll protoplasts of fenugreek *(Trigonella foenumgraecum):* isolation, culture, and shoot regeneration, *Plant Cell Rep.*, 2, 119, 1983.
198. **Gil, R. and Eapen, S.,** Plant regeneration from hypocotyl protoplasts of mothbean *(Vigna aconitifolia), Curr. Sci.*, 55, 100, 1986.
199. **Wilson, V. M., Haq, N., and Evans, P. K.,** Protoplast isolation, culture and plant regeneration in the winged bean (*Psophocarpus tetragonolobus* L.), *Plant Sci.*, 41, 61, 1985.
200. **Gill, R., Eapen, S., and Rao, P. S.,** Callus induction from protoplasts of *V. uniquiculata, V. sublobata* and *V. mungo, Theor. Appl. Genet.*, 74, 100, 1987.
201. **Pental, D., Cooper-Bland, S., Harding, K., Cocking, E. C., and Muller, A. J.,** Culture studies on nitrate reductase deficient *Nicotiana tabacum* mutant protoplasts, *Z. Pflanzen.*, 105, 219, 1982.
202. **Hamill, J. D., Ahuja, P. S., Davey, M. R., and Cocking, E. C.,** Protoplast-derived streptomycin resistant plants of the forage legume *Onobrychis viciifolia* Scop (Sainfoin), *Plant Cell Rep.*, 5, 439, 1986.
203. **Lu, C. V., Vasil, V., and Vasil, I. K.,** Isolation and culture of protoplasts of *Panicum maximum* Jacq. (Guinea Grass): somatic embryogenesis and plantlet formation, *Z. Pflanzenphysiol.*, 104, 311, 1981.
204. **Yamada, Y., Yang, Z. O., and Tang, D. T.,** Plant regeneration from protoplast-derived callus of rice (*Oryza sativa* L.), *Plant Cell Rep.*, 4, 85, 1986.
205. **Ogura, H., Kyozuka, J., Hayashi, Y., Koba, T., and Shimamoto, K.,** Field performance and cytology of protoplast-derived rice (*Oryza sativa*): high yield and low degree of variation of four Japonica cultivars, *Theor. Appl. Genet.*, 74, 670, 1987.
206. TCCP (Tissue Culture for Crops Project), Progress Report, Colorado State University, Fort Collins, 1987.
207. **Sticklen, M. B., Domir, S. C., and Lineberger, R. D.,** Shoot regeneration from protoplasts, of *Ulmus* × 'Pineer', *Plant Sci.*, 47, 29, 1986.
208. **Xu, X. H. and Davey, M. R.,** Shoot regeneration from mesophyll protoplasts and leaf explants of *Rehmannia glutinosa, Plant Cell Rep.*, 2, 55, 1983.
209. **Cameron, J. W. and Frost, H. B.,** Genetic breeding and nucellar embryony, in *The Citrus Industry,* Reuther, W., Batchelor, L. D., and Webber, H. J., Eds., Division of Agricultural Science, University of California, 1968, 325.
210. **Vardi, A., Hutchinson, D. J., and Galun, E.,** Protoplast-to-tree system in *Microcitrus* based on protoplasts derived from a sustained embryogenic callus, *Plant Cell Rep.*, 5, 412, 1986.
211. **Kobayashi, S., Uchimiya, H., and Ikeda, I.,** Plant regeneration from ''Trovita'' orange protoplasts, *Jpn. J. Breed.*, 33, 119, 1983.
212. **Kobayashi, S., Ikeda, I., and Nakatani, M.,** Induction of nucellar callus from orange *(Citrus sinensis* Osb) ovules and uniformity of regenerated plants, *Bull. Fruit Tree Res. Stn.*, E5, 43, 1984.
213. **Kobayashi, S., Ikeda, I., and Uchimiya, H.,** Conditions for high frequency embryogenesis from orange (*Citrus sinensis* Osb) protoplasts, *Plant Cell Tissue Organ Cult.*, 4, 249, 1985.
214. **Kobayashi, S.,** Uniformity of plants regenerated from orange *(Citrus sinensis* Osb) protoplasts, *Theor, Appl. Genet.*, 74, 10, 1987.
215. **Russell, J. A. and McCown, B. H.,** Recovery of plants from leaf protoplasts of hybrid-poplar and aspen clones, *Plant Cell Rep.*, in press.
216. **Schopke, C., Mueller, L. E., and Kohlenbach, H. W.,** Somatic embryogenesis and regeneration of plantlets in protoplast cultures from somatic embryos of coffee *(Coffea canephora* P. ex. Fr.), *Plant Cell Tissue Organ Cult.*, 8, 243, 1987.
217. **Ochatt, S. J., Cocking, E. C., and Power, J. B.,** Isolation, culture and plant regeneration of colt cherry *(Prunus avium × Peseuocerasus)* protoplasts, *Plant Sci.*, 50, 139, 1987.
218. **Canas, L. A., Wyssman, A. M., and Benbadis, M. C.,** Isolation culture and division of olive *(Olea europaea* L.) protoplasts), *Plant Cell Rep.*, 6, 369, 1987.
219. **Binding, H. and Mordhorst, G.,** Haploid *Solanum dulcamara* L: shoot culture and plant regeneration from isolated protoplasts, *Plant Sci. Lett.*, 35, 77, 1984.
220. **Berry, S. F., Lu, D. Y., Pental, D., and Cocking, E. C.,** Regeneration of plants from protoplasts of *Lactuca sativa* L., *Z. Pflanzenphysiol.*, 108, 31, 1982.

221. **Carlson, P. S.**, The use of protoplasts for genetic research, *Proc. Natl. Acad. Sci. U.S.A.*, 70, 598, 1973.
222. **Chellappan, K. P., Seeni, S., and Gnanam, A.**, The isolation, culture and organ differentiation from mesophyll protoplasts of *Mollugo nudicaulis* Lam., 1, *Experientia*, 36, 60, 1980.
223. **Cocking, E. C.**, Applications of protoplasts technology to agriculture, *Experientia Suppl.*, 46, 123, 1983.
224. **Davey, M. R.**, Recent developments in the culture and regeneration of plant protoplasts, *Experientia Suppl.*, 46, 19, 1983.
225. **Dodds, J. H. and Roberts, L. W.**, *Experiments in Plant Tissue Cultures*, Cambridge University Press, London, 1982.
226. **Engler, D. E. and Grogan, R. G.**, Isolation, culture and regeneration of lettuce leaf mesophyll protoplasts, *Plant Sci. Lett.*, 28, 223, 1982.
227. **Evans, D. A.**, Protoplast fusion and plant regeneration in tobacco, in *Plant Regeneration and Genetic Variation*, Praeger Press, New York, 1982, 303.
228. **Gosch, G. and Reinert, J.**, Nuclear fision in intergeneric heterokaryocytes and subsequent mitosis of hybrid nuclei, *Naturwissenschaften*, 63, 534, 1976.
229. **Klimaszewska, K. and Keller, W. A.**, Plant regeneration from stem cortex protoplasts of *Brassica napus*, *Plant Cell Tissue Organ Cult.*, 8, 225, 1987.
230. **Saxena, P. K., Gill, R., and Rashid, A.**, Isolation and culture of protoplasts from mesophyll tissue of the legume *Cyamopsis tetragonoloba*, *Plant Cell Tissue Organ Cult.*, 6, 173, 1986.
231. **Sikdar, S. R., Chatterjee, G., Das, S., and Sen, S. K.**, Regeneration of plant from mesophyll protoplasts of the wild crucifer *Eruca sativa* Lam., *Plant Cell Rep.*, 6, 486, 1987.
232. **Christou, P., Murphy, J. E., and Swain, W. F.**, Stable transformation of soybean by electroporation and root formation from transformed callus, *Proc. Natl. Acad. Sci. U.S.A.*, 84, 3962. 1987.
233. **Lin, W., Odell, J. T., and Schreiner, R. M.**, Soybean protoplast culture and direct gene uptake and expression by cultured soybean protoplasts, *Plant Physiol.*, 84, 856, 1987.
234. **Sun, Y., Heil, B. M., Kahl, G., and Kohlenbach, H. W.**, Plant regeneration from protoplasts of monocotylendonous *Haworthia magnifica* v. Polln, *Plant Cell Tissue Organ Cult.*, 9, 91, 1987.
235. **Schachakr, D. and Ducrenx, D.**, Plant regeneration from protoplast culture of sweet potato *(Ipomoea batatas* Lam.), *Plant Cell Rep.*, 6, 326, 1987.
236. **Koblitz, H. and Koblitz, D.**, Tissue and protoplast culture studies in *Lycopersicon esculentum* Miller var. *flammatum* Lehm. cv. Bonner Beste and its mutant Chloronerva, *Plant Cell Rep.*, 2, 194, 1983.
237. **Adams, T. L. and Quiros, C. F.**, Somatic hybridization between *L. peruvianum* and *L. pennellii*, Report of the Tomato Genetics Cooperative, Department of Horticulture, Purdue University, West Lafayette, IN, 1985, 1.
238. **Johnson, L. B., Stuteville, D. L., Higgins, R. K., and Skinner, D. Z.**, Regeneration of alfalfa plants from selected regen S clones, *Plant Sci. Lett.*, 20, 297, 1981.
239. **Bourgin, J. P.**, Valine resistant plants from *in vitro* selected tobacco cells, *Mol. Gen. Genet.*, 161, 225, 1978.
240. **Marse, J. L.**, Rice plants regenerated from protoplasts, *Science*, 235, 31, 1987.
241. **Toriyama, K., Hinata, K., and Sasaki, T.**, Haploid and diploid plant regeneration from protoplasts of anther callus of rice, *Theor. Appl. Genet.*, 73, 16, 1988.
242. **Power, J. B., Frearson, E. M., Hayword, C., and Cocking, E. C.**, Some consequences of the fusion and selection of *Petunia* and *Parthenocissus* protoplasts, *Plant Sci. Lett.*, 5, 197, 1975.
243. **Power, J. B., Frearson, E. M., Hayword, C., George, D., Evans, P. K., Berry, S. F., and Cocking, E. C.**, Somatic hybridization of *Petunia hybrida* and *Petunia parodii*, *Nature (London)*, 263, 500, 1976.
244. **Power, J. B., Berry, S. F., Frearson, E. M., and Cocking, E. C.**, Selection procedures for the production of interspecies somatic hybrids of *Petunia hybrida* and *Petunia parodii:* I. Nutrient media and drug sensitivity complementation selection, *Plant Sci. Lett.*, 10, 1, 1977.
245. **Power, J. B., Berry, S. F., Chapman, J. V., Cocking, E. C., and Sink, K. C.**, Somatic hybrids between unilateral cross-incompatible *Petunia* species, *Theor. Appl. Genet.*, 55, 97, 1979.
246. **Power, J. B., Berry, S. F., Chapman, J. V., and Cocking, E. C.**, Somatic hybridization of sexually incompatible petunia: *Petunia parodii, Petunia parviflora*, *Theor. Appl. Genet.*, 57, 1, 1980.
247. **Attree, S. M., Bekkaoui, F., Dunstan, D. I., and Fowke, L. C.**, Regeneration of somatic embryos from protoplast isolated from an embryogenic suspension culture of white spruce *(Picea glauca)*, *Plant Cell Rep.*, 6, 480, 1987.
248. **Binding, H. and Nehls, R.**, Somatic cell hybridizaton of *Vicia faba* and *Petunia hybrida*, *Mol. Gen. Genet.*, 164, 137, 1978.
249. **Shepard, J. F., Bidney, D., and Shahin, E.**, Potato protoplasts in crop improvement, *Science*, 208, 17, 1980.
250. **Saxena, P. K., Gill, R., Rashid, A., and Maheshwari, S. C.**, Plantlets from mesophyll protoplasts of *Solanum xanthocarpum*, *Plant Cell Rep.*, 1, 219, 1982.
251. **Okada, K., Hasezawa, S., Syono, K., and Nagata, T.**, Further evidence for transformation of *Vicia rosea* protoplasts by *Agrobacterium tumefaciens* spheroplasts, *Plant Cell Rep.*, 4, 133, 1985.

252. **Schilperoort, R. A. and Wullems, G. J.,** Protoplast transformation by Ti plasmid-whole plants and progeny, *Int. Rev. Cytol.*, 16 (Suppl.), 169, 1983.
253. **Krens, F. H., Molendijk, L., Wullems, G. J., and Schilperoort, R. A.,** In vitro transformation of plant protoplasts with Ti-plasmid DNA, *Nature (London)*, 296, 72, 1982.
254. **Lorz, H., Baker, B., and Schell, J.,** Gene transfer to cereal cells mediated by protoplast transformation, *Mol. Gen. Genet.*, 199, 178, 1985.
255. **Potrykus, I., Paszkowski, J., Saul, M. W., Petruska, J., and Shillito, R. D.,** Molecular and general genetics of a hybrid foreign gene introduced into tobacco by direct gene transfer, *Mol. Gen. Genet.*, 199, 169, 1985.
256. **Potrykus, I., Saul, M. W., Petruska, J., Paszkoski, J., and Shillito, R. D.,** Direct gene transfer to cells of a graminaceous monocot, *Mol. Gen. Genet.*, 199, 183, 1985.
257. **Meyer, P., Walgenbach, E., Bussman, K., Hombrecher, G., and Saedler, H.,** Synchronized tobacco protoplasts are efficiently transformed by DNA, *Mol. Gen. Genet.*, 201, 513, 1985.
258. **Uchimiya, H., Hirochika, H., Hashimoto, H., Hara, A., Masuda, T., Kasumimoto, T., Harada, H., Ikeda, J. E., and Yoshioka, M.,** Co-expression and inheritance of foreign genes in transformants obtained by direct DNA transformation of tobacco protoplasts, *Mol. Gen. Genet.*, 205, 1, 1980.
259. **Deshayes, A., Herrera-Estrella, L., and Cabache, M.,** Liposome-mediated transformation of tobacco mesophyll protoplasts by an *Escherichia coli* plasmid, *EMBO J.*, 4, 2731, 1985.
260. **Shillito, R., Saul, M., Paszkowski, J., and Potrykus, I.,** High efficiency direct gene transfer to plants, *Biotechnology*, 3, 1099, 1985.
261. **Chen, W. H., Gartland, K. M. A., Davey, M. R., Sotak, R., Gartland, J. S., Mulligan, B. J., Power, J. B., and Cocking, E. C.,** Transformation of sugarcane protoplasts by direct uptake of a selectable chimaeric gene, *Plant Cell Rep.*, 6, 297, 1987.
262. **Junker, B., Zimmy, J., Luhrs, R., and Lorz, H.,** Transient expression of chimaeric genes in dividing and non dividing cereal protoplasts after PEG induced DNA uptake, *Plant Cell Rep.*, 6, 329, 1987.
263. **Ecker, J. R. and Davis, R. W.,** Inhibition of gene expression in plant cells by expression of antisense RNA, *Proc. Natl. Acad. Sci. U.S.A.*, 83, 5372, 1986.
264. **Fromm, M., Taylor, L. P., and Walbot, V.,** Stable transformation of maize after gene transfer by electroporation, *Nature (London)*, 319, 791, 1986.
265. **Zimmerman, U. and Viernken, J.,** Electric field-induced cell-to-cell fusion, *J. Membr. Biol.*, 67, 165, 1982.
266. **Potter, H., Wier, L., and Leder, P.,** Enhancer dependent expression of human K-immunoglobulin genes introduced into mouse Pre-B lymphocytes by electroporation, *Proc. Natl. Acad. Sci. U.S.A.*, 81, 7161, 1984.
267. **Hauptmann, R. M., Ozias-Atkins, P., Vasil, V., Tabaeizadeh, Z., Rogers, S. G., Horsch, R. B., Vasil, I. K., and Fraley, R. T.,** Transient expression of electroporated DNA in monocotyledonous and dicotyledonous species, *Plant Cell Rep.*, 6, 265, 1987.
268. **Doughlas, M. W., Perani, L., Radke, S., and Bossert, M.,** The application of recombinant DNA technology toward crop improvement, *Physiol. Plant.*, 68, 560, 1986.
269. **Perani, L., Radke, S., Douglas, M. W., and Bossert, M.,** Gene transfer methods for crop improvement. Introduction of foreign DNA into plants, *Physiol. Plant.*, 68, 566, 1986.
270. **Crossway, A., Uakes, J. V., Irvine, J. M., Ward, W., Knauf, V. C., and Shewmaker, C. K.,** Integration of foreign DNA following microinjection of tobacco mesophyll protoplasts, *Mol. Gen. Genet.*, 202, 179, 1986.
271. **Lawrence, W. A. and Davies, D. R.,** A method for microinjection and culture of protoplast at very low densities, *Plant Cell Rep.*, 4, 33, 1985.
272. **Steinbiss, H. H. and Stabel, P.,** Protoplast derived tobacco cells can survive capillary microinjection of the fluorescent dye Lucifer Yellow, *Protoplasma*, 116, 223, 1983.
273. **Morikawa, H. and Yamada, Y.,** Capillary microinjection into protoplasts and intranuclear localization of injected materials, *Plant Cell Physiol.*, 26, 229, 1985.
274. **Linsefors, L. and Brodelius, P.,** Immobilization of protoplasts: viability studies, *Plant Cell Rep.*, 4, 23, 1985.
275. **Ou-Lee, T. M., Turgeon, R., and Wu, R.,** Expression of a foreign gene linked to either a plant virus or a *Drosophila* promotor, after electroporation of protoplasts of rice wheat and sorghum, *Proc. Natl. Acad. Sci. U.S.A.*, 83, 6815, 1986.
276. **French, R., Janda, M., and Alquist, P.,** Bacterial gene inserted in an engineered RNA virus: efficient expression in monocotyledonous plant cells, *Science*, 231, 1294, 1986.
277. **Fromm, M., Taylor, T., and Walbot, V.,** Expression of genes transferred into monocot and dicot plant cells by electroporation, *Proc. Natl. Acad. Sci. U.S.A.*, 82, 5824, 1985.
278. **Okada, K., Nagata, T., and Takabe, I.,** Introduction of functional RNA into plant protoplasts by electroporation, *Plant Cell Physiol.*, 27, 619, 1986.

279. **Nishiguchi, M., Langridge, W. H. R., Szalay, A. A., and Zaitlin, M.,** Electroporation-mediated infection of tobacco leaf protoplasts with tobacco mosaic virus and cucumber mosaic virus RNA, *Plant Cell Rep.,* 5, 57, 1986.
280. **Pullock, K., Barfield, D. G., Robinson, S. J., and Shields, R.,** Transformation of protoplast-derived cell colonies and suspension cultures by *Agrobacterium tumefaciens, Plant Cell Rep.,* 4, 202, 1985.
281. **Guerche, P., Charbonnier, M., Jouanin, L., Tourneur, C., Paszkowski, J., and Pelletier, G.,** Direct gene transfer by electroporation in *Brassica napus, Plant Sci.,* 52, 111, 1987.
282. **Odell, J. T., Nagy, F., and Chua, N. H.,** Identification of DNA sequences required for activity of the cauliflower mosaic virus 35 S promotor, *Nature (London),* 313, 810, 1985.
283. **Boston, R. S., Becwar, M. R., Ryan, R. D., Goldsbrough, P. B., Larkins, B. A., and Hodges, T. K.,** Expression from heterologous promotors in electroporated carrot protoplasts, *Plant Physiol.,* 83, 742, 1987.
284. **Pedersen, K., Devereux, J., Wilson, D. R., Sheldon, E., and Larkin, B. A.,** Cloning and sequence analysis reveal structural variation among related zein genes in maize, *Cell,* 29, 1013, 1982.
285. **Deka, P. C. and Sen, S. K.,** Differentiation in calli orginated from isolated protoplasts of rice (*Oryza sativa* L.) through plating technique, *Mol. Gen. Genet.,* 145, 239, 1976.
286. **Morikawa, H., Iida, A., Matsui, C., Ikegami, M., and Yamada, Y.,** Gene transfer into intact plant cells by electroinjection through cell walls and membranes, *Gene,* 41, 121, 1986.
287. **Maliga, P.,** Protoplast in mutant selection and characterization, *Int. Rev. Cytol.,* 16 (Suppl), 161, 1983.
288. **Krumbiegel, G.,** Mutagenic treatment of haploid and diploid protoplasts from *Datura innoxia* and *Petunia hybrida* L., *Environ. Exp. Bot.,* 19, 99, 1979.
289. **Aviv, D. and Galun, E.,** An attempt at isolation of nutritional mutants from cultured tobacco protoplasts, *Plant Sci. Lett.,* 8, 299, 1977.
290. **Galun, E. and Raveh, D.,** In vitro culture of tobacco protoplasts: survival of haploid and diploid protoplasts exposed to X-ray radiation at different times after isolation, *Radiat. Bot.,* 15, 79, 1975.
291. **Maliga, P., Breznovits, Sz., and Marton, L.,** Non-Mendelian streptomycin resistant tobacco mutant with altered chloroplasts and mitochondria, *Nature,* 255, 401, 1975.
292. **Maliga, P., Menczel, L., Sodorov, V., Marton, L., Cseplo, A., Medgyesy, P., Dung, T. M., Lazar, G., and Nagy, F.,** Cell culture mutants and their uses in plant improvement and somatic cell genetics, Vasil, I. K., Frey, K. J., and Scowcroft, W. R., Eds., Academic Press, New York, 1982, 221.
293. **Maliga, P.,** Isolation, characterization and utilization of mutant cell lines in higher plants, in *Perspectives in Plant Cell and Tissue Culture,* Academic Press, New York, 1980, 225.
294. **Vunsh, R., Aviv, D., and Galun, E.,** Valine resistant plants derived from mutated haploid and diploid protoplasts of *Nicotiana sylvestris* and *N. tabacum, Theor. Appl. Genet.,* 64, 51, 1982.
295. **Sung, Z. R.,** Mutagenesis of cultured plant cells, *Genetics,* 84, 51, 1976.
296. **Lawyer, A. L., Berlyn, M. B., and Zeltich, I.,** Isolation and characterization of glycine hydroxamate resistant cell lines of *Nicotiana tabacum, Plant Physiol.,* 66, 334, 1980.
297. **Brettell, R. I. S. and Ingram, D. S.,** Tissue culture in production of novel disease resistant crop plants, *Biol. Rev.,* 54, 379, 1979.
298. **Brettell, R. I. S., and Thomas, E.,** Reversion of Texas male sterile maize cytoplasm in culture, to give fertile T-toxin resistant plants, *Theor. Appl. Genet.,* 58, 55, 1980.
299. **Earle, E. D.,** Phytotoxin studies with plant cells and protoplasts, in *Frontiers of Plant Tissue Cultures,* Proc. 4th Int. Congr. Plant Tissue and Cell Culture, Thorpe, T. A., Ed., Calgary, Alberta, 1978, 363.
300. **Behnke, M. and Lonnendonker, N.,** Isolation and partial characterization of phytotoxic substances from culture filtrates of the fungus *Phytophthora infestans, Z. Pflanzenphysiol.,* 85, 17, 1977.
301. **Earle, E. D., Gracen, V. E., Yoder, O. C., and Gemmil, K. P.,** Cytoplasm-specific effects of *Helminthosporium maydis* race T-toxin on survival of corn mesophyll protoplasts, *Plant Physiol.,* 61, 420, 1978.
302. **Breiman, A. and Galun, E.,** Plant protoplasts as tools in quantitative assays of phytotoxic compounds from culture filtrates of *Phytophthora citrophthora, Physiol. Plant Pathol.,* 19, 181, 1981.
303. **Matern, V., Strobel, G., and Shepard, J.,** Reaction to phytotoxins in a potato population derived from mesophyll protoplasts, *Proc. Natl. Acad. Sci. U.S.A.,* 75, 4935, 1978.
304. **Behnke, M.,** Selection of potato callus for resistance to culture filtrates of *Phytophthora infestans* and regeneration of resistant plants, *Theor. Appl. Genet.,* 55, 69, 1979.
305. **Smith, H. H., Kao, K. N., and Combatti, N. C.,** Interspecific hybridisation by protoplast fusion in *Nicotiana.* Confirmation and extension, *J. Hered.,* 67, 123, 1976.
306. **Dudits, D., Hadlaczky, G. Y., Levi, E., Fefer, O., Haydu, Z. S., and Lazar, F.,** Somatic hybridization of *Daucus carota* and *Daucus capillifolus* by protoplast fusion, *Theor. Appl. Genet.,* 51, 127, 1977.
307. **Sundberg, E. and Glimelius, K.,** A method for production of interspecific hybrid within *Brassica* via somatic hybridization, using resynthesis of *Brassica napus* as a model, *Plant Sci.,* 43, 155, 1986.
308. **Kumar, P. M., Das, K., Sinha, R. R., Mukherjee, P., and Sen, S. K.,** Interspecific somatic protoplast fusion products in cultivated jute species, *Basic Life Sci.,* 22, 237, 1983.

309. **Toshiaki, K., Horn, M. E., and Widholm, J. M.,** Hybrid shoot formation from fused *Daucus carota* and *Daucus capillifolius,* protoplasts, *Z. Pflanzenphysiol.*, 104, 459, 1981.
310. **Cocking, E. C., Power, J. B., and Kinsara, A. M.,** Method of Producing Somatic Hybrids of Genus *Lycopersicon,* European Patent Appl. Ep. 6235697, Pub 9, September 1987.
311. **Teoule, E.,** Somatic hybridization between *Medicago sativa* L. and *Medicago falcata* L., *C. R. Acad. Sci. (Paris),* 297, 13, 1983.
312. **Gangadevi, T., Rao, P. N., Rao, B. H., and Satyanarayana, K.,** A study of morphology, cytology and sterility in interspecific hybrids and amphidiploids of *Nicotiana knightiana* × *N. umbratica, Theor. Appl. Genet.*, 70, 330, 1985.
313. **Mueller, G. E. and Schieder, O.,** Interspecific T-DNA transfer through plant protoplast fusion, *Mol. Gen. Genet.*, 208, 235, 1986.
314. **Aviv, D. and Galun, E.,** Restoration of male fertile *Nicotiana* by fusion of protoplasts derived from two different cytoplasmic male sterile cybrids, *Plant Mol. Biol.*, 7, 411, 1986.
315. **Aviv, D., Chen, R., and Galun, E.,** Does pretreatment of rhodamine 6-G affect the mitochondrial composition of fusion-derived *Nicotiana* cybrids?, *Plant Cell Rep.*, 3, 227, 1986.
316. **Nagao, T.,** Somatic hybridization by fusion of protoplasts. II. The combinations of *Nicotiana tabacum* and *N. glutinosa,* and *N. tabacum* and *N. alata, Jpn. J. Crop Sci.*, 48, 385, 1979.
317. **Gleba, Y. Y and Hoffmann, F.,** *Arabidobrassica:* plant genome engineering by protoplast fusion, *Naturwissenschaften*, 66, 547, 1979.
318. **Kosakovs'ka, I. V.,** Subunit structure of D-ribulose-1,5-diphosphate carboxylase of parasexual hybrids, *Nicotiana tabacum* and *Nicotiana debneyi, Ukr. Bot. Z.*, 37, 86, 1980.
319. **Uchimiya, H.,** Somatic hybridization between male sterile *Nicotiana tabacum* and *N. glutinosa* through protoplast fusion, *Theor. Appl. Genet.*, 61, 69, 1982.
320. **Evans, D. A., Flick, C. E., and Jensen, R. A.,** Disease resistance: incorporation into sexually incompatible somatic hybrids of the genus *Nicotiana, Science,* 213, 907, 1981.
321. **Evans, D. A., Flick, C. E., Kut, S. A., and Reed, S. M.,** Comparison of *Nicotiana tabacum* and *Nicotiana mesophilia* hybrids produced by ovule culture and protoplast fusion, *Theor. Appl. Genet.*, 62, 193, 1982.
322. **Hein, T., Przewozny, T., and Schieder, O.,** Culture and selection of somatic hybrids using an auxotrophic cell line, *Theor. Appl. Genet.*, 64, 119, 1983.
323. **Menczel, L., Nagy, F., Lazar, G., and Maliga, P.,** Transfer of cytoplasmic male sterility by selection for streptomycin resistance after protoplast fusion in *Nicotiana, Mol. Gen. Genet.*, 189, 365, 1983.
324. **Nagao, T.,** Somatic hybridization by fusion of protoplasts. III. Somatic hybrids of sexually incompatible combinations of *Nicotiana tabacum* + *Nicotiana repanda* and *Nicotiana tabacum* + *Salpiglossi sinuata, Jpn. J. Crop Sci.*, 51, 35, 1982.
325. **Nagao, T.,** Breeding by somatic hybridization based on protoplast fusion. I. The combination of *Nicotiana tabacum* and *Nicotiana rustica, Jpn. J. Crop Sci.*, 47, 491, 1978.
326. **Iwai, S., Nagao, T., Nakata, K., Kawashima, N., and Matsuyama, S.,** Expression of nuclear and chloroplastic genes coding for Fraction 1 protein in somatic hybrids of *Nicotiana tabacum* cultivar bright yellow and *Nicotiana rustica* var. *rustica, Planta,* 147, 414, 1980.
327. **Peitian, W., Jiayu, C., Shimin, Z., Jinziang, X., and Lianqing, W.,** Interspecific hybrid plants by protoplast fusion between *Nicotiana tabacum* and *N. rustica, Annu. Rep. Inst. Genet. Acad. Sin.*, 124, 1981.
328. **Melchers, G.,** Microbial techniques in somatic hybridization by fusion of protoplasts, in *International Cell Biology,* Brinkley, B. R. and Porter, K. R., Eds., Rockefeller University Press, New York, 1977, 207.
329. **Melchers, G., Sarcristan, M. D., and Holder, A. A.,** Somatic hybrid plants of potato and tomato regenerated from fused protoplasts, *Carlsberg Res. Commun.*, 43, 203, 1978.
330. **Izhar, S. and Power, J. B.,** Somatic hybridization in *Petunia:* a male sterile cytoplasmic hybrid, *Plant Sci. Lett.*, 14, 49, 1979.
331. **Izhar, S., Schlecter, M., and Swartzberg, D.,** Sorting out in somatic hybrids of *Petunia* and prevalence of the heteroplasmon through several meiotic cycles, *Mol. Gen. Genet.*, 190, 468, 1983.
332. **Cocking, E. C., George, D., Price-Jones, M. J., and Power, J. B.,** Selection procedures for the production of interspecies somatic hybrids of *Petunia hybrida* and *Petunia parodii.* II. Albino complementation selection, *Plant Sci. Lett.*, 10, 7, 1977.
333. **Cocking, E. C.,** Hybrid and cybrid production via protoplast fusion, in *Genetic Engineering: Application to Agriculture,* Owens, L. D., Ed., Rowman and Allanheld, Grananda, 1983, 257.
334. **Cocking, E. C., Devey, M. R., Pental, D., and Power, J. B.,** Aspects of plant genetic manipulation, *Nature,* 193, 256, 1981.
335. **Gleddie, S., Keller, W. A., and Setterfield, G.,** Production and characterization of somatic hybrids between *Solanum melongena L.* and *S. sisymbriifolium* Lam., *Theor. Appl. Genet.*, 71, 613, 1986.

336. **Binding, H., Jain, S. M., Finger, J., Mordhorst, G., Nehls, R., and Gressel, J.**, Somatic hybridization of atrazine resistant biotype of *Solanum nigrum* with *Solanum tuberosum*. I. Clonal variation in morphology and atrazine sensitivity, *Theor. Appl. Genet.*, 63, 273, 1982.
337. **Austin, S., Ehlenfedt, M. K., Baer, M. A., and Helgeson, J. P.**, Somatic hybrids produced by protoplast fusion between *Solanum tuberosum* and *Solanum brevidens:* phenotypic variation under field conditions, *Theor. Appl. Genet.*, 71, 682, 1986.
338. **Barsby, T. L., Shepard, J. F., Kemble, R. J., and Wong, R.**, Somatic hybridization in the genus *Solanum: Solanum tuberosum* and *Solanum brevidens, Plant Cell Rep.*, 3, 165, 1984.
339. **Butenko, R. G. and Kuchko, A. A.**, Somatic hybridization of *Solanum tuberosum* L. and *Solanum chacoense* Bitt. by protoplast fusion, in *Advances in Protoplast Research,* Ferenczy, L. and Farkas, G. L., Eds., Pergamon Press, Oxford, 1980, 293.
340. **Sidorov, V. A., Zubko, M. K., Kuchko, A. A., Komarnitsky, I. K., and Gleba, Y. Y.**, Somatic hybridization in potato: use of irradiated protoplast of *Solanum pinnatisectum* in genetic reconstruction, *Theor. Appl. Genet.*, 74, 364, 1987.
341. **Rennie, P. J., Weber, G., Constabel, F., and Fowke, L. C.**, Dedifferentiation of chloroplasts in interspecific and homospecific protoplast fusion products, *Protoplasma,* 103, 253, 1980.
342. **Gleba, Y. Y. and Hoffman, F.**, *Arabidobrassica:* a novel plant obtained by protoplast fusion, *Planta,* 149, 112, 1980.
343. **Gleba, Y. Y., Momot, V. P., Cherep, N. N., and Skarzynskaya, M. V.**, Intertribol hybrid cell lines of *Atropa belladonna* (×) *Nicotiana chinensis* obtained by cloning individual protoplast fusion products, *Theor. Appl. Genet.*, 62, 75, 1982.
344. **Gosch, G. and Reinert, J.**, Cytological identification and colony formation of intergeneric somatic hybrid cells, *Protoplasma,* 96, 23, 1978.
345. **Ohgawara, T., Kobayashi, S., Ohgawara, E., Uchimiya, H., and Ishii, S.**, Somatic hybrid plants obtained by protoplast fusion between *Citrus sinensis* and *Poncirus trifoliata, Theor. Appl. Genet.*, 71, 1, 1985.
346. **Krumbiegel, G. and Schieder, O.**, Selection of somatic hybrids after fusion of protoplasts from *Datura innoxia* Mill. and *Atropa belladonna* L., *Planta,* 145, 371, 1979.
347. **Krumbiegel, G. and Schieder, O.**, Comparison of somatic and sexual incompatibility between *Datura innoxia* and *Atropa belladonna, Planta,* 153, 466, 1981.
348. **Harms, C. T. and Oertli, J. J.**, Complementation and expression of amino acid analog resistance studied by intraspecific and interfamily protoplast fusion, in *Plant Tissue Culture,* Fujiwara, A., Ed., Japanese Association of Plant Tissue Culture, Tokyo, 1982, 467.
349. **Reinert, J. and Gosch, G.**, Continuous division of heterokaryons from *Daucus carota* and *Petunia hybrida* protoplasts, *Naturwissenschaften,* 63, 534, 1976.
350. **Potrykus, I., Jia, J., Lazar, G. B., and Saul, M.**, *Hyoscyamus muticus* + *Nicotiana tabacum* fusion hybrids selected via auxotroph complementation, *Plant Cell Rep.* 3, 68, 1984.
351. **Kartha, K. K., Gamborg, U. L., Constabel, F., and Kao, K. N.**, Fusion of rapeseed and soybean protoplasts and subsequent division of heterokaryocytes, *Can. J. Bot.*, 52, 2435, 1974.
352. **Constabel, F., Kirkpatrick, J. W., Kao, K. N., and Kartha, K. K.**, The effect of canavanine on the growth of cells from suspension cultures and on intergeneric heterokaryons of canavanine sensitive and tolerant plants, *Biochem. Physiol. Pflanz.*, 168, 319, 1975.
353. **Constabel, F., Weber, G., Kirkpatrick, J. W., and Pahl, K.**, Cell division of intergeneric protoplast fusion products, *Z. Pflanzenphysiol.*, 79, 1, 1976.
354. **Wetter, L. R. and Kao, K. N.**, Chromosome and isoenzyme studies on cells derived from protoplast fusion of *Nicotiana glauca* with *Glycine max* + *Nicotiana glauca* cell hybrids, *Theor. Appl. genet.*, 57, 273, 1980.
355. **Wetter, L. R.**, Isoenzyme patterns in soybean-*Nicotiana* somatic hybrid cell lines, *Mol. Gen. Genet.*, 150, 231, 1977.
356. **Chien, Y. C., Kao, K. N., and Welter, L. R.**, Chromosomal and isozyme studies of *Nicotiana tabacum-Glycine max* hybrid cell lines, *Theor. Appl. Genet.*, 62, 301, 1982.
357. **Constabel, F., Weber, G., and Kirkpatrick, J. W.**, Chromosome compatability of intergeneric cell hybrids of *Glycine max* × *Vicia hajastana, C. R. Acad. Sci. (Paris),* 285, 319, 1977.
358. **Tabaeizadeh, Z., Perennes, C., and Bergounious, C.**, Increasing the variability of *Lycopersicon peruvianum* Mill. by protoplast fusion with *Petunia hybrida, Plant Cell Rep.*, 4, 7, 1985.
359. **Kushnir, S. G., Shlumukov, L. R., Pogrebnyak, N. J., Berger, S., and Gleba, Y.**, Functional cybrid plants possessing a *Nicotiana* genome and an *Atropa* plastome, *Mol. Gen. Genet.*, 209, 159, 1987.
360. **Potrykus, I., Shillito, R. D., Jia, J., and Lazar, G. B.**, Auxotroph complementation via protoplast fusion in *Hyoscyamus muticus* and *Nicotiana tabacum,* in *Genetic Engineering in Eukaryotes,* Lurquin, P. F. and Kleinhofs, A., Eds., Plenum Press, New York, 1983, 253.
361. **Xianghui, L., Wenbin, L., and Meijuan, H.**, Somatic hybrid plants from intergeneric fusion between tumor B653 and *Petunia hybrida* W43 and expression of LPDH, *Sci. Sin. (B),* 25, 611, 1982.

362. **Nehls, R.,** The use of metabolic inhibitors for the selection of fusion products of higher plant protoplast, *Mol. Gen. Genet.*, 166, 117, 1976.
363. **Horst, B. and Nehls, R.,** Somatic cell hybridization of *Vicia faba* + *Petunia hybrida, Mol. Gen. Genet.*, 164, 137, 1978.
364. **Tabaeizadeh, Z., Ferl, R. J., and Vasil, I. K.,** Somatic hybridization in the graminea: *Saccharum officinarum* L. (sugarcane) and *Pennisetum americanum* (L) *K. Schum* (Peral nullet), *Proc. Natl. Acad. Sci. U.S.A.*, 83, 5616, 1986.
365. **Handley, L. W., Nickels, R. L., Cameron, M. W., Moore, P. P., and Sink, K. C.,** Somatic hybrid plants between *Lycopersicon esculentum* and *Solanum lycopersicoides, Theor. Appl. Genet.*, 71, 691, 1986.
366. **Brac, D. S., Rambold, S., Constabel, F., and Gamborg, O. L.,** Isolation, fusion and culture of *Sorghum bicolor* cultivar GRP-168 and corn *(Zea mays* cultivar Punjab Local) protoplasts, *Z. Pflanzenphysiol.*, 96, 269, 1980.
367. **Schieder, O. and Vasil, I. K.,** Protoplast fusion and somatic hybridization, *Int. Rev. Cytol.*, 11B (Suppl.), 21, 1980.
368. **Galun, E., Arzee-Gonen, P., Fluhr, R., Edelman, M., and Aviv, D.,** Cytoplasmic hybridization in *Nicotiana:* mitochondrial analysis in progenies resulting from fusion between protoplasts having different organelle constitutions, *Mol. Gen. Genet.*, 186, 50, 1982.
369. **Breiman, A., Vunsh, R., and Galun, E.,** Characterization of resistance to the proline analog azetidine carboxylic acid in a *Nicotiana sylvestris* cell line, *Z. Pflanzenphysiol.*, 105, 383, 1982.
370. **Galun, E.,** Somatic cell fusion for inducing cytoplasmic exchanges: a new biological system for cytoplasmic genetics in higher plants, in *Plant Improvement and Somatic Cell Genetics,* Vasil, I. K., Frey, K. J., and Scowcroft, W. R., Eds., Academic Press, New York, 1982, 205.
371. **Galun, E. and Aviv, D.,** Cytoplasmic hybridization — genetic and breeding application, in *Application of Plant Tissue Culture Methods for Crop Improvement,* Evans, D. A., Sharp, W. R., and Amirata, P. V., Eds., Macmillan, New York, 1983, 358.
372. **Galun, E., Bleichman, S., and Aviv, D.,** Development of an organelle genetics system by unilateral transfer of organelles through protoplast fusion and somatic hybridization, in Plant Tissue Culture, Proc. 5th Int. Congr. on Plant Cell and Tissue Culture, Tokyo, 1982, 645.
373. **Vardi, A., Spiegel-Roy, P., Ben-Hayyim, G., and Galun, E.,** Protoplast derived plants and fusion experiments in different *Citrus* species, in Plant Tissue Culture, Proc. 5th Int. Congr. on Plant Cell and Tissue Culture, Fujiwara, A., Ed., Japan, 1982, 619.
374. **Aviv, D. and Galun, E.,** Chondriome analysis in sexual progenies of *Nicotiana* cybrids, *Theor. Appl. Genet.*, 73, 821, 1987.
375. **Galun, E. and Aviv, D.,** Manipulation of protoplasts, organelles and genes: applicability to plant breeding, in Proc. 10th Congr. of the European Association for Research on Plant Breeding (EUCARPIA), Wageningen, The Netherlands, 1984, 228.
376. **Menczel, L., Polsby, L. S., Steinback, K. L., and Maliga, P.,** Fusion-mediated transfer of triazine resistant chloroplast: characterization of *Nicotiana tabacum* cybrid plants, *Mol. Gen. Genet.*, 205, 201, 1986.
377. **Kowalczyk, T. P., Mackenzie, I. A., and Cocking, E. C.,** Plant regeneration from organ explants and protoplasts of the medicinal plant *Solanum khasianum* C. B. Clarke Var., *Chatterjeeanum sengupta* (Syn. *Solanum viarum* Dunal), *Z. Pflanzenphysiol.*, 111, 55, 1983.
378. **Melchers, G.,** Topatoes and pomatoes, somatic hybrids between tomatoes and potatoes, in *Tissue Culture and Research,* Rohlich, P. and Bacsy, E., Eds., Hungarian Academy of Science, Budapest, 1984, 499.
379. **Austin, S., Baer, M. A., and Helgeson, J. P.,** Transfer of resistance to potato leaf roll virus from *Solanum tuberosum* by somatic fusion, *Plant Sci. Lett.*, 39, 75, 1985.
380. **Helgeson, J. P., Hunt, G. J., Haberlach, G. T., and Austin, S.,** Somatic hybrids between *Solanum brevidens* and *Solanum tuberosum:* expression of a late blight resistance gene and potato leaf roll resistance, *Plant Cell Rep.*, 3, 212, 1988.
381. **Ehlenfedt, M. K. and Helgeson, J. P.,** Fertility of somatic hybrids from protoplast fusion of *Solanum brevidens* and *Solanum tuberosum, Theor. Appl. Genet.*, 73, 395, 1987.
382. **Beversdorf, W. D., Weiss-Lerman, J., Erickson, L. R., and Souza Machado, V.,** Transfer of cytoplasmically-inherited triazine resistance from bird's rape to cultivated oilseed rape *(Brassica compestris* and *B. napus), Can. J. Genet. Cytol.*, 22, 167, 1980.
383. **Pelletier, G., Primard, C., Vedel, F., Chetrit, P., Remy, R., Rousselle, A., and Renard, M.,** Intergenetic cytoplasmic hybridization in Cruciferae by protoplast fusion, *Mol. Gen. Genet.*, 191, 244, 1983.
384. **Yarrow, S. A., Cocking, E. C., and Power, J. B.,** Plant regeneration from cultured cell derived protoplasts of *Pelargonium aridum, P.* × *hortorum* and *P. peltatum, Plant Cell Rep.*, 6, 102, 1987.
385. **Anon.,** Herbicide tolerant canola cybrids, *Biotechnol. News,* 5 (September), 9, 1986.
386. **Anon.,** New varieties, *Seed World,* 125, 22, 1987.
387. **Dudtis, D., Fejer, O., Hadlaczky, G., Koncz, C., Lazar, G. B., and Howath, G.,** Intergeneric gene transfer mediated from plant protoplast fusion, *Mol. Gen. Genet.*, 179, 283, 1980.

388. **Dudtis, D.,** Back fusion with somatic protoplasts, a method in genetic manipulation of plants, *Acta Biol. Acad. Sci. Hung.*, 32, 215, 1981.
389. **Schiller, B., Herrmann, R. G., and Melchers, G.,** Restriction endonuclease analysis of plastid from tomato, potato and some of their somatic hybrids, *Mol. Gen. Genet.*, 186, 453, 1983.
390. **Sears, B. B.,** Elimination of plastids during gametogenesis and fertilization in the plant kingdom, *Plasmid,* 4, 233, 1980.
391. **Koller, B., Fromm, H., Galun, E., and Edelman, M.,** Evidence for *in vivo* trans splicing of pre-mRNAs in tobacco chloroplasts, *Cell,* 48, 111, 1987.
392. **Sparks, R. B. and Dale, R. M. K.,** Characterization of ^{3}H-labelled supercoiled mitochondrial DNA from tobacco suspension culture cells, *Mol. Gen. Genet.*, 180, 351, 1980.
393. **Gressel, J., Cohn, N., and Binding, H.,** Somatic hybridization of an atrazine resistant biotype of *Solanum nigrum* with *Solanum tuberosum*. Segregation of plastomes, *Theor. Appl. Genet.*, 119, 1984.
394. **Shepard, J. F., Bidney, D., Barsby, T., and Kemble, R.,** Genetic transfer in plants through interspecific protoplast fusion, *Science,* 219, 688, 1983.
395. **Stumpf, P. K. and Pollard, M. R.,** *High and Low Crucic Acid Rapeseed Oils,* Krassel, J. K. G., Ed., Academic Press, New York, 1983, 131.
396. **Bingham, E. T.,** Better *Crops for Food,* 97th Ciba Foundation Symp., 1983, 130.
397. **Belliard, C., Pelletier, G., Vedel, F., and Quetier, F.,** Morphological characteristics and chloroplast DNA distribution in different cytoplasmic parasexual hybrids of *Nicotiana tabacum, Mol. Gen. Genet.*, 165, 231, 1978.
398. **Belliard, G., Vedel, F., and Pelletier, G.,** Mitochondrial recombination in cytoplasmic hybrids of *Nicotiana tabacum* by protoplast fusion, *Nature, (London),* 281, 401, 1979.
399. **Uchimiya, H., Ohgawara, T., Kato, H., Akiyama, T., and Harads, H.,** Detection of two different nuclear genomes in parasexual hybrids by ribosomal RNA gene analysis, *Theor. Appl. Genet.*, 64, 117, 1983.
400. **Nagy, F., Torok, I., and Maliga, P.,** Extensive rearrangements in the mitochondrial DNA in somatic hybrids of *Nicotiana tabacum* and *Nicotiana knightiana, Mol. Gen. Genet.*, 183, 437, 1981.
401. **Negy, F., Lazar, G., Menczel, L., and Maliga, P.,** A heteroplasmic state induced by protoplast fusion in a necessary condition for detecting rearrangements in *Nicotiana* mitochondrial DNA, *Theor. Appl. Genet.*, 66, 203, 1983.
402. **Aviv, D., Bleichman, S., Arzee-Gonen, P., and Galun, E.,** Intersectional cytoplasmic hybrids in *Nicotiana*. Identification of plastomes and chondriomes in *N. sylvestris* + *N. rustica* cybrids having *N. sylvestris* nuclear genomes, *Theor. Appl. Genet.*, 67, 499, 1984.
403. **Aviv, D., Gonen, A. P., Bleichman, S., and Galun, E.,** Novel alloplasmic *Nicotiana* plants by "donor-recipient" protoplast fusion: cybrids having *N. tabacum* or *N. sylvestris* nuclear genomes and either or both plastomes and chondriomes from alien species, *Mol. Gen. Genet.*, 196, 244, 1984.
404. **Bohnert, H. J., Crouse, E. J., and Schmitt, J. M.,** Organization and expression of plastid genomes, *Encycl. Plant Physiol. (New Ser.),* 14B, 475, 1982.
405. **Leaver, C. J. and Gray, M. W.,** Mitochondrial genome organization and expression in higher plants, *Annu. Rev. Plant Physiol.*, 33, 373, 1982.
406. **Cseplo, A. and Maliga, P.,** Lincomycin resistance, a new type of maternally inherited mutation in *Nicotiana plumbaginifolia, Curr. Genet.*, 6, 105, 1982.
407. **Durbin, R. D. and Uchytil, T. F.,** Cytoplasmic inheritance of chloroplast coupling factor I subunits, *Biochem. Genet.*, 15, 1143, 1977.
408. **Gasquez, J. and Barralis, G.,** Mise en evidence la resistance aux triazines chez *Solanum nigrum* et *Polygonum lapathifolium* par observation de la fluorescence de feuilles isolees, *C.R. Acad. Sci. (Paris),* D288, 1391, 1979.
409. **Neeman, M., Aviv, D., Degani, H., and Galun, E.,** Glucose and glycine metabolism in regenerating tobacco protoplasts, *Plant Physiol.*, 77, 374, 1985.
410. **Boeshore, M. L., Hanson, M. R., and Izhar, S.,** A variant mitochondrial DNA arrangement specific to *Petunia* stable sterile somatic hybrids, *Plant Mol. Biol.*, 4, 125, 1985.
411. **Fromm, H., Edelman, M., Koller, B., Goloubinoft, P., and Galun, E.,** The enigma of the gene coding for ribosomal protein S12 in the chloroplasts of *Nicotiana, Nucl. Acids Res.*, 14, 883, 1986.
412. **Perl-Treves, R. and Galun, E.,** The cucumis plastome: physical map, intrageneric variation and phylogenic relationship, *Theor. Appl. Genet.*, 71, 417, 1985.
413. **Green, R. M., Vardi, A., and Galun, E.,** The plastome of *Citrus,* physical map, variation among *Citrus* cultivars and species and comparison with related genera, *Theor. Appl. Genet.*, 72, 170, 1986.
414. **Bonnett, H. T. and Glimelius, K.,** Somatic hybridization in *Nicotiana:* behavior of organelles after fusion of protoplasts from male-fertile and male sterile cultivars, *Theor. Appl. Genet.*, 65, 213, 1983.
415. **Fluhr, R., Aviv, D., Galun, E., and Edelman, M.,** Generation of heteroplastic *Nicotiana* cybrids by protoplast fusion: analysis for recombinant types, *Theor. Appl. Genet.*, 67, 491, 1983.

416. **Gerstel, D. V.,** Cytoplasmic male sterility in *Nicotiana.* (a review), *Tech. Bull. No. 263,* North Carolina Agricultural Research Service, 1980, 1.
417. **Medgyesy, P., Fejes, E., and Maliga, P.,** Interspecific chloroplast recombination in a *Nicotiana* somatic hybrid, *Proc. Natl. Acad. Sci. U.S.A.,* 82, 6960, 1985.
418. **Chetrit, P., Mathieu, C., Vedel, F., Pelletier, G., and Primard, C.,** Mitochondrial DNA polymorphism induced by protoplast fusion in cruciferae, *Theor. Appl. Genet.,* 69, 361, 1985.
419. **Mathews, B. F. and Widholm, J. M.,** Organelle DNA compositions and isoenzyme expression in an interspecific somatic hybrids of *Daucus, Mol. Gen. Genet.,* 198, 371, 1985.
420. **Glimelius, K., Chen, K., and Bonnett, H. T.,** Somatic hybridization in *Nicotiana:* segregation of organellar traits among hybrid and cybrid plants, *Planta,* 153, 504, 1981.
421. **Clark, E. M., Izhar, S. and Hanson, M. R.,** Independent segregation of the plastid genome and cytoplasmic male sterility in *Petunia* somatic hybrids, *Mol. Gen. Genet.,* 199, 440, 1985.
422. **Chaleff, R. S.,** *Genetics of High Plants — Applications of Cell Culture,* Cambridge University Press, London, 1981, 24.
423. **Imumura, J., Saul, M. W., and Potrykus, I.,** X-ray irradiation promoted asymmetric somatic hybridization and molecular analysis of the products, *Theor. Appl. Genet.,* 74, 445, 1987.
424. **Gupta, P. P., Gupta, M., and Schuder, O.,** Correlation of nitrate reductase defect in auxotrophic plant cells through protoplast mediated intergeneric gene transfer, *Mol. Gen. Genet.,* 188, 378, 1982.
425. **Gupta, P. P., Schieder, O., and Gupta, M.,** Intergeneric nuclear gene transfer between somatically and sexually incompatible plants through asymmetric protoplast fusion, *Mol. Gen. Genet.,* 197, 30, 1984.
426. **Bates, C. W., Hasenkampf, C. A., Contolini, C. L., and Piastuch, W. C.,** Asymmetric hybridization in *Nicotiana* by fusion irradiated protoplasts, *Theor. Appl. Genet.,* 74, 718, 1987.
427. **Galun, E. and Aviv, D.,** Organelle transfer, *Methods Enzymol.,* 118, 595, 1986.
428. **Galun, E. and Vardi, A.,** Novel *Citrus* trees by protoplast manipulation, in *Building Nature — The Genetic Manipulation of Plants,* Costanitino, P., Ed., S.P.A., Rome, 1986, 30.
429. **Galun, E. and Aviv, D.,** Manipulation of protoplasts, organelles and genes: applicability to plant breeding, in *Efficiency in Plant Breeding,* Lange, W., Zeven, A. C., and Hogenboom, N. G., Eds., Poduce, Wageningen, 1984, 228.
430. **Frankel, R. and Galun, E.,** *Pollination, Mechanisms, Reproduction and Plant Breeding,* Springer-Verlag, Heidelberg, 1977, 281.
431. **Widholm, J. M.,** Isolation and characterizations of mutant plant cell cultures, in *Plant Cell Culture in Crop Improvement,* Sen, S. K. and Giles, K. L., Eds., Plenum Press, New York, 1983, 71.
432. **Sidorov, V., Menczel, L., and Maliga, P.,** Isoleucine requiring *Nicotiana* plant deficient in threonine deaminase, *Nature,* 294, 87, 1981.
433. **Pandy, K. K.,** Sexual transfer of specific genes without genetic fusion, *Nature,* 256, 310, 1975.
434. **Jinks, J. L., Caligari, P. D. S., and Ingram, N. R.,** Gene transfer in *Nicotiana rustica* using irradiated pollen, *Nature,* 291, 586, 1981.
435. **Snape, J. W., Parker, B. P., Simpson, E., Ainsworth, C. C., Payne, P. I., and Law, C. N.,** Irradiated pollen for differential gene transfer in wheat, *Triticum aestivum, Theor. Appl. Genet.,* 65, 103, 1983.
436. **Vasil, I. K.,** Progress in the regeneration and genetic manipulation of cereal crops, *Biotechnology,* 6, 397, 1988.
437. **Dodds, J. H.,** *Plant Genetic Engineering,* Dodds, J. H., Ed., Cambridge University Press, London, 1985, 5.
438. **Szabados, L. and Roca, W. M.,** Regeneration of isolated mesophyll and cell suspension protoplasts to plants in *Stylosanthes guianensis.* A tropical forage legume, *Plant Cell Rep.,* 5, 174, 1986.

Chapter 4

CELL SELECTION AND LONG-TERM HIGH-FREQUENCY REGENERATION OF CEREALS AND LEGUMES

I. INTRODUCTION

Tissue culture is a collection of methods of growing large numbers of cells in a sterile and controlled environment. The cells are obtained from embryos, stems, roots, or other plant parts which are encouraged to produce more cells in culture. Usually such cultures are set up in vials, small jars, or petri plates, containing a medium with mineral nutrients, vitamins, and hormones to encourage cell division and growth. For plant cell cultures it is frequently possible to encourage individual cells to express their totipotency, that is, their genetic ability to produce entire plants. A cubic centimeter-sized culture contains around 100,000 cells. Thus, many genetically identical copies of plants can be produced in a short time through the tissue culture technique. This group of similar plants is known as a "clone". The ability to clone unique stress-resistant, disease-resistant, or high yielding plants through tissue culture would shorten by years the time required to release them for use in the field. Also various types of useful mutant traits can be selected directly within the cultures. A cell carrying the useful characteristics can then be cloned and many identical plants can be regenerated.

Plant breeders and agriculturists worldwide realized the urgent need for stress-tolerant plants. In particular, dry areas of the world are frequently subjected to stress from drought and from sodium salts occurring naturally in many soils and gradually added by irrigation. A major stress problem in areas with more rainfall is the occurrence of soil acidity and the resulting increase in solubility of toxic metals such as aluminum. These stresses lower agricultural production in many areas which are severely affected by drought conditions. In terms of drought stress, the problem is complex; irrigation systems are increasingly expensive to install and operate; available water from wells and rivers is in shorter supply due to increased pressure from population centers and from agriculture itself; and irrigation systems inevitably add to the salt concentration of soils. There is, thus, a clear, often desperate need for crop varieties, with increased levels of stress tolerance. Such plants can extend the lifetime of irrigation systems, make water of low quality useful for irrigation, lower the need and expense involved in reducing soil acidity, and, in a general sense, increase agricultural production on land which is today unavailable, marginally available, or available only with expensive environmental modification for agricultural use. In order to achieve these goals through tissue culture techniques it is necessary to demonstrate (1) that tissue cultures of crop plants can be set up, maintained, and induced to regenerate plants and (2) that useful mutations can be selected in tissue cultures, carried to the regenerated plants, and passed to future generations in an inheritable fashion. The ability to control and manipulate plant tissue cultures is essential if their potential to help the plant breeder is to be realized. This is particularly important for cereal and legumes which form the major source of food for people. As with other techniques, the tissue culture technique is easy to be utilized in plants which can be propagated through asexual or vegetative means. Thus, it is difficult to apply tissue cultures to cereals and legumes. Scientists working on several aspects of tissue culture have now been successful in regenerating and selecting stress-resistant cereals and legumes. In this chapter, the major emphasis is laid on the long-term high-frequency regeneration of such plants and subsequent utilization through tissue culture for the production of stress-tolerant plants.

II. TISSUE CULTURE AND LONG-TERM HIGH-FREQUENCY REGENERATION OF CEREALS AND LEGUMES

Ideally, suspension cultures are best suited for selection procedures since virtually every cell is in contact with the selective agent. In cereal systems, however, reliable regeneration of liquid cultures has not yet been achieved, partly because embryogenic cells have been difficult to maintain in liquid cultures. While suspensions are relatively easy to initiate, the lack of complete regeneration is very common in this system. The alternate system is callus cultures. Most callus cultures are established using diploid tissues. In addition, plants regenerated from cell lines that have been selected for stress tolerance often do not exhibit stable, heritable tolerance when the selection cycle is too short. This might be because some cells are able to adapt to stress physiologically, without genetic changes. It might also indicate that populations of mutant cells must be given sufficient time to establish themselves and to overgrow a given cell culture before there is a high probability of regenerating plants composed of entirely mutant cells. The production of calli from plants and subsequent high-frequency regeneration from such calli subjected to stress for a prolonged period could, therefore, be a significant method for the production of several stress-tolerant plants. Thus, before discussing the cell selection strategies, a brief account of initiation of callus formation and subsequent long-term high-frequency regeneration in cereals which are difficult to raise through tissue culture is given.

The greatest advances in increasing regeneration frequency have come from the realization that cereal tissue cultures produce different types of calli (masses of undifferentiated cells) which may differ in their regenerative potentials. First, most cereals seem to produce callus tissue with and without green spots and a positive correlation between the presence of such spots in a callus and its regenerative potential has been observed.[1-7] Second, cereal callus is of at least two types (embryogenic and nonembryogenic) irrespective of the presence of green spots.[8-13] The embryogenic (E) callus has a smooth, white, knobby appearance and is composed of small isodiametric cells with an average of 31 μm in diameter (Figure 1). ''Nonembryogenic'' (NE) callus is yellow to translucent, wet, rough to crystalline in appearance and is composed of larger, elongated cells which average 52 μm in diameter and 355 μm in length. E Callus is produced on a small fraction of a given callus. The E callus frequently gives rise to shoots or roots by organogenesis. The term E and NE are used freely in literature to describe callus even though the existence of true embroids is not usually histologically demonstrated.[14]

E callus in cereals produces more regenerated plants than NE callus, and E callus production and plant regeneration can be increased by appropriate alterations of the medium. In addition, the maintenance and regeneration of calli depend on the major factors which are described below.

A. EXPLANT SOURCE

The undifferentiated mass of plant cells (callus) maintained in a sterile environment, with the use of an appropriate nutrient medium. By manipulating one or more medium components, whole plants can be regenerated from these calli. The conditions, i.e., medium, temperature, light, humidity, and air supply required to culture a plant tissue, often vary with each species. Suitable explant source and a medium have to be worked out before using tissue culture for the improvement of a particular crop. The variations in explant sources and culture conditions for a wide variety of plants have been discussed by Conger,[15] Evans et al.,[16] George and Sherrington,[17] Reinert and Yeoman,[18] Vasil,[19] and Wetter and Constabel.[20] Explant sources and culture media leading to long-term high-frequency regeneration of certain cereals and legumes which were difficult to culture have been worked out recently by the Tissue Culture for Crops Project.[14] The calli so raised have been exploited for the

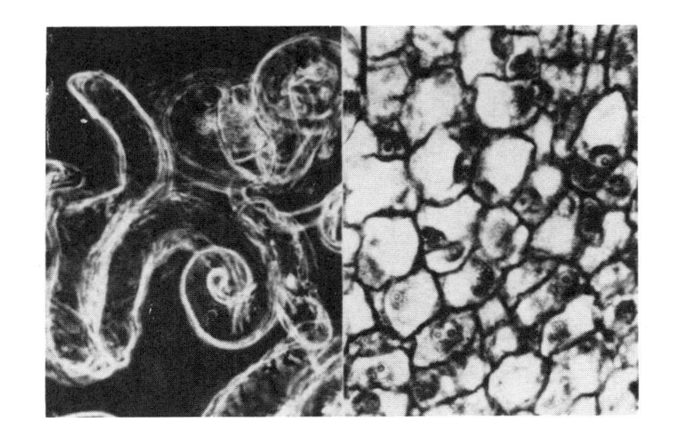

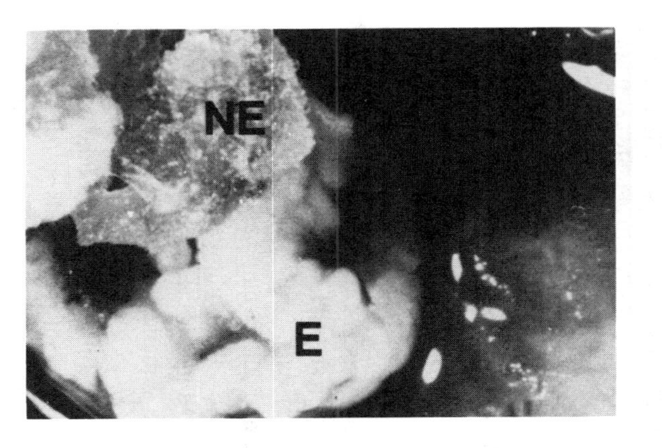

FIGURE 1. A highly regenerable form of callus which produces a high frequency of regenerated plants from tissue culture. Embryogenic (E) callus consists of nonvacuolated, meristematic cells in contrast to (NE) nonembryogenic callus which consists of highly vacuolated, elongated cells. (From TCCP [Tissue Culture for Crops Project], Progress Report, Colorado State University, Fort Collins, 1987. With permission.)

production of salt-tolerant plants. With this fortuitous beginning, reliable and consistent regeneration tissue culture methods have now become available for several cereals.[6,10-12,21-23]

Invariably long-term high-frequency regeneration of grasses has been associated with E callus cultures. Specific evidence for somatic embryogenesis has been presented for many cereals.[10,11,24-26] The most widely used explant for initiating embryogenic tissue cultures has been the immature embryo culture at a specific stage of development with scutellum facing away from the medium.[27] Other frequently deported sources of E callus in cereals are young influorescences[25,28-30] and unexpanded leaves.[9,31] It has been suggested that only explants still close to a meristematic state are reliable for E callus production[27,32] and the fully differentiated tissues of the Gramineae may not readily produce E callus.[33,34]

A more convenient source for starting tissue cultures in cereals is mature seeds. A number of workers have used seeds to start callus cultures.[6,35,36] Sears and Deckard,[37] using one defined series of media, achieved high rates from immature embryo-derived calli of some cultivars and found that both the rate and occurrence of regeneration were quite cultivar specific. Starting with mature seeds as the explants, scientists at Colorado State University working under the Tissue Culture for Crops Project[14] have produced long-term high-frequency regeneration calli for rice, wheat, sorghum, millet, and oats. They have achieved high-frequency, long-term regeneration for both Japonica and Indica cultivars of rice. Mature seeds have been very successful explant for establishing E callus cultures and the scutellum and occassionally the shoot apical meristems gave rise to embroids (Figure 2). Unlike wheat and corn, the potential of the rice scutellum to produce somatic embryos in tissue culture is very high as the embryo matures (Figure 3).

In wheat each callus is much more difficult to recognize than in rice and other cereals. Nabors et al.[6] found that smooth, compact callus was produced from seeds in culture and callus was moderately regenerable. Even attempts to regenerate high efficiency smooth-compact callus from bumpier callus derived from smooth-compact callus failed.[38] However, successful E calli were regenerated from developing scutella. Scutella of immature wheat embryos have the capacity to develop somatic embryos only during early stages of embryo development.

For oats, long-term totipotent tissue cultures have been obtained for a few varieties.[5] Cure and Mott[39] considered the proliferation of oat tissue in culture to be by growth of aberrant root-like tissues which had the external appearance of callus. Occasional shoot production in secondary callus cultures was considered by them to be a carryover from the primary culture which included meristem derived directly from the explant. Green spots have often been observed in cereal tissue cultures.[1,3,40-42] Nabors et al.[5] examined the relationship between green spot formation and shoot production in oat calli. Only calli with green spots grew and produced shoots. However, shoot production ceased after 38 weeks in callus selected for high green spot production. Calli initiated from mature seeds, mesocotyls, and immature embryos of oat (*Avena sativa*) produced E and NE callus and green spots.[11] E callus which is white, opaque, and convoluted in oat produced shoots, roots, or complete plantlet at much higher frequencies. Green spots did not initiate calli formation.

Several reports have been made on the *in vitro* culture and regeneration of grain sorghum.[43-45] The limitations of this technology are the inconsistent production of E callus and the lack of reliable, long-term whole plant regeneration. Mackinnon et al.[38] described the culturability of some sweet sorghums and outlined methods for the consistent production from developmentally mature embryos with the capability for plant regeneration. Tissue culture methods have now been developed for the induction, maintenance, and regeneration of E callus for sweet sorghum (*Sorghum bicolor*).[46] Callus production in sorghum, though dependent upon the type of the explant source, is also dependent on the donor plant.[14] Cultivar RT × 430 responded better if immature embryos were used as the explant source (95 of 102 immature embryos produced a total of 16 g of E callus). However, RT × 700 produced more E callus from mature than immature embryos.

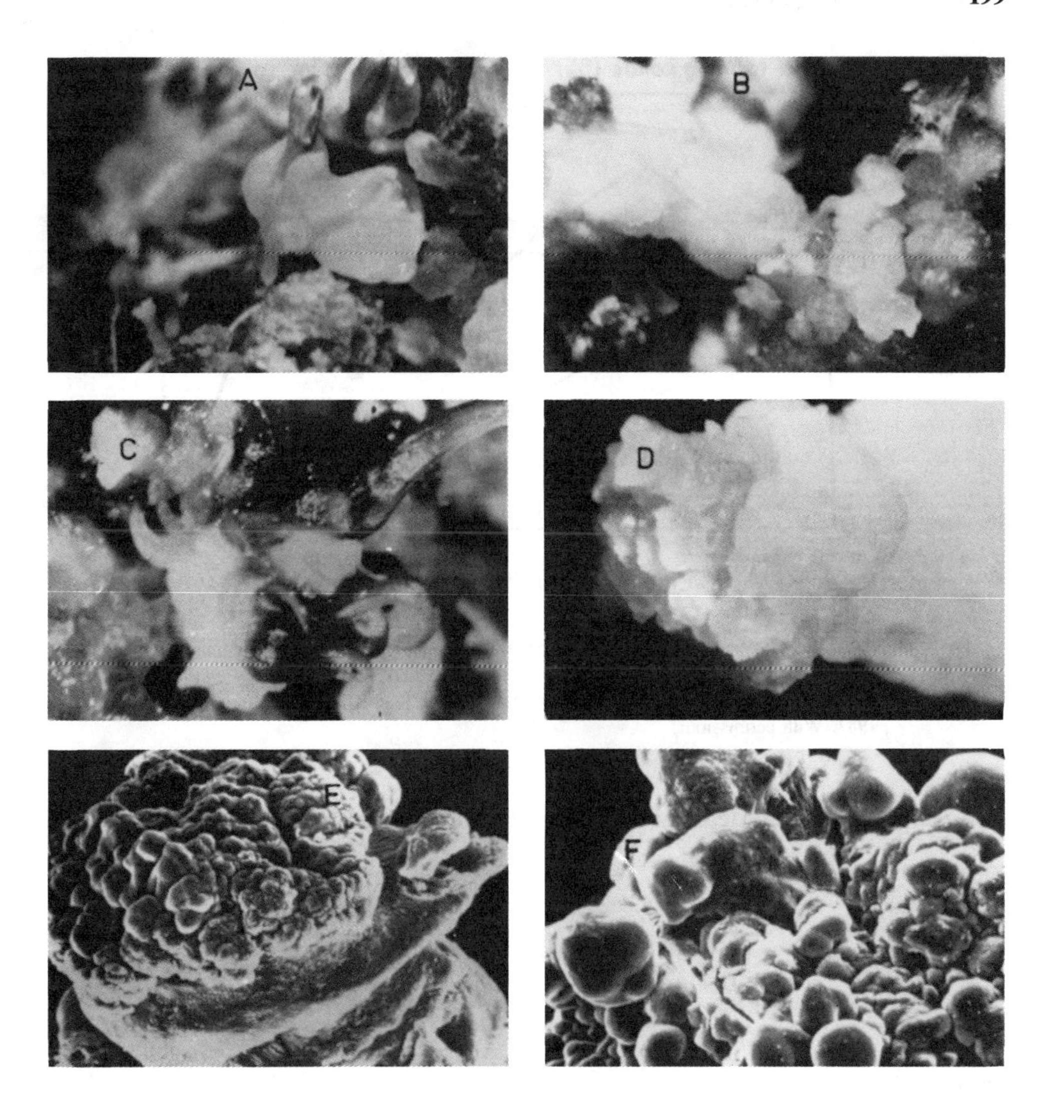

FIGURE 2. (A) A single somatic embryo on rice cultivar Giza 159 callus; (B) RD-23 E callus at second passage; (C) shoot forming on E callus of Mahsuri at third passage; (D) embryo forming on scutellum; (E) scanning EM micrograph of scutellum; (F) close-up of scutellum. (From TCCP [Tissue Culture for Crops Project], Progress Report, Colorado State University, Fort Collins, 1987. With permission.)

Rangan[47] reported the regeneration of proso millet (*Panicum milliaceum* L.) from calli initiated from dark grown mesocotyls which had been cultured for six to eight passages. These calli were apparently not embryogenic because scutellum formation which is characteristic in developing somatic embryos was not reported. A normal coleoptile was not formed during shoot initiation and newly initiated shoots were cultured for a second passage to induce adventitious roots. Heyser and Nabors[10] reported the retention of totipotency for over 24 weeks in proso millet calli through the selection and maintenance of embryogenic cell lines initiated from immature embryos, mature seeds, mesocotyls, and leaf and stem segments.

Somatic embryogenesis has been observed in cell suspension culture of soybean.[48] During induction of somatic embryogenesis from immature embryo of soybean a smooth shining-type callus developed from callus derived from cotyledon tissue and cultured on media containing 10 mg/l 2,4-dichlorophenoxy acetic acid (2,4-D). The rough type was derived

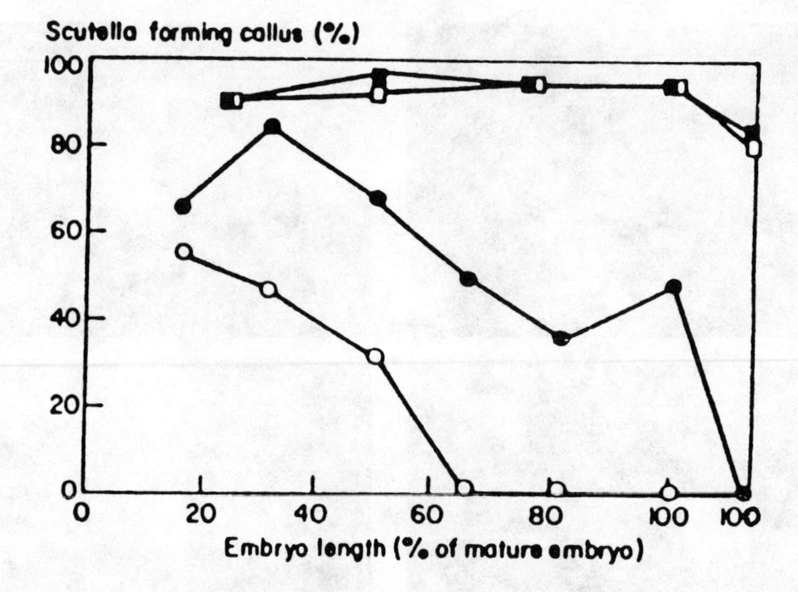

FIGURE 3. Formation of callus by rice (squares) and wheat (circles) scutella. Open symbols indicate E callus and closed symbols indicate total callus. (From TCCP [Tissue Culture for Crops Project], Project Report, Colorado State University, Fort Collins, 1987. With permission.)

TABLE 1
Optimal Media and Growing Conditions for Rice Cultivars

Japonica	Initiation	Maintenance	Regeneration (light)
Calrose 76	1 D + 0.2 K + 100 TRP (d)[a]	2 D + 0.2 K (I)	0.5 TIBA
Giza 159	1 D + 0.3 K (d)	1 D + 0.3 K (d)	0.5 A + 0.4 B
Indica			
IR-36	1 D + 0.5 K + 50 TRP (I)	0.5 D + 0.25 K (I)	0.5 A + 0.2 B
Pokkali	0.5 D + 0.2 K + 50 TRP (d)	0.5 D + 0.4 K (I)	0.5 B
(hybrid) Mahsuri	0.5 D + 0.5 K (I)	0.5 D + 0.5 K (I)	0.5 A + 0.48

Note: D = 2,4-D; K = KIN; B = BAP; N = NAA; A = IAA. Numbers preceding hormones denote concentration in mg/l.

[a] Culture conditions: callus grown in the light (I) or dark (d).

From TCCP (Tissue Culture for Crops Project), Progress Report, Colorado State University, Fort Collins, 1987. With permission.

from immature embryos, cotyledons, hypocotyls, and hypocotyl segments from germinated seedlings on a medium containing various growth regulators.

B. CULTURE CONDITIONS

1. Rice

Regeneration of shoots and plantlets from secondary rice tissue cultures (*Oryza sativa*) has been achieved for several varieties,[3,49] but many varieties do not initiate the formation of shoots[49] and the ability to initiate shoots is often lost after a few passages in culture.[3]

Optimal media for initiation, maintenance, and regeneration and growing conditions differ for each rice cultivar and do not seem to be consistent (Table 1).[14] In rice, for example,

TABLE 2
Regeneration Capacity of Calrose 76 E and NE Callus of Rice on Media Containing One of Three Different Growth Regulators

Concentration (mg/l)	% vials with shoots					
	BAP		TIBA		IAA	
	E	NE	E	NE	E	NE
0	25.0	0	25.0	0	25.0	0
0.1	53.0	0	60.0	0	20.0	0
0.2	68.8	0	73.3[a]	0	36.4	0
0.5	81.3[a]	0	75.0[a]	0	36.4	0

Note: Callus was initiated on medium containing 1 mg/l 2,4-D and 100 mg/l TRP. Sample size was 20 vials each containing 40 mg E callus.

[a] Significantly different from control (2 × 2 contingency test) at 0.05 level of probability.

From TCCP (Tissue Culture for Crops Project), Progress Report, Colorado State University, Fort Collins, 1987. With permission.

for cultivar Calrose 76, seeds form maximal E callus in the dark on LS (Linsmaier and Skoog) medium supplemented with 1 mg/l 2,4-D and 0.2 mg/l KIN (kinetin).[14] Addition of TRP (tryptophan) to the medium at 100 mg/l increased the proportion of E to NE callus significantly. It was also reported that IAA (indole-3-acetic acid) had a similar effect.[50] Plant regeneration in rice occurs only from E callus. A higher cytokinin-to-auxin ratio was required for plant regeneration than for E callus formation (Table 2). This appears to be attained by the addition of more cytokinin or by the inhibition of auxin transport. Cultures of rice cultivar Giza 159 were reported to be initiated and maintained in the dark on LS medium containing 1 mg/l 2,4-D and 0.3 mg/l KIN.[14] The medium requirement for another cultivar IR-36 was slightly different (1 mg/l 2,4-D, 0.5 mg/l KIN, and 50 mg/l sucrose). The TRP, though, enhanced E callus formation but did not help in its maintenance. Regeneration of IR-36 callus was obtained with medium containing 0.5 mg/l IAA and 0.2 mg/l BAP (benzylaminopurine). Initiation, maintenance, and regeneration media have also been determined for some other IR lines. Similarly, initiation, maintenance, and regeneration media have now been reported for several cultivars of rice including Mahsuri, Pokkali, Thia rice cultivar (RD-25), etc.

Studies carried out at different research stations under TCCP[14] showed that in Thia rice cultivar callus initiation rates were significantly higher when LS medium with 2.0 mg/l 2,4-D, 10% CW (coconut water), and 3% sucrose were used instead of a TCCP combination of LS medium with 1.0 mg/l 2,4-D, 0.4 mg/l KIN, and 4% sucrose (Figure 4). The Colorado State regeneration medium (LS + 0.5 mg/l IAA + 0.3 mg/l BAP + 4% sucrose) produced more plants per gram of callus than the Chulalongkorn University (CU) regeneration medium (Figure 5). The differences in callus initiation medium in two experimental stations (at Colorado State and Chulalongkorn University, Bangkok) were suggested to be due to the difference in hydrocarbon concentration in the air (Bangkok is a large metropolitan city as compared to Fort Collins, a small rural city). It was also discovered that exact replication of experimental methods at different locations is difficult. Chemical and water supplies as well as containers and culture room conditions would certainly vary. In addition, simple procedural differences may also profoundly affect the results. For instance, in the TCCP

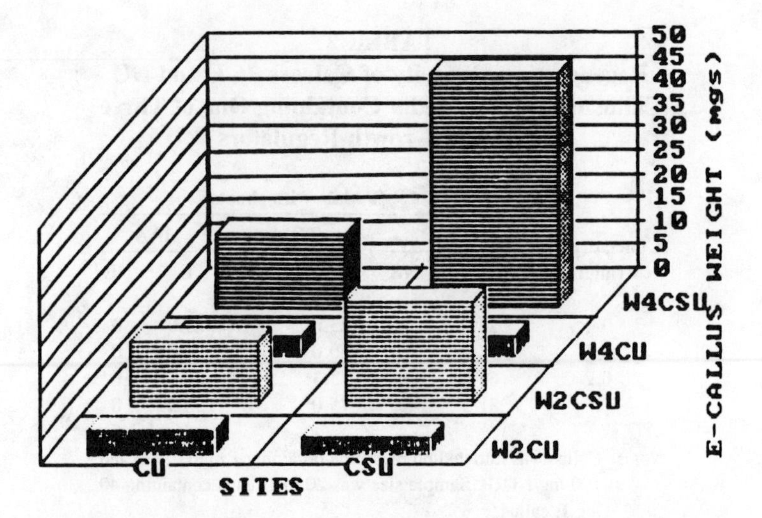

FIGURE 4. Callus initiation using CU and Colorado State methods at both sites. W2CU = 2-week culture using CU method; W2CSU = 2-week culture using Colorado State method; W4CU = 4-week-old culture using CU method; W4CSU = 4-week-old culture using Colorado State method. (From TCCP [Tissue Culture for Crops Project], Progress Report, Colorado State University, Fort Collins, 1987. With permission.)

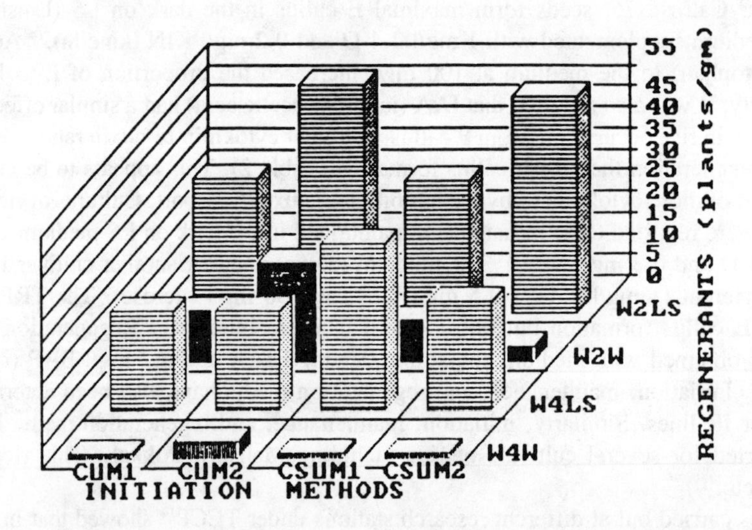

FIGURE 5. Plant regeneration in CU and Colorado State methods at Colorado State. CUM1 = CU initiation medium using CU method; CUM2 = Colorado State initiation medium using CU method; CSUM1 = CU initiation medium using Colorado State method;* CSUM2 = Colorado state initiation medium using Colorado state method;* W2W = 2-week-old calli in CU regeneration medium; W2LS = 2-week-old calli in Colorado State regeneration medium; W4W = 4-week-old calli in CU regeneration medium; W4LS = 4-week-old calli in Colorado State regeneration medium. *Each method was used for both initiation and regeneration. The number of regenerants from each treatment is significantly different at 0.01 probability level. (From TCCP [Tissue Culture for Crops Project], Progress Report, Colorado State University, Fort Collins, 1987. With permission.)

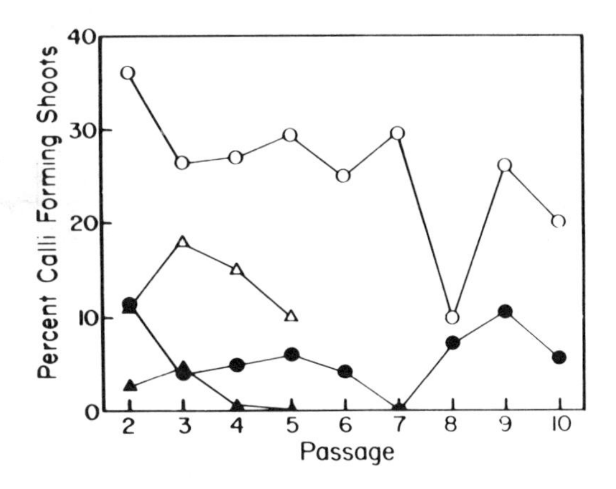

FIGURE 6. Percent of calli forming shoots in rice as a function of number of 5-week passages. The root-derived callus data are combined from all four varieties (Pokkali, IR-36, Mashuri, and Giza-159). The immature embryo-derived data are from Pokkali. Average sample size was 98 calli. ○, embryogenic root callus; ●, nonembryogenic root callus; △, embryogenic immature embryo; ▲, nonembryogenic immature embryo. (From Heyser, J. W., Dykes, T. A., DeMott, K. J., and Nabors, M. W., *Plant Sci. Lett.*, 29, 175, 1983. With permission.)

program at CU, scientists were able to obtain callus in rice with 100% E cells which had never been possible at Colorado State. These experiments also show the importance of complete description of the methods.

2. Wheat

Sears and Deckard,[37] using one defined series of media, achieved high rates from immature embryo-derived calli of some cultivars and found that both the rate and occurrence of regeneration were quite cultivar specific. In recent years, close visual observation of tissue in calli in a number of cereals has revealed that E callus is the source of most, if not all, plant regenerations.[6,24] The frequently low rates of plant regeneration in cereal tissue cultures, particularly those derived from mature embryos, are explained by the fact that E callus typically makes up a small fraction of the callus. Also, most media which select for rapidly growing callus usually, favor growth of larger nonembryogenic cells which form the friable, sometimes crystalline-appearing callus masses typical of cereal tissue cultures.[21]

Excised and subcultured calli for different varieties of wheat derived from roots of seeds produced more shoots than NE callus (Figures 6 and 7),[12] indicating that E callus from all passages produces high-frequency regenerated plants. This also suggests that maintaining continued production of embryogenic regions by subcultured callus is the key to obtain long-term regeneration.[12]

Series of experiments were conducted at the TCCP laboratory to maximize the amount of the smooth-compact callus produced in culture from seeds in wheat.[14] It was observed that 5 mg/l 2,4-D plus 20 mg/l TRP produced the maximum amount of smooth-compact callus.[21] It has also been determined that TRP does not consistently enhance E callus amounts in wheat and was eliminated from initiation and maintenance media. Isolated mature embryos also produce callus with embryos.[38] In cultivar Pavon, 2 mg/l 2,4-D was better than 5 mg/l 2,4-D at most KIN concentrations (Table 3).

Formation of E callus in wheat takes 1 to 2 months when immature embryos are used, as compared to 2 to 4 months when mature embryos are the explant source.[14] The composition of the regeneration medium does not appear to be critical, particularly in wheat, for long-

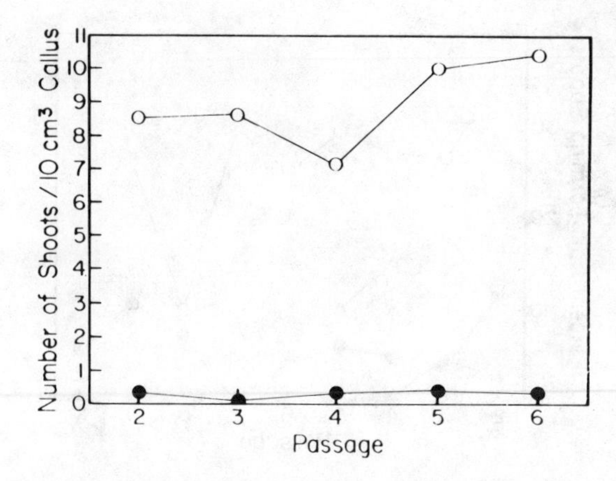

FIGURE 7. Number of shoots per 10 cc of rice callus as a function of number of 5-week passages. The data are combined from all four varieties. Average sample size is 102 shoots. ○, E callus; ●, NE callus. (From Heyser, J. W., Dykes, T. A., DeMott, K. J., and Nabors, M. W., *Plant Sci. Lett.*, 29, 175, 1983. With permission.)

TABLE 3
Effects of KIN,[a] Change of 2,4-D Level, and Selective Transfer on Percent of Cultures with Embryos, Wheat Cultivar Pavon

	Passage						
	1	2		3		4[b]	
Auxin (mg/l)	5D	5D	2D	5D	2D	5D	2D
Kinetin (mg/l)							
0	4	30	37	64	88	38	100
0.1	4	41	25	73	100	63	100
0.3	0	42	50	81	100	77	100
0.5	7	25	30	89	100	50	100
1	4	44	37	75	93	89	100
2	10	54	93	80	100	89	100

[a] Percentages within treatments at some passages varied significantly.
[b] At the start of fourth passage, only vials with embryos were transferred.

From TCCP (Tissue Culture for Crops Project) Progress Report, Colorado State University, Fort Collins, 1987. With permission.

term, high-frequency plant regeneration from the E callus. It was found that embryos were able to develop into plants on media devoid of any hormone, but more vigorous shoot and root formation occurred on media with hormones.[38] A critical factor for optimal regeneration of wheat was determined to be the inoculum size cultured on regeneration media.[14] When less than 0.01 g of E callus was used on regeneration medium (LS + 0.1 mg/l IAA + 0.5 mg/l BAP), plant formation was reduced (about 60% of the cultures formed plants). However, if larger pieces were used, approximately 90% of the cultures regenerated. Mackinnon[38] concluded from these results that an inoculum mass of 0.02 g of E callus per 10 ml of medium was optimal to obtain maximum regeneration. The majority of regenerates obtained from such long-term cultures were phenotypically normal and reached maturity under greenhouse conditions.

The success of callus initiation medium for wheat also depends more upon the developmental maturity of the explant than on the cultivar.[14] When whole seeds are used as explants, 5 mg/l 2,4-D is required to maximize the amount of smooth-compact callus and minimize the amount of NE callus. If isolated mature embryos are used, smooth-compact callus readily forms at 2 mg/l 2,4-D and produces significant amounts of E callus when maintained at this concentration.[38] While the majority of the embryos are formed during first passage, they are not (because of their small size, generally less than 1 mm) usually detectable. Thus, E callus is not evident in quantity until the second passage. With improved techniques, regenerative callus of spring wheat has been maintained for 22 passages with average regeneration rates of over 100 plants per gram.[51]

Some additional factors critical for the regeneration of wheat plants through tissue cultures are (1) visual selection of E and smooth-compact callus during the first and second subculture, (2) selective transfer of E callus to ensure long-term maintenance and (3) careful inoculation of regeneration media with E callus.

3. Sorghum

The major callus type produced by sorghum is a watery, sometimes crystalline, friable, large-celled NE callus.[6] A second type of NE callus formed in association with the friable callus is densely compact, nodular, and creamy yellow in color. This second type of NE callus should not be confused with E callus, because E callus is distinguished by being very compact, nodular, often cleft, and always creamy white in color. This callus type, produced by both immature and mature embryos, is recognized as E callus in grain sorghum[52] and other cereals.[6] Occasionally, in sweet sorghum cultures, a smooth, densely green, compact nodular callus is produced in close association with creamy white E callus from which regeneration can also be obtained.

A standard initiation and maintenance medium for sorghum was developed containing LS salts and vitamins, 2% sucrose, 1.1% agar, and 2 mg/l 2,4-D + 0.5 mg/l KIN.[14] The medium was adjusted to pH 5.5 prior to autoclaving.[46] Embryos were placed on this medium and subcultured for one 5-week passage in the dark. All additional passages consisted of 4 weeks in the light. During the first and second passages callus production was low (less than 0.05 g total callus per embryo). Visual selection of E callus did not occur until the third or fourth passage.

In sorghum, plant regeneration from E callus readily occurs on LS media supplemented with 1.0 mg/l IAA + 0.5 mg/l BAP.[14] Growth regulators are essential for the occurrence of long-term high-frequency plant regeneration. Occasionally, some E callus requires an additional passage on regeneration media before developing into plants. Regenerated sorghum plantlets are subcultured on a medium containing 3.0 mg/l IBA (indolebutyric acid) to complete root formation.[14]

Several reports have been made on the *in vitro* culture and regeneration of grain sorghum. The limitations of this technology are the inconsistent production of E callus, especially from developmentally mature explants and the lack of reliable, long-term whole plant regeneration. Mackinnon et al.[46] established methods in sweet sorghums for the consistent production of callus from developmentally mature embryos with capability for plant regeneration (Figure 8). They concluded that the key factors for obtaining highly regenerable callus in sweet sorghum are (1) culturing isolated mature or immature embryos on 2,4-D containing media for at least two 4-week passages, (2) identification of the compact nodular callus, and (3) selective transfer of the compact callus to ensure maximum growth while the production of E callus is initially low; recurrent subculture of this callus type results in increased production over time.

4. Millet

E callus produced in pearl millet cultures is similar to E callus produced by other cereal

FIGURE 8. Sweet sorghum regenerates from E callus established in a glass house. (From Mackinnon, C., Gunderson, G., and Nabors, M. W., *Plant Cell Rep.*, 5, 349, 1986. With permission.)

crops in culture. It is characteristically dense, compact, nodular, and creamy white in color.[14] Another type of callus commonly seen in millet appears yellowish in color and is also compact and nodular, but its ability to regenerate has not been established so far.

Early media experiments have determined that LS media supplemented with 5 mg/l 2,4-D, 1 mg/l IAA, and 0.3 mg/l KIN + 4% sucrose resulted in maximal E callus production from seed.[14] E callus production during the first subculture is low and tends to occur inside the callus mass; therefore, the entire mass of callus minus the shoot growth and the original explant source is subcultured for additional 4-week passages after which time callus types can be differentiated. E callus can then be separated from NE and is subcultured. This type of selective subculturing is necessary to maintain long-term cultures of E callus.

The E callus is placed on LS media with either 0.2 mg/l IAA and 0.5 mg/l BAP for millet variety Hays Population or 0.5 mg/l BAP for Senegal Bulk to obtain plantlets of millet.[14] Callus is subcultured for a minimum of two passages to allow the development of shoots. Following the formation of well-developed shoots they are transferred to LS media containing 3 mg/l IBA to induce root formation. Usually one 4-week passage is sufficient for a well-established root system, although two passages may be required at times.

5. Corn

The long-term tissue culture and regeneration of corn callus have been a challenge for many investigators. Part of the difficulty with corn culture is the inability of mature explants to yield regenerable callus. Immature embryos, though, yield E callus, but long-term maintenance and regeneration is not readily achieved.[27,53,54]

Plant regeneration capability has been demonstrated with several hybrid cultivars and inbred lines of corn.[14] When immature embryos of corn are cultured, the tissue gives rise to two separate types of calli. Plant regeneration is achieved by somatic embryogenesis from what is identified as type II callus after a period of up to 4 months. Type II callus is very friable, almost thread-like, with well-defined somatic embryos on suspensorlike structures (Figure 9A). It is formed from immature embryos cultured on LS media + 1 mg/l 2,4-D, 1 m*M* L-asparagine, and 2% sucrose. Type I callus is more abundant and grows at a more rapid rate, but plant regeneration is not successful beyond a 2-month culture period (Figure 9B).

Whole seeds, when cultured on LS media containing 2 to 10 mg/l 2,4-D, produced masses of roots; but the callus produced was NE.[14] The best responding explant source for corn is immature embryos. Seeds are planted in large pots approximately 9 in. in diameter to obtain the embryos. In 2 to 3 months depending on the time of the year and the cultivar, ear will appear. Once the ears have been ready, they are sterilized before the embryos are excised and initiated. The excised embryo is placed scutellum up on the media and grown in the dark for 4 to 6 weeks until callus develops.

Scientists at TCCP have been successful in extending regeneration from type I callus to over 5 months. To obtain regeneration from this callus five different types of media were tried.[14] These are initiation, maintenance, budding, regeneration, and rooting. Type I callus produced in this way is translucent, convoluted, and compact similar to E callus found in rice and sorghum. It is produced from immature embryos cultured in small petri dishes on 15 ml of LS media + 1 mg/l 2,4-D and 12% sucrose. Maintenance of the callus is on 2 ml/l 2,4-D with 4% sucrose. Calli have been successfully maintained for up to 90 d before moving them to the budding media. Only after buds are formed are they moved to regeneration media containing 1 mg/l BAP, 0.3 mg/l IAA, and 4% sucrose. The shoots are then placed on a rooting media of LS + 3 mg/l IBA and 4% sucrose. A total of 49 plants have been regenerated by the program with 7 plants set selfed seed.

6. Oats

Oat tissue culture is primarily obtained from mature seeds, although mesocotyls and immature embryos are also used in some experiments. In earlier reports[5,11] it was found that if seeds were used as explant, the better callus was obtained only from the root region. In subsequent investigations, however, it was recognized that callus derived from roots is exclusively NE callus. Thus, oat callus which was observed to be regenerable was derived from the scutellum or shoot regions of mature seeds.

Investigations on the better media required for the maintenance of oat callus revealed that the greatest amount of total callus was produced on LS medium containing 1mg/l 2,4, 5-T when cultures were grown in light. However, media containing 2,4-D produced significantly more E callus than media containing 2,4,5-T. [11] The E callus produced far more plants than either NE callus or callus with both E and NE regions (Figure 10).

7. Soybean

The induction of somatic embryogenesis and plant regeneration from immature embryos of soybean, *Glycine max* (L.) cv. Prize, has been achieved. [48] Soybean has been henceforth a very difficult plant for tissue culture. Scientists at the TCCP laboratory investigated the procedure for the regeneration of callus which was used for the production of plants with increased salt tolerance. [14] In this research, two morphological types of E calli, the origin and duration of their differentiation, and subsequent plant regeneration from soybean cultures were also investigated.

In this project whole immature embryos were obtained from pods (2 to 10 cm long) produced on plants grown in the greenhouse. Hypocotyls (from seedlings), embryos, and cotyledonary segments were planted on a solid basal medium (0.7% agar) containing LS major and minor salts, vitamins, 2% sucrose, and various concentrations of growth regulators. Variables investigated included hormone type and concentration, nitrogen source, proline requirement, and developmental stage of the explant. In addition, 2,4-D and proline com-

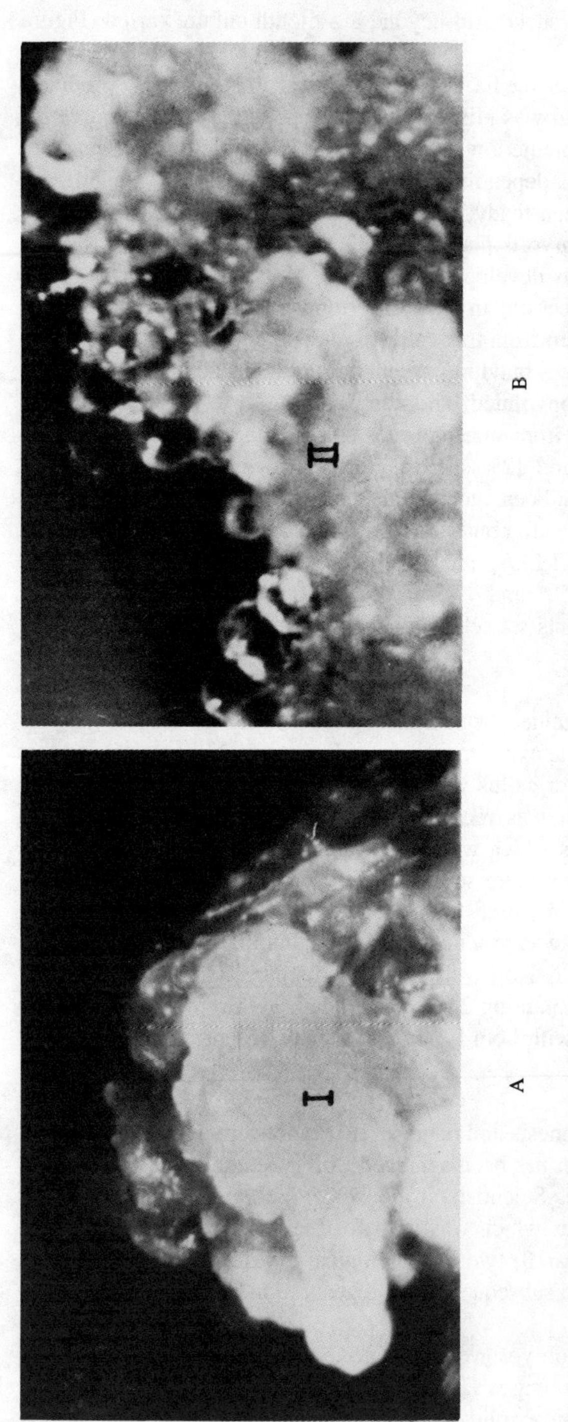

FIGURE 9. Type I (A) and II (B) callus from corn. (From TCCP [Tissue Culture for Crops Project], Progress Report, Colorado State University, Fort Collins, 1987. With permission.)

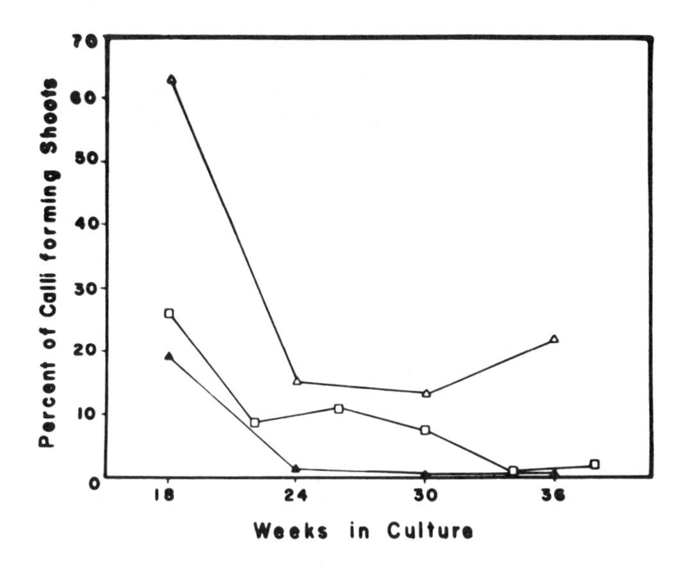

FIGURE 10. Shoot formation over time in oat culture. (△) E callus; (▲) NE callus; and (□) callus selected for high green spot production. For E callus there were 30 to 78 samples per point, while NE callus represented by 39 to 759 samples per point. (From TCCP [Tissue Culture for Crops Project], Progress Report, Colorado State University, Fort Collins, 1987. With permission.)

binations, nitrogen source (NH_4NO_3 and NH_4Cl), GA, zeatin, IBA, and ABA, were tested as components of maintenance and regeneration media (Table 4). At the end of the 4-week initiation passage, callus obtained from different sizes of cotyledonary segments, cultured on media containing 10 mg/l 2,4-D or 10 mg/l 2,4-D in combination with 0.132 to 0.264 mg/l ABA, were transferred to fresh medium of the same composition. In addition, 15 mg calli were transferred to a series of media for studies (Table 4) in which the effect of nitrogen source and proline on callus maintenance was tested.

A total of 33 different initiation and maintenance media have been tested. It was observed that callus from all explant sources produced on media containing 2,4-D was friable and creamy in color. [14] Callus initiated from whole embryos, cotyledons, hypocotyl segments, and hypocotyls from germinated seedlings on media containing IAA was greenish and more compact. These four explant sources, on media containing NH_4Cl as a nitrogen source, produced more vigorous growing callus compared with media containing NH_4NO_3. Proline had no apparent effect on callus growth.

Two types of E callus were observed. One type appeared smooth-shiny and was produced from cotyledonary segments. Media containing 10 mg/l 2,4-D in combination with 0, 0.132, or 0.264 mg/l ABA were not required for embryo formation.[14] The other type of E callus was consistently rough and was produced on cotyledons, hypocotyls, and hypocotyls from germinated seedlings on all initiation media except those containing 10 mg/l 2,4-D. The smooth-shiny E callus was greenish and translucent, while the rough E callus was opaque and creamish in color. The greatest amount of smooth-shiny E callus was produced from cotyledons 6 mm long (Table 5). When 2,4-D was used in combination with 0.264 mg/l ABA, the best development of somatic embryos from smooth-shiny E callus occurred.

When smooth-shiny E callus type was incubated in the dark in regeneration medium for 2 to 3 weeks, three different patterns of developing embryos were observed.[14] A few plantlets were also produced in some of the cultures. In the medium supplemented with 0.104 mg/l GA or 0.110 mg/l zeatin, some of the cultures differentiated and produced roots, shoots, or complete plantlets. Moreover, shoots or roots alone were obtained from these cultures on

TABLE 4
Hormonal Composition of LS Media for Embryo Initiation, Maintenance, and Plant Regeneration from Various Explants of Soybean Cv. Prize

Initiation Media (mg/l)

IAA	2 2,4-D
1 IAA + 0.5 BAP	2 2,4-D + 0.5 BAP
1 IAA + 5 BAP	2 2,4D + 5 BAP
1 2,4-D	10 2,4-D + 0.132 ABA
1 2,4-D + 0.5 BAP	5 IAA
1 2,4-D + 0.5 BAP	5 IAA + 0.5 BAP
10 2,4-D	5 IAA + 0.5 BAP
2 IAA	5 2,4-D
2 IAA + 0.5 BAP	5 2,4-D + 0.5 BAP
2 IAA + 5 BAP	5 2,4-D + 1 BAP
	10 2,4-D + 0.264 ABA

Maintenance Media (mg/l)

	+0.2 2,4-D	+0	Proline
	+0.2 2,4-D	+25 m*M*	Proline
	+0.5 2,4-D	+0	Proline
	+0.5 2,4-D	+25 m*M*	Proline
Minus	+0.2 2,4-D	+0	Proline
NH_4NO_3	+0.2 2,4-D	+25 m*M*	Proline
	+0.5 2,4-D	+0	Proline
	+0.5 2,4-D	+25 m*M*	Proline
Minus NH_4NO_3	+0.2 2,4-D	+0	Proline
Plus 800	+0.2 2,4-D	+25 m*M*	Proline
NH_4Cl	+0.5 2,4-D	+0	Proline
	+0.5 2,4-D	+25 m*M*	Proline

Regeneration Media (mg/l)

No growth regulators (control)
0.104 GA
0.103 GA + 0.132 ABA
0.103 GA + 0.132 ABA + 0.102 IBA
0.110 Zeatin
0.110 Zeatin + 0.132 ABA
0.110 zeatin + 0.132 ABA + 0.102 IBA
0.110 zeatin + 0.132 ABA + 0.103 GA

From TCCP (Tissue Culture for Crops Project), Progress Report, Colorado State University, Fort Collins, 1987. With permission.

media containing ABA plus either GA or zeatin. When such shoots were transferred to B5 medium[55] supplemented with 0.122 mg/l IBA or LS media combined with 5 mg/IBA, rooting occurred and whole plants were obtained. About 30 plants were successfully established in the greenhouse (Figure 11). There was an inverse relationship between induction frequency of rough E callus and length of hypocotyl segments obtained from seedlings. Low concentrations of 2,4-D produced less rough E callus and the callus required a longer time to develop into discrete structures. The higher the concentration of 2,4-D, the more smooth-shiny E callus was produced and the less time was required for its subsequent development (Figure 12).

TABLE 5
Formation of Smooth-Shiny E Callus in Soybean from Developmentally Different Cotyledonary Segments on Media Containing 10 mg/l

Immature embryo lengths (mm)	Cultures with embryos (%)	Embryos per culture (no.)
2	—	—
3	6	0.06
4	36	2.04
5	59	2.00
6	39	2.42
7	38	1.23
8	44	2.00
9	25	1.00
10	13	0.13

From TCCP (Tissue Culture for Crops Project), Progress Report, Colorado State University, Fort Collins, 1987. With permission.

FIGURE 11. Regenerated soybean plants from callus culture. (From TCCP [Tissue Culture for Crops Project], Progress Report, Colorado State University, Fort Collins, 1987. With permission.)

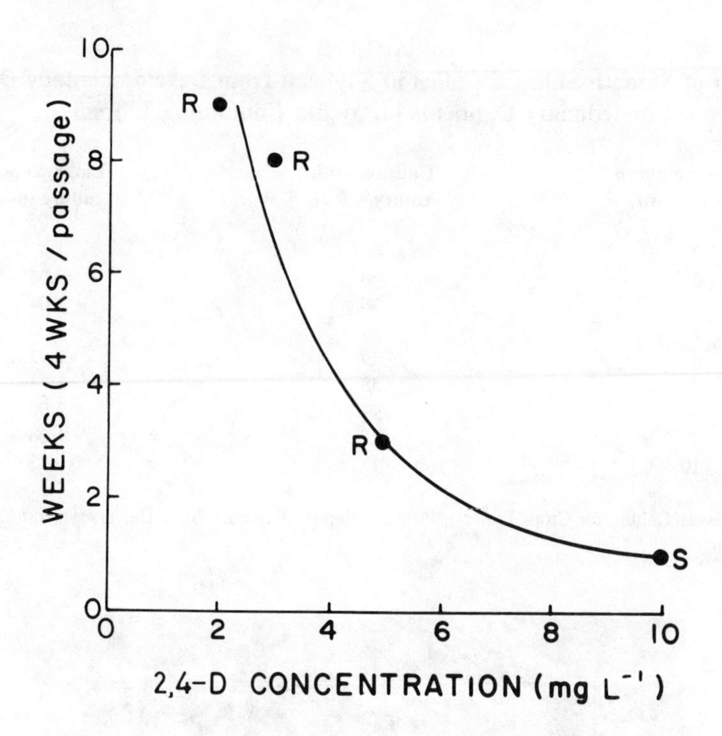

FIGURE 12. Relationship between time interval for E callus formation and 2,4-D concentration in soybean. Each coordinate represents at least 30 cultures. "S" represents smooth-shiny and "R" represents rough. (From TCCP [Tissue Culture for Crops Project], Progress Report, Colorado State University, Fort Collins, 1987. With permission.)

When immature leaves of soybean cultivar FER 335 (a line from India) were cultured on 2,4-D media, the E callus remained at the globular stage without further development. After 4 months on a media with 0.1 mg/l 2,4-D and 1 mg/l KIN, shoots differentiated in 4% of the cultures, but further developments did not occur. Stem segments, including nodes, were cultured on media with 4 or 5 mg/l 2,4-D. Callus production was observed on both ends and small, proembryonic structures appeared.

8. Pigeon Pea

A reliable regeneration system was developed from primary explant and first passage callus of seedling leaves and excised cotyledons of pigeon pea by Kumar et al.[56] A systematic approach has been undertaken at TCCP to produce E callus and to achieve high frequency of plantlet regeneration from long-term cultures of pigeon pea.[14] The mature and immature embryos and excised cotyledons of three genotypes of pigeon pea (ICP-7182, ICP-6917, and ICP-7128 obtained from the germplasm collections of ICRISAT) were used for tissue culture studies. When cultured on MMS (modified Murashige and Skoog) medium containing auxin, callus was initiated from both embryo and cotyledonary segments within a week of inoculation. At the end of 3 weeks, the E and NE calli were separated. Brown, loose callus was produced from the explants cultured on MMS media with 0.1 to 4.0 mg/l 2,4-D, picloram, dicamba, or NAA. This callus turned dark brown and became necrotic within one subculture on the media tested. A whitish green, nodular callus was observed on MMS media either with 1.8 mg/l 2,4-D or 1.1 mg/l NAA. Three-week-old callus was subcultured onto maintenance media (MMS with 1.1 mg/l NAA or 1.8 mg/l 2,4-D), which in turn produced organogenic callus. This organogenic capacity of the callus was retained until third passage. The whitish green, nodular callus turned brown after three subcultures and plant

TABLE 6
The Proportion of E Callus to Total Callus in Pigeon Pea on MMS Medium with 1.1 mg/l BAP + 200 mg/l CH and 2% Sucrose

Media type	Total weight of callus (mg)	Weight of E callus
MMS-1[a]	436.00	22.40
MMS-1 + 1% mannitol	421.06	124.36
MMS-1 + 2% mannitol	439.50	92.40
MMS-1 + 1% sorbitol	468.00	172.20
MMS-1 + 2% sorbitol	412.00	84.00

[a] MMS-1: Modified MS medium with 1.1 mg/l BAP + 200 mg/l CH and 2% sucrose.

From TCCP (Tissue Culture for Crops Project), Progress Report, Colorado State University, Fort Collins, 1987. With permission.

regeneration did not occur. To prevent browning, media was incorporated with polyvinylpyrrolidone, charcoal, or ascorbic acid at 0.1 to 1.0 g/l. The callus was also subjected to cold treatment at 4 to 6°C for several hours. The above treatments failed to inhibit browning in the callus; however, increasing sucrose concentration and organic nitrogen sources of either amino acids or CH were observed to prevent browning of callus until fourth passage.

The primary and subcultured calli were transferred to MMS basal medium for 4 to 5 d before being placed on regeneration medium. Shoot bud regeneration was observed from primary and subcultured callus up to fourth passge initiated and maintained on MMS medium with 1.1 mg/l NAA or 1.8 mg/l 2,4-D, and transferred onto the same medium supplemented with 5.6 mg/l BAP, 0.2 mg/l NAA, and 0.03 mg/l GA. The frequency of regeneration from embryo and cotyledonary calli varied from 38 to 42%, with four to nine shoot buds occurring on MMS medium containing 0.2 mg/l NAA and 0.02 mg/l KIN. The regenerated shoots were rooted on Blaydes medium containing 1.1 mg/l NAA and 0.1 mg/l KIN within 3 weeks. The regenerated plantlets were established in pots containing vermiculite and soil mixture.

All the auxin concentrations (1.0 to 25.0 μM), either alone or in combination with cytokinins, failed to induce somatic embryos either from explants or from callus; however, two to three proembryos developed on the surface of shiny green callus induced on MMS medium with 1.1 mg/l BAP + 200 mg/l CH. In 4 weeks of culture, the somatic embryos were further differentiated with distinct radicles, hopocotyls and cotyledons. These embryos developed into plantlets on hormone-free MMS medium and the regenerated plantlets were established in the greenhouse.

To increase the number of embryos per unit mass of callus, experiments were conducted with a different source of nitrogen, increasing sugar concentration, using mannitol and sorbitol, and reducing the sulfate concentration.[14] The MMS media with reduced $MgSO_4$ (70.7 mg/l) + 2.2 mg/l BAP + 200 mg/l CH, and 1 and 2% of mannitol or sorbitol were found to be optimum to induce proembryos from excised cotyledon and first- and second-passage callus (Table 6). Further growth of these proembryos into fully differentiated embryos was observed on MMS medium with 1.1 mg/l BAP + 200 mg/l CH. The proportion of E callus was highest in MMS with 1.0% sorbitol, and lowest without mannitol or sorbitol. Experiments are in progress to enhance E callus production and to retain the E potential in long-term cultures.[14]

III. CELL SELECTION

Cell selection, in its broadest sense, is the isolation and characterization of cells that differ from their parental cells in some observable manner. Several recent reviews have described plant cell selection and the use of culture-derived variant cells.[57-60] For the mutation to be passed on from one generation to another, it must occur either in meristem or germ cell level. It is difficult to achieve this through traditional methods. In addition, mutations in somatic cells will be lost when the tissue containing that cell dies. This problem appears to be solved through the tissue culture technique. Plant regeneration from tissue culture does not depend on whether a cell is in a meristem or whether it is a gamete;[61] instead it depends on the totipotency of a cell, i.e., the ability of any plant cell to develop into a whole plant. As described earlier, several variations appear in regenerants raised through tissue culture. For instance, Evans et al.[62] conducted several somaclonal variation experiments using tomato. Callus derived from sterile leaves was regenerated and the resultant plants (Ro) were self-pollinated and their progeny (R_1) were examined for plants that differed from seed-grown controls. From 230 plants, 13 nuclear-gene mutations were observed. Selected plants were self-fertilized (R_2) for analysis of inheritance. The mutations observed included recessive traits such as flower and fruit color, jointless pedicel, and virescence, and dominant mutations controlling fruit ripening and growth habit. Such cells through mutations may acquire the capacity to tolerate some type of stress. While the term "mutant" is frequently and casually used in reports of selection in plant cell cultures, "variant" is usually considered to be a more accurate word. This is because selected cells represent altered phenotypes, but may not necessarily represent altered genotypes, thus confusing the whole issue. It is further made difficult to define mutant with some certainty. It is now agreed that the term mutant should be exclusively kept for those situations where genetic stability and sexual transmission of a trait have been demonstrated. In situations where it is not possible to differentiate the mutant from variant with certainty, molecular and biochemical evidences should be gathered to resolve the issue.

A. SELECTION STRATEGIES

For selection to be successful for a particular crop it is necessary to develop a tissue culture procedure for successful regeneration of the plant. This has already been achieved for several crops, the selection of which is underway. Cereals and legumes, which until recently were difficult to regenerate through tissues cultures, have now been regenerated successfully. The methods for raising such crops through tissue culture have already been discussed in the previous section. Following is the brief description of procedures which are commonly used for cell selection.

The description of several selection procedures has been discussed from time to time.[63-67] In general, strategy as outlined in Figure 13 is used to obtain the desired characters in plants through tissue culture. The essential steps in the protocol are

1. An inhibitory level of the selection agent must be determined at which 100% of the cultured cells die or do not grow.
2. A selection medium (liquid or agar solidified) is produced at two to three times the predetermined inhibitory concentration.
3. The cell culture is plated on or suspended in the inhibitory medium.
4. The mutant cells which may grow at this inhibitory level are then detected.

In some tissues the protocol outlined above may be too harsh and lead to the death of even resistant cells due to toxic metabolites exuded from dying, susceptible cells. An alternative, multistep selection procedure that uses sublethal concentrations of an inhibitor and

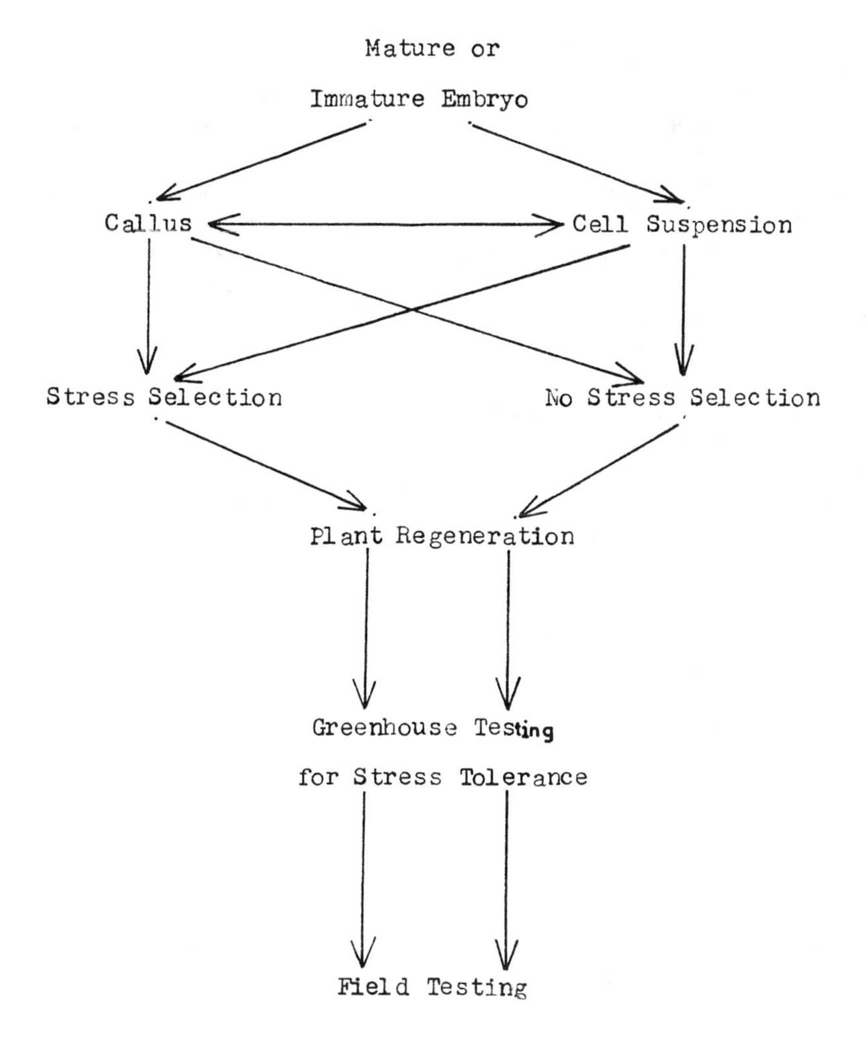

FIGURE 13. Tissue culture selection of stress-resistant plants. (From TCCP [Tissue Culture for Crops Project], Progress Report, Colorado State University, Fort Collins, 1987. With permission.)

subsequent subculturing onto media of higher inhibitor concentrations has also been suggested.[68] Glyphosate resistance in carrot cell cultures has been obtained from such a gradual selection.[68]

The procedure mentioned above is not enough to release the stress-tolerant plants to the fields or to hand over the plants to the farmers. After selection, once a variant is grown, the culture must be retested for resistance to the inhibitor and its stability is tested by growing the line away from the inhibitor for several generations and retesting it for resistance. This is then followed by the regeneration of the whole plant from the culture. Subsequent efforts should be made to determine the means of inheritance.[69] Finally, the mechanism of resistance and the manner of its expression in the whole plant (e.g., is the mutation a dominant or recessive trait, is it expressed throughout all stages of growth, is it expressed in all or few plant tissues) must be investigated before the variant is given to the breeders for development.

Specific variants may be identified and isolated by the application of some selection pressure that permits the preferential survival and/or growth of a desired phenotype, but selection is not essential. A great deal of genetic variation (usually called somaclonal variation) may be expressed in plants regenerated from cell cultures to which no selection has been applied. Some phenotypes may not be expressed in cultured cells, and for others no effective selection strategy may yet be conceivable. Those phenotypes are better isolated by simply screening regenerated plants. Nonselective screening can also be applied directly to cultured cells. For example, pigment variants can be easily identified by visually scanning large numbers of cultured cells.[70] However in those cases in which the phenotypes are expressed in culture and a selection strategy can be conceived, the effective use of selection pressure is likely to greatly increase the efficiency of recovery of new phenotypes. The particular selection method employed can greatly influence the nature of variant obtained and so should be chosen with care.

The length of the selection is an important criterion. Since selection agents allow for the preferential growth of variants, several generation times are required for the few mutant cells to become the dominant cell type in the population. In callus cultures this usually involves 10 to 12 months.[14] In suspension culture in which cell doubling times are usually faster and the selective agent is more extensively distributed, 3 to 4 months is minimum. Some investigators have combined both liquid and solid phase selection systems in order to take advantage of the more dispersed state of the selective agent in liquid and the regenerative capacity of the callus culture. Long-term cultures can be maintained on one concentration level of selective agent, or the concentration increased stepwise over time. The time required before the selective agent is increased in concentration is usually determined by the time required for doubling of fresh weight of the callus mass.

An alternate selection route as described is short-term exposure to very high, essential, lethal concentrations of selective agents. This procedure allows for the survival of those cells which possess an altered genotype and minimize the risks of selecting cells with physiologically adapted phenotypes.

The level of selective agents used must be determined empirically; i.e., for any crop, E callus must be initiated/maintained on a level of stress which reduces growth. The initial reduction in growth should be at least half of the controls. In some cereal systems that require 2 to 3 months for production of significant quantities of E callus, it may be more efficient to initiate callus first, than transfer callus to selective media. To increase stress tolerance in such crops the tissue culture process involves several steps. These steps as outlined by scientists working at the TCCP laboratory are[14] the following.

1. Callus Initiation and Culture

A large population of cells sufficient to contain desired mutants or produce a desired number of regenerated plants must be obtained. This is accomplished by removing an explant and placing it in liquid or on solid medium. The media contain various nutrients and hormones to allow the cells to divide and reproduce, eventually forming a large collection of cells (suspensions, if liquid medium; calli, if solid medium). After 1 to 4 months, depending on species and genotype, this primary suspension callus is large enough to be subdivided into identifiable cell types and transferred to a second medium where stress selection can take place.

2. Screening

Stress-tolerant, mutant (variant) cells must be identified. Once a sufficiently large callus has been produced, it is ready to be subjected to a specific stress. In the case of salt tolerance, the callus will be subdivided and placed on a medium containing a level of NaCl sufficient to severely retard the growth of nontolerant cells (overall growth reduction should be about 10% of controls). If mutant cells are present in the callus, they will grow and divide more rapidly than nontolerant cells and will eventually comprise the majority of the population. If the level of NaCl is not sufficient to eliminate nontolerant cells, these cells may divide and regenerate along with tolerant cells. During the selection process, other beneficial mutations are also identified. In nature, many salt-tolerant plants are also drought tolerant. Thus, if one is selecting for salt-tolerant mutants in culture, one might also discover drought-tolerant mutants. In addition, salt-tolerant plants in nature tend to be shorter than related nontolerant varieties. Therefore, in selecting for salt tolerance, one might also be selecting for resistance to lodging.

Crucial to the effectiveness of plant tissue culture as a means of decreasing environmental alterations is the genetic stability of the variant cells. Several types of alterations are possible, for example, point mutations and chromosomal alterations. Increase in duration of culture also increases the amount of chromosomal alterations. Long periods of selection may allow for occurrence of cells with two separate genetic mechanisms of tolerance. Such cells would be less affected by reverse selection pressure.

3. Plant Regeneration

To validate the usefulness of *in vitro* manipulations, plants must be regenerated from long-term, stress-selected cultures as well as from control cultures. In tobacco, cultures have been regenerated after 10 years or more in culture; however, it has been widely accepted that for most types of cultured cells, the rate of plant regeneration declines as the time in culture increases. Results from TCCP experiments have revealed that in wheat, rice, and millet, plant regeneration was dependent on the duration to maintain E callus in culture. In the case of wheat, regeneration has been obtained even after 20 months in culture. Some rice cultivars still produce plants with high frequency after 4 years in culture. Long-term regeneration is important because selection for stable variants requires up to 12 months of growth *in vitro*. In addition, such regeneration must be of high frequency to ensure that enough plants are produced for seed production and subsequent whole plant screening. Undesirable chromosomal abnormalities and chromosomal loss also increase with length of time in culture. Thus, the correct period of selection before plant regeneration must be long enough to ensure stable variants and short enough to maintain agronomic chromosomal integrity.

4. Testing of Regenerated Plants and Progeny

Plants obtained from stress-tolerant cultures must be shown to retain the tolerance and to consistently pass it on to future generations. The TCCP laboratory was first to demonstrate that NaCl tolerance obtained in cells was also apparent in regenerated plants;[71] however, it is possible that resistance, shown in other cultures, will not be passed on to regenerated plants. Most *in vitro* selection procedures isolate dominant or codominant variants; therefore, a generation of inbreeding may be required to obtain a stable, tolerant line of plants. It must be demonstrated, through whole plant testing, that stress-tolerant plants obtained from cell cultures also show increased tolerance during germination, seedling growth, maturity, flowering, and seed set. In addition, it is also important to determine that tolerance is stable in the absence of stress. Care should also be taken that an increase in stress tolerance does not favorably alter other desirable characters.

***In vitro* selection** — Because plantlets could be regenerated from pigeon pea only from third-passage callus, a slightly different screening and selection strategy was followed at the TCCP laboratory.[14] Cultures were grown on media supplemented with NaCl, Na_2SO_4, and mixed salts. The schematic representation of the strategy used for screening salt-tolerant pigeon pea is shown in Figure 14. First-passage callus was placed on saline MMS maintenance medium with 2.5 to 30 g/l of NaCl, 2.5 to 30 g/l Na_2SO_4, and mixed salts (17 to 68 m*M* Cl salts). In the initial experiments the pigeon pea callus was found to be highly sensitive to salt; callus became brown on the MMS saline medium with more than 10 g/l. However, the callus remained green and nodular, hence, this callus was subcultured to MMS

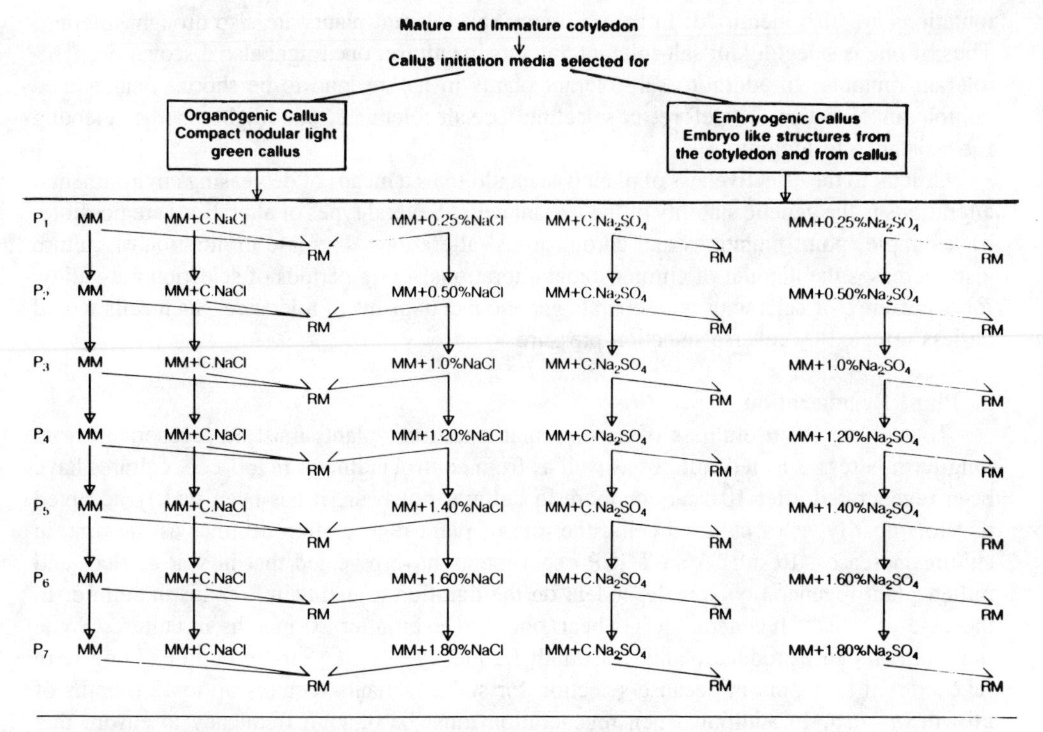

FIGURE 14. Schematic representation of the salt selection strategy followed in pigeon pea tissue culture. (From TCCP [Tissue Culture for Crops Project], Progress Report, Colorado State University, Fort Collins, 1987. With permission.)

media with 25 to 5 g/l of salts and by increasing the concentrations of salt at 0.50 to 1.0 g/l for every passage. At the end of every passage the salt-stressed callus was placed on regeneration media.

B. SELECTION METHODS

In selection methods either negative or positive selection procedures are used to produce stress-tolerant plants. Negative selection, in which growing cells are preferentially killed and nongrowing cells survive, is appropriate for obtaining auxotrophic mutants, those lacking a metabolic function such that their growth depends on a nutritional supplement. While negative selection methods have been employed with plant cells,[72,73] these methods have proven difficult and most auxotrophs have been obtained by nonselective screening.[74,75] Positive selection, in which only the desired cell type grows, favors the isolation of resistant mutants. Most mutants obtained in plant cell cultures to date are the results of positive selection.[76] Whether the selection is negative or positive the culture methods employed in selection experiments are also of critical importance. Selection can be imposed on virtually all kinds of cultures, but the results can vary markedly.

1. Callus Selection

Callus selection in which the selection pressure is imposed on pieces of calli[77] can be very efficient. This method for stress selection has been extensively used for raising stress tolerance plants of rice, wheat, oats, corn, etc. in TCCP.[14]

Raghava Ram and Nabors[78] suggested several convenient steps for the selection of cultivars with increased salt tolerance through tissue culture. They also suggested that callus culture should be preferred for cereals and legumes.

The starting material or the explant source in cereals and legumes is seedling roots or immature embryos. These are encouraged to grow and divide to form large, unorganized cell masses known as calli. The calli are then placed in one of the two groups: those to be stressed to obtain resistant variant, and those not stressed from which nonresistant controls are selected. Cell suspensions are produced when appropriate for either cell proliferation or selection. Selection is maintained for a long-enough period so that the entire culture comes to be composed of resistant cells. In some cases resistant cells are subjected to additional selection at increased stress levels. Frequently, selection begins with and continues through the callus initiation stage. In this case stressed cells grow and divide slowly. A callus which increases rapidly in size is identified as one which may be stress resistant. Four different selection strategies were followed for *in vitro* stress selection in E calli of grain and sweet sorghums (vs. Hegari, Rio, and Keller) at the TCCP laboratory.[14] These strategies are

1. Direct selection on 9 g/l NaCl on solid media for 4 months
2. Direct selection on 12 g/l NaCl in solid media for up to 3.5 months
3. Stepwise selection: 2 to 3 months on 6 g/l NaCl, then stepped up to 9 g/l NaCl for 3 to 4 months
4. Liquid agar selection: callus cultured in liquid media containing 1.5, 3, 6, or 9 g/l NaCl followed by either continuous or stepwise selection for up to 2 weeks

E callus produced from the above selection strategies was subsequently used to regenerate stress-tolerant plants. Callus which was stressed on 12 g/l NaCl, showed substantial decrease in fresh weight, whereas callus stressed on 12 or 9 g/l NaCl showed decrease in fresh weight over time, with trace amounts of callus remaining in fresh weight. However, callus stressed first on 6 g/l NaCl, then stressed on 9 g/l NaCl decreased in fresh weight only by 50%. In this way scientists at TCCP were able to regenerate sorghum from NaCl-stressed cultures (Table 7).

In the third strategy E callus was successfully grown in liquid cultures. Every 2 to 3 d medium was poured off and fresh medium added. The callus recovered from liquid cultures was frequently completely black, but within a few days formed regions of E clusters and, eventually, regenerated plants. The greatest number of plants were recovered after the first few days in liquid culture; however, plants were recovered from callus cultured for up to 12 d in liquid media.[46] Thus, experiments were carried with liquid saline cultures and the callus recovered from liquid saline cultures was also completely black. When plated directly on solid regeneration media, some calli produced E regions which regenerated into plants. If the callus was plated first on solid saline media and then cultured on regeneration media, fewer E calli were recovered; hence, regeneration was poor. In one experiment, however, callus recovered from liquid saline media was cultured on media containing 9 g/l NaCl, then placed on 10 g/l NaCl. Upon transfer to regeneration media, 24 plants from 0.3 g callus were recovered.

Sorghum callus is somewhat unique to cereals in that phenolic compounds are exuded by some cultivars of sorghum. Hegari, a grain type, produces a purplish-black exudate, while Rio, a sweet sorghum, exudes a reddish-orange color.[14] The phenols do not appear to interfere with E callus production or development; however, there is a correlation between the amount of phenols produced and level of stressing agents in media. The callus stress response associated with successfully higher NaCl concentrations could be seen by the extent of phenolics exuded into the medium. Rarely did callus growing on salt-containing media fail to exude phenolics. It has been suggested that the phenolics produced by the calli indicate stresses associated with the differentiated state of tissue.[79] Callus stressed on 6 g/l or higher concentrations of NaCl eventually became intensely pigmented with the black phenolics,

TABLE 7
Number of Sorghum Plants Regenerated from Salt-Stress Experiments Which Matured and Set Seed

Cultivar	Range of total passages	NaCl (g/l)	Range of passages on stress	Number of plants regenerated
Hegari	4—23	0	—	284
	11—15	9	4—7	301
	12	10	1	2
	8—22	12	1—4	55
	7—14	15	1—3	36
Keller	3—18	0	—	177
	10	9	4	7
	13—14	10	2—6	23
	13—16	12	1—5	74
	13—15	15	1—3	65
Rio	3—15	0	—	178
	21	9	2	11
	13—14	10	3—4	21
	13—18	12	1—2	51
	14	15	1	132
RTx-430	3—8	0	—	132
	14	10	2	1
	15—16	12	3—4	5
RTx-7000	4.5	0	—	22

From TCCP (Tissue Culture for Crops Project), Progress Report, Colorado State University, Fort Collins, 1987. With permission.

and it was difficult to determine if the callus was still viable. Once transferred to regeneration media, some of the stressed calli produced regions of embryoid clusters. These embryoid regions subsequently regenerated into plants.

An indirect method for *in vitro* selection to enhanced stress tolerance was attempted at the TCCP laboratory by altering concentration of proline (an amino acid involved with osmoregulation and essential to cell survival).[14] Hydroxyproline, a proline analogue, was chosen as the selective agent because it causes feedback inhibition of proline biosynthesis. Cells that survive may have an altered proline enzyme system that might also have enhanced tolerance through cellular adaptation to osmotic stress. Hegari E callus was inoculated onto maintenance media containing 0, 2, 5, and 10 m*M* hydroxyproline. After first passage, 10 m*M* hydroxyproline was determined to be severely growth restrictive. Upon testing the cultivars Keller, Rio, and Hegari at 10 m*M* hydroxyproline, differences in callus growth were observed between cultivars.

Screening for drought tolerance generally involves the use of polyethylene glycol (PEG). This polymer will interfere with the solidification of the agar media unless special techniques are used. A filter paper bridge method has been used with higher concentration of osmotica. This technique uses liquid media with the callus supported by the paper bridges. To make the bridges, a double-layer filter paper (Whatman No. 1) is cut and folded to make a 130 × 30-mm strip. This strip is folded into three sections with the sections being 50 mm long and the center being 20 mm. The bridges are placed in jars (75 mm tall by 35 mm in diameter) and 30 ml of medium containing PEG is added prior to autoclaving (Figure 15). Callus grows more rapidly on filter paper bridges, but setup time is considerably lengthened. Initial stages of stress studies with $AlCl_3$ have also been completed with wheat and 100 mg/l $AlCl_3$ was found to be lethal, as no callus survived at this concentration.

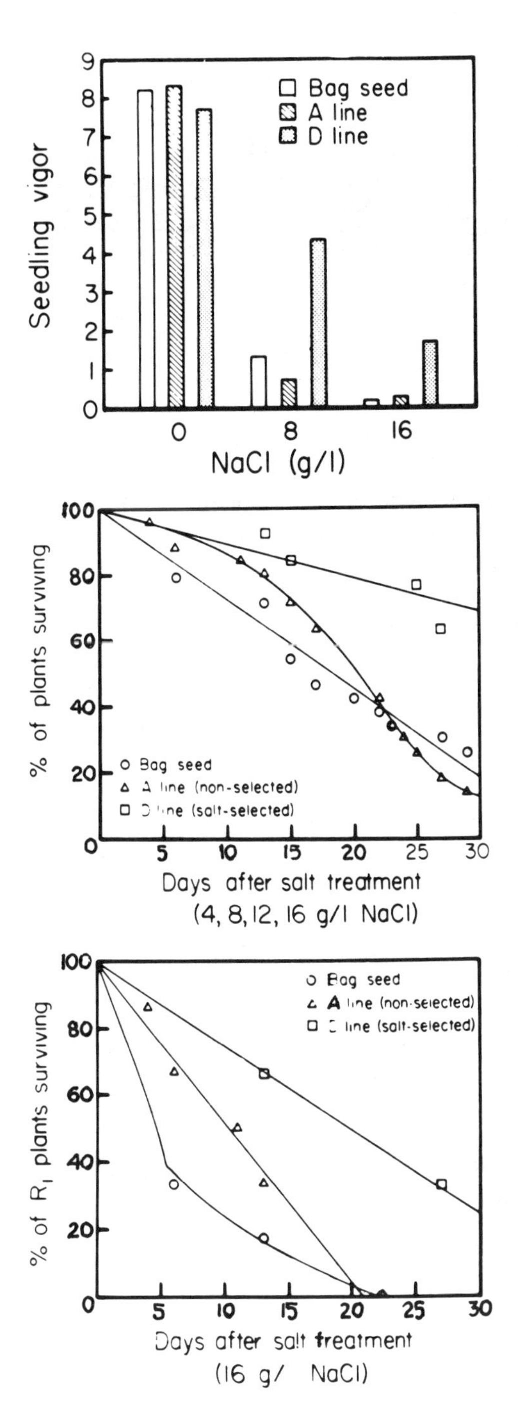

FIGURE 15. Data from hydroponic studies of regenerated oats. (From TCCP [Tissue Culture for Crops Project], Progress Report, Colorado State University, Fort Collins, 1987. With permission.)

2. Selection via Suspension Culture

Cells within a callus piece are not uniformly exposed to the selective agent, since not all are in direct contact with the medium. Callus cells are also in close contact with each other, thus permitting cross-feeding and promoting escape from the selection. Those rare cells of the desired phenotype that may exist in a callus piece can easily go undetected if they are physically surrounded by nongrowing cells all around them. Therefore, not only are the desired cells less likely to be detected with callus selection, but the surviving sectors may consist of a mixed population of cells which includes many cells that simply escaped the selection pressure. For these reasons the use of suspension cultures has been suggested. This method has been used by several workers to produce stress-tolerant plants.[80-88] Such suspension in cereals and legume cultures is usually difficult to obtain. This, however, has been successfully established for tepary bean (*Phaseolus acutifolius*) and moth bean (*Vigna aconitifolia*).[83] Both of these grain legumes are exceptionally drought tolerant.

Selection in suspension cultures[89] is also complicated by a number of factors. The selection pressure is applied to an entire flask of cells, and eventual recovery of growth in the flask is taken as an indication that the flask population is now enriched for the desired variant cell type. However, with this method it is not possible to discriminate between the rapid growth of a rare variant cell type and the slow adaptation to the selective agent of the entire cell population. In addition, it is not possible with this method to separate distinct variants arising from indepenedent events. If the original cell population contains several variants of the desired phenotype, the recovered cell population will be a mixture and further characterization will certainly be confused.

Chemicals which induce mutations have been used by some workers to increase the frequency of variants. Raghava Ram and Nabors[78] found that in tobacco suspension cultures, salt-tolerant variant occurred spontaneously in 1 of every 10^5 to 10^6 cells. Simons et al.[90] have produced selection in plant cell suspensions which can be used to estimate mutation rates from growth data. This model may also apply to display a prolonged adaptation period after initiation during which successive subcultures grow progressively faster. After a long period, the culture's growth rate stabilizes.[91] The resulting cells unavoidably selected for rapid growth in culture may have accumulated mutations which will affect plants regenerated from them. The application of this model to test the stability of tolerance has been suggested. Cells selected for salt tolerance are often grown in the absence of salt for a few passages and then returned to saline media to test the stability of the salt tolerance phenotype. Such experiments are the reverse of selection of salt tolerance.[80] It is assumed in the model that the growth rate of a few salt-sensitive cells (which are maintained at a low equilibrium frequency by back mutation in the selected culture) in stressful media is substantially higher than the growth rate of the salt tolerance cells when grown in the absence of salts. This will lead to a selection for salt-sensitive cells. In this case, the result will be a culture very similar to one before selection, predominently wild-type cells with mutants maintained at a low frequency by mutations. Thus, phenotypic instability after long periods in nonselective media should not be considered as proof that the tolerant cells are not true mutants.

3. Selection via Plated Cells

Selection via plated cells overcomes the problems associated with selection in callus and suspension cultures.[92] The cells can be plated directly on agar or modifications that permit the use of liquid medium and/or feeder cells can be incorporated to increase the efficiency of variant recovery.[93-95] Selection with plated cells offers the advantage that all cells are uniformly exposed to the selective agent. The cells are not in contact, so cross-feeding is minimized. Rare variants can easily be distinguished from general adaption of the cell population. Rather than being mixed, distinct variants can be recognized as separate colonies on a plate, although some may certainly be chimeras since the plating units are often not single cells. Additionally, selected colonies are not necessarily genetically distinct, though they may be clonally derived from the same mutation event. Selection with plated cells is somewhat more laborious than either callus selection or selection in suspensions. For this reason plated protoplasts for selection experiments to minimize the selection of chimerical colonies have been choosen by some workers. Protoplasts are, by definition, single cells and would be expected to give rise to a genetically homogenous colony. The possibility of a chimerical colony still exists, even in protoplasts. An advantageous feature of employing protoplasts is that they can be isolated from differentiated tissues, such as leaf mesophyll or root tissue, thereby suggesting the possibility of imposing selection pressures on differentiated functions that may not be present in established cell cultures. Protoplast selection is by far the most laborious of all the techniques; however, its major advantage, that of insuring the single cell origin of selected variants, is not so absolute to eliminate the necessity of cloning variants to completely eliminate the possibility of chimeras.

Greenhouse and field testing of plants regenerated from salt, drought, or acid-tolerant cell cultures is another important aspect.[96] In such testing, it is usually not possible to use Ro plants (those regenerated directly from tissue culture), because such plants need special care during the early stage of growth in the greenhouse, and it is in some cases difficult to obtain a sufficient number of plants of the same age. It has been suggested, therefore, to carry greenhouse testing in R_1 plants (plants from the seeds of selfed R_0) or later generation plants. Scientists at the TCCP laboratory[96] have sought to develop simple hydroponic methods which meet the criteria of allowing testing methods throughout the complete life cycle of the plants. These methods are now available for almost all cereals. In these methods small plastic tubes (24 × 30 × 13 cm) are filled with nutrient solution which is changed every 7 d. A piece of styrofoam, in which 12 × 2 cm holes are bored for plants, is then placed over the tub. Control plants (from field-grown seeds) are included in each tub along with the plants to be tested. Seeds are germinated in vermiculite filled in 12.5-cm-wide-pots. The pots are placed in trays with 5 cm of standing water. No nutrients are supplied. Enough seeds are planted to obtain eight seedlings per pot. At 14 d when the seedlings have two leaves, they are transplanted to tubs. The pH is adjusted to 5.5 in tubs containing wheat, sorghum, corn, or oats. Except for rice all tubs are supplied with bubbled air from a small aquarium pump. Tap water can be used to mix the nutrient solutions for the tubs if an ionic analysis is available so that nutrient concentration can be corrected. The next step is to transplant seedlings from pots to tubs. The pots are submerged in a container of water to easily remove all vermiculite from the roots. The seedlings are then moved to the tubs, held in place in the holes in the styrofoam with non absorbent glass wool. Salt damage can be compounded by excessive sunlight, heat, or wind; therefore, experiments should be randomized to eliminate potential problems with these variables within a greenhouse.

IV. APPLICATIONS

Plant-breeding goals according to Day[97] include (1) increasing crop yield, (2) improving crop quality, and (3) reducing crop production costs. Crop improvements through tissue culture could easily include, for example, developing herbicide tolerance, developing tolerance to salt or drought stress, and altering the nutrient content of the marketed crop part. Day,[97] for example, has suggested that the following approaches might be used to improve crop production: (1) increase in crop yield by altering photosynthesis to increase biomass production; (2) improve crop quality by altering high-molecular-weight glutonins to improve the dough elasticity and leaf volume properties of wheat flour; and (3) improve efficiency of crop production by developing salt tolerance, drought, and disease resistance and inserting nitrogen fixation capabilities. For the convenience of discussion, this section has been divided arbitrarily into four areas: salt tolerance, mineral tolerance, herbicide resistance, and disease resistance. Since there are a large number of reports in the literature describing work in these areas,[59,60,69] the following discussion is limited to a few examples of successful cell selections, particularly those examples where plants have been regenerated. Major emphasis will be laid on the production of stress-tolerance plants.

A. SALT TOLERANCE

Salinity problems in agriculture are primarily found in the wasteland, arid, and semiarid regions of the world, which comprise more than 25% of the earth surface.[98] Salinity poses problems in almost every irrigated area of the world as well as in the coastal regions. Salt concentration becomes a problem as water evaporates from the soil. The salt concentration in the soil after evapotranspiration (water loss from transpiration and direct evaporation from soil) may become four to ten times the salt concentration in irrigation water.[98] A level of 800 ppm salt is generally recognized as the threshold of severe agriculture and nonagriculture damage.

Crop productivity on about a third of the world land area is limited by water availability, high soil salt content, and high or low temperature extremes.[71,99] Furthermore, the need to irrigate much of this land has increased the problems of high concentrations of salts in the soil. Consequently, many stress-related selections in tissue culture have been aimed at selecting resistance to drought, elevated salt concentrations, and abnormally high or low temperatures.

The first reported success in selecting for stress tolerance was the selection of tobacco cell lines tolerant to elevated levels of NaCl in the culture medium.[86] With a similar selection procedure, Nabors et al.[71] developed NaCl-tolerant tobacco lines capable of plant regeneration. The tolerance was expressed through two generations after plant regeneration, but the inheritance pattern was not explainable by Mendelian genetics.

Nabors[4] has extended these studies to wheat, oats, rice, pearl millet, and proso millet and has developed cultures of each of these species tolerant to elevated NaCl concentrations in their culture media. Similar selections have been conducted with alfalfa[43] and rice.[100] However, in these cases regenerated plants showed either no tolerance (alfalfa) or the tolerance was unstable (rice).

A multistep selection procedure was used by Nabors et al.[71] to produce variant tobacco cultures tolerant to higher levels of NaCl. They also showed that tobacco plants derived from seeds of NaCl-stressed plants were more tolerant to NaCl stress than plants derived from seeds from unstressed plants.[71] These reports, along with those of Smith and McComb[43] and Yano et al.,[100] suggest that the selected tolerances required a high level of NaCl to be present for continued expression or the degree of tolerance was gradually reduced or lost. This tolerance appears to be similar to that reported by Handa et al.[101] for the adaptation of tomato cultures to high levels of PEG in the culture medium. The PEG was used to induce water stress. Handa et al.[99,102] showed that the loss of the selected tolerance was very gradual, requiring 20 cell mass doublings. They attributed the selected tolerance to osmotic adjustments made by the cultured cells.

Ben-Hayyim and Kochba,[103] using cultures of Shamouti orange (*Citrus sinensis*), selected what appeared to be a stable alternative form of resistance to elevated NaCl concentrations in the culture medium. They demonstrated that NaCl uptake was reduced substantially in the variant cultures and indicated that chloride was the toxic agent inhibiting the parental cells.

Genetic variability of salt tolerance has been characterized for certain agronomic crops such as barley.[104] Some wild species exhibit tolerance to extremely high salinity, up to 2 to 3 % NaCl (w/v).[105,106] It is theoretically possible to introgress salt tolerance genes from wild into cultivated gene pools. For example, high salt tolerance has been characterized in a wild relative of the tomato, *Lycopersicon cheesmanii*.[106] Hybrids between this species and tomato were sythesized and advanced tomato lines exhibiting salt tolerance were successfully developed. The likelihood of achieving this goal depends on the complexity of the trait.

Scientists at the TCCP laboratory[14] have conducted several long-term selection studies through callus culture of different cereals and legumes. They have been successful in raising salt tolerance plants from E callus of several cultivars of rice including Calrose, Giza 159, IR-36, Mahsuri, and Pokkali. Eighty-one plants of Calrose, 535 of Giza 159, 519 of IR-36, 258 of Mahsuri, and 146 of Pokkali, tolerant to NaCl, have been regenerated and grown to maturity (Table 8). Callus cultures of these plants were maintained at respective levels

TABLE 8
Selected Salt Levels and Their Influence on Length of Culture and Number of Rice Plants Regenerated and Grown to Maturity in the Greenhouse

Cultivar	Salt level (g/l)	Maximum total passage length in culture	Plants which set seed	Total
Calrose	0	11	29	
	3	8	47	
	6	7	5	81
G-159	0	20	108	
	3	18	130	
	6	11	183	
	8	10	13	
	9	5	38	
	10	10	63	535
IR-36	0	22	258	
	3	11	173	
	5	2	1	
	6	21	70	
	8	5	7	
	9	15	7	
	10	2	3	519
Mahsuri	0	22	15	
	1	4	3	
	3	6	209	
	6	8	26	
	8	4	4	
	9	3	1	258
Pokkali	0	13	31	
	3	13	38	
	6	22	65	
	9	7	12	146

From TCCP (Tissue Culture for Crops Project), Progress Report, Colorado State University, Fort Collins, 1987. With permission.

of NaCl until the callus was moved to regeneration media. Though cultures did not survive seven passages at higher levels, such cultures were successfully maintained on 6 g/l NaCl for at least seven passages for all cultivars and have retained regeneration ability to maturity and seed set. With rice cultivar Giza 159 and Mahsuri (3 g/l NaCl), it was even possible to maintain long-term cultures for more than 12 passages under NaCl stress.

Preliminary experiments to obtain reduced callus growth sufficient to obtain tolerant mutants in the presence of PEG and $AlCl_3$ in rice revealed that callus grown in the presence of 10, 20, and 50 g/l regenerated and set seed from PEG-stressed culture of cv. Pokkali. Plants were recovered under $AlCl_3$ stress (250, 500, 1000, and 2000 μ g/l) after a minimum of 4 months in culture for Pokkali, Giza 159, and IR-36. A total of 170 fertile plants from IR-36, 38 from Giza 159, and 1 from Pokkali were regenerated.

R_1 seedlings of Calrose derived from callus that had been selected for tolerance at 6 g/l NaCl were approximately 70% more tolerant than the R_1 progeny of plants derived from callus cultures which were not selected for salt tolerance.[14] At the highest level of salt tested (15 g/l NaCl), the R_1 progeny of callus selected at 6 g/l NaCl were approximately 100% more tolerant than the controls. IR-36 plants in R_1 regenerated from callus selected for tolerance at 9 g/l NaCl did approximately 100% better than R_1 plants derived from nonselected callus. With 15.0 g/l NaCl in the test solution, R_1 progeny from the 9 g/l NaCl-selected cultures did almost 200% better than the controls. It is interesting to note that at a low *in*

TABLE 9
Response of Sorghum E Callus and Regeneration to One Passage on 50 mg/l $KClO_3$

Cultivar	Culture period (months)	Mean weight E callus		Plants/g	Total number of plants
		Inoculated	Recovered		
Hegari	1	3.4	0.33	2	9
	2	8.0	0.17	2	15
	3	11.2	0.19	1	10
RTx-430	1	0.2	0.09	0	0
	2	0.1	0.05	0	0
Keller	1	1.0	0.23	15	74
	2	1.5	0.05	16	58
	3	1.8	0.09	10	44
Rio	1	0.4	0.22	45	11
	2	0.7	0.23	19	57
	3	0.6	0.16	10	17

From TCCP (Tissue Culture for Crops Project), Progress Report, Colorado State University, Fort Collins, 1987. With permission.

vitro selection pressure (3 g/l NaCl), tolerance of the R_1 progeny of several rice cultivars did not appear to be significantly greater than the controls when the selection period was for 7 months or less. When the selection period was more than 7 months, however, some NaCl tolerance was expressed in the R_1 progeny. Thus, rice callus, selected for salt tolerance *in vitro,* can produce salt tolerance seedlings when given a selection period of at least 7 months. R_1 seedlings (61%) derived from callus culture stressed at 3 g/l NaCl, 71% of the R_1 seedlings derived from callus stressed at 6 g/l NaCl, and 92% of the R_1 seedlings derived from callus stress at 9 g/l did better in seedling tests than did R_1 plants derived from callus that was never stressed. This suggests that the total stress in culture is a function of both the level of NaCl in the medium and the length of the selection period. Several experiments investigated the effects of NaCl on long-term callus growth and fresh weight measurements in wheat.[14] Total callus amounts grew reasonably well on media containing 9 g/l NaCl; however, E callus was more sensitive to salt in that fresh weight amount decreased over time. Therefore, for further screening cultures were initiated from mature embryos of spring wheat cultivars Pavon and Glennson and maintained on 2 mg/l 2,4-D with 3, 6, or 9 g/l NaCl. Green and yellow compact E calli were maintained for five passages on their respective selection media. When transferring to the sixth passage, one half of the E calli from each group was divided into small pieces and placed on regeneration medium containing 0.1 mg/l IAA and 0.5 mg/1, and 0 plants from 9 g/l NaCl. In another experiment on Pavon, cultures were maintained for four passages on nonstress media, then moved to 12 g/l NaCl for four additional passages. After a total of eight passages, all of the calli were moved to regeneration media and 22 plants were regenerated. In similarly designed experiments with Glennson, after four passages on nonstressed media the cultures were moved to either 10 or 11 g/l NaCl. This has resulted in the production of over 2200 plants from cultures containing NaCl and 1650 from control cultures.

Embryogenic calli of four sorghum cultivars, Keller, Rio, Hegari, and RTx-430, were screened for tolerance to $KClO_3$.[14] Potassium chlorate was added to the maintenance media at a rate of 50 mg/l. Data indicated a significant decrease in callus growth after 1 month (Table 9). The cultivar Rio showed a similar growth response after exposure to chlorate. Regeneration rates of sweet sorghums (Keller and Rio) showed a different response to chlorate stress as compared to the grain sorghums (Hegari and RTx-430).

TABLE 10
Regenerated Pearl Millet Plants Which Have Set Seed from NaCl-Stressed Cultures

Cultivar	Range of total passage	NaCl level (g/l)	Range of passages on stress	Plants regenerated
HMP-559	5—9	0	—	29
	4—6[a]	3	—	21
	4—8[a]	6	—	15
	13—14	8	2—6	4
	3[a]	9	—	2
SB	9—15	0	—	19
	9—11[a]	3	—	7
	7[a]	6	—	3
	12—15	8	3	13

[a] Cultures started on media containing NaCl.

From TCCP (Tissue Culture for Crops Project), Progress Report, Colorado State University, Fort Collins, 1987. With permission.

tolerance plants using NaCl and $AlCl_3$.[14] The number of cultures initially stressed on 6 g/l NaCl and then stressed on 8 g/l NaCl was drastically reduced from 250 cultures in the first passage to 43 in the second passage. A continual decline in the number of cultures was observed until tenth passage, while those remaining turned brown and necrotic. Callus stressed on 9 g/l NaCl also showed a decrease in growth over time until it was transferred to regeneration media without NaCl. After the callus was removed from the stress, the growth rate of E callus remained almost constant. Several plants were regenerated which set seed (Table 10).

Oat tissue cultures were used as the model cereal system to select *in vitro* for NaCl-resistant variants at the TCCP. In initial experiments, it was observed that as the NaCl concentration increased, the quantity of callus initiated decreased (Figure 16); however, callus formation occurred on media containing up to 10 g/l NaCl. In several later experiments, additional callus cultures were stressed on media containing 3 and 6 g/l NaCl for periods of one to five subcultures. Several hundred plants were regenerated from some of the short-term, salt-stressed cultures, and grown to maturity. Plants from the R_3 generation (Figure 17A) whose genetic heritage began with salt-tolerant cell lines were considerably more tolerant in pot experiments than those from salt-sensitive lines. In addition, the salt-tolerant R_4 progeny flowered and set seed 2 weeks earlier than plants from salt-sensitive cultures (Figure 17B). The oat data represented the second demonstration in the literature, and the first for cereals, that traits (1) selected in cell cultures also appear in regenerated plants, (2) are stable in the plants in the absence of stress, and (3) are inherited to subsequent generations. This does not mean that all cell-selected stress tolerance will behave identically. In fact, whole plant testing of such selected traits is required to make any valid conclusion.

The early flowering and seed set of the salt-tolerant oats are of interest for several reasons. First of all, this phenotypic trait could be of considerable value since oats are a cold weather crop and are frequently subject to production losses due to early frost. Second, the genetic relations between tolerance and earliness need to be determined; they could be pleiotropic effects. In the latter case, earliness would represent a somaclonal variation not selected for. There is some concern among breeders that selections in cell culture might produce the desired alteration in a cultivar as well as other undesired changes. While this is a clear danger, it can largely be controlled by close scrutiny of the plants produced.

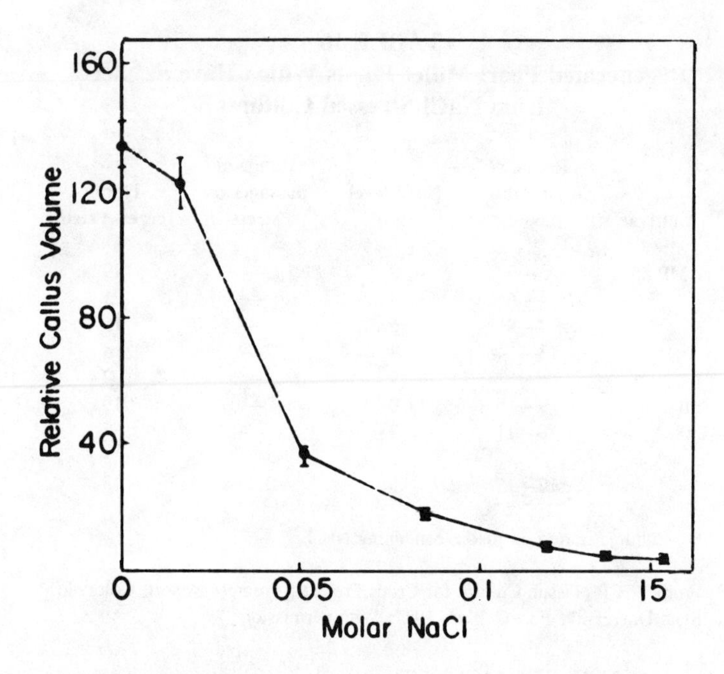

FIGURE 16. Callus initiation in oat and a function of NaCl concentration when 2,4,5-T concentration was 7.8×10^{-6} *M*. Each sample consisted of 20 vials. The no-growth percentages are 0, 0, 0, 4, 8, 21, and 36 for data points from left to right. (From Nabors, M. W., Kroskey, C. S., and McHugh, D. M., *Z. Pflanzenphysiol.*, 105, 341, 1982. With permission.)

A hydroponic system, similar to that described for rice, was developed to determine exact levels of salt tolerance in oats. Data from hydroponic studies were used to answer a question frequently posed by plant breeders. Does tissue culture selection simply pick out stress-tolerant recombinants naturally occurring in field-grown seed? Salt tolerance, for example, could arise by mutation or by meiotic recombination. Most cereal cultivars are highly inbred and homozygous for selected agronomic data, and hydroponic experiments (Figure 17) indicate that, in Park oats at least, salt tolerance does not occur to a high degree in field-grown seed.

A "decay curve"-type response to increasing NaCl concentration was obtained in soybean (Figure 18).[14] Since the F test was highly significant for the regression, the negative influence of salt on production of embryo-like structures is suggested. The curve also shows that when Prize cultivar is used in future studies, NaCl concentrations greater than 3 g/l are unnecessary. An optimum explant size of 6 mm resulted in maximum production of embryo-like structures; however, as salt concentration increased, influence of explant size on the number of embryo-like structures disappeared. Thus, a certain amount of salt tolerance is accomplished by choosing embryos (explants) at an intermediate stage of development.

Twenty-five mature soybean plants were raised from 30 cultures, each culture containing 1 g/l NaCl. The five plants obtained from 30 cultures containing NaCl at 3 g/l did not survive. At 5 g/l, a few embryo-like structures were produced but plants did not develop. Seven and 9 g/l of NaCl failed to yield embryo-like structures.

If the technique of cellular selection is to be successfully applied in crop improvement programs, the efficacy of the selection program, in terms of progeny performance in soil, will have to be demonstrated. Two critical questions need to be addressed: (1) is the trait, selected at the cellular level, active at the whole plant level? and (2) is the trait genetically stable? It is interesting to note that in a recent report Raghava Ram and Nabors[107] cite 37 cases of cellular selection *in vitro* for salt tolerance, but in only 6 were plantlets regenerated.

FIGURE 17. (A) R_3 oats watered with 0, 5, and 10 g/l NaCl solution. Front row: plants from nontolerant cultures; back row: plants from cultures tolerant to 3 g/l NaCl. Twenty-three seeds planted per pot. (B) R_4 plants whose parents were regenerated from tissue cultures tolerant (right) and nontolerant (left) to NaCl. Note that tolerant plants are setting seed, whereas nontolerant plants set seed after photograph was taken. Tolerant plants are 2 weeks earlier than nontolerant plants. (From TCCP [Tissue Culture for Crops Project], Progress Report, Colorado State University, Fort Collins, 1987. With permission.)

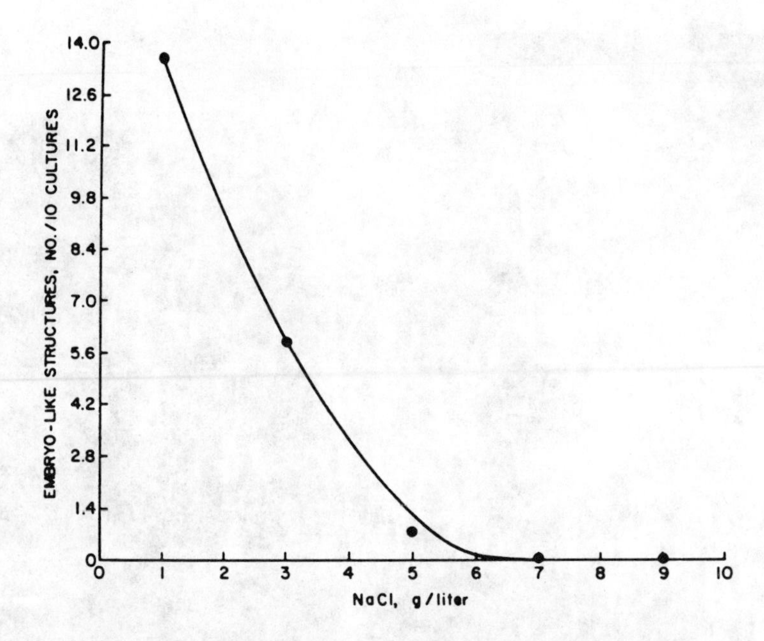

FIGURE 18. Influence of NaCl concentration on production of embryo-like structures, based on 2100 cultures. The F value obtained for regression, 191, was highly significant with 2 and 4° of freedom. (From TCCP [Tissue Culture for Crops Project], Progress Report, Colorado State University, Fort Collins, 1987. With permission.)

Putative salt-tolerant cell lines of crop plants include, for example, tobacco,[86,108] alfalfa,[109] flax,[110] and chickpea,[111] but often plants were not regenerated from the selected cell lines and the progeny typically were not analyzed for salt tolerance with some notable exceptions. Scientists at the TCCP laboratory used a hydroponics system to test progeny of a cereal regenerated from a salt-tolerant cell lines *in vitro*.[96] The progeny of a flax (*Linum usitatissimum*) plant (selected *in vitro* for salt tolerance) was tested for its performance over two generations in normal and in saline soil against the parent variety.[112] The putative salt-tolerant line was superior in saline soil for all parameters measured indicating that the mechanism in selected cells *in vitro* was also active in whole plants. This indicates that cellular selection can be a useful adjunct to traditional breeding programs. Selection of stable, NaCl-tolerant alfalfa (*Medicago sativa*) was established by a step-up procedure.[113] These cell lines retained tolerance following four subcultures. Plants regenerated from NaCl tolerance showed an extensive amount of somaclonal variation. Whole plant NaCl tolerance was expressed only in two plants.

The physiological effects of salts on plants are not clear; however, it is now well-accepted that differential salt resistance in plants is under genetic control.[114] Salt resistance is generally regarded as a polygenic trait,[115,116] although monogenic control has been reported.[117] The effects of salinity on plants are complex. Not only does salinity expose plants to osmotic stress, thus restricting water uptake, but it can also produce specific ion toxicities (e.g., Na^+, SO_4^{-2}, Cl^-, CO_3^{-2}). Some crops are more sensitive to specific ion effects than others.[118] Efforts to generate salt-resistant crops should take both these effects into account.

Most *in vitro* selection programs have used NaCl as the selective agent.[25,71,109,119-122] While NaCl is clearly a very important factor in salt-affected soils, there may well be other toxic ions that play an important role in certain agricultural situations. NaCl selection is not likely to produce genotypes with resistance to toxic ions other than Na^+ or Cl^-. While it may result in genotypes with resistance to osmotic stress, NaCl selection may also produce a halophytic type of genotype that is highly tolerant to the ionic stresses, and not so much

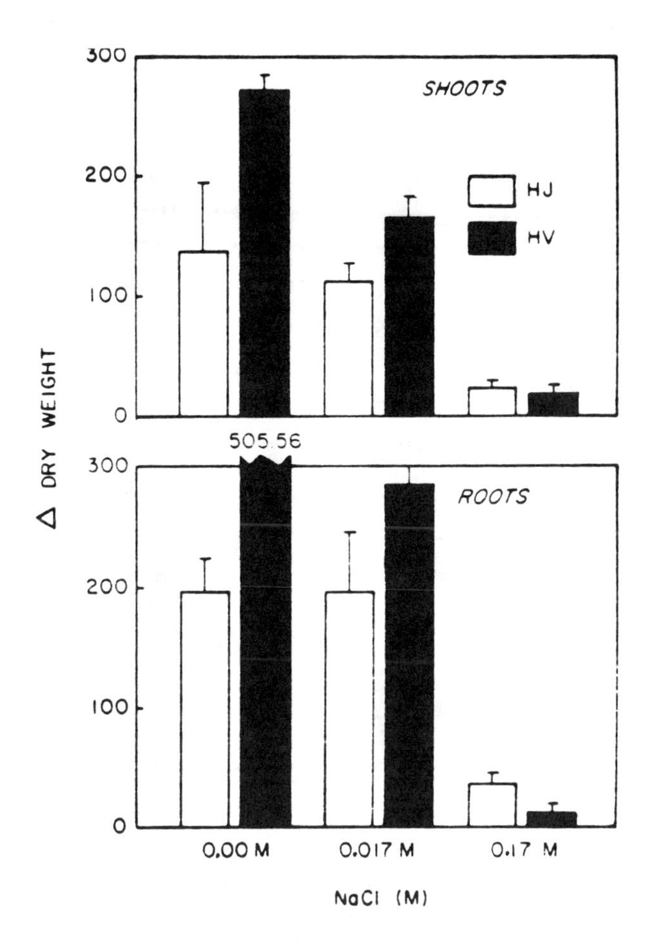

FIGURE 19. Relative growth (ΔDW) of shoots and roots of whole HV and HJ plants after 40 d on various NaCl concentrations. (From Orton, T. J., *Z. Pflanzenphysiol.*, 98, 105, 1980. With permission.)

to osmotic stress (since it might take up the Na^+ and Cl^- freely and, thus, never encounter the full osmotic stress). Such a genotype would not be expected to perform well in a saline situation in which other ions were contributing to salinity if the species were particularly sensitive to any of the lesions. NaCl selection may, thus, not be an appropriate selective agent in all cases. Several investigators have employed salt mixtures[100,123] and might thereby be more closely simulating the agricultural situation.

Genetic variability of salt tolerance has been characterized for certain agronomic crops, such as barley.[104] Some wild species exhibit tolerance to extremely high salinity up to 2 to 3% NaCl (w/v).[105,106] Salt tolerance has been characterized in the wild relative of the tomato *Lycopersicon cheesmanii*.[106] Hybrids between this species and tomato were synthesized, and advanced tomato lines exhibiting salt tolerance have been produced. To achieve these goals successfully much depends on the genetic complexity of the trait and the ability to isolate fertile hybrid intermediates. Orton[85] studied and contrasted *Hordeum vulgare* (cultivated barley) with a wild perennial relative *H. jubatum* with respect to vegetative growth and development, relative water content, ion uptake, and transport under different salt regimes. Both shoots and roots of *H. vulgare* and *H. jubatum* exhibited similar growth responses over NaCl (Figure 19). Relative water content remained constant over NaCl for shoots of both *H. vulgare* and *H. jubatum* (Figure 20). Sodium increased concomitantly with increasing NaCl in roots with no differences between *H. vulgare* and *H. jubatum*. In shoots, however,

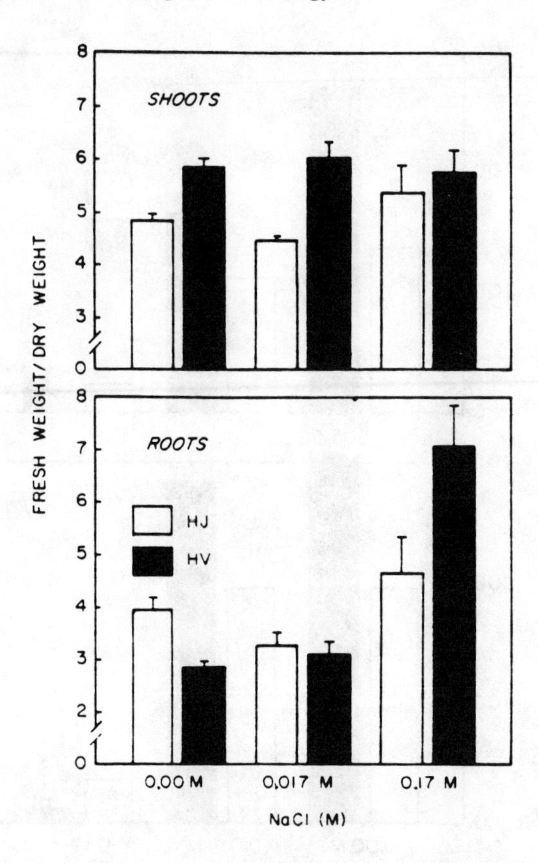

FIGURE 20. Relative water content (FW/DW) of shoots and roots of whole HV and HJ plants after 40 d on various NaCl concentrations. (From Orton, T. J., *Z. Pflanzenphysiol.*, 98, 105, 1980. With permission.)

H. vulgare exhibited a much greater increase in Na^+ and Cl^- accumulation at 0.017 and 0.17 *M* NaCl than *H. jubatum* (Figure 21). Whole plants of *H. vulgare* showed decreased growth and ion imbalance at 0.17 *M* and were severely stressed at 0.17 *M*, while *H. jubatum* exhibited no inhibition. A depression of callus growth rate was observed in salt stressed compared to control, independent of NaCl or genotype (Figure 22). However, *H. vulgare* cultures failed to recover after 3 days on 0.17 *M* NaCl, while all *H. jubatum* cultures recovered. *H. vulgare* cultures were also unable to osmotically adjust to the salt stress while HJ appeared capable of adjustment at all concentration. These experiments clearly demonstrated differential salt tolerance between cultivated barely and wild relative. The proposed mechanism suggests a relatively simple genetic basis which may be transferrable by hybridization and selection. The salt tolerance exhibited by the whole plants of these species is also manifested in callus cultures originated from these plants. This further provides theoretical support for the hypothesis that salt-tolerance in plants may be acquired through the regeneration of salt-selected cells.

Differential ion toxicities have been observed in salinity studies with cultured cells. Carrot (*Daucus carota*) callus has been shown to be more sensitive to inhibition by a salt mixture than by a mannitol solution of equal osmotic potential. These cells were also much more sensitive to K^+ than other ions[124] (one with salt glands and the other with succulence), demonstrating that their resistance is not cellular, but dependent upon higher levels of organization, whereas the resistance of *Beta vulgaris* is cellular.[87] Another halophyte, *Distichlis spicata*, is also more resistant in culture than to the glycophytic species, *Nicotiana*

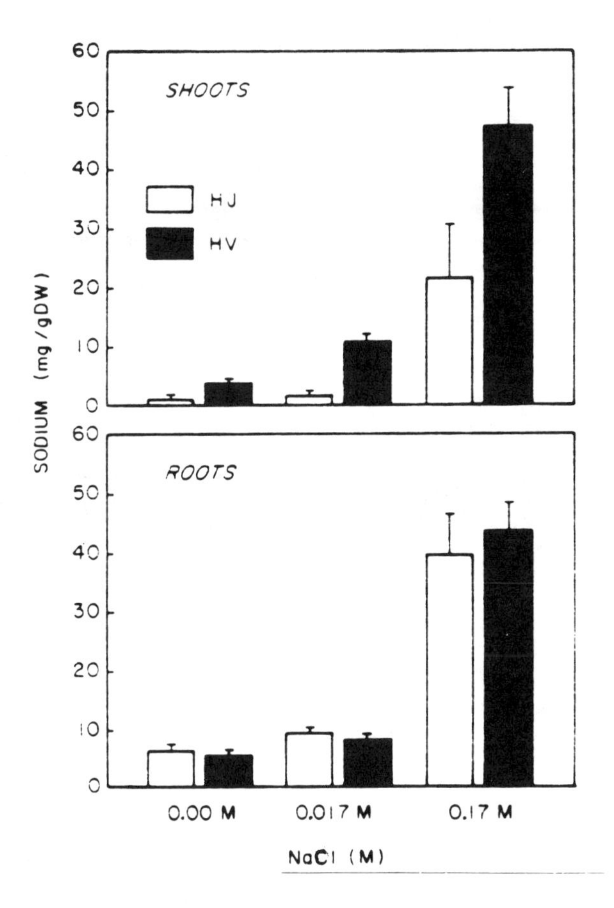

FIGURE 21. Relative sodium content (mg/g DW) in shoots and roots of whole HV and HJ plants after 40 d on various NaCl concentrations. (From Orton, T. J., *Z. Pflanzenphysiol.*, 98, 105, 1980. With permission.)

sylvestris and *Zea mays*.[125] Two barley species, *H. jubatum* and *H. vulgare*, were compared both as callus and as plants; *H. jubatum* was more resistant in both states.[85] Likewise, callus cultures of the salt-resistant wild tomato species *L. peruvianum* and *L. pennellii* are more resistant than callus of the salt-sensitive cultivated tomato *L. esculentum*.[126] Finally, the salt resistance of four alfalfa (*M. sativa*) genotypes was compared both in callus and in plants. Resistance was found in only one of the genotypes and it was expressed both in the plant and the callus.[88]

The facts that salinity affects cultured cells and that salt resistance mechanisms do operate in cultured cells strongly suggest that salt-resistant plants can be obtained by *in vitro* selection. To our knowledge only one case of citrus cells which are differentially sensitive to ions (Cl^- being particularly toxic) has been reported. The resistance of a selected citrus line for Cl^- was greatly influenced by the cation, with resistance being expressed to NaCl, but lost when Cl^- was provided as KCI.[121]

Salinity is clearly inhibitory to cultured cells and thus can be studied in culture. However, are salt-resistance mechanisms also of a cellular nature? Can salt resistance be selected at the cellular level? Plants can resist salinity both by excluding the toxic ions and compensating osmotically via the synthesis of compatible solutes, and/or by freely taking up the ions so that osmotic stress is avoided.[127,128] The latter mechanism required that ion toxicities be overcome. Both mechanisms of resistance could conceivably be expressed by cells in culture.

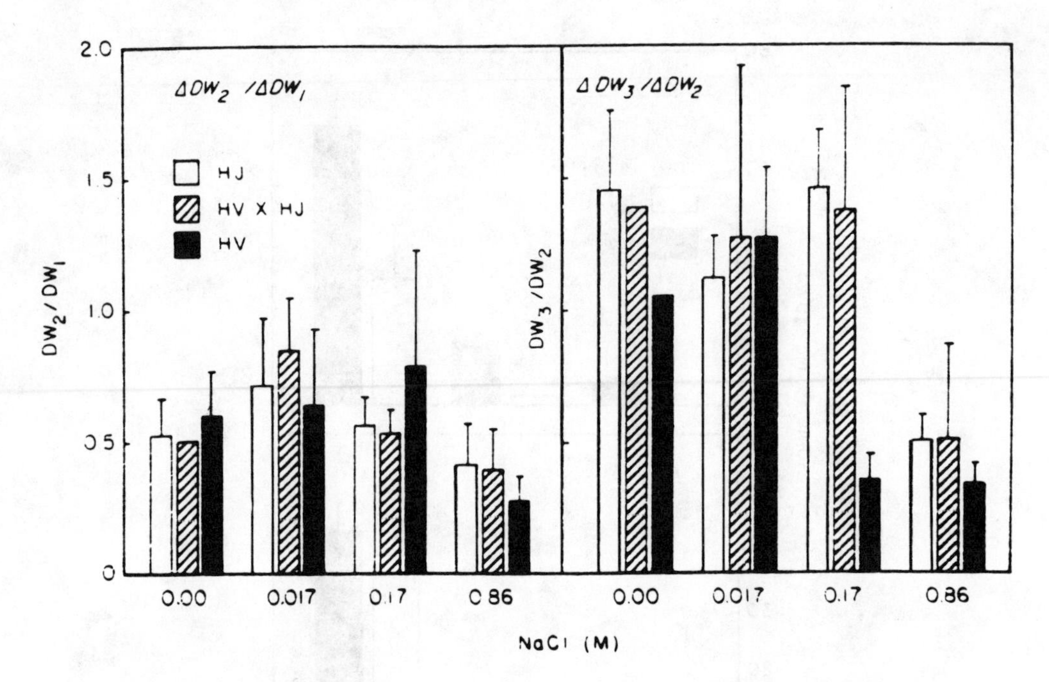

FIGURE 22. Ratio of relative growth (IDW) of HV, HJ, and HV × HJ callus cultures between periods B/A and C/B after 30-d growth. (From Orton, T. J., *Z. Pflanzenphysiol.*, 98, 105, 1980. With permission.)

In intergeneric, interspecific, and intraspecific comparisons, genetic differences in salt resistance have been shown to be expressed in cultured cells, thus at least some resistance mechanisms are cellular in nature. Callus of *B. vulgaris*, a salt-resistant species, is more resistant than callus of *Phaseolus vulgaris*, a salt-sensitive species. However, the salt resistance of two halophytes and sexual transmission of salt resistance so obtained have been reported.[71] By repeated selection for improved growth in the presence of NaCl, a salt-resistant cell line of *N. tabacum* was obtained from which plants were regenerated. The plants exhibited greater NaCl resistance than plants regenerated from unselected cells and an elevated level of salt resistance was also expressed in two subsequent generations of seed progeny from these plants. The inheritance of the increased resistance did not fit a recognizable pattern and the genetic basis for this trait has not yet been elucidated.

There are a number of other reports in which selected salt resistance is retained through the plant regeneration process. Salt resistance selected in citrus cell culture is expressed in somatic embryos regenerated from the cells.[121] Some degree of resistance is expressed in rice (*Oryza sativa*) plants regenerated from salt-resistant callus. In other cases, resistance has been shown to be retained in callus derived from plants regenerated from salt-resistant callus.[120,129] It may be possible to obtain salt-resistant plants by selecting for resistance to only the osmotic stress component of salinity. Tomato cells resistant to elevated concentrations of PEG also show enhanced resistance to NaCl.[80] Such cells presumably are resistant by virtue of their ability to accumulate osmoregulatory compounds.[130]

A problem that may limit selection for PEG resistance, and perhaps also NaCl resistance, is the remarkable capacity of cultured cells to adapt to osmotic stress. While cell populations probably do contain genetic variants with enhanced resistance, it may be difficult to distinguish them from wild-type cells that have adapted to the stress.[130] It may be necessary to screen a large number of resistant selections in order to identify a few stable variants.

The resistance mechanisms of selected cell lines have been elucidated in some cases. Salt-resistant cell lines selected in alfalfa, *N. tabacum*, *N. sylvestris*, and *N. tabacum/gossii*, have been shown not to exclude salt ions and are thus able to maintain the osmotic gradient necessary for continued function.[109,131,132] In alfalfa, the resistant cell line actually exhibits a halophytic mode of resistance, in that it accumulates higher ion concentrations than the unselected line and actually seems to require elevated salt for normal growth. With *N. tabacum/gossii,* both resistant and sensitive cells take up Na^+ from a saline medium, but the resistant cells, as in the case of alfalfa, are able to maintain high K^+ concentrations, a characteristic of many true halophytes.[118] In these four cases, the selected cell lines are truly salt tolerant in that they can withstand high cellular salt concentrations. The situation is different in a salt-resistant citrus cell line in which resistance seems to be due to the exculsion of Na^+ and Cl^-.[103] In this case salt stress is actually avoided. It is likely that the cells osmoregulate via the synthesis of some compatible solute to maintain their water status. That such osmoregulation can occur in cultured cells is strongly suggested by the studies of Bressan et al.[133] with cells adapted to conditions of very negative water potential created by PEG. In this case the medium contained insufficient solutes to account for the degree of osmoregulation necessary for the observed continued growth of these cells, but their adaption could be accounted for by osmoregulation via the synthesis of organic solutes.

As with intact plants, proline has been implicated in salt resistance in selected cell lines; but as is also the case with intact plants, there is no consensus on whether or not its role is adaptive. In *N. sylvestris*, a salt-resistant line did not accumulate as much proline as the unselected line, arguing for proline synthesis as a stress response rather than an adaptive mechanism.[131] However, evidence to the contrary is provided by observations with carrot and *Kickxia*. Several carrot cell lines selected to overproduce proline (by selection for resistance to azetidine-2-carboxylic acid, a proline analogue) demonstrated increased resistance to NaCl, although the degree of resistance was not proportional to proline concentration.[134] In *Kickxia* cultures, the addition of exogenous proline was shown to overcome inhibition by NaCl.[119]

Salt-resistant variants can clearly be isolated in plant cell cultures and the resistance mechanisms observed in them are similar to those known to operate in intact plants. This is encouraging evidence in that salt-resistant plants can be obtained from such variant cells. The polygenic nature of salt resistance might be expected to limit the frequency with which such a genetic change could occur in cultured cells, but modifications in polygenic traits have been observed in plants regenerated from cultured cells.[135] A difficulty is presented by the physiological plasticity of cells in readily adapting to osmotic stress. Selection strategies must be employed that facilitate the identification of stable variants among adapted, but genetically unchanged, cells. It is difficult to evaluate the importance of specific ions in *in vitro* selection, since no field evaluation of plants regenerated from salt-resistant cells has yet been performed. However, it is reasonable to expect that the greater likelihood of eventual success in the field will be improved if the salt stress employed for selection reflects the ionic composition of the soil salts in the area for which the crop is intended.

B. TOLERANCE TO MINERAL STRESS

Mineral stresses include both deficiencies of essential nutrient elements and excesses of toxic elements. Such stresses are widespread and represent a major limitation to crop production worldwide.[136,137] The genetic control of plant response to mineral stress is well established,[118] and in a number of cases genotypic differences have been attributed to a single major gene.[138] The roles played by mineral nutrients are generally of a fundamental cellular nature (e.g., enzyme activation, components of fundamental molecules) and so are just as critical to cultured cells as to whole plants. Likewise, the primary lesions of toxic ions are also of a fundamental cellular nature (e.g., competition with essential elements for uptake, inactivation of enzymes, displacement of essential elements from functional sites).[118] It is not surprising that the cultured plant cells are sensitive to artificially applied mineral stresses, both nutrient deficiencies and mineral toxicities.

The selection strategies necessary to isolate variants resistant to mineral stresses are relatively straight forward to conceive, consisting simply of subjecting cultured cells to the deficiency or toxicity of interest by modifying the culture medium. It is critical, however, that interactions with other elements, such as pH, be considered in the design of the inorganic nutrient mixture used to produce the desired stress. Oversimplification is to be avoided. While it may lead to quick results and even new plant genotypes, these plants are likely to fail in the agricultural setting in which the mineral stress occurs unless the selection strategy employed closely simulates the conditions that exist in the soil environment.

Just as the primary lesions caused by mineral stresses are cellular, so are at least some resistance mechanisms by which plants withstand these stresses. For example, more efficient uptake of a deficient nutrient element may result from an altered membrane transport mechanism, or more efficient utilization can be achieved via higher enzyme affinity for an element. Ion toxicity resistance mechanisms can include exclusion at the plasmalemma, detoxification via binding to an organic molecule, or altered target sites.[139]

Genetic differences in response to mineral stress have been shown to be expressed in cultured cells. This has been reported in *Anthoxanthum* genotypes differing in resistance to zinc and lead[140] and soybean (*Glycine max*) genotypes differing in resistance to iron deficiency.[141] Christianson,[142] however, did not observe the expected differences in zinc sensitivity in *Phaseolus* genotypes.

The cellular nature of mineral nutrition suggests that mineral stresses applied to cultured cells can be expected to have an effect similar to that in a whole plant. Likewise, the cellular nature of many mechanisms by which plants resist mineral stresses suggests that variants possessing resistance mechanisms that will function in whole plants can be selected. This has yet to be proven; however, there have been few reports of selection for resistance to mineral stresses in cultured plant cells. Phillips and Collins[143] have obtained red clover (*Trifolium pratense*) cells capable of growth on low phosphorus. While plants have been regenerated, they have not yet been fully characterized.[209] Meredith[144] obtained a number of stable aluminum-resistant tomato cell lines, but plants could not be regenerated from the cultures. Mercury-resistant petunia cells have also been selected, but no information regarding regenerated plants has yet been reported.[145] Cadmium-resistant *Datura* cells[146] and carrot cells resistant to both aluminum and manganese[147] have also been selected. While the nature of the regenerated *Datura* has not yet been reported, there are preliminary indications that the carrot plants both retain and transmit the aluminum resistance.[148]

There is ample evidence to expect that new crop genotypes with resistance to important mineral stresses can be produced by *in vitro* selection. The selection strategies are clear-cut, known resistance mechanisms operate in cultural cells, and resistant cells lines have been isolated. It is most encouraging that many cases response to mineral stress are under monogenic control. Lack of success in this area to date probably reflects the small number of investigators with interests in this area. This, in turn, may be related to the only recent general recognition of the agricultural importance of mineral stress. The potential for the genetic manipulation of plant response to mineral stress is not yet fully appreciated at the whole plant level, let alone in cultured cells.

The quality of a crop is often measured in terms of its capacity to supply the nutritive needs of the animals that eat it. In this respect many crops are deficient, cereals often lack sufficient lysine, tryptophan, and threonine, while legumes often lack sufficient methionine.[149,150]

As previously mentioned, subtle changes in a plant (e.g., an elevated concentration of a single amino acid) can be very difficult to obtain at the whole plant level by classical breeding techniques such as recurrent selection, because such a change cannot be readily detected. A potential technique to circumvent the difficulties of whole plant selections for altered nutritional quality is to use metabolic inhibitors. The inhibitor is used in a cell culture system to exert a selection pressure such that any nonresistant cells eventually die and only resistant cells survive.

There are numerous ways to inhibit the metabolism of a cell; inhibitors may act at sites of metabolic regulation, they may compete with normal substrates for incorporation into cellular metabolites, or they may block the uptake of nutritive materials. Similarly, there are numerous ways for a plant cell to overcome the effects of an inhibitor. One means of inhibitor reversal is the overproduction of a metabolite, which can be the end result of a hyperactive isozyme, an elevated level of an enzyme in a biosynthetic pathway, or the deregulation of a site of metabolic control. Typically, the overproduced metabolite competes with the inhibitor for its site of action and dilutes the inhibitor's effect to a tolerable level. It is this overproduction of a metabolite, by whatever means, that can improve the nutritive quality of a crop.

Amino acid analogues are commonly used to select for the overproduction of amino acids in culture. Ranch et al.[151] reported the successful use of 5-methyltryptophan to select TRP overproduction in *D. innoxia*, a close relative of Jimson weed. Five of the 79 resistant cell strains isolated were regenerated to whole plants, and a varying degree of TRP overproduction was noted in the leaves of these plants. Anthranilate synthase was shown to be less sensitive to inhibition by TRP in both the resistant calli and plants.[152] The resistance (i.e., the overproduction of TRP) has been shown to be heritable by amino acid analysis of seed and leaf material or progeny from regenerated plants, but the mode of inheritance has not been determined to date. [67] Carlson[153] similarly reported success with the selection of tobacco (*N. tabacum*) resistant to methionine sulfoximine. Plants regenerated from the resistant cultures had six times more methionine in their leaves than did susceptible plants.

Another example involves the aspartate biosynthetic pathway, which is regulated through the feedback inhibition of aspartokinase by lysine and threonine and of homoserine dehydrogenase by threonine. Large quantities of both amino acids can prevent the production of methionine, an end product of the aspartate pathway, and subsequently, can induce methionine starvation in cultured cells. Consequently, elevated levels of lysine and threonine provided adequate selection pressure to identify variants with alterations in the asparate pathway.[150] Hibberd et al.[149] took advantage of this and selected for isoleucine, lysine, methionine, and threonine overproduction in cultures of maize (*Z. mays*). Aspartokinase was found to be less sensitive to lysine inhibition in the resistant line than in the susceptible material. When Hibberd[154] carried out the same selection using a maize line capable of plant regeneration, he isolated variants that overproduced free threonine. The overproduction of threonine was noted in the whole plant tissue. In the kernel, the homozygous mutants had 33 to 59% greater total threonine (free and protein associated) per kernel than did the wild-type plants. Progeny from selfs and backcrosses with the wild-type parent overproduced threonine, demonstrating that the mutation was heritable and dominant. However, the inheritance pattern was not discernible in terms of simple Mendelian genetics. In this mutant lysine and methionine levels were not significantly different from those of the wild type.

C. HERBICIDE RESISTANCE

Herbicides are used widely for improving crop production by reducing competition for light and nutrients from weeds and also by limiting weeds that complicate harvesting. However, problems arise with the use of herbicides. In some cases, they are ineffective at concentrations that crop plants tolerate; in other cases, they are safe to use only on a limited range of crops. Carryover of herbicides from one season to the next in fields where crop rotation is practiced can severely damage the following crop which increases some of the problems with their use. Much effort has been put into applying cell selection in tissue culture for developing herbicide-resistant plants.

Chaleff and Parsons [92] reported the selection of diploid tobacco (*N. tabacum* var. *xanthi*) suspension cell culture resistant to picloram (4-amino-3,5,6-trichloropicolinic acid). Five stable lines were developed that were capable of regenerating plants. The plants were selfed and backcrossed and the progeny tested for resistance to picloram. The results indicated that cell resistance was associated with a single gene; in four of the lines the gene was dominant, while in one line it was semidominant. Similarly, Swanson and Tomes[155] selected birdsfoot trefoil (*Lotus corniculatus* cv. Leo) callus and suspension cultures tolerant to elevated levels of 2,4-D. Crosses of regenerated tolerant plants produced plants with tolerant calli and tolerant plants themselves yielded resistant progeny showing selection for 2,4-D. This tolerance was considered possibly of complex genetic origin, unlike the picloram resistance developed by Chaleff and Parsons.[92] Moreover, the regenerated plants did not show as great a 2,4-D tolerance as did plants derived from a 5-year recurrent selection program carried out in the field.

Miller and Hughes,[156] using tobacco (*N. tabacum* cv. Wisconsin 38 grown in the dark, and Thomas and Pratt,[157] using tomato (*Lycopersicon esculentum* × *L. peruvianum*) also grown in the dark, selected diploid calli resistant to paraquat (1,1′-dimethyl-4,4′-dipyridylium dichloride, also known as methyl violgogen). In both cases, several resistant or slightly resistant plants were regenerated from the cultures. Thomas and Pratt[157] selfed and outcrossed plants derived from resistant calli to sensitive plants. Several of the progeny were as paraquat tolerant as the parent, indicating that the tolerance was dominant and heritable, but the number of alleles involved was not known.

Paraquat acts by reducing the electron transport system in chloroplasts and subsequent reoxidization by molecular oxygen. This reoxidization yields superoxide radicals, which are the actual lethal agents in the herbicide treatment. Although paraquat reacts in the chloroplast, selection for resistance in the above studies was only in callus grown in the dark. The resistance produced was due to the overproduction of superoxide isomutase, or peroxidase, which destroys the superoxide radicals.

D. DISEASE RESISTANCE

The most obvious selection strategy to identify and isolate disease-resistant cells is to challenge a cell population with the pathogen. This approach, however, may involve several difficulties. Since plant tissue culture media are nutrient rich, most pathogens will grow readily on them. A pathogen introduced into a plant cell culture will often overgrow the plant cells.[158] A means of avoiding this is to introduce the pathogen onto the plant cells only and not onto the medium, for example, by inoculating callus pieces.[64] Since a plant cell culture is an unnatural situation, it is also possible that a pathogen introduced into such a culture may produce unspecific killing that is not governed by mechanisms that determine pathogenicity in the whole plant.[159]

In contrast, obligate pathogens may not grow at all in plant cell cultures. This has been the case with several rust fungi (*Puccinia antirrhini, P. helianthi, Gymnosporangium, Juniperivirginianae*) on cultured tissues of snapdragon (*Antirrhinum majus*), sunflower (*Helianthus annuus*), and red cedar (*Juniperus virginiana*)[160] powdery mildew (*Erysiphe graminis* f. sp. hordei) on barley;[161] and *Plasmodiphora brassicae* on callus of *Brassica campestris*.[162] There is evidence that the presence of organized plant structures may be essential for infection with certain pathogens. Such is the case with *P. brassicae* on *B. napus* where callus cannot be infected, but somatic embryos can be infected.[163]

To avoid the inherent difficulties when introducing a pathogen into a culture, some have looked to toxins as selective agents. There are, however, limitations regarding the use of toxins, not the least of which is that there must be a toxin associated with the disease in question. Only very few plant diseases are known to be associated with a toxin at present.[164] To be effective in selecting disease resistance, the toxin must be a significant determinant of disease, that is, it must play an important role in either the development or the severity of the disease. While the toxin need not be the sole cause of the disease, toxin resistance can only be expected to result in increased disease resistance to the extent that the toxin contributes to the production of the disease. It is not necessary that a toxin be host specific to be effective as a selection agent.[159] Even if other factors control the host specificity of the pathogen, as long as the toxin plays a significant role in the disease, toxin resistance can be expected to confer a significant degree of resistance to the disease.

In those diseases where a toxin is suspected to play a significant role but has not been isolated, crude culture filtrates have sometimes been used as selective agents.[158,165-167] A danger inherent in this approach is that the toxicity to which resistance is being selected may be due to components of the pathogen culture medium and not to a pathogen-produced toxin. Additionally, a pathogen may produce a toxic principle in culture that is insignificant or absent in infected plants.[168] These doubts can be erased by using purified toxins whose role in disease has been confirmed.

The requirements for the use of toxin to select disease resistance are met by very few plant diseases. Relatively few diseases are associated with toxins. Of these toxins, even fewer have been purified and conclusively shown to play a significant role in plant disease. While many more toxins will undoubtedly be implicated in plant disease in the future,[164] the number is now so small that the use of toxin selection can by no means be considered a generally applicable approach to disease resistance.

There are alternatives to the use of either the pathogen or its toxin to select resistant cells. When the mechanism of the toxin is known, other compounds with the same effect that might perhaps be more readily available can be considered as selection agents. This was the case with the selection of tobacco cells for resistance to tobacco wildfire disease (*Pseudomonas tobaci*) using methionine sulfoximine, a compound with similar effects to those of the wildfire toxin produced by the pathogen.[153] In those cases in which disease resistance is associated with the biosynthesis of a specific compound by the plant,[169] selection strategies that permit the isolation of variants that overproduce the critical compound might also result in increased disease resistance. As with any phenotype, to be selectable in culture, disease resistance must be expressed in the cultured cells to which the selection pressure is being applied. It is reasonable to expect such a phenotype to be expressed in cultured cells.

Resistance to disease may be associated with certain organized structures and certainly would not be expected to operate in *in vitro* systems where these structures may not be present. The absence of cuticle and wax in rice callus, for example, has been associated with increased susceptibility to infection by several fungal pathogens.[170] Since many secondary compounds are often absent or present in reduced concentrations in cultured cells,[171] the expression of disease resistance that is associated with secondary compounds would also be expected to be reduced. The reduced resistance of rice callus to infection by fungal pathogens, for example, has also been attributed to the absence of lignin in this tissue.[172]

While there are a number of cases where resistance has not been expressed in cultured tissue derived from plants of a resistant genotype, there have been several well-documented instances in which disease resistance is clearly expressed in cultured cells. A number of these cases involve *Phytophthora* species. Monogenic resistance to *P. infestans* tomato race 0 is expressed in callus cultures of several resistant tomato cultivars.[173] Growth of the fungus was not supported by callus from resistant cultivars to nearly the extent that it was supported by callus from susceptible cultivars. Similar results were obtained with calli of two near-isogenic soybean genotypes differing in monogenic resistance to race 1 of *P. megasperma* var. *sojae*.[174]

Expression of resistance to *P. parasitica* var. *nicotianae* in *Nicotiana* callus has been investigated in several laboratories. Monogenic, race-specific resistance to race 0 is expressed in callus of *N. tabacum* genotypes whether the resistance is derived from *N. plumbaginifolia*[175] or *N. longiflora*.[176,177] Comparisons of plant and callus resistance over several segregating generations have clearly shown that, at least with the *N. plumbaginifolia*-derived resistance, the callus resistance and the plant resistance represent the expression of the same gene.[175] Polygenic resistance to both races 0 and 1 (derived from yet another source) has also been shown to be expressed in tobacco callus.[177]

In maize (*Z. mays*), cytoplasmic resistance to the toxin of *Helminthosporium maydis* is also expressed in callus cultures. Calli of genotypes with N cytoplasm (resistant) were not inhibited by toxin concentration lethal to calli of genotypes with T cytoplasm (susceptible).[77,178,179]

The expression or nonexpression of disease resistance should not be considered an immutable condition. Plant cell cultures are not fixed in their developmental state, since morphological, physiological, and biochemical characteristics will vary in response to physical, nutritional, and hormonal factors. It stands to reason, therefore, that even in those cases where resistance does not seem to be expressed in cultured cells of resistant genotypes, it might be possible to elicit the expression of resistance by manipulating one or more of these factors. This has been clearly demonstrated by Haberlach et al.[180] with regard to the expression of resistance to *P. parasitica* var. *nicotianae* race 0 in tobacco callus. At low kinetin concentrations, calli from resistant cultivars exhibited resistance of a hypersensitivity type, while susceptible cultivars did not. At higher kinetin concentrations, however, the expression of resistance was lost and calli from all cultivars were equally susceptible. Benzyladenine was equally effective in eliminating resistance, while two other cytokinins were not. This work should serve as a clear example that investigations concerning the expression of disease resistance in cultured cells should not be discontinued should resistance not be initially apparent, but should include the manipulation of culture variables that might permit the expression of the resistance phenotype in resistant genotypes. In selection programs, to obtain resistant cells, preliminary studies with known resistant and susceptible genotypes, if available, should be undertaken so as to optimize culture conditions for the maximum expression of known resistance mechanisms.

While the availability of suitable selection strategies and the expression of resistance in cultured cells seriously limit the *in vitro* selection of disease resistance, there have been some successes in this area. Carlson[153] selected tobacco cells resistant to methionine sulfoximinc, which has similar effects to those of tabtoxin, a broad-sprectum toxin produced by *Pseudomonas tabaci*, the causal organism of tobacco wildfire disease.[181] Plants regenerated from the selected cells did not develop chlorotic halos (one of the disease symptoms) but did develop necrotic lesions. This partial resistance was genetically transmissable and was inherited as a monogenic semidominant trait in one case and as recessive alleles at two loci in each of two other selections.

The purified toxin of *H. maydis* Race T has also been successfully employed as a selective agent.[77,179] Callus from a susceptible maize genotype was subjected to several cycles of sublethal exposure to the toxin and resistant callus sectors were isolated. Plants regenerated from resistant callus retained resistance to the toxin (but lost their original male sterility) and the resistance was inherited as a cytoplasmic trait. Resistance to the toxin was closely correlated as was resistance to the pathogen.[178] Partially purified toxins from *P. syringae* pv. *tabaci* and *Alternaria alternata* pathotype tobacco have been used to select resistant protoplast-derived calli of tobacco. With each toxin, some plants regenerated from selected calli were also resistant to the pathogen and resistance was sexually transmitted.

Behnke[165,167] has used crude culture filtrates of two potato (*Solanum tuberosum*) pathogens, *Phytophthora infestans* and *Fusarium oxysporum*, to select resistant callus. In the case of *P. infestans*, selected callus was resistant to all four pathotypes, although the crude filtrate came from only one pathotype. The resistance was stable in the absence of the filtrate and also retained in callus derived from some regenerated plants.[165] Plants from resistant selections also exhibited some general resistance to a mixture of races of the pathogen itself.[166] Potato callus was similarly selected for resistance to crude culture filtrate of *F. oxysporum*. Resistance to the filtrate was retained in some regenerated plants. Plant response to the pathogen itself was not reported.[167] Callus and embryogenic cultures of *B. napus* have been selected for growth in the presence of a crude culture filtrate of *Phoma lingam*. Some generated plants exhibited increased resistance to the pathogen. Preliminary studies of the progeny of regenerated plants suggested a genetic basis for the resistance.[158]

Limited resistance to tobacco mosaic virus has been obtained by Murakishi and Carlson[183] by selecting green callus from leaf strips of gamma-irradiated *N. sylvestris* plants infected with the virus. Some of the selected callus colonies were virus free. Some of the plants regenerated from those displayed limited resistance that was transmitted to seedling progeny. Critical factors in the success of this experiment were the careful control of experimental conditions to insure a uniform infection rather than the typical mosaic and the use of a yellow strain of the virus as a marker for infected tissue.

The preceding examples all employed a selective agent to deliberately select a resistant phenotype. There is a rapidly growing body of evidence, however, to indicate that disease-resistant individuals can be recovered at a high frequency in populations of plants regenerated from cultured cells even without selection pressure. The earliest indications of this phenomenon were observed in sugarcane (*Saccharum officinarum*). Increased resistance to a number of diseases has been noted in sugarcane plants regenerated from cultured cells.[184-188] Individuals with increased resistance to both early blight (*A. solani*) and late blight (*Phytophthora infestans*) have been identified in populations of potato regenerated from mesophyll protoplasts.[189,190] Resistance to *F. oxysporum* f. sp. *apii* has been observed in celery (*Apium graveolens*) plants regenerated from suspension cultures.[191]

In two of the cases of deliberate selection described above, resistance was also recovered from unselected cultures. With callus cultures of maize that had never been challenged with T-toxin, more than half of the regenerated plants could be shown to be toxin resistant.[28,29] Similarly, some regenerants from *B. napus* cultures not exposed to culture filtrate showed an increased resistance to *Phoma lingam*.[158]

These observations of increased frequency of disease resistance in plants regenerated from cultured cells are most encouraging in view of the limitations associated with *in vitro* selection for disease resistance. The exploitation of such somaclonal variation may permit the recovery of resistant variants in cases where no selective agent is available or where resistance is not expressed in culture. Selection is still advisable in those few cases where it is possible, because it increases the recovery of resistance by enriching the culture for resistant cells.

Several attempts have been made to select disease resistance expressed at the whole plant level by using tissue culture systems. Experimental approaches for such selections have been described by Daub,[192] Grout and Weatherhead,[63] Helgeson and Haberlach,[64] and Ingram.[193] The use of a pathogen directly in a selection experiment can present difficulties depending on the type of host-parasite interaction that is involved. For instance, if the pathogen is a facultative saprophyte, once it makes contact with the nutrient medium on which the callus is growing, it often overgrows the medium and never produces a true parasitic interaction with the callus. In such a case no selection pressure is applied to the callus and no selection will be made. Consequently, the diseases to which resistance has been developed so far are those that are caused by pathogens producing a toxin. In this case a toxic substance (the toxin) can be incorporated into the medium, and the toxin supplies the selection pressure against which resistance can be developed. Tabtoxin is such a toxin and is produced by *Pseudomonas tabaci*, the pathogen of wildfire of tobacco. Tabtoxin is the causative agent of the yellow halo formed around wildfire leaf lesions. Methionine sulfoximine (MSO) produces, in appearance, a similar yellowing of whole leaf tissue as does tabtoxin. Using MSO, which was thought at the time to be an analogue of tabtoxin,

Carlson[153] selected resistance to wildfire of tobacco. Plants regenerated from resistant cultures were crossed to nonresistant plants and the progenies were selfed. F_2 plants showed a loss of the yellow halo around wildfire. Lesions still formed to varying degrees, indicating that the yellowing was suppressed, but not the actual disease. The regenerated plants also showed resistance to MSO. While Carlson[153] selected resistant tobacco lines without using a toxin, other researchers have carried out selection experiments using the actual toxin associated with the disease. Brettell et al.[28] and Gengenbach and Green[77] have selected maize callus resistant to the T-toxin produced by race T of Bipolaris (*H. maydis*), the causative agent of Southern corn leaf blight (SCLB). The selections were done in lines possessing the disease-sensitive, male-sterile T cytoplasm in an attempt to produce SCLB-resistant, male-sterile maize lines. The sensitivity to the disease and the male sterility are mitochondrial traits, and the resistant calli were shown to have toxin-insensitive mitochondria typical of wild-type, SCLB-resistant, male-fertile N cytoplasm. Plants were regenerated from the resistant calli and were shown to be resistant to SCLB, but the plants were also male fertile.[178] The resistance was inherited through female as would be expected for a cytoplasmic trait. Restriction endonuclease analysis of mitochondrial DNA (mtDNA) showed that the mtDNA from the selected resistant plants had a T-cytoplasm mtDNA origin, and was not N-cytoplasm mtDNA. There were, however, some distinct differences in the restriction patterns of the original T cytoplasm and the selected resistant material. One cannot prove that the resistance obtained was due to the observed changes in the mtDNA, but the correlation seems very good. Since male fertility and toxin resistance were not separated, the selection has not proved useful in terms of crop improvement.

Brettell et al.[28] found that resistance to SCLB with a concomitant loss of male sterility was produced in plants derived from callus that had been in culture for several transfers and that had not been subjected to T-toxin selection pressure (somaclonal variants). A similar event was noted by Larkin and Scowcroft[188] in the selection of sugarcane plants resistant to eyespot (caused by *H. sacchari*). Larkin and Scowcroft[188] also attempted to select for resistance by using the toxin produced by *H. sacchari*. In this case the toxin was added to the medium used for the last callus maintenance transfer and also to the plant regeneration medium. Regenerated plants showed tolerance to eyespot that was passed through several vegetative propagation cycles. In both the experiments of Brettell et al.[28] and of Larkin and Scowcroft,[188] the recovery of resistant material was greater when toxin was used.

Behnke[166] was able to select potato callus resistant to the crude culture filtrate of *Phytophthora infestans*, the causative agent of late blight of potato leaflet. Similarly, Hartman et al.[194] selected resistance to fusarium wilt of alfalfa in cultures of alfalfa (*M. sativa*) using culture filtrates of *F. oxysporum* f. *sp. medicaginis*. The resistance produced was expressed in cultures placed on inhibitory medium after having been subcultured repeatedly on non-inhibitory medium for 5 months. The resistance was also expressed in plants regenerated from the cultures and in callus established from the regenerated plants. Genetic transmission of the trait was not examined. The disease discussed by Behnke,[165] Brettell et al.[28] Carlson,[153] Gengenbach and Green,[77] Gengenbach et al.,[178] and Larkin and Scowcroft[188] is a leaf disease. Hartman et al.,[194] having selected for resistance to a vascular wilt, illustrated that resistance to several types of diseases can be selected in tissue culture.

Many toxins produced by pathogens are unidentified or difficult to purify; consequently, crude filter extracts may be the only source of the material. The work of Behnke[166] and Hartman et al.[194] suggests that the purity of the toxic material may not limit its use in selecting resistance in a tissue culture. Also, the use of culture filtrates may be a good first step in attempting to select for resistance to pathogens where the identity and mode of action of a toxin are unknown. Regretfully, few diseases are known to have a pathogen-derived toxin associated with symptom expression or involved in pathogenesis. Consequently, selection for disease resistance in tissue culture is presently very limited.

V. CONCLUSIONS AND FUTURE PROSPECTS

The advisability of employing *in vitro* selection in a crop improvement program is dependent upon both the objective of the program and the crop. The objective should involve a phenotype that is readily amenable to an *in vitro* approach. In addition to *in vitro* data available, the knowledge of the conventional breeding technology for a crop is also essential. These data are certainly helpful in deciding whether selection with cultured cells is worthy of consideration. With many agronomic crops, the conventional breeding technology is extremely sophisticated and powerful, while cell culture manipulations are still difficult. On the other hand, for some woody perennial crops the power of conventional breeding technology is severely limited by such factors as generation time and plant size, making even the current primitive state of cell culture technology associated with these crops attractive by comparison, and justifying further efforts to improve *in vitro* methods for these crops.

There are many limitations to the application of *in vitro* selection for crop improvement, the major ones being that cell culture technology is not sufficiently advanced for many crops and that most agriculturally significant phenotypes are as yet too poorly understood to permit the design of effective selection strategies. It is clear, however, that novel genotypes can be obtained with this approach and that some contributions can be made to crop improvement. Along with established plant breeding methods and other newly emerging genetic technologies, selection in plant cell cultures has both strengths and limitiations. For certain crop species and certain agricultural objectives, a thorough analysis of the considerations presented here may reveal *in vitro* selection to be an effective genetic tool in crop improvement. Apart from the development of useful somaclonal variants obtained by random chance, useful traits selected in tissue culture have yet to be incorporated into general crop production.[195-198] These traits stem from the need for basic research into the physiology and genetic regulation of tissue cultures and plants in general. Also, a clear understanding of the phenotypes desired from a culture is needed.

Until recently, the application of cell selection had limited success in regenerating plants from many economically important crops.[199,200] In order to produce mutant plants from stress-tolerant tissue cultures, reliable and consistent regeneration techniques are important. Clearly, the production of cell lines tolerant to external stressing agents is not of much practical value, unless whole plants can be regenerated. Sporadic, short-term plant regeneration is of little value. Valuable *in vitro* mutations may be relatively rare and may require several months of culturing to stabilize painstaking selection of promising cell culture. The difficulty in raising whole plants from such lines has frustrated the efforts of many tissue culture researchers. This has been specially true for those working with cereals. Considerable strides have been made in developing long-term high-frequency regeneration techniques for rice, wheat, proso millet, pearl millet, oats, and legumes at the TCCP laboratory.[14] For some cultivars, regeneration rates have exceeded even most optimistic expectations. Significant progress has been made in the production of E callus, its maintenance, and regeneration into whole plant in these species. The most dramatic increase in plant regeneration to date at the TCCP has occurred in certain rice cultivars.

Another difficulty with present tissue culture systems is that not all variants produced in culture are stable or are expressed in the whole plant.[58,69,92,155,156,201] Mein and Binn's[201]work on cytokinin habituation of tobacco cultures showed that variants could result from epigenetic changes rather than from genetic mutations. Such epigenetic changes are considered to be alterations in gene expression and typically revert to the parental state upon plant regeneration.[69] Such changes could, for instance, be due to subtle changes in isozyme patterns, as seen in potato cultures[197] and tobacco cultures[67] resistant to 5-methyltryptophan. In the case of tobacco, resistance was expressed in culture but not in the whole plant, yet cultures derived from the regenerated plants were resistant.[202] The appearance and disappearance of an isozyme of anthranilate synthase have been correlated to the resistance pattern seen in the culture and the plant.[67] Whether these changes in isozymes are epigenetic or mutational is unknown at this time,[69] but they illustrate the problem of phenotypic changes being expressed in culture and not at the whole plant level. This lack of expression is probably an uncontrollable consequence of using relatively undifferentiated tissue to select genetic alterations that are to be expressed in a highly differentiated plant.[203]

Despite reports of extensive amounts of heritable somaclonal variation, mass production of regenerates — some which might be variant — might not be the most efficient means of selecting variants with specifically altered traits. A more appropriate means of obtaining mutants is selection *in vitro*; that is, applying some selective agent that permits the preferential survival and/or growth of a desired phenotype in an otherwise lethal environment.

On the other hand, traits expressed at the whole plant level may not be expressed at the cultured cell level.[204] Radin and Carlson[205] demonstrated that the leaves of some tobacco cultivars were susceptible to the herbicides bentazone and phermedipham, inhibitors of the Hill reaction, but callus derived from those leaves was not susceptible. In this instance, selection for resistance to the herbicides was accomplished by irradiating haploid plants with gamma-radiation, spraying them with the herbicides, and then initiating cultures from the surviving sectors of leaf tissue. After a colchicine treatment to double the chromosome number, homozygous-recessive resistant plants were regenerated.

The reason the initial tobacco cultures of Radin and Carlson[205] were not sensitive to bentazone or phermedipham was that the callus was not photosynthesizing and, hence, had no functional Hill reaction to be inhibited. Most plant tissue cultures lack chlorophyll and active chloroplasts and, thus, lack photosynthesis. This lack of photosynthesis is due, in most cases, to the high levels of sucrose and 2,4-D in the media used to grow the cultures.[206] The possibility exists that a well-founded understanding of desired selection would permit the circumvention of the photosynthesis problem, as illustrated by the previously mentioned case of selecting paraquat resistance in cultures grown in the dark. However, it seems reasonable to expect that, in most cases, cultures with functional chloroplasts will be needed to select chloroplast mutants or, as in the case of Radin and Carlson,[205] selection will have to be at the whole plant level with the recovery of mutants via tissue culture.

Such a selection system would be useful in soybean to select for atrazine resistance, as described by Horn and Widholm.[206] Soybean and corn rotation is a common planting strategy in the corn belt. Atrazine is a good broadleaf herbicide for corn cultivation, yet carryover to the next year's soybean crop can be very damaging. In the years since atrazine was first used, no herbicide-resistant soybean plant has been found in areas where atrazine carryover has been a problem. Arntzen et al.[207] demonstrated that in weed biotypes, the only known resistance to triazine herbicides (e.g., atrazine) resulted from the direct alteration of electron transport properties of the chloroplast. The failure to find resistant soybean plants in the field and the knowledge of where and how atrazine affects soybean plants suggest that photosynthesis-dependent (photoautotrophic) cell cultures may be the most probable system in which to select triazine resistance. To date, success with such a selection has not been realized.

In addition to herbicide tolerance, salt tolerance, drought tolerance, aluminum tolerance, tolerance to high and low temperatures, tolerance to specific plant diseases, and efficient fertilizer utilization, increased growth rates are some of the characteristics which could be selected in tissue culture. Research work at the TCCP has already made sufficient progress in achieving these goals in rice, wheat, corn, oats, barley, millet, sorghum, corn, and soybean. Several salt-tolerant lines have been selected and are being field tested. It is hoped that these efforts will lead to the production of stress-tolerant varieties of cereals in the near future.

Another difficulty with utilizing the variants produced in tissue culture stems from the genetic background from which these variants originate. One of the comments often aired in discussion concerning crop improvement and tissue culture is that breeding new lines of a crop will be faster using tissue culture than using conventional breeding techniques.[195,198] However, Bingham[195] has argued that too often the background in which a variant is found is not the best background to optimize the expression of the variant, even if the line used for the selection was an elite line.

In addition, one of the benefits of tissue culture (somaclonal variation) can also be a hindrance, if upon regenerating a variant plant, more variation appears than just that which was selected. Also, the commercial life expectancy of an elite line can be very brief because of competition with other lines, breakdown of disease resistance, and the like. Consequently, Bingham[195] suggests that little time will be saved by mutant selection in tissue culture, because for a cell-selected variant to be of use in crop improvement, it will eventually (if not immediately) have to be moved into new and improved lines of the proper genetic background. One could argue the reverse, however, and suggest that if the culturing, selecting, and plant regenerating period could be as short as 6 months, then rapid production of selected elite plant lines would be possible. In addition, even if *in vitro* selection does not speed up the development of new lines, it may make the process easier. Cell selection and plant regeneration would eliminate most plants lacking a desired trait. Consequently, the number of plants the breeder must examine to find the trait can be greatly reduced to a small group of plants with a high probability of success.[44]

Although still encumbered by many difficulties, cell selection has the potential to be a powerful tool for the breeder. Depending on the crop in question, use of tissue culture can (1) significantly increase the variation available to breeders and (2) help them select for unique and specific modifications in a crop. The use of cultured cells or protoplasts permits one to select from billions of entities easily and also permits the use of biochemical selection systems that cannot be easily applied to whole plants, including the use of rare or expensive compounds or conditions difficult to control. So far, some success has been realized in selecting for qualitative traits such as resistance to antimetabolites, herbicides, toxins, and stresses (e.g., heavy metals). One can predict that success is also possible in selecting useful quantitative traits such as resistance to heat, cold, high or low soil pH, flooding, or drought. Indirect methods could be used to select for resistance to pests like insects and nematodes if a plant metabolite can be identified that would confer resistance; selection for increase in that metabolite could then be carried out. Also, cultures that photosynthesize may be used to screen for more efficient photosynthesis by selecting, for instance, cultures with reduced light requirements but with growth comparable to the parental cells.

The major difficulty in the immediate use of cell selection for crop improvement arises from a fundamental lack of knowledge in the areas of plant physiology and plant genetic regulation. A focus of research on several basic questions such as those listed above would ultimately speed up the development of stress-tolerant plants. There is a need for further technical advances in the production of all culture types (suspension, callus, and protoplast) in all crops and plant regeneration from those cultures. Until recently, for example, this was the case with cereals; it is still the case for most legumes. This is further exemplified by corn and soybean cultures. Scutellum-derived callus of maize can readily regenerate plants, but protoplast or suspension cultures are not reproducibly formed and plants are seldom regenerated from these types of cultures when they have been established.[23,53,208] On the other hand, all culture types can readily be produced in soybean; however, plants cannot be regenerated from any culture type in a reproducible manner.[200]

Understanding the location and metabolic effect of inhibitors in a plant cell is essential to produce salt-tolerant plants. This would help the researcher to assess and predict the results of a particular selection scheme. Similarly, a clearer understanding is needed of gene expression and of the genetic and biochemical basis of complex phenotypic traits such as yield, stem strength, hybrid vigor, winter hardiness, and host-pathogen recognition. As our knowledge increases in these areas, cell selection can be optimized by the development of biochemical systems that will facilitate the production of unique, specific, and desirable modifications in complex whole plant traits. Despite several gaps, sufficient evidence exists to justify that tissue culture can be used to obtain stress-tolerant plants.

REFERENCES

1. **Ogura, H. and Shimada, T.,** Preliminary report on shoot redifferentiation from wheat callus, *Wheat Inf. Serv.*, 45, 26, 1978.
2. **Shimada, T. and Yamada, Y.,** Wheat plants regenerated from embryo cell cultures, *Jpn. J. Genet.*, 54, 379, 1979.
3. **Inoue, M. and Maeda, E.,** Stimulation of shoot bud and plantlet formation in rice callus cultures by two-step culture method using abscisic acid and kinetin, *Jp. J. Crop Sci.*, 50, 318, 1981.
4. **Nabors, M. W.,** Progress report: tissue culture for crops project, Agency for Internal Development, Department of State, Washington, D.C., 1982.
5. **Nabors, M. W., Kroskey, C. S., and McHugh, D. M.,** Green spots are predictors of high callus growth rates and shoot formation in normal and salt stressed tissue cultures of oat (*Avena sativa* L.), *Z. Pflanzenphysiol.*, 105, 341, 1982.
6. **Nabors, M. W., Heyser, J. W., Dykes, T. A., and Demott, K. J.,** Long duration, high frequency plant regeneration from cereal tissue cultures, *Planta,* 157, 385, 1983.
7. **Nabors, M. W. and Dykes, T. A.,** Obtaining cereal cultures with increased tolerance to salt, drought and acid stressed soils through tissue culture, in Intercentre Semin. on IARICs and Biotechnology, IRRI Los Banos, Laguna, Philippines, 1984.
8. **Thomas, E., King, P. J., and Potrykus, I.,** Shoot and embryolike structure formation from cultured tissues of *Sorghum bicolor, Naturwissenschaften,* 64, 587, 1977.
9. **Lu, C. and Vasil, I. K.,** Somatic embryogenesis and plant regeneration from leaf tissues of *Panicum maximum* Jacq, *Theor. Appl. Genet.*, 59, 275, 1981.
10. **Heyser, J. W. and Nabors, M. W.,** Regeneration of proso millet from embryogenic calli derived from various plant parts, *Crop Sci.*, 22, 1070, 1982.
11. **Heyser, J. W. and Nabors, M. W.,** Long term plant regeneration, somatic embryogenesis and green spot formation in secondary oat *(Avena sativa)* callus, *Z. Pflanzenphysiol.*, 107, 153, 1982.
12. **Heyser, J. W., Dykes, T. A., DeMott, K. J., and Nabors, M. W.,** High frequency long term regeneration of rice from callus culture, *Plant Sci. Lett.*, 29, 175, 1983.
13. **Heyser, J. W. and Nabors, M. W.,** Osmotic adjustment of cultured tobacco cells *(Nicotiana tabacum* var. *Samsun)* grown on sodium chloride, *Plant Physiol.*, 67, 720, 1981.
14. TCCP (Tissue Culture for Crops Project), Progress Report, Colorado State University, Fort Collins, 1987.
15. **Conger, B. V.,** Agronomic crops, in *Cloning Agricultural Plants via* in Vitro *Techniques,* Conger, B. V., Ed., CRC Press, Boca Raton, FL, 1981, 165.
16. **Evans, D. A., Sharp, W. R., Ammirato, P. V., and Yamada, Y.,** *Handbook of Plant Cell Culture,* Macmillan, New York, 1983.
17. **George, E. F. and Sherrington, P. D.,** Plant propagation by tissue culture, in *Handbook and Directory of Commercial Laboratories,* Exegetics, Eversley, England, 1984.
18. **Reinert, J. and Yeoman, M. M.,** *Plant Cell and Tissue Culture: A Laboratory Manual,* Springer-Verlag, New York, 1982.
19. **Vasil, I. K.,** *Cell Culture and Somatic Cell Genetics of Plants,* Vol. 1, Academic Press, New York, 1984.
20. **Wetter, L. R. and Constabel, F.,** Plant Tissue Culture Methods, National Council Canada, Saskatoon, 1982.
21. **Heyser, J. W., Nabors, M. W., Mackinnon, C., Dykes, T. A., De Mott, K. J., Kautzmann, D. C., and Mujeeb-Kaji, A.,** Long-term, high frequency plant regeneration and the induction of somatic embryogenesis in callus cultures of wheat *(Triticum aestivum), Z. Pflanzenphysiol.*, 94, 218, 1985.
22. **Raghva Ram, N. V. and Nabors, M. W.,** Cytokinin mediated long term, high frequency plant regeneration in rice tissue cultures, *S. Pflanzenphysiol.*, 113, 315, 1984.
23. **Vasil, I. K.,** Isolation and culture of protoplasts of grasses, *Int. Rev. Cytol.*, 16 (Suppl.), 79, 1983.
24. **Czias-Akins, P. and Vasil, I. K.,** Plant regeneration from cultured immature embryos and influorescences of *Triticum aestivum* L. (wheat): evidence for somatic embryogenesis, *Protoplasma,* 110, 95, 1982.
25. **Rangan, T. S. and Vasil, I. K.,** Sodium chloride tolerant embryogenic cell lines of *Pennisetum americanum* (L.) K. Schum, *Ann. Bot.*, 52, 59, 1983.

26. **Vasil, V. And Vasil, I. K.,** The ontogeny of somatic embryos of *Pennisetum americanum* (L.) K. Schum. I. In cultured immature embryos, *Bot. Gaz.*, 143, 454, 1982.
27. **Green, C. E. and Phillips, R. L.,** Plant regeneration from tissue cultures of maize, *Crop Sci.*, 15, 417, 1975.
28. **Brettell, R. I. S., Thomas, E., and Ingram, D. S.,** Reversion of Texas male-sterile cytoplasm maize in culture to give fertile, T-toxin resistant plants, *Theor. Appl. Genet.*, 58, 55, 1980.
29. **Brettell, R. D., Ingram, S., and Thomas, E.,** Selection of maize tissue cultures resistant to Drechslera *(Helminthosporium maydis)* T-toxin, in *Tissue Culture Methods for Plant Pathologists*, Ingram, D. S. and Helgeson, J. P., Eds., Blackwell Scientific, London, 1980, 223.
30. **Vasil, V. and Vasil, I. K.,** Somatic embryogenesis and plant regeneration from tissue cultures of *Pennisetum americanum*, and *P. americanum* × *P. purpureum* hybrid, *Am. J. Bot.*, 68, 864, 1981.
31. **Wernicke, W. and Brettell, R.,** Somatic embryogenesis from *Sorghum bicolor* leaves, *Nature*, 287, 138, 1980.
32. **Wernicke, W., Potrykus, I., and Thomas, E.,** Morphogenesis from cultured leaf tissue of *Sorghum bicolor* — the morphogenetic pathways, *Protoplasma*, 111, 53, 1982.
33. **Vasil, I. K.,** Somatic embryogenesis and plant regeneration in cereals and grasses, in *Plant Tissue Culture 1982*, Fujiwara, A., Ed., Japanese Association for Plant Tissue Culture, Tokyo, 1982, 101.
34. **Botti, C. and Vasil, I. K.,** Plant regeneration by somatic embryogenesis from parts of cultured embryos of *Pennisetum americanum* (L.) K. Schum., *Z. Pflanzenphysiol.*, 111, 319, 1983.
35. **Nishi, T., Yamada, Y., and Takahashi, E.,** The role of auxins in differentiation of rice tissues cultured *in vitro*, *Bot. Mag. (Tokyo)*, 86, 183, 1973.
36. **Tamura, S.,** Shoot formation in calli originated from rice embryo, *Proc. J. Acad.*, 44, 544, 1968.
37. **Sears, R. G. and Deckard, E. L.,** Tissue culture variability in wheat: callus induction and plant regeneration, *Crop Sci.*, 22, 546, 1982.
38. **Mackinnon, C.,** Morphogenic Responses of Wheat Calli Derived from Developmentally Mature Embryo Explants, Ph.D. dissertation, Colorado State University, Fort Collins, 1985.
39. **Cure, W. W. and Mott, R. L.,** A comparative anatomical study of organogenesis in cultured tissues of maize, wheat and oats, *Physiol. Plant.*, 42, 91, 1978.
40. **Brenneman, F. B. and Galston, A. W.,** Experiments on the cultivation of protoplasts and calli of agriculturally important plants. I. Oat (*Avena sativa* L.), *Biochem. Physiol. Pflanz.*, 168, 453, 1975.
41. **Nakano, H. and Moveda, E.,** Shoot differentiation in callus of *Oryza sativa* L., *Z. Pflanzenphysiol.*, 93, 449, 1979.
42. **Gresshoff, P. M. and Doy, C. H.,** *Zea mays:* methods for diploid callus culture and the subsequent differentiation of various plant structures, *Aust. J. Biol. Sci.*, 26, 505, 1973.
43. **Smith, M. K. and McComb, J. A.,** Selection for NaCl tolerance in cell cultures of *Medicago sativa* and recovery of plants from a NaCl-tolerant cell line, *Plant Cell Rep.*, 2, 126, 1983.
44. **Tomes, D. T.,** An assessment of the impact of biotechnology on plant breeding, *Newsl. Int. Assoc. Plant Tissue Cult.*, 42, 9, 1984.
45. **Tomes, D.,** Cell culture, somatic embryogenesis and plant regeneration in maize, rice, sorghum and millet, in *Cereal Tissue and Cell Culture, Advances in Agriculture Biotechnology*, No. 15, Bright, S. W. S. and Jones, M. G. K., Eds., Dr. W. Junk Publishers, The Hague, 1985, 176.
46. **Mackinnon, C., Gunderson, G., and Nabors, M. W.,** Plant regeneration by somatic embryogenesis from callus cultures of sweet sorghum, *Plant Cell Rep.*, 5, 349, 1986.
47. **Rangan, T. S.,** Morphogenic investigations on tissue cultures of *Panicum miliaceum, Z. Pflanzenphysiol.*, 72, 456, 1974.
48. **Ghazi, T. D., Cheema, H. V., and Nabors, M. W.,** Somatic embryogenesis and plant regeneration from embryogenic callus of soybean, *Glycine max* L., *Plant Cell Rep.*, 5, 452, 1986.
49. **Bhattacharya, S., Chatterjee, S., Biswas, P., and Mukherjee, B.,** Role of inoculum weight on physiology of growth and development of *Corchorus olitorius* and *Nigella sativa* cultured in vitro, *Indian J. Exp. Biol.*, 19, 1030, 1981.
50. **Sirwardana, S. and Nabors, M. W.,** Tryptophan initiated somatic embryogenesis and plant regeneration in rice, *Plant Physiol.*, 73, 142, 1983.
51. TCCP (Tissue Culture for Crops Project), Newsl. No. 7, Colorado State University, Fort Collins, 1987.
52. **Wernicke, W. and Brettell, R.,** Somatic embryogenesis from *Sorghum bicolor* leaves, *Nature*, 287, 138, 1980.
53. **Green, C. E., Armstrong, C. L., and Anderson, P. C.,** Somatic cell genetic systems in corn, in *Advances in Gene Technology: Molecular Genetics of Plants and Animals*, Downey, K., Voellmy, R. W., Ahmed, F., and Schultz, J., Eds., Academic Press, New York, 1983, 147.
54. **Hedges, T. K., Kamo, K. K., Imbrie, C. W., and Becwar, M. R.,** Genotype specificity of somatic embryogenesis and regeneration in maize, *Biotechnology*, 4, 219, 1986.
55. **Gamborg, O. L., Miller, R. A., and Ojima, K.,** Nutrient requirements of suspension cultures of soybean root cells, *Exp. Cell Res.*, 50, 151, 1968.

56. **Kumar, A. S., Reddy, T. P., and Reddy, G. M.,** Plantlet regeneration from different callus cultures of pigeonpea *(Cajanus cajan* L.), *Plant Sci. Lett.*, 32, 271, 1983.
57. **Chaleff, R. S.,** Isolation of agronomically useful mutants from plant cell culture, *Science,* 219, 676, 1983.
58. **Maliga, P.,** Isolation and characterization of mutants in plant cell culture, *Annu. Rev. Plant Physiol.*, 35, 519, 1984.
59. **Widholm, J. M.,** Isolation and characterization of mutant plant cell cultures, in *Plant Cell Culture in Crop Improvement,* Sen, S. K. and Giles, K. L., Eds., Plenum Press, New York, 1983, 71.
60. **Widholm, J. M.,** Selection and characterization of plant cell mutants for molecular biology studies, *Newsl. Int. Assoc. Plant Tissue Cult.*, 44, 2, 1984.
61. **Chaleff, R. S.,** Genetics of higher plants, applications of cell culture, Cambridge University Press, London, 1981.
62. **Evans, D. A., Sharp, W. R., and Medina-Filho, H. P.,** Somaclonal and gametoclonal variation, *Am. J. Bot.*, 71, 759, 1984.
63. **Grout, B. W. W. and Weatherhead, M. A.,** A strategy for the production of disease-resistant mutants, in *Tissue Culture Methods for Plant Pathologists,* Ingram, D. S. and Helgeson, J. P., Eds., Blackwell Scientific, London, 1980, 249.
64. **Helgeson, J. P. and Haberlach, G. T.,** Disease resistance studies with tissue cultures, in *Tissue Culture Methods for Plant Pathologists,* Ingram, D. S. and Helgeson, J. P., Eds., Blackwell Scientific, London, 1980, 179.
65. **Strauss, A., Gebhardt, G., and King, P. J.,** Methods for selection of drug-resistant plant cell cultures, in *Tissue Culture Methods for Plant Pathologists,* Ingram, D. S. and Helgeson, J. P., Eds., Blackwell Scientific, London, 1980, 241.
66. **Widholm, J. M.,** Selection of plant cell lines which accumulate certain compounds, in *Plant Tissue Culture as a Source of Biochemicals,* Staba, E. J., Ed., CRC Press, Boca Raton, FL, 1980, 100.
67. **Duncan, D. R. and Wildholm, J. M.,** Cell selection for crop improvement, *Plant Breed. Rev.*, 4, 153, 1986.
68. **Nafziger, E. D., Widholm, J. M., Steinrucken, H. C., and Kilmer, J. L.,** Selection and characterization of a carrot cell line tolerant to glyphosate, *Plant Physiol.*, 76, 571, 1984.
69. **Meins, J. R. F.,** Heritable variation in plant cell culture, *Annu. Rev. Plant Physiol.*, 34, 327, 1983.
70. **Schieder, O.,** Isolation of mutants with altered pigments after irradiating haploid protoplasts from *Datura innoxia* Mill. with x-rays, *Mol. Gen. Genet.*, 149, 251, 1976.
71. **Nabors, M. W., Gibbs, S. E., Bernstein, C. S., and Meis, M. E.,** NaCl tolerant tobacco plants from cultured cells, *Z. Pflanzenphysiol.*, 97, 13, 1980.
72. **Street, H. E.,** Cell cultures: a tool in plant biology, in *Higher Plants,* Dudits, D., Farkas, G. L., and Maliga, P., Eds., Akadamiai Kiado, Budapest, 1976, 7.
73. **Polacco, J. C.,** Arsenate as a potential negative selection agent for deficiency variants in cultured plant cells, *Planta,* 146, 155, 1979.
74. **King, J., Horsch, R. B., and Savage, A. D.,** Partial characterization of two stable auxotrophic cell strains of *Datura innoxia* Mill., *Planta,* 149, 480, 1980.
75. **Gebhardt, C., Schnebli, V., and King, P. J.,** Isolation of biochemical mutants using haploid mesophyll protoplasts of *Hyoscymus muticus.* II. Auxotrophic and temperature-sensitive clones, *Planta,* 153, 81, 1981.
76. **Maliga, P.,** Isolation, characterization and utilization of mutant cell lines in higher plants, in *Perspectives in Plant Cell and Tissue Culture,* Vasil, I. K., Ed., Academic Press, New York, 1980, 225.
77. **Gengenbach, B. G. and Green, C. E.,** Selection of T-cytoplasm maize callus cultures resistant to *Helminthosporium maydis* Race T pathotoxin, *Crop Sci.*, 15, 645, 1975.
78. **Raghava Ram, N. V. and Nabors, M. W.,** Plant regeneration from tissue cultures of pokkali rice is promoted by optimizing callus to medium volume ratio and by a medium conditioning factor produced by embryogenic callus, *Plant Cell Tissue Organ Cult.*, 4, 241, 1985.
79. **Oberthur, E. E., Nicholson, R. L., and Butler, R. G.,** Presence of polyphenolic materials, including condensed tannins, in sorghum callus, *J. Agric. Food Chem.*, 31, 660, 1983.
80. **Bressan, R. A., Hasegawa, P. M., and Handa, A. K.,** Resistance of cultured higher plant cells to polyethylene glycol-induced water stress, *Plant Sci. Lett.*, 21, 23, 1981.
81. **Fincham, J. R. S.,** Localized instabilities in plants — a review and some speculations, *Gen. Suppl.*, 73, 195, 1973.
82. **Heyser, J. W. and Nabors, M. W.,** Growth, water content and solute accumulation of two tobacco cell lines cultured on sodium chloride, dextran and polyethylene glycol, *Plant Physiol.*, 68, 1454, 1981.
83. TCCP (Tissue Culture for Crops Project), Newsl. No. 6, University of Colorado, Fort Collins, 1986.
84. **Maliga, P.,** Isolation, characterization and utilization of mutant cell lines in higher plants, *Int. Rev. Cytol. Suppl.*, 11A, 225, 1980.
85. **Orton, T. J.,** Comparison of salt tolerance between *Hordeum vulgare* and *H. jubatum* in whole plants and callus cultures, *Z. Pflanzenphysiol.*, 98, 105, 1980.

86. **Orton, T. J.,** Somaclonal variation: theoretical and practical considerations, in *Gene Manipulation in Plant Improvement,* Gustafson, J. P., Ed., Plenum Press, New York, 1984, 427.
87. **Smith, M. K. and McComb, J. A.,** Effect of NaCl on the growth of whole plants and their corresponding callus cultures, *Aust. J. Plant Physiol.,* 8, 267, 1981.
88. **Smith, M. K. and McComb, J. A.,** Use of callus cultures to detect NaCl tolerance in cultivars of three species of mature legumes, *Aust. J. Plant Physiol.,* 8, 437, 1981.
89. **Widholm, J. M.,** Selection and characterization of cultured carrot and tobacco cells resistant to lysine, methionine, and proline analogs, *Can. J. Bot.,* 54, 1523, 1976.
90. **Simons, R. A., Nabors, M. W., and Lee, C. W.,** A model of mutant selection in plant suspension cultures, *J. Plant Physiol.,* 116, 95, 1984.
91. **Gamborg, O. L. and Shyluk, P.,** Nutrition, media and characteristics of plant cell and tissue cultures, in *Plant Tissue Culture,* Thorpe, T. A., Ed., Academic Press, New York, 1981, 21.
92. **Chaleff, R. S. and Parsons, M. F.,** Direct selection in vitro for herbicide-resistant mutants of *Nicotiana tabacum, Proc. Natl. Acad. Sci. U.S.A.,* 75, 5104, 1978.
93. **Conner, A. J. and Meredith, C. P.,** An improved polyurethane support system for monitoring growth in plant cell cultures, *Plant Cell Tissue Organ Cult.,* 3, 59, 1984.
94. **Horsch, R. B. and Jones, G. E.,** A double filter paper technique for plating cultured plant cells, *In Vitro,* 16, 103, 1980.
95. **Weber, G. and Lark, K. G.,** An efficient plating system for rapid isolation of mutants from plant cell suspensions, *Theor. Appl. Genet.,* 55, 81, 1979.
96. TCCP (Tissue Culture Crop Project), Newsl. No. 5, Colorado State University, Fort Collins, 1986.
97. **Day, P. B.,** Long-term goals in agricultural plant improvement, in *Genetic Manipulation: Impact on Man and Society,* Arber, W., Illmensee, K., Peacock, W. J., and Starlinger, P., Eds., Cambridge University Press, Cambridge, 1984, 139.
98. **Carter, D. L.,** Problems of salinity in agriculture, in *Plants in Saline Environments,* Poljakoff-Mayber, A. and Gale, J., Eds., Springer-Verlag, New York, 1975, 25.
99. **Handa, A. K., Bressan, R. A., Handa, S., and Hasegawa, P. M.,** Tolerance to water and salt stress in cultured cells, in *Plant Tissue Culture, Proc. 5th Int. Congr. Plant Tissue Cell Culture,* Fujiwara, A., Ed., Maruzen Co., Tokyo, 1982, 471.
100. **Yano, S., Ogawa, M., and Yamada, Y.,** Plant formation from selected rice cells resistant to salts, in *Plant Tissue Culture,* Fujiwara, A., Ed., Japanese Association for Plant Tissue Culture, Tokyo, 1982, 495.
101. **Handa, A. K., Bressan, R. A., Handa, S., and Hasegawa, P. M.,** Clonal variation for tolerance to polyethylene glycol-induced water stress in cultured tomato cells, *Plant Physiol.,* 72, 645, 1983.
102. **Handa, S., Bressan, R. A., Handa, A. K., Carpita, N. C., and Hasegawa, P. M.,** Solutes contributing to osmotic adjustment in cultured plant cells adapted to water stress, *Plant Physiol.,* 73, 834, 1983.
103. **Ben-Hayyim, G. and Kochba, J.,** Aspects of salt tolerance in a NaCl-selected stable cell line of *Citrus sinensis, Plant Physiol.,* 72, 685, 1983.
104. **Iyengar, E. R. R., Patobia, J. S., and Kurian, T.,** Varietal differences in barley to salinity, *Z. Pflanzenphysiol.,* 84, 355, 1977.
105. **Waisel, Y.,** *Biology of Halophytes,* Academic Press, New York, 1972.
106. **Rush, D. W. and Epstein, E.,** Genotypic responses to salinity, *Plant Physiol.,* 57, 162, 1976.
107. **Raghava Ram, M. V. and Nabors, M. W.,** *Salinity Tolerance in Biotechnology: Applications and Research,* Cheremisinoff, P. N. and Culette, R. P., Eds., Technomic Publ. Lan., 1985, 623.
108. **Dix, P. J. and Street, H. E.,** Sodium chloride resistant cultured cell lines from *Nicotiana sylvestris* and *Capsicum annuum, Plant Sci. Lett.,* 50, 151, 1988.
109. **Croughan, T. P., Stavarek, S. J., and Rains, D. W.,** Selection of a NaCl tolerant line of cultured alfalfa cells, *Crop Sci.,* 18, 959, 1978.
110. **McHughen, A. G. and Swartz, M.,** A tissue culture derived salt tolerant line of flax *(Linum usitatissimum), J. Plant Physiol.,* 117, 109, 1984.
111. **Pandey, R. and Ganapathy, P. S.,** Isolation of sodium chloride tolerant callus line of *Cicer arietinum* L.cv.BG-203, *Plant Cell Rep.,* 3, 45, 1984.
112. **McHughen, A. G.,** Salt tolerance through increased vigor in a flax line (STS-II) selected for salt tolerance *in vitro, Theor. Appl. Genet.,* 74, 727, 1987.
113. **McCoy, T. J.,** Characterization of alfalfa *(Medicago sativa* L.) plants regenerated from selected NaCl tolerant cell lines, *Plant Cell Rep.,* 6, 417, 1987.
114. **Epstein, E., Norlyn, J. D., Rush, D. W., Kingsbury, R. W., Kelley, D. B., Cunningham, G. A., and Wrona, A. F.,** Saline culture of crops: a genetic approach, *Science,* 210, 399, 1980.
115. **Venables, A. V. and Wilkins, D. A.,** Salt tolerance in pasture grasses, *New Phytol.,* 80, 613, 1978.
116. **Humphreys, M. O.,** The genetic basis of tolerance to salt spray in populations of *Festuca rubra* L. *New Phytol.,* 91, 287, 1982.
117. **Abel, G. H.,** Inheritance of the capacity for chloride inclusion and chloride exclusion by soybeans, *Crop Sci.,* 6, 697, 1969.

118. **Epstein, E.,** *Mineral Nutrition of Plants: Principles and Perspectives,* John Wiley & Sons, New York, 1972.
119. **Mathur, A. K., Ganapathy, P. S., and Johri, B. M.,** Isolation of sodium chloride-tolerant plantlets of *Kickxia ramosissima* under in vitro conditions, *Z. Pflanzenphysiol.,* 99, 287, 1980.
120. **Tyagi, A. K., Rashid, A., and Maheshwari, S. C.,** Sodium chloride resistant cell line from haploid *Datura innoxia* Mill.: a resistance trait carried from cell to platelet and vice versa *in vitro, Protoplasma,* 105, 327, 1981.
121. **Kochba, J., Ben-Hayyim, G., Spiegel-Roy, P., Saad, S., and Neumann, H.,** Selection of stable salt-tolerant callus cell lines and embryos in *Citrus sinensis* and *C. aurantium, Z. Pflanzenphysiol.,* 106, 111, 1982.
122. **Wong, C. K., Ko, S. W., and Woo, S. C.,** Regeneration of rice plantlets on NaCl-stressed medium by anther culture, *Bot. Bull. Acad. Sin.,* 24, 59, 1983.
123. **Nyman, L. P., Gonzales, C. J., and Arditti, J.,** In vitro selection for salt tolerance of taro *(Colocasia esculenta* var. *antiquerum), Ann. Bot.,* 51, 229, 1983.
124. **Goldner, R., Umiel, N., and Chen, Y.,** The growth of carrot callus cultures at various concentrations and composition of saline water, *Z. Pflanzenphysiol.,* 85, 307, 1977.
125. **Warren, R. S. and Gould, A. R.,** Salt tolerance expressed as a cellular trait in suspension cultures developed from the halophytic grass *Distichlis spicata, Z. Pflanzenphysiol.,* 107, 347, 1982.
126. **Tal, M., Heikin, H., and Dehan, K.,** Salt tolerance in the wild relatives of the cultivated tomato: responses of callus tissue of *Lycopersicon esculentum, L. peruvianum* and *S. pennellii* to high salinity, *Z. Pflanzenphysiol.,* 86, 231, 1978.
127. **Flowers, T. J., Troke, P. F., and Yeo, A. R.,** The mechanism of salt tolerance in halophytes, *Annu. Rev. Plant Physiol.,* 28, 89, 1977.
128. **Greenway, H. and Munns, R.,** Mechanisms of salt tolerance in nonhalophytes, *Annu. Rev. Plant Physiol.,* 31, 149, 1980.
129. **Dix, P. J.,** Cell culture manipulations as a potential breeding tool, in *Low Temperature Stress in Crop Plants,* Lyons, J. M., Raison, J. K., and Steponkus, P. L., Eds., Academic Press, New York, 1979, 463.
130. **Handa, A. K., Bressan, R. A., Handa, S., and Hasegawa, P. M.,** Clonal variation for tolerance to polyethylene glycol-induced water stress in cultured tomato cells, *Plant Physiol.,* 72, 645, 1983.
131. **Dix, P. J. and Pearce, R. S.,** Proline accumulation in NaCl resistant and sensitive cell lines of *Nicotiana sylvestris, Z. Pflanzenphysiol.,* 102, 243, 1981.
132. **Watad, A. A., Reinhold, L., and Lerner, H. R.,** Comparison between a stable NaCl-selected *Nicotiana* cell line and the wild type, *Plant Physiol.,* 73, 624, 1983.
133. **Bressan, R. A., Handa, A. K., Handa, S., and Hasegawa, P. M.,** Growth and water relations of cultured tomato cells after adjustment to low external water potentials, *Plant Physiol.,* 70, 1303, 1982.
134. **Riccardi, G., Cella, R., Camerino, G., and Ciferri, O.,** Resistance to azetidine-carboxylic acid and sodium chloride tolerance in carrot cell cultures and *Spirulina platensis, Plant Cell Physiol.,* 24, 1073, 1983.
135. **Larkin, P. J., Ryan, S., Brettell, R. I. S., and Scowcroft, W. R.,** Heritable somaclonal variation in wheat, *Theor. Appl. Genet.,* 67, 443, 1984.
136. **Sanchez, P. A., Bandy, D. E., Villachica, J. H., and Nicholaides, J. J.,** Amazon Basin soils: management for continuous crop production, *Science,* 216, 821, 1982.
137. **Swaminathan, M. S.,** Biotechnology research and third world agriculture, *Science,* 218, 967, 1982.
138. **Devine, T. E.,** Genetic fitting of crops to problem soils, in *Breeding Plants for Less Favorable Environments,* Christiansen, M. N. and Lewis, C. F., Eds., John Wiley & Sons, New York, 1982, 143.
139. **Foy, C. D., Chaney, R. L., and White, M. C.,** The physiology of metal toxicity in plants, *Annu. Rev. Plant Physiol.,* 29, 511, 1978.
140. **Qureshi, J. A., Collin, H. A., Hardwick, K., and Thurman, D. A.,** Metal tolerance in tissue cultures of *Anthoxanthum odoratum, Plant Cell Rep.,* 1, 80, 1981.
141. **Sain, S. L. and Johnson, G. V.,** Iron utilization by iron efficient and inefficient soybean cultivars in cell suspension culture (abstr.), *Plant Physiol.,* 72 (Abstr. Suppl.), 5, 1983.
142. **Christianson, M. L.,** Zinc sensitivity in *Phaseolus:* expression in cell culture, *Environ. Exp. Bot.,* 19, 217, 1979.
143. **Phillips, G. C. and Collins, G. B.,** Growth and selection of red clover *(Trifolium pratense* L.) cells on low levels of phospate, *Agron. Abstr.,* 187, 1981.
144. **Meredith, C. B.,** Selection and characterization of aluminum resistant variants from tomato cell cultures, *Plant Sci. Lett.,* 12, 25, 1978.
145. **Colijn, C. M., Kool, A. J., and Nijkamp, H. J. J.,** An effective chemical mutagenesis procedure of *Petunia hybrida* cell suspension cultures, *Theor. Appl. Genet.,* 55, 101, 1979.
146. **Jackson, P. J., Roth, E. J., and McClure, P. R.,** Coinduction of synthesis of two metallotheionein-like, cadmium binding proteins in cadmium resistant suspension cell cultures of *Datura innoxia,* Abstract, ARCO Solar-UCLA Symp. on Plant Molecular Biology, Keystone, CO, April 16 to 22, 1983.

147. **Ojima, K. and Ohira, K.,** Characterization of aluminum and manganese tolerant cell lines selected from carrot cell cultures, *Plant Cell Physiol.*, 24, 789, 1983.
148. **Ojima, K., and Ohira, K.,** Characterization and regeneration of an aluminum-tolerant variant from carrot cell cultures, in *Proc. 5th Int. Congr. Plant Tissue,* Fujiwara, A., Ed., Maruzen Co., Tokyo, 1982, 475.
149. **Hibberd, K. A., Walter, T., Green, C. E., and Gengenbach; B. G.,** Selection and characterization of a feedback-insensitive tissue culture of maize, *Plant,* 143, 183, 1980.
150. **Cattoir-Reynaerts, A., Degryse, E., and Jacobs, M.,** Selection and analysis of mutants over-producing amino acids of the aspartate family in barley, *Arabidopsis* and carrot, in *Induced Mutations, a Tool in Plant Breeding,* 1981, 353.
151. **Ranch, J. P., Rick, S., Brotherton, J. E., and Widholm, J. M.,** Expression of 4-methyltryptophan resistance in plants regenerated from resistant cell lines of *Datura innoxia, Plant Physiol.*, 71, 136, 1983.
152. **Widholm, J. M.,** Control of tryptophan biosynthesis in plant tissue culture, lack of repression of anthranilate and tryptophan synthetases by tryptophan, *Physiol. Plant.*, 25, 75, 1971.
153. **Carlson, P. S.,** Methionine sulfoximine-resistant mutants of tobacco, *Science,* 180, 1366, 1973.
154. **Hibberd, K. A.,** Induction, selection and characterization of mutants in maize cell cultures, in *Cell Culture and Somatic Cell Genetics of Plants,* Vol. 1, Vasil, I. K., Ed., Academic Press, New York, 1984, 571.
155. **Swanson, E. B. and Tomes, D. T.,** Evaluation of birds foot trefoil regenerated plants and their progeny after in vitro selection for 2,4-dichlorophenoxyacetic acid, *Plant Sci. Lett.*, 29, 19, 1983.
156. **Miller, Q. K. and Hughes, K. K.,** Selection of paraquat-resistant variants of tobacco from cell culture, *In Vitro,* 16, 1085, 1980.
157. **Thomas, B. R. and Pratt, D.,** Isolation of paraquat-tolerant mutants from tomato cell cultures, *Theor. Appl. Genet.*, 63, 169, 1982.
158. **Sacristan, M. D.,** Resistance responses to *Phoma lingam* of plants regenerated from selected cell and embryogenic cultures of haploid *Brassica napus, Theor. Appl. Genet.*, 61, 193, 1982.
159. **Shepard, J. F.,** Protoplasts as sources of disease resistance in plants, *Annu. Rev. Phytopathol.*, 19, 145, 1981.
160. **Maheshwari, R., Hildebrandt, A. C., and Allen, P. J.,** Factors affecting the growth of rust fungi on host tissue cultures, *Bot. Gaz.*, 128, 153, 1967.
161. **Franzone, P. M., Foroughi-Wehr, B., Fischbeck, G., and Friedt, W.,** Reaction of microspore callus, androgenetic albino platelets and roots of barley to *Erysiphe graminis* f. sp. *hordei, Phytopathol. Z.*, 105, 170, 1982.
162. **Dekhuijzen, H. M.,** The enzymatic isolation of secondary vegetative plasmodia of *Plasmodiophora brassicae* from callus tissue of *Brassica campestris, Physiol. Plant Pathol.*, 6, 187, 1975.
163. **Sacristan, M. D. and Hoffmann, F.,** Direct infection of embryogenic tissue cultures of haploid *Brassica napus* with resting spores of *Plasmodiophora brassicae, Theor. Appl. Genet.*, 54, 129, 1979.
164. **Scheffer, R. P. and Briggs, S. P.,** Introduction: a perspective of toxin studies in plant pathology, in *Toxins in Plant Disease,* Durbin, R. D., Ed., Academic Press, New York, 1981, 1.
165. **Behnke, M.,** Selection of potato callus for resistance to culture filtrates of *Phytophthora infestans* and regeneration of resistant plants, *Theor. Appl. Genet.*, 55, 69, 1979.
166. **Behnke, M.,** General resistance to late blight of *Solanum tuberosum* plants regenerated from callus resistant to culture filtrates of *Phytophthora infestans, Theor. Appl. Genet.*, 56, 151, 1980.
167. **Behnke, M.,** Selection of dihaploid potato callus for resistance to the culture filtrate of *Fusarium oxysporum, Z. Pflanzenzu.*, 85, 254, 1980.
168. **Yoder, O. C.,** Assay, in *Toxins in Plant Disease,* Durbin, R. D., Ed., Academic Press, New York, 1981, 45.
169. **Keen, N. T.,** Evaluation of the role of phytoalexins, in *Plant Disease Control: Resistance and Susceptibility,* Staples, R. C. and Toenniessen, G. H., Eds., John Wiley & Sons, New York, 1981, 155.
170. **Uchiyama, T. and Ogasawara, N.,** Disappearance of the cuticle and wax in outermost layer of callus cultures and decrease of protective ability against microorganisms, *Agric. Biol. Chem.*, 41, 1401, 1977.
171. **Bohm, H.,** The formation of secondary metabolites in plant tissue and cell cultures, in *Perspectives in Plant Cell and Tissue Culture,* Vasil, I. K., Ed., Academic Press, New York, 1980, 183.
172. **Uchiyama, T., Sata, J., and Ogasawara, N.,** Lignification and qualitative changes of phenolic compounds in rice callus tissues inoculated with plant pathogenic fungi, *Agric. Biol. Chem.*, 47, 1, 1983.
173. **Warren, R. S. and Routley, D. G.,** The use of tissue culture in the study of single gene resistance of tomato to *Phytophthora infestans, J. Am. Soc. Hortic. Sci.*, 95, 266, 1970.
174. **Holliday, M. J. and Klarman, W. L.,** Expression of disease reaction types in soybean callus from resistant and susceptible plants, *Photopathology,* 69, 576, 1979.
175. **Helgeson, J. P., Haberlach, G. T., and Upper, C. D.,** A dominant gene conferring disease resistance to tobacco plants is expressed in tissue cultures, *Photopathology,* 66, 91, 1976.
176. **Moronek, D. M. and Hendrix, J. W.,** Resistance to race O of *Phytophthora parasitical* var. *nicotianae* in tissue cultures of a tobacco breeding line with black shank resistance derived from *Nicotiana longiflora, Phytopathology,* 68, 233, 1978.

177. **Deaton, W. R., Keyes, G. J., and Collins, G. B.,** Expressed resistance to black shank among tobacco callus cultures, *Theor. Appl. Genet.,* 63, 65, 1982.
178. **Gengenbach, B. G., Green, C. E., and Donovan, C. M.,** Inheritance of selected pathotoxin resistance in maize plants regenerated from cell cultures, *Proc. Natl. Acad. Sci. U.S.A.,* 74, 5113, 1977.
179. **Brettell, R. I. S., Goddard, B. V. D., and Ingram, D. S.,** Selection of Tms-cytoplasm maize tissue cultures resistant to *Drechslera maydis* T-toxin, *Maydica,* 24, 203, 1979.
180. **Haberlach, G. T., Budde, A. D., Sequira, L., and Helgeson, J. P.,** Modification of disease resistance of tobacco callus tissues by cytokinins, *Plant Physiol.,* 62, 522, 1978.
181. **Yoder,, C. C.,** Toxins in pathogenesis, *Annu. Rev. Phytopathol.,* 18, 103, 1980.
182. **Thanutong, P., Furusawa, I., and Yamamoto, M.,** Resistant tobacco plants from protoplast-derived calluses selected for their resistance to *Pseudomonas* and *Alternaria* toxins, *Theor. Appl. Genet.,* 66, 209, 1983.
183. **Murakishi, H. H. and Carlson, P. S.,** *In vitro* selection of *Nicotiana sylvestris* variants with limited resistance to TMV, *Plant Cell Rep.,* 1, 94, 1982.
184. **Krishnamurthi, M. and Tlaskal, J.,** Fiji disease resistant *Saccharum officinarum* var. Pindar Sub-clones from tissue cultures, *Proc. Int. Soc. Sugarcane Technol.,* 15, 130, 1974.
185. **Heinz, D. J.,** Sugarcane improvement through induced mutations using vegetative propagules and cell culture techniques, in *Induced Mutations in Vegetatively Propagated Plants,* International Atomic Energy Commission, Vienna, 1973, 53.
186. **Heinz, D. J.,** Tissue culture in breeding, *Annu. Rep. Hawaiian Sugar Planteers Assoc. Exp. Stn.,* p. 9, 1976.
187. **Lie, M. C. and Chen, W. H.,** Improvement in sugarcane using tissue culture methods, in *Frontiers of Plant Tissue* Culture, Abstract, Thorpe, T. A., Ed., University of Calgary, Canada, 1978, 515.
188. **Larkin, P. J. and Scowcroft, W. R.,** Somoclonal variation and eyespot toxin tolerance in sugarcane, *Plant Cell Tissue Organ Cult.,* 2, 111, 1983.
189. **Mattern, U., Strobel, G., and Shepard, J.,** Reaction to phytotoxins in a potato population derived from mesophyll protoplasts, *Proc. Natl. Acad. Sci. U.S.A.,* 75, 4935, 1978.
190. **Shepard, J. F., Bidney, D., and Shahin, E.,** Potato protoplasts in crop improvement, *Science,* 208, 17, 1980.
191. **Pullman, G. S. and Rappaport, L.,** Tissue culture-induced variation in celery for *Fusarium* yellows, *Phytopathology,* 73 (Abstr.), 818, 1983.
192. **Daub, M. E.,** A cell culture approach for the development of disease resistance: studies on the phytotoxin cerosporin, *Hortic. Sci.,* 19, 382, 1984.
193. **Ingram, D. A.,** Tissue culture methods in plant pathology, in *Tissue Culture Methods for Plant Pathologists,* Ingram, D. S. and Helgeson, J. P., Eds., Blackwell Scientific, London, 1980, 1.
194. **Hartman, C. L., McCoy, T. J., and Knous, T. R.,** Selection of alfalfa *(Medicago sativa)* cells and regeneration of plants resistant to the toxin(s) produced by *Fusarium oxysporum* f.sp. *medicaginis, Plant Sci. Lett.,* 34, 183, 1984.
195. **Bingham, E. T.,** Molecular genetic engineering vs. plant breeding, *Plant Mol. Biol.,* 2, 222, 1983.
196. **Carlson, P. S., Conrad, B. F., and Lutz, J. S.,** Sorting through the variability, *Hortic. Sci.,* 19, 388, 1984.
197. **Carlson, J. E. and Widholm, J. M.,** Separation of two forms of anthranilate synthetase from 5-methyltryptophan-susceptible and resistant cultured *Solanum tuberosum* cells, *Physiol. Plant.,* 44, 251, 1978.
198. **Duvick, D. N.,** Plant breeding with molecular biology, *Plant. Mol. Biol.,* 2, 221, 1983.
199. **Cocking, E. C. and Riley, R.,** Application of tissue culture and somatic hybridization to plant improvement, in *Plant Breeding II,* Frey, K. J., Ed., Iowa State University Press, Ames, 1981, 85.
200. **Christianson, M. L., Warnick, D. A., and Carlson, P. S.,** A morphogenetically competent soybean suspension culture, *Science,* 222, 632, 1983.
201. **Meins, F. and Binns, A. N.,** Cell determination in plant development, *BioScience,* 29, 221, 1979.
202. **Widholm, J. M.,** Differential expression of amino acid biosynthetic control isoenzymes in plants and cultured cells, in *Plant Cell Cultures: Results and Perspective,* Sala, F., Parisi, B., Cella, R., and Ciferri, O., Eds., Elsevier/North-Holland, Amsterdam, 1980, 157.
203. **Chaleff, R. S.,** Considerations of developmental biology for the plant cell geneticist, in *Genetic Engineering of Plants: An Agricultural Perspective,* Kosuge, T., Meredith, C. P., and Hollaender, A., Eds., Plenum Press, New York, 1983, 257.
204. **Gressel, J., Zilkah, S., and Ezra, G.,** Herbicide action, resistance and screening in cultures vs. plants, in *Frontiers of Plant Tissue Culture,* Thorpe, T. A., Ed., University of Calgary, Canada, 1978, 427.
205. **Radin, D. N. and Carlson, P. S.,** Herbicide tolerant tobacco mutants selected in situ and recovered via regeneration from cell cultures, *Genet. Res.,* 32, 85, 1978.
206. **Horn, M. E. and Widholm, J. M.,** Aspects of photosynthetic plant tissue cultures, in *Applications of Genetic Engineering to Crop Improvement,* Collins, G. B. and Petolino, J. F., Eds., Martinus Nijhoff/Dr. W. Junk, New York, 1985, Chap. 5.

207. **Arntzen, C. J., Pfister, K., and Steniback, K. E.,** The mechanism of chloroplast triazine resistance: alterations in the site of herbicide action, in *Herbicide Resistance in Plants,* Lebaron, H. M. and Gressel, J., Eds., John Wiley & Sons, New York, 1982, 185.
208. **Chourey, P. S. and Zurawski, D. B.,** Callus formation for protoplasts of a maize cell culture, *Theor. Appl. Genet.,* 59, 341, 1981.

Chapter 5

AGROBACTERIA-MEDIATED GENE TRANSFORMATIONS AND VECTORS FOR GENE CLONING IN PLANTS

I. INTRODUCTION

A plant gene cloning vector may be defined as an agent which will facilitate one or more steps in the overall process of placing foreign genetic material, from whatever source, into plants or their constituent parts. The steps of gene cloning in plants include uptake, incorporation, transcription, translation, maintenance, and passage through mitosis and meiosis of the exogenous material. Thus, ''plant gene vector'' applies to potential vectors both for the transfer of genetic information between plants and also from other organisms (bacteria, fungi, and animals) to plants. In short, a vector is a go-between, transferring genetic information from the donor to the recipient.[1]

Several plant DNA viruses have been proposed as candidates for such a role, but they are probably much more complicated. The plant cell wall is a barrier to plant DNA uptake experiments and the problem has been solved to some extent by the use of protoplasts which can directly take up foreign genes with ease. Another difficulty which is encountered in this type of experiment is the rapid degradation of the genetic material inside the host cells. Thus, natural vectors such as *Agrobacterium,* tumor-inducing (Ti) plasmids, viruses, and transposable elements may prove quite useful in overcoming the problems generally encountered during the introduction of new genetic material. The list of cloned plant genes is lengthening rapidly, and a new situation has arisen leading to the transfer of desired genes through such vectors to plants without much difficulty. In this chapter major groups of gene vectors for plants, the Ti plasmids of *A. tumefaciens,* the caulimo viruses, and the gemini viruses are considered in detail. In addition, the applications which have emerged during recent years are also discussed in detail.

II. PLANT GENE VECTORS

A. *AGROBACTERIUM* Ti PLASMID SYSTEM

Crown gall disease is the infection inflicted mainly on young fruit trees by *Agrobacterium tumefaciens*. Infection occurs at wound sites on the plant, although wounding may not be an absolute requirement for infection; rather, access to a constituent of the primary cell wall may trigger establishment of the disease. Though the bacterium is required for infection, subsequent maintenance of the gall is not dependent on its continued presence. The disease has been known and studied for a long time, but a proper understanding of the bacterial interaction came about through an intense study of the bacterium itself, and the development of extremely sensitive techniques for the detection of specific pieces of DNA. Material from an established gall may be removed and kept in sterile culture, free of the bacteria, for considerable periods of time up to several years. Furthermore, unlike normal plant cells, these tumorous cells can grow on a chemically defined medium lacking added auxins and cytokinins (plant growth substances).

The ability to transform plant cells is correlated with the presence of either a tumor-inducing *(A. tumefaciens)* Ti or root-inducing *(A. rhizogenes)* Ri plasmid. There are two families of bacteria, and the crown galls they established differed in that they produced two different amino acid derivatives, subsequently called octopine and nopaline. The ability to produce these ''opines'' was specified by the transferred bacterial DNA and not by the host plant genome.

Bacteria unable to elicit crown galls on susceptible plants do not carry a Ti or Ri plasmid. These bacteria are known as avirulent. Avirulent strains become virulent and start utilizing specific opines produced in the crown gall tissue on receiving these plasmids from virulent strain. The ability of virulent strains to transfer the Ti plasmid is found to be dependent on the presence of a particular opine encoded by that plasmid. Only a small part of the Ti plasmid, the transferred DNA (T-DNA), is transferred to and integrated into the host plant nuclear genome.[2,3]

Ti plasmids have been studied in detail using molecular techniques such as restriction mapping and insertional mutagenesis. The Ti plasmids of *Agrobacterium* are large, up to 200 kbp (kilobase pairs) in length. The two families of Ti plasmids differ in their properties. The octopine group of Ti plasmids are closely related and show extensive homology to one another, while the nopaline family are less closely related. Comparison of the octopine and nopaline groups shows that overall they exhibit about 30% homology, this homology residing in specific regions of the Ti plasmid. Restriction endonuclease digestion and *in vitro* DNA/DNA hybridization studies established the size and location of the T-DNA on the Ti plasmids. Insertional transposon mutagenesis revealed localized regions on the Ti plasmid concerned with the major functions specified by the plasmids, for example, virulence, origin of replication, oncogenicity, and catabolism of opines. These functions, with the exception of some of those controlling tumorigenicity, do not reside within the T-DNA.

1. Incorporation, Localization, and Organization of T-DNA

Two important regions in the Ti plasmid are essential for transformation by T-DNA. These are the T-DNA itself and the virulence region (vir). The T-DNA is defined by flanking 25-base pair (bp) of direct repeated sequences.[4-6] These sequences are required in *cis* for T-DNA transfer and are the recognition sequences for a site-specific endonuclease encoded by the virD operon.[7,8] Cleavage results in a linear, single-stranded molecule that is presumed to be an intermediate in T-DNA transfer to plant cells. What is critical from a perspective of vector design is that the border sequences must flank the DNA to be transferred and that they are the only *cis*-acting elements required for T-DNA transfer. No other T-DNA genes play a necessary role in transfer, and all of the essential transfer functions act in *trans* on the border sequences. Any DNA placed between the borders will be transferred to a plant.

The T-DNA is capable of inducing tumor formation in a transformed plant cell (Figure 1). This is accomplished by the products of three genes, iaaM (tms1), iaaH (tms2), and ipt (tmr). It has now been clearly established by biochemical means that these genes are involved in the synthesis of phytohormones. The iaaM and iaaH genes encode, respectively, a tryptophan monooxygenase[3,9] and an indoleacetamide hydrolase[9,10] which, together, synthesize the auxin indoleacetic acid. The ipt gene encodes an isopentenyl transferase,[11] which uses isopentenyl pyrophosphate and adenosine monophosphate to synthesize the cytokinin isopentenyl adenosine. The combination of auxin and cytokinin biosynthesis directed by the T-DNA leads to the tumorous morphology of transformed plant tissue. These phytohormone biosynthetic genes are almost certainly bacterial in origin. The auxin biosynthetic pathway utilized by *Agrobacterium* is not normally used by plants. These T-DNA genes, while bacterial in origin, are powerful tools to the plant biologist trying to understand the roles of phytohormones in plant development.

The other major portion of the Ti plasmid involved in T-DNA transfer is the vir region. This region of about 35 kb encompasses at least six operons.[12,13] While most of these operons are essential for T-DNA transfer,[14,15] they need not be physically linked to the T-DNA.[16,17] In practice this has led to the development of binary transformation vectors. The virulence functions, since they encompass such a large segment of DNA, are left intact on the Ti plasmid; and the T-DNA is placed on a much smaller, easily manipulable plasmid that replicates autonomously.

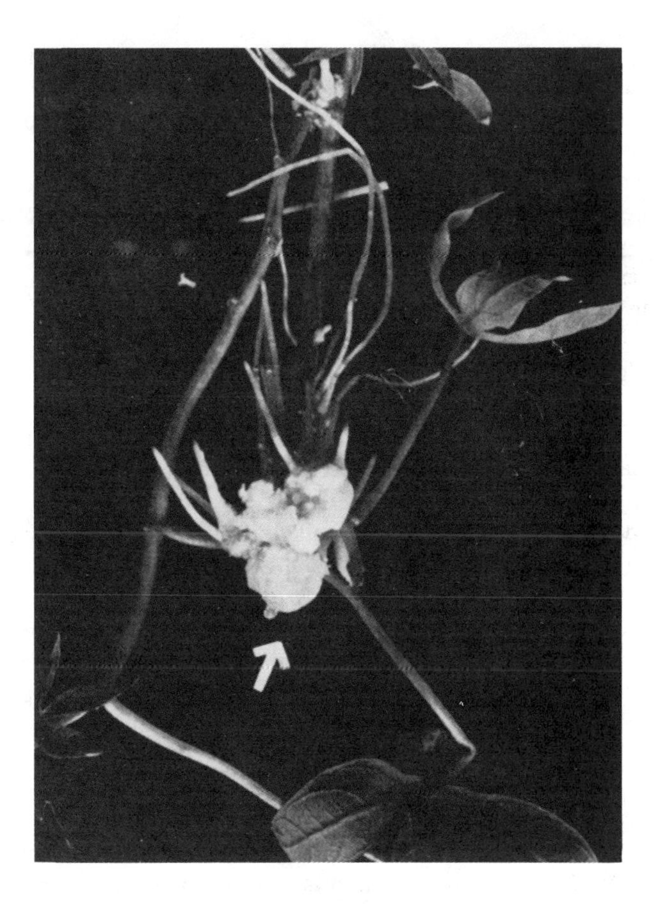

FIGURE 1. Tumor induction of *in vitro* propagated cassava plantlet, 3 weeks after having been inoculated with a wild-type *A. tumefaciens,* strain 1182 CIAT. This cassava-specific strain is being engineered at LSU to harbor genes of interest to cassava, including the shooting genes for plant regeneration. (From Roca, W. M., Szabados, L., and Hussain, A., Annual Report, Biotechnology Research Unit, CIAT, Colombia, 1987. With permission.)

The regulation of the vir genes has turned out to be an interesting and potentially useful story. These genes are expressed at very low levels in *Agrobacterium* under normal conditions. When they are exposed to plant cells or exudates they are turned on to varying degrees.[18,19] This induction is a slow process, taking 8 to 16 h to reach maximal levels of expression. Several phenolic compounds capable of induction have been identified and can be used in pure form to stimulate expression of the various vir operons. Two of the vir operons, virA and virG, have been implicated in the induction process.[18]

Thus, there is an intimate interaction between bacteria and the plant cell. The bacteria sense the presence of plant cells, and only then do they express the genes necessary for T-DNA excision and transfer. The T-DNA of Ti plasmids is transferred and stably incorporated into the nuclear genome of the infected plant cells. However, there is no incorporation into plastid or mitochondrial DNA. The exact mechanisms of transfer and subsequent incorporation of T-DNA into the plant host are not fully understood. The site of integration of the T-DNA on the plant chromosomes appears to be random. However, integration of T-DNA in any plant chromosome which is constantly actively expressed would be the appropriate selection.

In the crown gall, T-DNA has been shown to contain recognizable left- and right-hand borders. The left border of octopine T-DNA is fairly constant, whereas the right border is more variable, and there may, in fact, be duplications of all or part of the T-DNA. In T-DNA from nopaline Ti plasmids these borders consist of 14-bp direct repeats, overlapped by 25-bp imperfect repeats. T-DNA of nopaline tumors has a simpler arrangement with fixed borders and colinearity with T-DNA of the Ti plasmid. It is interesting to note that genes on the T-DNA are linked to typical eukaryotic control sequences and T-DNA is transcribed in the plant cell by host DNA polymerase II, the polyadenylated transcript being translated on plant ribosomes. This information has helped in the design and construction of Ti vector plasmids carrying foreign genes.

2. Vectors for *Agrobacterium*-Mediated Plant Transformation

Plant transformation vectors based on *Agrobacterium* can generally be divided into two categories: those that cointegrate into a resident Ti plasmid and those that replicate autonomously (the binary vectors). All of the vectors have several common features imposed upon them by the requirements of *Agrobacterium.*

As mentioned above, thc T-DNA is delimited by direct 25-bp imperfect repeats. No other element is required in *cis* to cause T-DNA excision and transfer. A transformation vector must, then, include at least one border or be capable of cointegrating in such a manner that it becomes flanked by borders. There is good evidence for the existence of a second sequence, termed "overdrive", which is found in the vicinity of the border sequence.[20] While this sequence is not essential for excision and transfer, it does appear to stimulate the process. Most vectors utilize a segment of a Ti plasmid that includes both the border and the overdrive sequences. Some vectors, however, use only the 25-bp border sequence and these seem to work efficiently for plant transformations.

Besides a border, all vectors should include a selectable marker for identification of transformed plant cells and a bacterial selectable marker essential for introduction of the vector into *Agrobacterium.* Some vectors also include a marker such as nopaline synthase. This allows for early verification of transformation by a simple assay for nopaline.

a. *Cointegrating Vectors*

Cointegrating transformation vectors must include a region of homology between the vector plasmid and the Ti plasmid. This requirement for homology means that the vector is capable of integrating into a limited number of Ti plasmids. The vector is usually designed to cointegrate into one or a few specific Ti plasmids. Two cointegrating, commonly known as disarmed and nondisarmed vectors are in extensive use.

i. *Disarmed Vectors*

This vector utilizes the disarmed *Agrobacterium* Ti plasmid pGV3850.[21] In this plasmid the phytohormone genes of the C58 plasmid have been excised and replicated by pBR322 sequence (Figure 2).[22] Any plasmid containing the pBR322 sequence homology can be cointegrated into the disarmed Ti plasmid. The border sequences as well as a nopaline synthase gene are part of the Ti plasmid, and the cointegration places the new sequence between the T-DNA borders.

A different approach to a cointegrating vector was used by Fraley et al.[23] In this system, the right border and all of the phytohormone genes are removed from the Ti plasmid. A left border and a small part of the original T-DNA, referred to as the limited internal homology (LIH), remain intact. The vector to be introduced into *Agrobacterium* contains the LIH region for homologous recombination as well as a right border. The cointegrated DNA reconstructs a functional T-DNA with a right border and a left border. This system has been used extensively for introduction of many genes into plants.

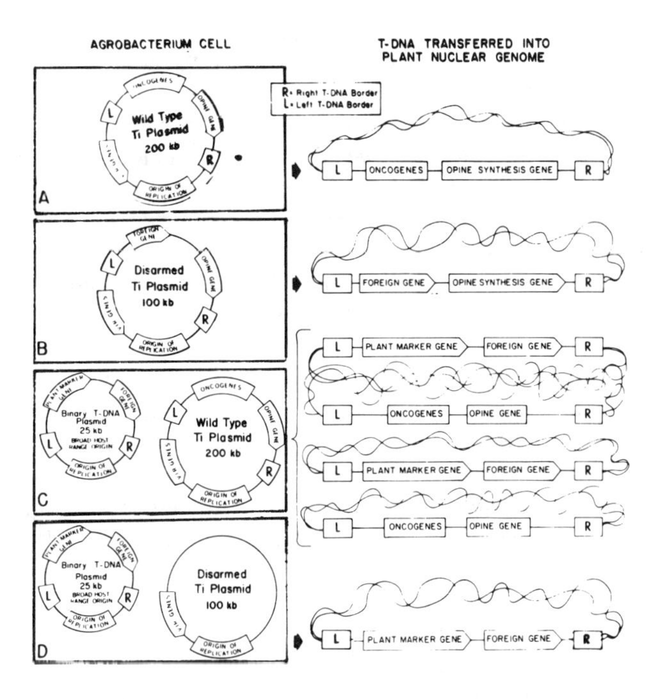

FIGURE 2. Ti plasmid-mediated T-DNA transfer. (A) Natural wild-type *Agrobacterium* Ti plasmid; (B) "disarmed" Ti plasmid (oncogenes have been removed and foreign gene inserted) containing the opine synthesis genes; (C) binary T-DNA plasmid in conjunction with a wild-type *Agrobacterium* Ti plasmid. With this arrangement it is possible to obtain three different integration events. The *vir* genes on the wild-type Ti plasmid may act either to transfer the binary T-DNA in addition to its own T-DNA (two T-DNAs are transferred), or may act to transfer each T-DNA separately; (D) binary T-DNA plasmid in conjunction with a disarmed Ti plasmid. Since the disarmed Ti plasmid does not contain a T-DNA region, the *vir* genes can only act to transfer the T-DNA on the binary T-DNA. The wavy lines represent helices in the plant genome.

ii. Nondisarmed Vectors

An alternative approach to separating the hormone biosynthetic genes from the gene of interest can be used with binary vectors. When two independent T-DNAs are present within a single *Agrobacterium,* as in the case of a cell containing a wild-type Ti plasmid and a binary vector, there is a high probability that the two T-DNAs will become integrated into a single plant cell.[24-26] However, some transformed cells will contain only one of the T-DNAs. Since some selective agent such as kanamycin is always used during the transformation/regeneration protocol, only the cells transformed with the binary T-DNA will survive. Some of these cells will contain only the binary T-DNA, while others will contain both T-DNAs. If fertile, transformed plants can be recovered, any wild-type T-DNA can be eliminated by outcrossing, assuming that the two T-DNAs are not linked.

The problem with this approach is that the cells transformed with the wild-type T-DNA will synthesize phytohormones that can still interfere with the regeneration protocol. The main advantage is related to host-range effects observed with different *Agrobacterium* strains. For the most part, disarmed Ti plasmids have been constructed from only a few common laboratory strains such as pTiC58[21] and pTiT37.[23] Construction of a disarmed Ti plasmid, while straightforward, requires a detailed knowledge of the plasmid. If one wishes to utilize an *Agrobacterium* strain that happens to transform a particular plant species very well, one need not do all of the work to construct a disarmed Ti plasmid. The cointegrate system, while more difficult to use, does offer advantages. Once the cointegrate has been formed, the plasmid is stable in *Agrobacterium* and is virtually impossible to lose. Binary vectors, on the other hand, are not completely stable in *Agrobacterium* in the absence of drug selection. There is also evidence that a cointegrating vector can transform tomato at a higher frequency than a binary vector.[27]

b. Binary Vectors

Binary transformation vectors are somewhat different from cointegrating vectors. Instead of a region of homology with the Ti plasmid, they contain origins of replication from a broad host-range plasmid (Figure 2). These replication origins permit autonomous replication of the vector in *Agrobacterium*. Since the plasmid does not need to form a cointegrate, these plasmids are considerably easier to introduce into *Agrobacterium*. The frequency of introduction into *Agrobacterium* is about 10^{-1}, whereas cointegrate formation is usually about 10^{-5}. Since these vectors do not cointegrate, they must contain the T-DNA border sequence(s). Many vectors contain two borders that delimit the portion of the plasmid transferred to the plant. However, a single border is sufficient for transfer[14] and the transfer initiates at the single border.

A growing body of evidence indicates that the tumorigenicity exhibited by different strains of *A. tumefaciens* and *A. rhizogenes* is highly dependent on the plant host.[28,29] This means that different strains may vary significantly in their abilities to infect a cultivar of a given plant species. It should be noted that tumorigenicity is not the same as T-DNA transformation. For example, Facciotti et al.[30] have observed transformed cells in wounded soybean seedlings despite the lack of tumor production. Most of the assays done to date have, by necessity, used tumor formation as an assay for infectibility. Many factors are involved in tumor induction. Most of these are probably related to T-DNA phytohormone production and the sensitivity of the plant to those hormones. Nonetheless, the observations related to *Agrobacterium* host range are important to consider, especially in plants not traditionally used as *Agrobacterium* hosts.

The problem of host range is probably best dealt with the use of binary vectors. These vectors can be readily mobilized into any *Agrobacterium* capable of providing vir functions in *trans*. This allows the experimenter to assay for T-DNA transfer by selecting the marker on the binary vector. With this approach, one should be able to screen a large number of *Agrobacterium* strains on a plant host and determine their capability of transformation.

A major advantage with binary vectors is their lack of dependence on a specific Ti plasmid. The vector may be introduced into virtually any *Agrobacterium* host containing any Ti or Ri plasmid, as long as the vir helper functions are provided. This may be important in the transformation of some plant species, since different *Agrobacterium* strains exhibit major differences in their abilities to infect different plant species.[31]

3. Limitations in Vector Construction and Plant Regeneration

Several problems are associated with the use of Ti plasmids as vectors. The plasmids are large and this does not allow for easy manipulation of their DNA; also, they have a large number of restriction sites which are not usefully distributed. In addition, the tumor cells with plasmid T-DNA (which specify production of substances which effectively convert the infected cells into tumor cells) cannot be regenerated into whole plants. Further, *Agrobacterium* infects only dicotyledonous plants, whereas many of the major food crops are monocotyledonous. The development and use of plant protoplast systems may go some way towards circumventing this problem and recent reports on successful regeneration of protoplasts derived from cereals are encouraging.[32,33]

In the beginning foreign DNA in the form of a bacterial transposon Tn7 was inserted at random sites on the Ti plasmid. This method has also been used to construct a functional map of genes on the Ti plasmid within the T-region because insertion of foreign genes may inactivate certain genes on the Ti plasmids. It has also revealed that none of the genes of the T-region are essential for the transfer and integration of the T-DNA. In addition, Tn7 integrated into the T-region is transferred and integrated into the plant nuclear genome. Thus, increase in the size of T-DNA by inserting certain foreign genes does not affect its integration into the plant nuclear genome. This has also permitted the study of expression of the transposon-encoded functions in the infected plant. The Tn7 gene transcription has been reported to transcribe and translate the methotrexate-resistant dihydrofolate reductase. Several workers have studied the precise organization of octopine and no paline synthase genes and numerous foreign genes have now been inserted into the T-region of *Agrobacterium*.[34-37]

The large size of the Ti plasmids and the lack of suitable restriction endonuclease target sites in the T-region for the cloning of foreign DNA have been the impetus for the development of smaller Ti cloning vectors. Allied to this is the requirement for other functions on the Ti plasmid which are essential for the transfer and integration of T-DNA into the plant genome. These virulence genes must therefore be present on any vector plasmid or their functions available in *trans*. Also, there is the situation that plant tissue, transformed with T-DNA with or without inserted sequences, will give rise to tumorous plant cells with grossly altered growth and differentiation characteristics. Ideally, one requires transformed cells to be capable of normal growth and development. Each of these problems has been overcome to a greater or lesser extent. Perhaps the easiest problem to overcome is that of tumorous growth of the plant cells. This was achieved by deleting from the T-region of the bacterial Ti plasmids all the genes concerned with tumorous growth, while conserving those concerned with opine synthesis.[38-40] Normal plants develop from infection with *Agrobacterium* strains containing such a deletion plasmid, and the presence of a functional octopine synthase gene has been demonstrated in their tissues. Moreover, they have been shown to be fertile and to sexually transmit the introduced gene as a single dominant Mendelian locus.[35]

The Ti plasmid is large and contains numerous functions not essential for transfer and integration of the T-region. In an attempt to reduce the overall size of the Ti plasmid and yet retain an infective agent capable of gene transfer to plant cells, Chilton and collaborators have set out placing the minimum number of essential functions onto smaller plasmids—the mini Ti strategy.[2] Any mini plasmid must contain the T-region, and by a series of *in vitro* manipulations, de Framond et al.[41] have placed the T-region on a plasmid which is capable of replication both in *Escherichia coli* and *A. tumefaciens*. This mini Ti by itself is avirulent, i.e., it will not induce the formation of tumors or production of opines in infected plant cells. However, in the presence of a second plasmid which carries the vir genes (responsible for transfer of the T-region from bacteria to plant) the mini Ti is virulent. de Framond et al.[41] pointed out that even this mini Ti is too large to allow easy manipulation. Similarly, Zambryski et al.[21] also proposed a modified Ti plasmid with the following T-region characteristics: only the border recognition sequences, no gene functions, and a marker gene (preferably dominant and selectable, and a sequence derived from a widely used cloning vector, such as pBR 322). It is envisaged that a single *in vivo* crossover event between an intermediate vector, a pBR322 plasmid carrying the gene of interest, and the acceptor Ti plasmid described above will generate a Ti plasmid with the inserted gene within the T-region. The plasmid may then be transferred to the plant via normal *Agrobacterium* infection.

4. Selectable Markers for Plant Transformation

Several requirements must be considered in the development of a truly useful selectable marker system. It is most critical that the selective agent be inhibitory to plant cells. However, not all compounds toxic to plant cells are necessarily useful as selective agents. Cells that are not transformed can be killed in such a manner that they become toxic to adjacent transformed cells. This presumably happens because of leakage of toxic compounds, such as phenols, from the dying cells. If this occurs, even high-level expression of a resistance gene in the transformed cells is insufficient to rescue these cells. The best selective agents are compounds that arrest growth of nontransformed cells or slowly kill them.

By far the most widely used selectable marker has been the neomycin phosphotransferase, type II (NPTII) enzyme, which was originally isolated from the prokaryotic transposon Tn5.[42] This enzyme detoxifies aminoglycoside compounds such as kanamycin and G418 by phosphorylation. This gene, fused to constitutive plant transcriptional promoters, has been used successfully to transform a large number of plant species and has been incorporated into numerous plant transformation vectors.[43-55]

For several reasons a single antibiotic resistance gene, even one as versatile as NPTII, has not fulfilled all the needs of plant molecular biologists. Probably the single most important reason is that this marker does not work in all plant species. This can be the consequence either of the lack of toxicity of kanamycin (G418) or of the failure of the enzyme to confer selectability in transformed cells. A good example of this lack of selectivity is *Arabidopsis thaliana* var. Columbia.[56] While the NPTII gene is clearly expressed and can, in fact, be used to select transformed, germinating seeds, it does not work as an efficient selectable marker for primary transformation. Another good reason for development of alternative selectable markers is related to the need for introducing more than one gene into a plant. If one desires to introduce multiple genes into a single plant, one simple approach is retransformation. This requires additional independent selectable markers.

The need for alternative markers has led to the development of two useful systems. The first of these is also a bacterial phosphotransferase, one encoding resistance to hygromycin.[53,54,57,58] This selectable marker has been demonstrated in a number of plant species. It has been particularly useful in the development of a transformation system for *Arabidopsis,* where the NPTII gene has not worked well.[56] The second generally useful marker system is based on the enzyme dihydrofolate reductase (DHFR). In this system the activity of DHFR is blocked by methotrexate. A DHFR enzyme with a 500-fold lower affinity for methotrexate has been isolated from mouse.[57] When the resistant enzyme is fused to the CaMV 35S promoter, transformed plant cells become highly resistant to methotrexate. This marker has worked well in a number of plant species.[31] Not only do these selectable markers extend the transformation of plants already resistant to kanamycin, but it is likely that other selectable marker systems will be developed in the near future. One promising system should be the acetohydroxy acid synthase enzyme described by Haughn and Somerville.[59] This mutant enzyme is tolerant to a 300-fold higher level of chlorsulfuron that is required to inhibit wild-type enzyme.

5. Transformation of Plants

a. Cocultivation

Two basic approaches have been used to obtain transgenic plants: cocultivation of regenerating protoplasts[48,60] and the leaf disk procedure.[49] Cocultivation was the first procedure used successfully to generate a transgenic plant. The procedure involves *Agrobacterium* transformation of regenerating protoplasts, followed by selection and regeneration to plants. For solanaceous plants such as *Petunia* and tobacco this has proved to be an efficient way to generate large numbers of independent transformants. Because it requires a good regeneration protocol for protoplasts of the plant species, the procedure is not useful for many important plant species.

b. Leaf Disk Transformation

In the leaf disk procedure, surface-sterilized leaf pieces, or other axenic explants, are cocultured on regeneration medium for 2 to 3 d with *Agrobacterium.* The best choice of explant is usually one that regenerates well in tissue culture for the species of interest. For example, in tomato, cotyledons are used as the explant source,[27] while in *Brassica napus,* stem segments seem to work well.[61] During cocultivation, the vir genes are induced in *Agrobacterium,* bacteria bind to plant cells around the wounded edge of the explant, and T-DNA transfer occurs. A nurse culture of tobacco cells is usually used during cocultivation to increase transformation frequency. This presumably results in better induction of the vir genes. Following coculture, the explants are transferred to regeneration/selection medium. This medium contains carbenicillin to kill the *Agrobacterium* and the appropriate antibiotic to select for transformed plant cells. During the next several weeks, transformed callus grows and differentiates into shoots. The shoots are then excised, rooted on an appropriate medium in the presence of the selective agent, and transferred to soil.

The leaf disk procedure and variants are a great improvement over cocultivation for several reasons. The leaf disk procedure requires far less tissue culture expertise than does protoplast preparation. It is far more generally applicable because a protoplasting procedure is not required. Also, transgenic plants are obtained more quickly and often, within 4 to 6 weeks of coculture.

One technical issue concerning leaf disk transformation should be considered, especially with new plant species. Some of the shoots that regenerate on selective medium do not contain T-DNA. The reasons for these escapes are not clear, but may include loss of T-DNA or incomplete selection due to cross-protection of wild-type cells by nearby transformed cells. The problem of escapes is most efficiently dealt with by a second selection for the ability to form roots in the presence of the selective agent. A scorable marker such as nopaline synthase, if present, can also be useful for identification of transformed shoots. Once a transformed shoot has been rooted on selective medium, the T-DNA insertion appears to be stable. T-DNAs have been characterized genetically for a number of generations and behave as normal Mendelian traits.[62]

B. CAULIMOVIRUS SYSTEM

Cauliflower mosaic virus (CaMV) is often cited as the most likely potential vector for introducing foreign genes into plants. This is mainly because caulimoviruses are unique among plant viruses in having a genome composed of double-stranded DNA, and this double-stranded DNA can be readily manipulated in recombinant DNA technology.

1. Localization and Organization of CaMV

The development of CaMV as a vector is as a consequence of its pathogenic activities on susceptible plants. The symptoms of infection vary, depending on the virus isolate, time of inoculation, and condition of the plant, from mild vein clearing to more severe stunting. In nature, this virus is transmitted by aphids. It is interesting to note that both the virus and the isolated DNA are infectious and can be easily transmitted by abrasion of the leaves. The infection becomes systemic yielding very high replication of the virus—10^5 virions per cell. Virus accumulates in the cytoplasm in inclusion bodies which consist of a protein matrix with embedded virus particles. It is rare to see free virus in the cytoplasm, and the inclusion body may be the site of virion assembly. The virus particle is spherical, isometric, about 50 nm in diameter, and may be isolated from the inclusion body using urea and nonionic detergents. Extraction of DNA from the virus particles is particularly difficult, involving the use of ionic detergents and proteolytic enzymes. The DNA molecule is about 8 kb long and several varieties (totaling 50,000 bases) have been sequenced. The DNA exists in linear, open circular, and twisted or knotted forms; however, none of the circular forms is covalently closed due to the presence of site-specific single-strand breaks (S1). These S1 nuclease-sensitive single-strand breaks are not true gaps, but short oligonucleotide overlapping regions, having complementary sequence and thus forming short triple-stranded structures with a fixed 5′ end. There are three such sites, one in the minus (coding or transcribed) strand yielding the large fragment and overlapping by eight residues. The other two are in the plus (noncoding or nontranscribed) strand yielding the fragments having 18 and 15 residue overlaps, respectively. One of the plus strand discontinuities is dispensable[63] and none are required for infection, as virus DNA previously cloned in bacteria and lacking the "gaps" is as infectious as native DNA.[64]

The sequence data obtained from CaMV have revealed (by analysis of open reading frames and stop codons) six major and two minor, tightly packed, potential coding regions distributed between the three reading frames. On either side of coding region VI there are two "intergenic regions" (IR), one large of approximately 1000 bp and one small of roughly 100 bp. The large IR flanks the α-strand break. The projected amino acid composition of the protein derived from the DNA sequence coding region IV has the capacity to code for a protein of 57,000 MW and correlates closely with the viral coat protein. Region VI codes for a nonviral protein of 61,000 to 66,000 MW thought to be involved in the viral inclusion body found in the cytoplasm. The proposed protein from region II is implicated in aphid transmissibility, as mutations in this region can abolish this capacity.

Viral-specific RNA isolated from infected tissue is analyzed using Northern blots. Transcription of CaMV is found to be asymmetric, with only the α-strand producing stable transcripts. The map location of only two major transcripts is known. The smaller of these is a 1.9-kb (19S) RNA derived from coding region VI, proposed to code for the 62,000-MW inclusion body protein, the 5′ end of which originates in the small IR between coding region V and VI. The other transcript is an 8-kb (35S) RNA of the whole α-strand, transcribed in a clockwise direction with the 5′ end in the large IR near the end of coding region VI. Upstream of the 5′ end of both transcripts are found typical eukaryotic promotor signals, "TATA" and "CAT" boxes, and at the 3′ end a polyadenylated signal is present; the two transcripts have a common 3′ end. The large 35S RNA transcript also has terminal repeats of 180. The accumulation of start and stop signals before the first open reading frame makes the transcript an unsuitable messenger and, unlike the 19S RNA, it is not translated *in vitro*.

The mechanisms of replication of CaMV have been worked out by Hull and Covey[65] and Pfeiffer and Hohn.[66] The infecting CaMV DNA enters the plant nucleus, where the single-stranded overlaps are digested and the gaps ligated to give supercoiled minichromosome. The function of this minichromosome is to act as a template for plant nuclear RNA polymerase II. The transcript thus formed is transported to the cytoplasm where it is either translated or replicated by reverse transcription. A site 6000 bp downstream of the promotor of the large transcript binds the proposed primer of reverse transcription, methionine tRNA. The RNA transcript is then copied into minus strand DNA. Synthesis of the plus strand DNA starts at two primer binding sites near "gaps" 2 and 3. From gap 2 synthesis proceeds to the 5′ end of the minus strand DNA, whereas synthesis from gap 3 continues to gap 2. Template switching and displacement of 5′ termini are proposed to account for the terminal repeats/overlaps at the three strand irregularities. This DNA molecule may then be packed into virus particles, or reenter the nucleus and undergo another round of transcription and/or translation/replication.[66] This model of DNA replication has certain implications for the use of CaMV as a possible gene vector. Transcription takes place in the nucleus, but unlike retrovirus, where integration is a prerequisite for transcription, there is no evidence that CaMV DNA integrates into the plant genome; indeed, the symptoms of infection are not transmitted through seed.

2. Use as a Cloning Vector

To date, the infectivity of the virus particle, and its naked DNA, are the most useful assets as regards the use of CaMV and its development as a gene cloning vector for plants. While some cloned viral DNAs are infectious following mechanical inoculation of leaves, others are not. The natural route for viral infection of plants is through insects that have fed on virus-infected plants, which is highly inconvenient. Stocks of insects must be maintained under strictly contained conditions, and the completely controlled introduction of insects to the plant is difficult. Although it is possible to clone and study a variety of viral genomes, for example, single-stranded DNA viruses by cloning double-stranded replicative-intermediate forms, no way had been found of reintroducing these genes into plants. Thus, the use of these viruses for *in vitro* mutagenesis or other applications of recombinant DNA technology was precluded.

The T-DNA of the crown gall bacterium *Agrobacterium tumefaciens* can be transferred to plant cells, and a method has been developed to transfer genes through this system to dicotyledons. This system uses the cloned viral DNA from bacteria to plants via the Ti plasmid using CaMV as a well-characterized model system. This is potentially a very flexible technique, because there is some other work that suggests that at least some monocotyledonous plants may be transferred by this bacterium.[67] Cloned CaMV DNA is not infectious when inoculated onto test plants, but viral DNA is infectious if it is excised from the bacterial vector at the cloning site, either as a monover or as subgenomic fragments showing that *in vivo* ligation within the plant cell can generate circular, infectious, viral DNA. The insertion of one genome of CaMV in the T-DNA would, therefore, not be a suitable way to test transfer of infectious virus to plants, as no mechanism for specific excision of the virus is likely to exist. On the other hand, it has been found that bacterial plasmids containing tandemly duplicated CaMV genomes are infectious.[68,69] Infection arises either as a result of intramolecular recombination or via products of 35S transcript. This transcript, which spans the entire CaMV genome and bears a 180-base-long terminal direct repeat, has been implicated as an intermediate in replication.[70,71] Thus, vectors with tandem repeated sequences of CaMV in the T-DNA might allow the virus to escape, giving rise to systemic infection when transferred to plants (Figure 3).[72] Grimsley et al.[72] tested the introduction of CaMV into plants by using a variety of constructs, including pCa305 and pGV96414D which contain 1.4 and 2 tandem genomes of CaMV, respectively (Figure 4). These constructs were able to produce large transcripts, including terminal repeats. Since naked CaMV DNA is usually inoculated onto wounded leaves of turnip plants and *A. tumefaciens* is inoculated by wounding at the crown, both routes of mechanical inoculation were tested with naked plasmid DNA, plasmids in *A. tumefaciens,* and total agrobacterial DNA containing Ti plasmid construct. Naked DNA of pCa305, pGV96414D, and pEA1 was found to be infectious when rubbed on leaves, but not when applied to wounded crown (Table 1). Inoculation with C58 (pGV3850::pCa305) or C58 (pGV3850::pGV96414D) bacteria rapidly led to viral infection following leaf or crown inoculation, whereas total bacterial DNA from these strains was only infectious when applied to the leaves in very large amounts. Inoculation with C58 (pEA1) bacteria did not give rise to symptoms. It was, therefore, concluded that the Ti plasmid has promoted the transfer of the infectious cloned viral DNA to the recipient plants. Inoculation with C58 (pGV3850::pCa305) and C58 (pGV3850::pGV96414D) gave rise to systemic infection more rapidly through the leaves than crown and also the symptoms on the host plants, including mustard, rape, and radish.

There are several problems associated with the use of CaMV as a gene cloning vector for plants. First, the genome is so tightly packed with coding regions that there is little room to insert foreign DNA. The question then arises as to whether any other genome is dispensable for viral functions. Most deletions of any significant size destroy virus infectivity, except for small modifications in coding region II. Random insertion of 8-bp restriction site linkers also destroys infectivity, except in coding region II and the large IR. Coding region II is highly polymorphic and has natural deletions and insertions; however, inserts up to 0.4 kbp long are tolerated, but those over 1.3 kbp destroy infectivity of the DNA.

Attempts have been made to sidestep this size limitation problem by using a helper virus

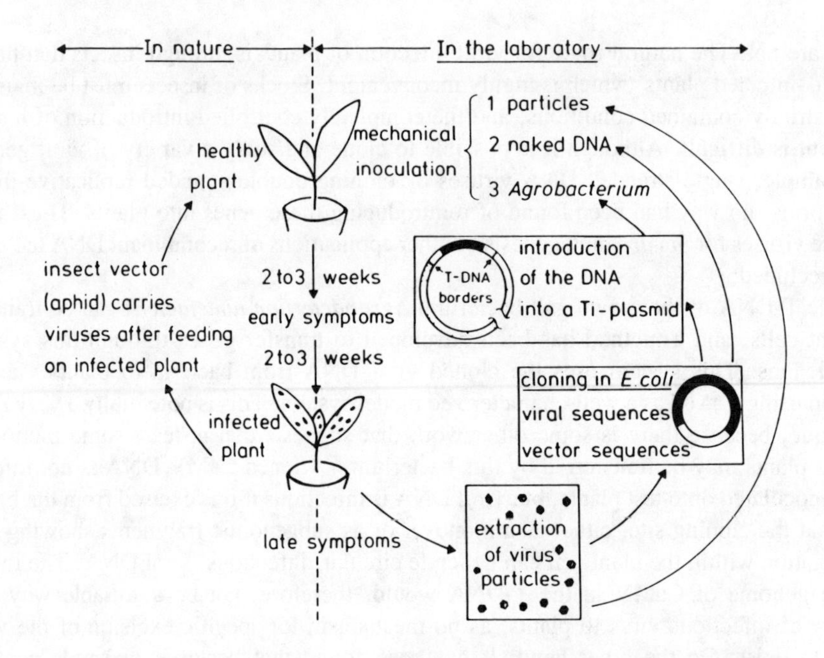

FIGURE 3. (Not to scale.) Routes of viral infection. Normal life cycle of CaMV (left side). In turnip, characteristic vein clearing develops 2 to 3 weeks after the inoculation, and, after a total of 6 to 10 weeks, stunting and premature senescence of leaves occur. In the laboratory (right side) crushed leaf tissue from infected plants or naked viral DNA suitably cloned in a bacterial vector can give rise to infection following mechanical inoculation of leaves (routes 1 and 2). Route 3 shows an alternative method of inoculation, ''agroinfection''. (From Grimsley, N., Hohn, B., Hohn, T., and Walden, R., *Proc. Natl. Acad. Sci. U.S.A.*, 83, 3282, 1986. With permission.)

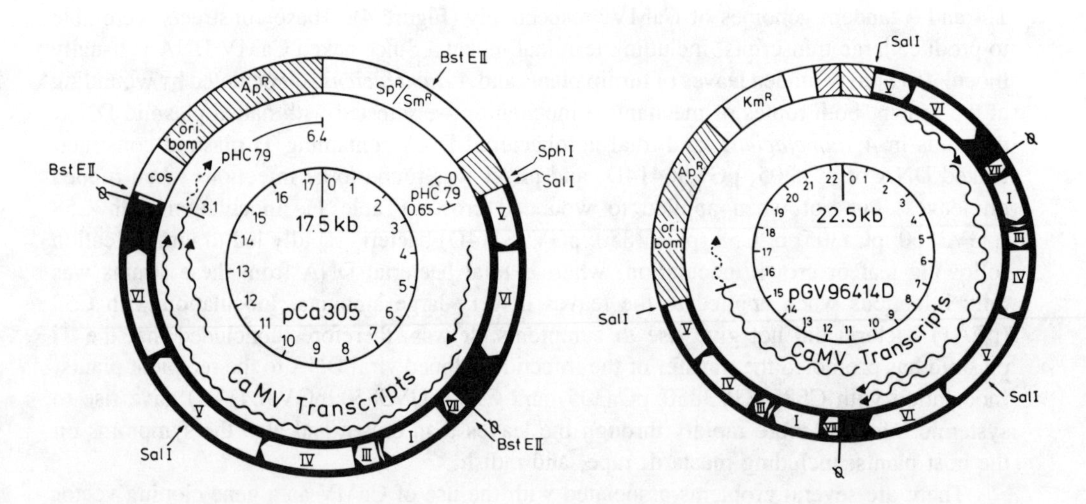

FIGURE 4. Vectors used for inoculation of plants either as naked DNA or in bacterial cells. CaMV sequences are shown as a black background with white open reading frames containing roman numerals, and regions of homology to pBR322 are shown hatched. ''Ø'' indicated the position of base pair 1 on a standard CaMV genome map. (From Grimsley, N., Hohn, B., Hohn, T., and Walden, R., *Proc. Natl. Acad. Sci. U.S.A.*, 82, 3282, 1986. With permission.)

TABLE 1
Appearance of Symptoms of Turnips after Inoculation

DNA or bacterial strain used	Whole plants showing symptoms/total plants inoculated (no./no.) Leaf inoculated	Crown inoculated
Plasmid DNA		
pCa305	8/12	0/4
pEA1	5/12	0/4
A. tumefaciens		
C58	0/12	0/8
C58 (pGV3850::pCa305)	8/8	14/14
C58 (pGV3850::pGV96414D)	8/8	8/8
C58 (pEA1)	0/24	0/16
Total bacterial DNA		
C58 (pGV3850::pCa305)(1.5 mg DNA per plant)	8/10	Not tested
C58 (pGV3850::pCa305)(100 μg DNA per plant)	0/10	Not tested
C58 (pGV3850::pGV96414D)(10 μg DNA per plant)	0/4	0/4

Note: Symptoms were assessed 6 weeks after inoculation

From Grimsley, N., Hohn, B., Hohn, T., and Walden, R., *Proc. Natl. Acad. Sci. U.S.A.*, 83, 3282, 1986. With permission.

system, where a substantial proportion of the viral genome is deleted and replaced with foreign DNA. The loss of function could be complemented by coinfection with a normal viral DNA, or viral DNA deleted for a different function. However, the rescue of viral functions in all cases occurred by recombination between the inactive viral genomes, and only normal infectious virus was recovered. For this system to be of any use the recombinational rescue of altered genomes must be suppressed, although the "retroviral-like" mode of replication produces a high recombination frequency and alteration of this would affect viral replication.

The second potential problem with CaMV is that infection, once established, becomes systemic, spreading throughout the whole plant. This lack of inheritance through the germ line may be advantagous, in that the CaMV DNA and any inserted gene sequence would be highly amplified in the host plant cells, potentially permitting the expression of large quantities of the foreign gene product. However, it appears that to propagate CaMV and to allow its movement throughout the vasculature of the plant, the DNA must be encapsidated, and this would impose serious constraints on the side of foreign DNA which can be inserted into the viral genome.

The problems outlined above have come to light through a detailed study of CaMV and its mode of replication and transcription, and there may be as yet unrecognized properties of the virus which will be exploited in its future development as a cloning vector. Also, one can envisage the possibility of introducing into the CaMV DNA short regions of host plant DNA, or the border regions from the T-region of the *Agrobacterium* Ti plasmid, to aid integration of CaMV DNA into the plant nuclear genome through these regions of homology. Subsequent infection using CaMV, containing integrated DNA, would allow integration via the homologous CaMV DNA already in the plant nuclear genome. Such inserts, stably maintained, may also be inherited through the seed.

C. GEMINI VIRUS SYSTEM

As with the Ti plasmid of *Agrobacterium,* and CaMV, the potential of gemini viruses as gene cloning vectors for plants stems from work on several plant diseases now recognized as being caused by these agents. Both curly top virus (CTV)—which causes major disease in several crops in the western U.S. and the Mediterranean—and maize streak disease (MSV)—one of the most important diseases of maize in Africa—have been major economic factors in the development of agriculture in areas where they occur. The gemini viruses recognized by the International Commission on the Taxonomy of Viruses in 1978 are characterized on the bases of their unique virion morphology and possession of single-stranded DNA.

1. Structural Features

The most surprising features of this virus group are the small capsid size, 19 to 20 $\times$ 30 nm, their geminate (paired particles) morphology, which sets them apart from all other classes of viruses, and the unexpected covalently closed circular (ccc) topography of the single-stranded DNA which is in the molecular weight range 7 to 9 $\times$ 10^5. All gemini viruses recognized so far have a single major coat protein subunit in the range 2.7 to 3.4 $\times$ 10^4 Da.

These features raised questions about the genome size of gemini virus and the biological or genetic significance of the geminate structure. Bean golden mosaic virus (BGMV) DNA was found to be 2510 nucleotides long, and if this was the complete genome it would be less than half the length of any other known autonomously replicating plant virus. By comparing the single-stranded DNA of the virus particle with the viral double-stranded DNA found in infected plants, it was found that the nucleotide sequence had a complexity twice that expected on the physical size of the viral DNA.[73] This indicates that the BGMV DNA is heterogenous, the virus having a divided genome consisting of two DNA molecules of approximately the same size, but different genetic content. It would appear that the gemini viruses consist of two populations of paired particles, differing only in the nucleotide sequence of the DNA molecules they contain.

Little is known about the DNA replication of gemini viruses. DNA replication occurs in the nuclei of infected plant cells. Viral assembly also takes place here and may be associated with the fibrillar rings seen in the nucleolus. A double-stranded DNA synthesis may occur on a circular template using, at least in part, host enzymes. No DNA/RNA hybrid molecules have been observed. Transmission of this virus in nature occurs by leaf hoppers or the tropical whitefly.

2. Use as a Cloning Vector

The potential of gemini viruses as vectors for the transfer of genes to plants is only beginning to be considered. As with other potential vectors, particularly viruses, the first step is to determine which genes carried by the virus are essential for their use as vectors. Tumor formation, though, provides a sensitive assay of T-DNA transfer to host cells, because transformants acquire the capacity for persistent proliferation. But this method does not appear to be valid for monocotyledonous plants, as integration of the transforming DNA, expression of the tumor genes, and response of the host plant to the hormones expressed are conditions which may not be fulfilled. However, opine test of whole plant tissue inoculated with *Agrobacterium* is also applicable in monocotyledonous plants.[67,74,75] The only exception to this is best CTV (BCTV) which is a monopartite virus transmitted by leafhoppers.[76] Cloned versions of several of these viruses infecting dicotyledonous plants have been shown to be infectious whereby coinoculation with both units of the bipartite viruses is required for symptom formation, but the single unit of BCTV DNA is infectious by itself.

However, the cloned and sequenced isolate of MSV could not be demonstrated to be infectious. In fact, neither native viral DNA nor virus particles could be shown to infect maize plants and the only successful method for infection remained was the natural. Agroin-

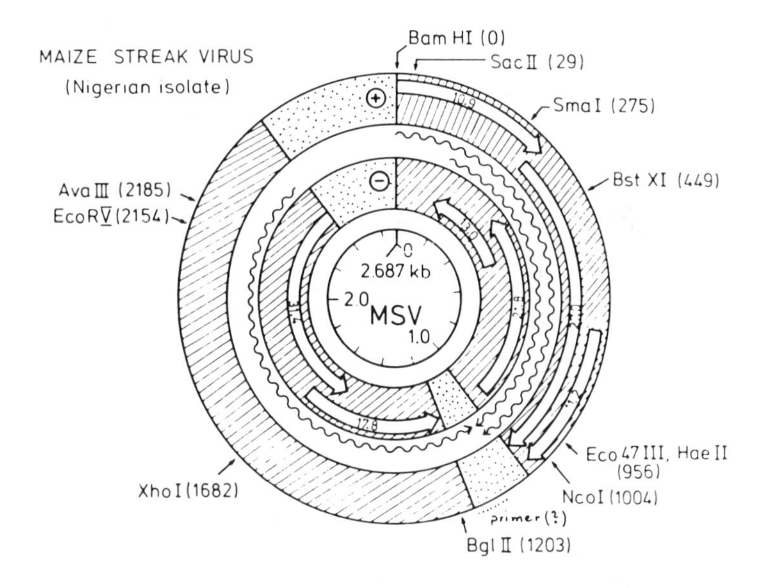

FIGURE 5. Map of MSV genome (Nigerian isolate) in its double-stranded form. (From Hohn, B., Hohn, T., Boulton, M. I., Davies, J. W., and Grimsley, N., *Plant Molecular Biology*, von Wettstein, D. and Chua, N. H., Eds., Plenum Press, New York, 1987, 1. With permission.)

fection of maize plants with dimer constructions therefore posed two questions: (1) is maize a host for *Agrobacterium?* and (2) is the MSV isolate used for experiment infectious? Grimsley et al.[77] found positive experimental results. They constructed strains of *A. tumefacines* in which dimers of MSV DNA were inserted in the T-DNA of a binary vector. Recently, a genetic map of MSV has been completed by Hohn and co-workers[78] and is shown in Figure 5. Transfer of intact natural T-DNA genes is not required for the transfer of MSV containing T-DNA. However, T-DNA containing MSV plasmid does not persist indefinitely and an average rate of plasmid loss of 1.7% per generation can be calculated (Figure 6).[78]

Through agroinfection, symptoms of T-DNA MSV infection appear between 4 to 15 d after inoculation with *Agrobacterium* (Figure 7). Although it is not clear why native and cloned MSV DNA is not infectious, *Agrobacterium*-mediated transformation should now allow the study of such constructs in whole maize plants and possibly in other members of Graminae.

A series of plasmids containing tandemly repeated dimer of MSV DNA was constructed by Grimsley et al.[77] and maize plants were infected with *A. tumefaciens* containing MSV dimer in the T-DNA (Figure 8a). When such bacteria were inoculated on the stem and leaves of young maize plants, symptoms consisting of yellow-white streaks appeared. Control experiments involving binary vectors lacking the T-DNA border sequences or using an agrobacterial virulence mutant, both of which gave negative results, indicated that most probably the T-DNA transfer mechanism, as described for dicotyledonous plants,[72,79] was operating in the maize system. This conclusion was further strengthened by the fact that naked DNA of MSV dimer was not infectious. Thus, lysis of the employed agrobacterial strains and infection of maize plants with the released DNA cannot be the route of infection.[77] DNA was extracted from infected and uninfected plants and infected plants only showed the presence of MSV DNA (Figure 8b).

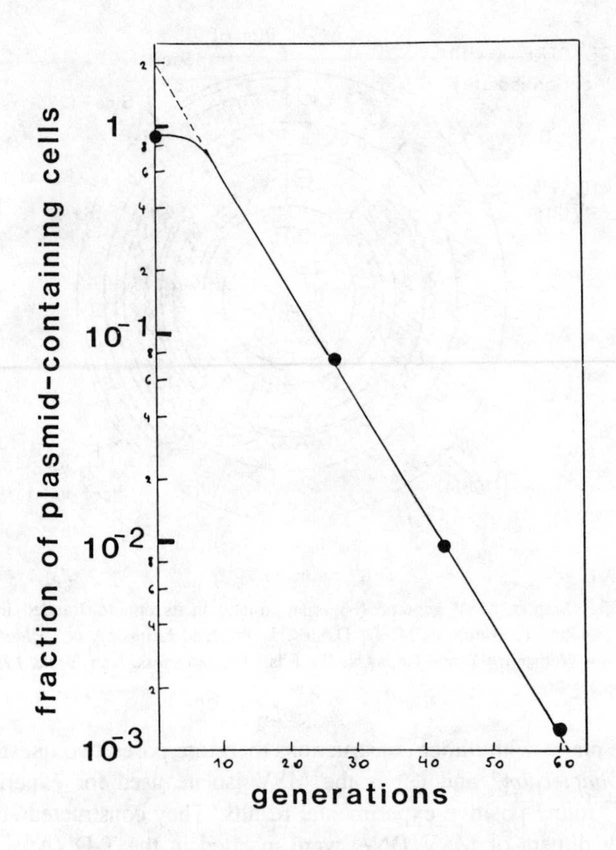

FIGURE 6. Maintenance of pEAP37 in C58, under nonselective conditions. Serial passages of C58 (pTiC58, pEAP37) were conducted in L-broth containing 100 μg/ml rifampicin, at 28°C. At intervals the culture was titered on plates containing rifampicin only or rifampicin and 25 μg/ml kanamycin. The segregation kinetics indicates a plasmid copy number of about 2 per cell under these conditions. (From Hohn, B., Hohn, T., Boulton, M. I., Davies, J. W., and Grimsley, N., *Plant Molecular Biology*, von Wettstein, D. and Chua, N. H., Eds., Plenum Press, New York, 1987, 1. With permission.)

III. APPLICATIONS

Agrobacteria carrying the recombinant Ti plasmid may use either of two approaches to transform plant cells via: (1) infection[80] or (2) uptake of Ti plasmid by plant protoplasts.[81-83] The end result of both approaches is to get cultured tumor callus, in the first instance, from excised crown gall freed of the inciting bacteria or, in the second instance, from transformed cells selected by their ability to grow on medium lacking phytohormones. Generally, however, the problem of deriving transformed fertile plants from tumor callus is a serious impediment to the practical use of both approaches to transformation.

The research efforts on *Agrobacterium*-mediated gene transformation are beginning to bear fruits. Several examples of transgenic plants have now become available.

A. PEST RESISTANCE TRANSGENIC PLANTS

Modern agriculture uses a wide variety of insecticides to control insect damage. Most of them are chemically synthesized. Notable exceptions are the insect toxins produced by *Bacillus thuringiensis*. Spore preparation of this Gram-positive bacterium has been used for

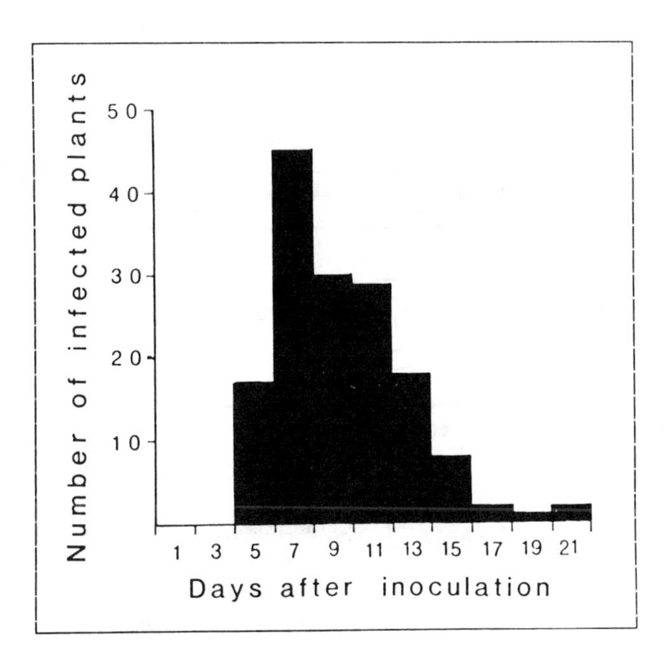

FIGURE 7. Kinetics of symptom appearance of MSV following agroinfection. The number of new plants showing symptoms, scored at 2-d intervals, is shown. (From Hohn, B., Hohn, T., Boulton, M. I., Davies, J. W., and Grimsley, N., *Plant Molecular Biology*, von Wettstein, D., and Chua, N. H., Eds., Plenum Press, New York, 1987, 1. With permission.)

more than 20 years as a biological insecticide. The insecticidal activity resides in crystalline inclusion bodies produced during sporulation of the bacteria which are composed of proteins (termed delta endotoxins) specifically toxic against a variety of insects.

Different strains of *B. thuringiensis* differ in their spectra of insecticidal activity. Most are active against Lepidoptera, but some strains are specific to Diptera and Coleoptera. The crystalline inclusion bodies which have an insecticidal property dissolve in alkaline conditions of the insect midgut and release proteins which are proteolytically processed by midgut proteases to yield smaller toxic fragments. *B. thuringiensis* insect toxins are highly specific, in that they are not toxic to other organisms. Hence, they are safe insecticides and present an interesting alternative to chemical control agents. Their commercial use, however, is limited by high production costs and the instability of the crystal proteins when exposed in the field.

Vaeck et al.[84] used *Agrobacterium*-mediated T-DNA transfer to express chimeric *B. thuringiensis* toxin genes in tobacco plants with the objective of protecting the plants from insect attack. These authors first characterized bt_2 gene (responsible for the production of protein which has insecticidal property) and cloned it from *B. thuringiensis* and recombinant polypeptide expressed in *Escherichia coli*. This protein, termed Bt_2, is 1155 amino acids long and is a potent toxin to lepidopteran larvae including those of *Manducta sexta*, a pest on tobacco. The smallest fragment of bt_2 gene that is still fully toxic was mapped. In plant transformation experiments, Vaeck et al.[84] used chimeric genes containing the entire coding sequence of bt_2 as well as truncated genes. Four chimeric genes containing modified *Bacillus* toxin genes under the control of 2′ promotor of the *Agrobacterium* Ti DNA were transferred to the tobacco plant. The intact and modified toxin genes of *B. thuringiensis* were inserted between the T-DNA borders of plant expression vector pGSH160 or pGSH150. The resulting plasmids were mobilized into *Agrobacterium*. Transgenic tobacco plants were obtained by

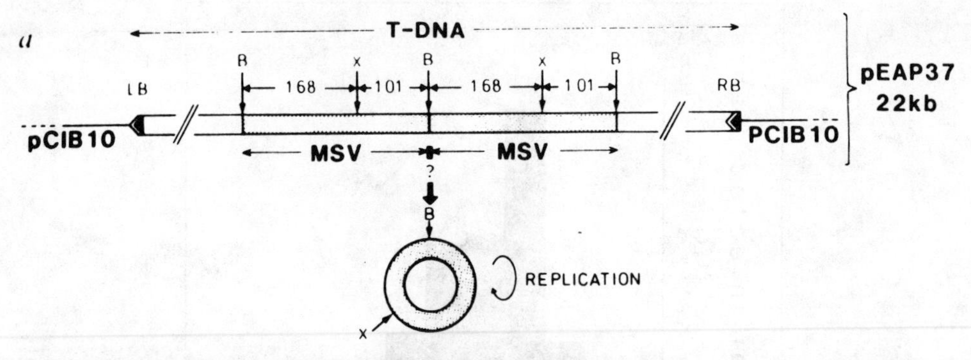

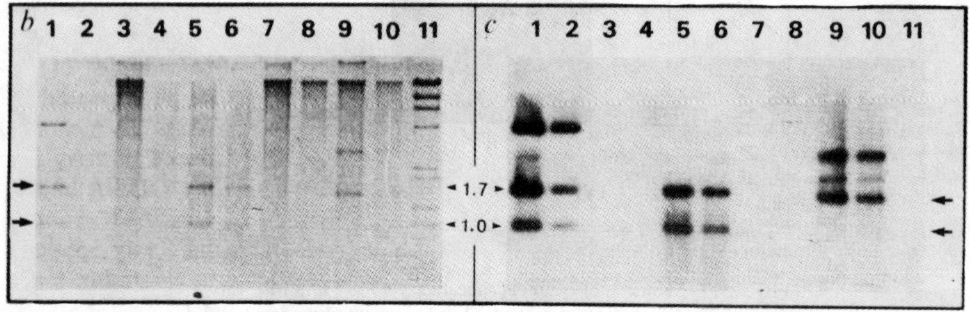

FIGURE 8. (a) Schematic representation of the input and output molecules of MSV agroinfection. LB, left border; RB, right border. Restriction enzyme sites: B, Bam HI; X, XhoI, Sizes are in kb·b, electrophoretic analysis of DNA of infected and control maize plants. Lanes 1 and 2: 500 and 50 ng of pMSV 12 digested with BamHI + XhoI; lanes 3 to 10: aliquots (1 μl) of DNA preparations of uninfected B73 (lanes 3 and 7), GB (lanes 4 and 8), infected B73 (lanes 5 and 9), and GB (lanes 6 and 10) plants. Samples in lanes 3 to 6 were restricted with Bam HI and XhoI and those in lanes 7 to 10 were left undigested. Lane 11, marker DNA (λHindIII + φX174 RF HaeIII). (c) Southern transfer of the gel in (b) and hybridization to nick-translated pMSV12 DNA. (From Grimsley, N., Hohn, T., Davies, J. W., and Hohn, B., *Nature*, 325, 177, 1987. With permission.)

leaf disk infection of *Nicotiana tabacum*. Leaves of transgenic plants containing the four types of *B. thuringiensis* gene construct were fed to *M. sexta* larvae to evaluate the expression of toxin genes in plants. High toxicity of insects, resulting in 75 to 100% mortality of the larvae, was observed in about 25% of the plants that expressed longer fusion protein Bt:NPT23 and 66% of those with shorter Bt:NPT860. Thus, shorter bt:ne0860 gene provided high level of biologically active protein. None of the plants transformed with the full length of bt_2 gene produced insect-killing activity above levels obtained in NPTII-expressing control plants, indicating that only truncated bt_2 genes gave rise to expression levels that are strongly insecticidal in transgenic tobacco. Selected transgenic plants, when grown in greenhouse and infected with larvae of *M. sexta,* were highly protected, whereas control plants were severely damaged. Further quantitative analysis of toxic *Bacillus* proteins revealed ten times more toxic proteins in plants transformed with truncated bt_2 gene or the fusion constructs. The failure to obtain insect-resistant plants using the intact bt_2 gene may be due to inefficient protein synthesis in the transformed plants. The plants that were recognized as producing a large amount of *Bacillus* protein generated exclusively kanamycin-resistant F_1 seedlings. E_1 progeny showed similar insecticidal activity that was observed in parental plants. These experiments thus illustrate the feasibility of engineering plants that defend themselves against lepidopteran larvae, which are sensitive to *B. thuringiensis* toxin. However, some species such as *Heliothis* and *Spodoptera* (important pests) are less sensitive to common *Bacillus* strain. Vaeck et al.[84] found that plants can be protected against these pests, but a higher level of expression of transformed genes in transgenic plants will be required. This may be achieved through 35S promotor of CaMV which directs 10- to 50-fold higher expression than the regular T-DNA promotors in plants. However, in a pilot experiment where the gene encoding the *B. thuringiensis* toxin was incorporated into a tobacco plant, the toxin produced by the newly introduced gene was inactivated by the ultraviolet component of sunlight.[85] This unexpected result warns us that biological systems can be very subtle, and it will often be easier to dream up revolutionary applications of recombinant DNA technology than to put these ideas into practice.

Recently, Fischhoff et al.[86] have determined the structure of an insect control protein gene from *B. thuringiensis* and generated a truncated gene that expressed a functional insecticidal protein. These truncated genes were incorporated into a plant vector for *Agrobacterium*-mediated transformation. Transgenic tomato plants containing the chimeric genes expressed the insect control gene which conferred tolerance to lepidopteran larvae on the transgenic plants and their progeny. These engineered tomato plants represented a significant step to increase selectivity, specificity, and efficiency to insect control. The transfer of insecticidal genes has now gained a significant importance. Realizing this the U.S. Department of Agriculture has given the go-ahead to Rohm and Huas to field test a tobacco plant containing a *B. thuringiensis* insecticidal protein gene.[87] The test will evaluate the plants' resistance to tobacco hornworm and tobacco budworm.

Another area where gene transfer technology can make a significant contribution is the integration of viral-resistant character into plants protecting them from several pathogenic viruses. A chimeric gene containing a cloned cDNA of the coat protein (CP) gene of tobacco mosaic virus (TMV) was introduced into tobacco cells on a Ti plasmid of *A. tumefaciens* from which tumor-inducing genes had been removed.[88] Plants regenerated from transformed cells expressed TMV mRNA and CP as a nuclear trait. Seedlings from self-fertilized transgenic plants that expressed the CP gene were delayed in symptom development, indicating that plant can be genetically transformed for resistance to virus disease development. *Agrobacterium*-mediated transfer from a binary vector was used to produce transgenic *N. tabacum* that expressed coat protein of alfalfa mosaic virus.[89] Plants that expressed the highest levels of coat protein developed fewer primary infections following inoculation with two strains of the virus, but infection by viral RNA was not reduced.

B. HERBICIDE RESISTANCE TRANSGENIC PLANTS

Transfer of different chimeric genes into a variety of crops may provide a new and environmentally safer method of controlling destructive pests. The example described above relates to the transformation of genes producing proteins with insecticidal property. On the contrary, De Block et al.[91] have been successful in engineering herbicide resistance in plants by the expression of a detoxifying enzyme. The use of herbicides to reduce loss in crop yield due to weeds is a very common practice with agriculturists. A new class of herbicides that fulfills these needs acts by inhibiting specific amino acid biosynthesis pathways in plants. Most of these herbicides are nonselective as they do not distinguish between weeds and crops. Modifying plants to become resistant to such broad-spectrum herbicides would allow their selective use for crop protection. As a consequence, a major effort has been devoted in several laboratories to engineer herbicide-resistant plants. Two approaches have been followed. In the first, a mutant form of target enzyme is produced which is still active, but less sensitive to the herbicide. In this way, mutant plants producing an altered form of the enzyme acetoacetate synthase have been selected which are resistant to sulfonylurea and imidazolinone herbicides.[92] In another example, a mutant form of the bacterial aroA gene was transferred to tobacco where it conferred tolerance to herbicide glyphosate.[93] The second approach involves the overproduction of target enzyme. It has been demonstrated that overexpression of the plant enzyme 5-enol-pyruvylshikimate-3 phosphate synthase conferred glyphosate tolerance in transgenic *Petunia* plants.[94]

Bialaphos and phosphinothricin (PPT) are new potent herbicides for weed control. Bialaphos is a tripeptide antibiotic produced by fermentation of *Streptomyces hygroscopicus* (Herbiace®, Meiji Seika Ltd.). It consists of the PPT analogue of L-glutamic acid and two L-alanine residues. PPT is chemically synthesized under the tradename Basta® (Hoechst AG). Upon removal of L-alanine residues by peptidases, PPT becomes a potent inhibitor of glutamic synthatase (GS). Inhibition of GS by PPT causes rapid accumulation of ammonia which leads to death of plant cells. De Block et al.[91] cloned and characterized bialaphos resistance (bar) genes from *S. hygroscopicus* which is envolved in the biosynthesis pathway. It encodes a phosphinothricin acetyltransferase (PAT) which acetylates the free NH_2 group of PPT and thereby prevents autotoxicity in the producing organism. The strategy followed by De Block et al.[91] in producing transgenic plants required the isolation of a bar gene from a streptomycete vector subcloned into an *E. coli* vector yielding pBG195. An ATG initiation codon through the N-terminal end of bar coding region was substituted for two complementary synthetic oligonucleotides. The bar gene in the pGSFRI with ATG initiation codon was inserted between cauliflower mosaic 35S promotor and the termination and polyadenylation signal of octopine T-DNA gene 7. This chimeric gene and a kanamycin resistance gene under the control of the nopaline synthase promoter were inserted between the octopine T-DNA borders of plant transformation vector pGV1500. The resulting plasmid, pGSFR280, was mobilized into the *Agrobacterium* recipient C58CI Rif® (PGV2260) to generate strain C58CI Rif ® (pGSFR1280). PAT was used as a selectable marker in protoplast cocultivation. The chimeric bar gene was expressed in tobacco, potato, and tomato plants. Low doses of 0.5 mg/l PPT were sufficient to inhibit the growth of protoplasts. The bar gene proved to be an efficient dominant selectable marker in potato protoplast transformation. Tobacco calli expressing the PAT enzyme grew on a medium containing 500 mg/l PPT. In greenhouse spraying tests transgenic plants expressed complete resistance to high doses of Basta® and Heblace®. The resistance was also inherited in the F_1 progeny as a single dominant trait. These results have clearly shown that engineering herbicide resistance using a detoxification or degradation process holds much promise. The successful engineering of a detoxification pathway will be largely independent from the plant species used. The same gene will thus be useful to engineer a variety of crops.

In addition to tobacco, extensive work is going on to transfer herbicide resistance genes through *Agrobacterium* in crops like potato and tomato. However, not much has been done to genetically engineer tree species for herbicide resistance. Methods such as microinjection[95] and direct DNA uptake[96] have been used to introduce foreign genes into herbacious crop species, whereas the most effective method of gene transfer through *Agrobacterium* has not received much attention.[37,47,97,98] Fillatti et al.[99] worked out a transformation system for a hybrid poplar clone ***Populus alba*** × ***grandidentata*** using *A. tumefaciens* as a vector. A major factor limiting the establishment and management of short rotation *Populus* plantations is the lack of a broad spectrum herbicide which effectively controls weeds.[100,101] Production of a *Populus* variety resistant to such herbicides would thus be economically attractive. Introducing the chimeric genes for glyphosate tolerance offers a unique opportunity for weed control. Secondly, *Populus* has been known to be a natural host for *A. tumefaciens* for over 10 years. Wild-type strains such as strain 27[102] and strain At 181[103] have been isolated from galls of *Populus* sp. Recently, *Populus* was confirmed as a host for *A. tumefaciens* by demonstrating that T-DNA sequences were present in gall tissue.[104] Fillatti et al.[99] for the first time successfully used *A. tumefaciens* as a vector to transfer foreign genes of agronomic importance into *Populus* plants. Leaf explants from stabilized shoot cultures of *Populus* hybrid were cocultivated with *A. tumefaciens* on a tobacco nurse culture. Both an oncogenic and disarmed strain of *A. tumefaciens* harboring a binary vector containing two neomycin phosphotransferase II (NPTII) and one bacterial enolpyruvylshikimate 3-phosphate (EPSP) synthase (aroA) chimeric gene fusion were used. Shoots did not develop when leaf explants were cocultivated with the binary disarmed strain of *A. tumefaciens*. In most Ti plasmid vector systems the oncogenes that encode for the production of auxin and cytokinin were deleted to ensure that species-specific hormonal regimes for shoot regeneration can be maintained. Most of these vectors, however, have been developed using solanaceous species such as tobacco,[49] tomato,[27] or petunia[49] as host tissue. Poplar appears to be an exception in this regard, since the hormones produced by the oncogenes or the TZS gene enhance shoot regeneration in combination with the hormones in the regeneration medium.[99] Parsons et al.[104] were successful, however, in obtaining evidence for gene transfer by *Agrobacterium* to cells of *Populus* hybrid in culture, but they were unable to recover transformed plants. This may be due to several reasons: (1) stabilized shoot cultures were not employed, (2) the strains of *Agrobacterium* used provided a different hormonal environment from the strains used by Fillatti et al.,[99] and (3) different genotype of *Populus* was used (*P. trichocarpa* × *P. delotides* hybrid). Further success with more transformation experiments in *Populus* hybrid colonies (*P. nigra* cv. *betulifolia* × *P. trichocarpa, P. nigra* × *P. laurifolia, P. maximowiczii* × *P. trichocarpa*) by Fillatti et al.[99] suggests the use of this regeneration and transformation technique to a wide range of *Populus* species. Combined research at Calgene (University of Wisconsin U.S. Forest Service) also led to the insertion of gene conferring resistance to glyphosate into poplar.[105] The gene was introduced via *A. tumefaciens* in plant tissue culture.

Chloroplasts are thought to be derived from symbiotic photosynthetic prokaryotes that colonize the cytoplasm of primitive eukaryotes. As predicted by the endosymbiont hypothesis, chloroplasts resemble bacteria in many ways. For example, ribosomes found in the cytoplasm of eukaryotes are resistant to the antibiotic chloramphenicol, but both prokaryotic and chloroplast ribosomes are sensitive. The differing antibiotic sensitivity of cytoplasmic and chloroplast ribosomes was used in an ingenious experiment designed to test whether T-DNA can insert into chloroplast as well as into nuclear genomes.[85] A gene fusion was constructed between the promoter of the nopaline synthetase gene and the coding region of the bacterial chloramphenicol resistance gene. Because the nopaline synthetase promoter was already known to function in bacteria, and the chloramphenicol resistance gene was itself derived from bacteria, it seemed possible that the gene fusion might function in chloroplasts. When the gene fusion was introduced into chloramphenicol-sensitive tobacco cells using a recombinant T-DNA vector, mature tobacco plants could be regenerated in the presence of chloramphenicol. The enzyme chloramphenicol transacetylase (which degrades chloramphenicol and is the product of the chloramphenicol resistance gene) was found exclusively in chloroplasts of the resistant plants; moreover, the chloramphenicol resistance trait was maternally inherited. These observations confirmed that the T-DNA had integrated directly into the chloroplast genome.

The ability of T-DNA to insert into the chloroplast genome has important agricultural implications. Many herbicides work by uncoupling light-driven electron transfer from production of high-energy chemical bonds during photosynthesis, and resistance to these herbicides often resides in chloroplast DNA.[85] The macromolecular complexes responsible for photosynthesis are generally put together from gene products encoded by both the nuclear and chloroplast genomes. Thus, it should prove technically possible to transfer resistance to various kinds of herbicides from one species of plant to another, using T-DNA as the vector to move the appropriate region of chloroplast DNA from the resistant to the sensitive chloroplasts. Thus, herbicide-resistant crop plants may soon be a reality.

Another potential use for recombinant T-DNA vectors stems from the observation that many soil bacteria are able to degrade various herbicides enzymatically. In principle, these bacterial genes could be cloned and expressed in new plant lines, so that herbicides could replace mechanical cultivation as a method for weed control. Still other bacteria may be found to produce antifungal agents, the biosynthetic gene(s) which could be incorporated into crop plants. However, ecosystems are often balanced, and the environmental impact of introducing transgenic plants will sometimes require serious scrutiny.

C. TRANSGENIC PLANTS WITH MARKER GENES (PRELIMINARY SUCCESS)

A system was established for introducing cloned genes into white clover *(Trifolium repens)* using binary *Agrobacterium* vectors containing a chimeric gene conferring kanamycin resistance.[106] The kanamycin resistance phenotype was stable in cell and regenerated shoots. Genetically engineered plants of commercial cotton varieties were obtained by *Agrobacterium*-mediated transformation.[107] Inoculated tissues selected for kanamycin-containing medium gave rise to calli that were resistant to the antibiotic and expressed the neomycin phosphotransferase enzyme. Somatic embryos germinated and whole plants also expressed the marker enzymes.

An attempt has been made recently to transfer the T-DNA of *A. tumefaciens* previously induced into plant cells via protoplast fusion from one species into another.[108] T-DNA was first transferred to *N. paniculta* which was fused with *N. tabacum*. Novel Ti plasmid-derived vectors produced by double recombination into the wild-type Ti plasmid of genetic information flanked by two DNA fragments on a wide host range plasmid or by the binary vector strategy produced transgenic tobacco plants.[52]

More detailed understanding of T-DNA organization in transgenic plants is important for two reasons. First, patterns of T-DNA sequence organization in transgenic plants are the end products of T-DNA transfer, and an integration process must be accounted for any complete description of the mechanisms by which these processes occur. Second, T-DNA structures might influence the genetic stability of introduced genes in transgenic plants. Jorgensen et al.[109] reported an extensive analysis of T-DNA organization in parallel transgenic tomato plants. The analysis revealed the previously unrecognized fact that inverted repeats are the predominant T-DNA organization pattern.

It is not clear in many cases whether foreign genes that have been added to a plant genome through *Agrobacterium*-mediated transformations would be maintained with high degree of meiotic stability required for commercial use. Relevant data are not available for several tobacco transformants produced by direct gene transfer. In most of the *Agrobacterium*-mediated transformations genetic instability has been observed.[110] Mueller et al.[111] transformed *N. tabacum* with a chimeric gene that consisted of a nopaline synthase promotor and coding sequence of Tn5 neomycin phosphotransferase II gene *(npt)* specifying resistance to kanamycin. The gene transfer was accomplished with the aid of *A. tumefaciens* and a novel binary vector, pC22. Mueller et al.[111] reported most of the transformants having monogenic inheritance of the kanamycin resistance and the others being two or three locus transformants. The integrated genes were found to be transmitted to progeny with high degree of meiotic stability. It was also suggested that cointegration should be avoided and nonhomologous integration should be preferred to achieve meiotically stable transformants.

Genetic transformation of oilseed rape *(Brassica napus)* was reported by Ooms et al.[112] They regenerated plants from roots induced by *A. rhizogenes* and described phenotypic alterations which indicated the presence of Ri (root-inducing) T-DNA. Direct evidence and the utilization of Ri T-DNA for transformation have been presented only recently.[113] Genetically transformed rapeseed roots were obtained by *in vitro* inoculation of *A. rhizogenes* and transformed cells raised to plants. The resulting plants synthesized opines, contained Ri T-DNA, and transmitted the transformed phenotype to their progeny. Longitudinal stem sections of *B. napus* were cultivated with *A. tumefaciens* A 208-E, carrying disarmed plasmid TiT37-SE and the binary vector p-MON 809.[114] The vector contained a mouse mutant dihydrofolate reductase coding sequence derived by the cauliflower mosaic 35S promotor. The transgenic plants were phenotypically normal and resistant to methotrexate. This gene was transmitted and expressed in seed progeny as a dominant Mendelian trait. A reproducible system to produce transgenic *B. napus* plants has been developed through *A. tumefaciens* containing a disarmed tumor-including plasmid carrying a bacterial chimeric gene encoding for kanamycin resistance. The regenerated transgenic plants were normal and fertile. T-DNA was stably integrated and transmitted to the progeny in a Mendelian fashion. Genetic transformation of flax *(Linum usitatissimum)* has also been achieved.[115] Fertile and transformed shoots were most easily obtained from kanamycin resistance callus.

Factors involved in the optimization of recovery of plant cells have been little studied and the importance of the organized tissue in the regeneration process has not been fully addressed. Chyi and Phillips[110] suggested that a successful *Agrobacterium*-mediated transformation system involving a disarmed Ti plasmid is composed of two stages: transformation of cells and recovery of transformed plants. A tissue transformation system with 34% efficiency was developed using stem segments of the interspecific hybrid *Lycopersicon esculentum* × *L. pennellii*. This transformation system emphasized three factors favoring the recovery of transformed plants: (1) promotion of cell division activity at the inoculation site with the kinetin in the incubation medium, (2) promotion of adventitious regeneration, but initiation by using organized tissue explants in culture, and (3) application of selection at the shoot development stage of adventitious regeneration.

A. rhizogenes has also been used to mediate gene transfer in *Lycopersicon* through a binary vector of *Agrobacter-lycoperiscon esculentum.*[116] Using the integrated T-DNA as a molecular marker transgenic plants of *L. esculentum* were produced. These transgenic plants contained one or two copies of the integrated vector T-DNA. The kanamycin trait in the progeny of most of the transgenic plants segregated in the ratio of 3:1, suggesting that vector T-DNAs were integrated at a single site on a tomato chromosome.

Potato cells have been transformed using *Agrobacterium* and plant regenerated with specific DNA.[112] An et al.[43,117] have reported the isolation of potato shoots which were resistant to kanamycin. The shoots were transformed with DNA from a binary vector plasmid. DNA binary vectors have also been characterized by Ooms et al.[118] and they described a technically simple mixed-infection approach for isolation of a transformed derivative of a potato cultivar which was morphologically and cytologically normal.

What new genes would we want to introduce to important crop plants? The seed proteins of such staple crops as wheat, beans, and corn are often deficient in particular amino acids; corn is especially poor in lysine, thereonine, and methionine. Genes encoding the major seed storage proteins have already been identified and cloned, and it should not prove difficult to use recombinant DNA technology to insert nucleotide sequences encoding the deficient amino acids into the correct reading frames. The structure of seed proteins (which are broken down to feed the germinating plant embryo) are presumably not so critical that insertion of a few new amino acids would interfere with function. Since abundant seed proteins are generally encoded by large multigene families, the genetically engineered seed protein genes will have to be introduced into the plant genome in relatively high copy number in order to have a significant effect on the nutritional characteristics of the seed. This presents no technical problem because T-DNA can integrate in the form of long, tandem repeats. However, because T-DNA integrates randomly into the plant genome (as to P elements in *Drosophila* and microinjected DNA in fertilized mouse eggs), one initial worry was the position effects which might prevent genes on T-DNA vectors from being properly regulated in their new chromosomal environment. Fortunately, as is the case for transgenic flies and mice, the immediate chromosomal environment usually has little effect on regulation of the transgene. For example, soybean seed storage proteins are subject to proper developmental regulation when introduced into tobacco plants. The transgenes are expressed in seeds, but not in roots, shoots, or leaves.

Gene-encoding maize seed storage proteins of 19 and 15 kDa have been inserted at various sites in the T-DNA of the Ti plasmid pTiA6.[119] *A. tumefaciens* strains harboring these plasmids were used to incite tumors on sunflower stem sections. Some of the resulting tumors were found to contain mRNAs transcribed from zein genes. Vanstogteren et al.[67] developed a process of incorporation of foreign DNA into the genome of monocotyledonous plants through *Agrobacterium* strains containing one or more Ti plasmids.

D. GENETIC TRANSFORMATION OF MARKER GENES THROUGH *AGROBACTERIUM* FOR SELECTION OF HYBRID SEEDS

The primary barrier to the use of Mendelian male-sterile (ms) genes in hybrid seed production is that pure stands of ms plants are difficult to obtain. This is because ms (ms/ms) plants must be propagated by backcrossing to male-fertile (+/ms) plants, yielding a mixed population of ms and +/ms plants in 1:1 ratio. Consequently, pure stands of ms plants can only be obtained by means of precise and expensive procedures for removing +/ms segregants.

Singleton and Jones[120] first presented a solution to this problem suggesting that a seed color gene could be used as a marker for a linked ms gene. By using such a marker, fertile plants could be removed before planting to produce a nearly pure stand of ms plants. Unfortunately, sufficient tight linkage has not been found between suitable marker and ms genes; thus, this approach has never been implemented on a significant scale despite many investigations into various markers and linkages.[121-124]

Gene transformation through *Agrobacterium* can be used to synthesize tight linkage which can allow the introduction of marker gene to a distinct genetic location in each of a large number of individual plants. Standard linkage analysis of the progeny of independent transgenic plants can then be used to identify a sufficiently tight linkage between the marker and the ms genes. A beginning has already been made in this direction and research in coming years will harvest the fruits of this technology.[125]

There are two principal requirements for implementation of this proposal: a genetic transformation system that permits approximately random introduction of the marker to the genome and a suitable marker gene.[125] The ease with which transgenic plants can be produced has made the *Agrobacterium* genetic transformation system a best choice for efficient production of large numbers of normal flowering transgenic plants carrying introduced genes that are expressed and inherited stably.[49,126-128] To determine whether the locations of genes introduced via *Agrobacterium* are random, Jorgensen[125] performed linkage study of the marker genes introduced to ten independent transgenic tomatos (transgenote) plants. Each transgenote carried the introduced gene insertions at a different location, indicating that gene insertion via *Agrobacterium* in tomato-like *Petunia*[129] occurred at random. Jorgensen[125] also determined the gene location of nopaline synthase (NoS) in relation ms10 and ms2 linkage genes. The results indicated that the NoS gene was tightly linked. This indicates that genetic transformation via *Agrobacterium* can be used to synthesize tight, novel linkages between markers and ms genes. However, the NoS gene cannot be used as a marker because a principal requirement of a proposed system for a suitable marker phenotype is that it should be expressed in the seed or seedling to permit effective removal of fertiles. Two main classes of markers have been suggested by Jorgensen:[125] (1) seed or seedling color and (2) dominant chemical sensitivity. A color marker will permit removal of fertiles prior to planting or transplanting. A chemical sensitivity marker will permit destruction of fertile seedlings either in the field or in the greenhouse prior to transplanting in the field. Though unusual, such genes exist that direct the conversion of noninhibitory chemical to an inhibitory chemical. Such a bacterial gene proposed for this purpose by Jorgensen[125] is indoleacetamide hydrolase (IaaH). IaaH determines the final enzymatic step in bacterial auxin (IAA) biosynthesis and is responsible for the conversion of indoleacetamide (IAM) to IAA. It has been shown that tobacco plants carrying this gene are normal except when treated with the IAM analogue, naphthalene acetamide (NAM).[125] NAM is efficiently converted by the IaaH gene product to naphthalene-acetic acid resulting in severe inhibition of the growth of the plants. Jorgensen[125] showed that three populations of tobacco growing in the presence of NAMs will be formed: (1) a normal wild-type tobacco population that is unaffected by NAM; (2) a population homozygous for IaaH that is uniformly inhibited by NAM; and (3) a backcross population of heterozygotes (IaaH/) and homozygotes ( )/( ) (i.e., wild type) segregating in a 1:1 ratio. This segregating population is equivalent to populations that in the proposed hybrid seed production system would be grown in maintainer fields and that would segregate 1:1 for ms and +/ms plants. Jorgensen[125] explained the manner in which a synthetic marker-ms linkage would be used in hybrid seed production with an IaaH-ms linkage system as an example. The maintainer field would consist of two genotypes planted in separate rows, a ms line (ms+/m+) and a +/ms, NAM-sensitive maintainer line [+ IaaH/ms ( )] with heterozygotes at both the ms locus and the marker locus. The maintainer is produced by crossing the ms line with a fertile, a NAM-sensitive homozygous (+ IaaH/+IaaH) line. These lines would otherwise be isogenic. Seed would be harvested only from ms rows and when replanted would yield 1:1 mixture of fertile and sterile plants. Fertile plants carrying the marker phenotype would be removed and could also be used as the maintainer if sterile recombinants were removed. Plants without the marker phenotype are sterile with the exception of infrequent fertile recombinants, and would serve as the female parent in the hybrid seed production when interplanted with the intended male parent.

An obvious requirement in the proposed system for hybrid seed production is that suitably stable and female-fertile ms genes must be available. This should not limit implementation of the method because many such genes exist in several plants. At least 20 such stable genes have been identified in corn, barley, and tomato.[121] This method proposed by Jorgensen[125] should be applicable to any hybrid breeding program given that a suitable marker and a suitable transformation system can be identified in that crop. The *Agrobacterium* transformation system is suitable for the synthesis of the required linkage in a number of dicotyledonous species. In species in which the *Agrobacterium* transformation system proves inefficient or impossible, an alternate transformation system would need to be found. The IaaH gene should suffice as a suitable marker in solanaceous and cucurbit crops, although its usefulness remains to be shown in practice. For other crops suitable markers still need to be identified.

E. VECTOR DESIGN TO FACILITATE THE EXPRESSION OF FOREIGN GENES

To test whether virulence mutants of *A. tumefaciens* are capable of promoting T-DNA transfer into plant cells, a tandem array of CaMV DNA was cloned between T-region border sequence on a wide-host-range plasmid and introduced into various virulence mutants.[130] The resulting strains were used to infect *B. napus* cv. Just Right. Mutants in the loci *vir* A, B, and G, which were avirulent on turnip, failed to induce viral symptoms. Mutants in *vir* E, C, and F, which induced, respectively, no, small, and normal tumors on turnip, all induced viral symptoms.

Schardl et al.[131] have built a series of vectors designed to facilitate the expression of foreign genes in plants. These vectors carry plant gene expression cassettes containing either a highly active, constitutive promotor (CaMV 35 promotor) or a light regulated promotor (the pea *rbc* S-E9 promotor), the M13umM20 MCS, and plant polyadenylation signal also derived from the *rbc* S-E 9 gene. These expression cassettes are carried in contexts that facilitate direct DNA uptake (as pBR322-derived replicon) or *A. tumefaciens*-mediated gene transfer into plant cells. As such, they can be adapted to a wide range of plant cell systems.

Several promotors carried by the expression cassettes were characterized in transgenic plants.[131] The CaMV 35S promotor is known to function in a number of different plant

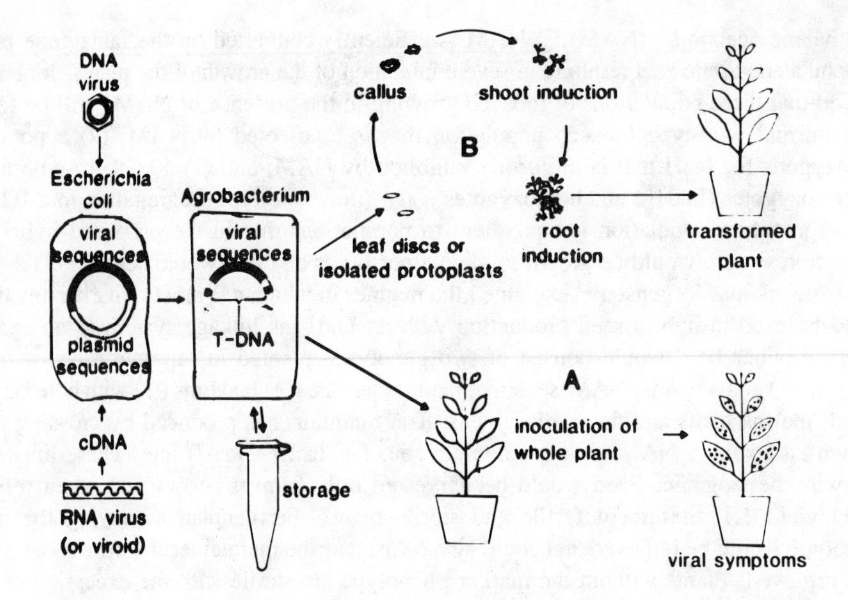

FIGURE 9. Agroinfection: transfer of viral or viroidal to plants via *Agrobacterium*. *Agrobacterium* can conveniently be stored at −70°C, and used for inoculation of whole plants (A) or used to obtain plants transgenic for viral sequences (B). (From Grimsley, N. and Bisaro, D., *Plant DNA Infectious Agents,* Hohn, T. and Schell, J., Eds., Springer-Verlag, New York, 1987, 87. With permission.)

species, including tobacco, petunia, maize, wheat, sorghum, and rice.[49,96,132-134] In tobacco, the promotor initiates transcripts at the same site as in CaMV-infected turnip.[133] This promoter has been shown to be constitutive and highly active and is thus well suited for high-level predictable initiation of transcription of chimeric genes in transformed plants.

The *rbc* S-E9 promotor has been shown to function in a light-regulated manner in transgenic tobacco and petunia.[135] In such instances, it directs transcription initiation at a site identical to that seen in pea.[131] Schardl et al.[131] assembled a series of modular expression cassettes and Ti plasmid binary vectors that allowed rapid cloning and expression of foreign genes in plants. The pKYLX vector series are currently in use to express a wide variety of foreign genes.[136-139]

F. AGROINFECTION

Development of *A. tumefaciens* Ti plasmid vectors permits the delivery of heritable foreign DNA into plant cells and regeneration of transformed cells to intact plants.[48,126] Requirements for T-DNA transfer are functional virulence regions on Ti plasmid and the bacterial chromosome as well as short (25 bp long), imperfect, direct repeat sequences which delineate the T-DNA.[140,141] This unique plant transformation occurs in many dicotyledonous as well as monocotyledonous plants. However, the presence of T-DNA in the recipient plant has not been confirmed in most monocotyledonous genera, including the grasses; in fact, the monocotyledons include several major food crops. Thus, a need was always felt to develop a system through *Agrobacterium*-T-DNA to transfer the foreign genes to monocotyledons. The reports on the regeneration of several cereals and legumes from tissue culture are encouraging.[142]

Grimsley and associates reported a system called agroinfection to transfer foreign genes to monocotyledons.[72,143] Agroinfection was broadly defined as the use of *Agrobacterium* to introduce viral or viroidal genetic information into plants. Grimsley and Bisaro[143] suggested the methods through which agroinfection can be performed (Figure 9). The first route (A) is by mechanical inoculation of intact plants with bacterial suspensions and does not require the regeneration of transformed plants. In the second route (B) plant cells are incubated with bacteria. In this case the use of a nononcogenic *Agrobacterium* vectors allows the plant species to contain viral genome in every cell regenerated from transformed cell.

1. *Agrobacterium* as an Organism for the Experimental Storage and Transmission of Plant Viruses

Agroinfection offers an alternative to the propagation of viruses by natural vectors. The procedure generally requires the placement of an oligomer with more than one complete unit of viral genetic information into T-DNA of a strain of *Agrobacterium* in such a way that infectious viral molecules can be produced by recombination, replication, or transcription during or after the process of plant transformation. Preparation of viral DNA oligomer is done by routine cloning techniques using a combination of *in vitro* nucleic acid reactions and *in vivo* bacterial manipulations. In the case of RNA virus, synthesis of cDNA is required from viral RNA prior to cloning. Suspensions of bacteria carrying T-DNA with inserted viral DNA are applied to parts of intact plants by wounding or abrasion in much the same way as mechanical inoculation of plants with virus preparations.

a. Storage Efficiency and Flexibility

Genetically engineered strains of *Agrobacterium* containing viral DNA are stored at −70°C until further use. Before inoculating such bacteria to plants, a large amount of culture is raised from the stock already stored at low temperature. The handling and maintenance of bacterial culture for the transfer of viral genome to plants is much easier as compared to insect vectors.

The transfer of viral genome through *Agrobacterium* T-DNA is highly efficient because it involves a natural system delivery of DNA into the plants. As described earlier, strains of *Agrobacterium* containing oligomers of complete viral or viroidal genomes in T-DNA are used to inoculate the plants. Grimsley and Bisaro[143] described three prominent systems that have been used to transfer the viral genome through T-DNA plasmids. These systems are (1) CaMV,[79] (2) potato spindle tuber viroid (PSTV),[144] and (3) MSV.[77]

CaMV as described earlier is an 8-kb-long double-stranded DNA virus which is transmitted in nature by aphids. It replicates via reverse transcription of an RNA molecule which spans the entire viral genome and bears terminal direct repeats.[64,71,145,146] The unit length genomes of viral DNA-cloned plasmid vectors are not infectious when inoculated as naked DNA onto the host plant leaves; oligomers of viral genomes cloned in the same way do reproducibly give rise to systemic infection when 5 to 10 μg DNA per plant is inoculated.[68,69,72] Thus, relatively large amounts of plasmid DNA must be prepared for inoculation following a growth of isolate of interest in *E. coli*. In order to test the use of *Agrobacterium* for introduction of viral DNA into host plant, an oligomer comprising 1.4 genomes of CaMV (known to be infectious as naked DNA) was placed in the T-DNA of a nononcogenic Ti plasmid.[143] The resulting agrobacterial strain was used for inoculation at the crown of turnip plants (*B. compestris* var. *rapa*). Within 2 weeks of inoculation, systemic infection appeared a week earlier than in plants inoculated simultaneously with naked DNA. Even the plants agroinfected on leaves showed more pronounced symptoms on the primary inoculated leaves.

Agroinfection by leaf inoculation was extremely efficient, since inoculation of only 10^4 *Agrobacterium* cells containing approximately 0.1 pg of CaMV DNA was sufficient to produce systemic infection. As a control the same viral oligomer was placed in a broad-host-range plasmid lacking T-DNA borders, and plants were inoculated following introduction of this construct into *Agrobacterium*. No viral symptoms were observed in recipient plants, showing that infection did not occur by lysis of bacterial cells and release of DNA. Grimsley and Bisaro[143] proved from these experiments that in terms of the amount of DNA

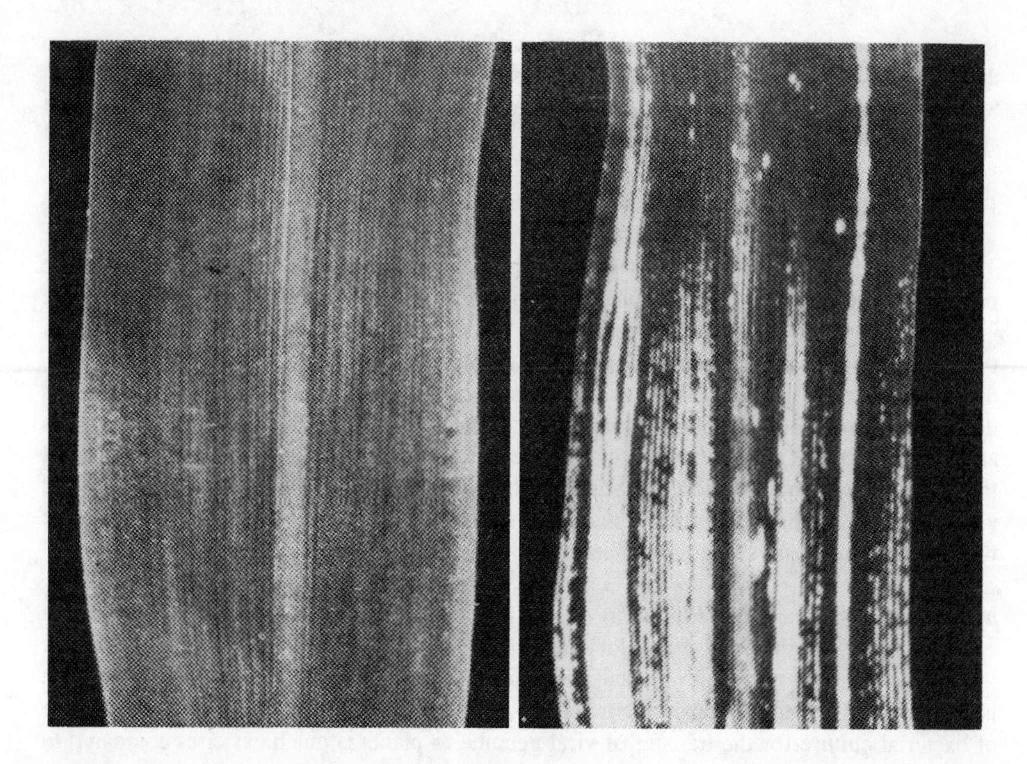

FIGURE 10. Infected and uninfected leaves of maize. Photographs were taken 12 d after inoculation with (left) C58 (pTiC58, pEAP37) and (right) C58 (pTiC58, pEA2). (From Grimsley, N., Hohn, T., Davies, J. W., and Hohn, B., *Nature,* 325, 177, 1987. With permission.)

required to produce infection, agroinfection (DNA in *Agrobacterium*) is several times more efficient than inoculation of leaves with naked DNA.

Hille et al.,[130] in an independent series of experiments using the same system (*Agrobacterium* cells containing a viral oligomer in the T-DNA were mixed with a similar strain of *Agrobacterium* which did not contain viral DNA), inoculated tubers of growing turnips just below the crown of the plant with *Agrobacterium* containing viral DNA in T-DNA. They also found agroinfection at least 100 times more efficient than transfer through crown gall tumor.

Agroinfection as pointed out by Grimsley and Bisaro[143] will also prove very helpful in situations where it is difficult or impossible to infect plants with cloned viral nucleic acids. In addition, *Agrobacterium* can colonize only dicotyledonous plants by transferring a portion of their T-DNA into the host plant cell, but most of the monocotyledonous species including the economically important plants of family Graminae are thought not to be susceptible. MSV has now been successfully transferred through agroinfection into whole maize plants by Grimsley et al.[77] (Figure 10). Mullineaux et al.[147] found it difficult to transfer directly viral DNA or cloned DNA into maize, though viral particle has been suggested to be transferred to maize plant through electroendosmosis.[148] The drawback associated with this technique is that cloned viral DNA is not infectious, making it difficult to study viral DNA manipulations. Transfer of MSV DNA via *Agrobacterium* prompted Grimsley et al.[77] to suggest that agroinfection can permit the analysis of *in vitro*-produced mutant strains of MSV and allow the potential of MSV as a vector for foreign DNA to be evaluated.

Viroids are small (250 to 370 nucleotides) circular RNA molecules which can replicate autonomously in host plants. Their replication cycle probably involves both tandem linear and circular DNA intermediates.[149,150] Full-length cDNA copies of PSTV have been cloned in plasmid vectors,[151,152] and whereas monomeric PSTV cDNA has little or no infectivity when inoculated as naked DNA, tandem linear cDNA copies are infectious. A variety of PSTV constructions were made, comprising monomeric or multimeric genomes (either not adjacent to a plant promotor or downstream of a plant promotor), and were introduced to whole tomato plants by agroinfection.[153] In agreement with the experiments done with naked DNA, the efficiency of agroinfection was shown to increase with increasing numbers of tandem PSTV units.

b. Viral Genome Release from T-DNA

Viral DNA, once transferred to a plant genome through T-DNA, is released into the plant cells followed by its systemic infection. The mechanisms of release and subsequent spread have not been studied in detail. Grimsley and Bisaro[143] discussed the possible mechanism of viral release from T-DNA transfer; this could either be used directly as a substrate for the production of an infectious viral molecule (as in the case of [+] strand of single-stranded DNA) or synthesis of complementary strand could occur.[7,153] In any case, escape can occur by the genome inserted into the vector, or by a replication or transcription mechanism in which one strand of the genome is initially copied. Grimsley et al.[79] investigated the mechanism of transfer in detail using CaMV. CaMV oligomer used in the transfer contains one genome of CaMV isolate "CM4-184" (CaMV CM4-184) and 0.4 genome of CaMV isolate "S" (CaMV S). These two isolates showed differences in their DNA sequences and in certain restriction enzyme sites. Genetic analysis of the progeny viruses produced by agroinfection made it possible to analyze the mechanisms of infectious viral escape from T-DNA.[79] The 1.4 kb viral genomes were arranged in such a way that the long transcript thought to be an intermediate in viral replication by reverse transcription could be produced with its 5′ end starting in one viral strain and its 3′ end finishing in the other. If there is reverse transcriptional replication, this would give rise to one kind of virus molecule, whereas homologous DNA recombination is expected to produce progeny viruses with a variety of genetic constitutions. The latter event depends on the place where recombination takes place. Grimsley et al.[79] found that 96% of the progeny viruses in agroinfected plants had arisen by reverse transcription. The remaining 10% were thought to arise due to homologous DNA recombination. These authors were not able to gather data to determine whether the escape of the viruses occurred before or after integration of the T-DNA into the host plant cell.

Escape of other viruses like MSV and PSTV from T-DNA is not clear. Gardner and Knauf[153] suggested that escape of PSTV may involve an RNA intermediate which is processed in the same way to yield an infectious PSTV genome. When a monomer of PSTV cDNA was placed adjacent to a CaMV 35S promotor, agroinfection was successful only if the PSTV genome was in orientation required to produce a (+) strand of RNA, but not if the CaMV promotor directed synthesis of a PSTV (−) strand.[143] Thus, PSTV might also escape from the transforming DNA by an RNA polymerase II-derived transcript, which is subsequently processed to produce an RNA replicative intermediate, and this then undergoes RNA-RNA replication. On the other hand, all constucts which contained multimers of the PSTV genome were infectious and infectivity was shown to increase with increasing numbers of PSTV units. This was true even of constructs in which a plant promotor was absent, as well as those in which PSTV (−) strand RNA would be produced. Thus, it appears that in this case infectious RNA results from (+) transcript produced by "background" transcription from adjacent T-DNA promotors in the plant. This, however, does not explain the noninfectious nature of the (−) strand monomer construct. An alternative explanation for the escape of PSTV as suggested by Grimsley and Bisaro[143] is that the homologous DNA recombination event occurs first to produce a circular DNA molecule which is then transcribed.

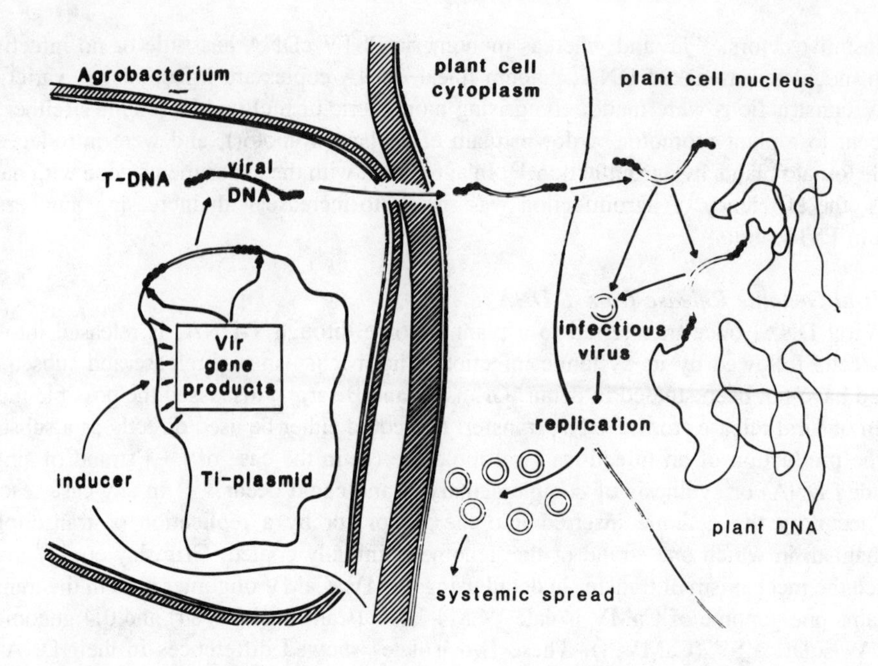

FIGURE 11. Systemic viral infection as an assay for T-DNA transfer. Agroinfection proves a sensitive assay, since theoretically only one infectious molecule is required, and it may not require integration of the transferred DNA. (From Grimsley, N. and Bisaro, D., *Plant DNA Infectious Agents,* Hohn, T. and Schell, J., Eds., Springer-Verlag, New York, 1987, 87. With permission.)

c. T-DNA Transfer and Its Analysis

Replication of viral molecules in recipient cells can be used as a sensitive essay for the transfer of the *Agrobacterium* T-DNA. Unlike other assays for T-DNA transfer, such as proliferation of crown gall tumor tissue or growth of antibiotic-resistant cells, integration of the T-DNA into the recipients nuclear genome may not be required for escape of replicating viral progeny[11,143] (Figure 11). Early events in the T-DNA transfer process can thus be studied using replicating viral sequences as a marker, and this approach may also permit the rescue and study of transfer intermediates. In addition, the role of T-DNA border sequences and Ti-plasmid-encoded *vir* genes, which are required for normal tumor formation and are known to play important roles in the transfer of the T-DNA from *Agrobacterium* cells to plant cells,[140] have been the first aspect of this process to be analyzed by agroinfection.

Gardner and Knauf[153] placed a PSTV cDNA trimer in *cis* with left and/or right T-DNA border sequences in a broad-host-range plasmid and this plasmid was introduced into a variety of *Agrobacterium* strains. In *trans* with a Ti plasmid bearing intact *vir* genes, systemic infection of agroinfected tomato plants occurred when a left border or right or a left plus right border was present, but not in the absence of a border sequence. A right border was more efficient in agroinfection than a left border as judged by the time of appearance of infection symptoms. The observation is in accordance with the recent discovery of an ''overdrive'' sequence in dose proximity to the right border, which was present in constructions used. The ''overdrive'' sequence allows expeditious transfer of the T-DNA.[20] Surprisingly, PSTV cDNA that was located outside the T-DNA borders was also transferred to the plant.

The binary vector containing a right border and the PSTV trimer were also tested for agroinfection in pairwise combinations in *trans* with Ti plasmids bearing a variety of *vir* gene mutations. Whereas *Agrobacterium* strains with mutations in the virulence loci A, B, C, and D could not agroinfect tomatoes, those carrying a mutation in vir E could do so.

Similarly, A CaMV oligomer in a binary vector was also tested in *trans* in a variety of Ti plasmid vir mutant strain.[72,130] In this system, using turnip as the host plant, a similar result was obtained. *Agrobacterium* strains carrying the vir E mutation could not induce tumors but were effective in agroinfection, however, at a reduced efficiency compared to wild-type strains.

The agroinfection of whole maize plants with MSV conclusively demonstrates that *Agrobacterium* can transfer T-DNA to graminaceous monocot.[79] This also provides an invaluable substitute for the insect vector. Grimsley et al.[77] thus developed a sensitive test for T-DNA transfer to maize, which requires the delivery of only one infectious MSV molecule to a recipient plant. A dimer of the MSV genome in two constructs, differing only in the orientation of the MSV sequences with respect to T-DNA borders, was also tested in an attempt to use this system to analyze the mechanism of T-DNA transfer. Either a (+) or a (−) strand of the virus would therefore have been transferred, but since both orientations were equally agroinfectious, no information about the single- or double-stranded nature of the T-DNA intermediate could be gained.

2. Transformation of Plant Cells with Viral Genetic Information

In cases where regeneration is possible, the *Agrobacterium*-mediated transformation of cultured plants cells (such as protoplasts or leaf disks) with viral genomes may be used to obtain transgenic plants.[49] Further, the agroinfection approach has the potential to facilitate studies of viruses which may be limited to particular tissues. Through agroinfection, it is also possible to use *Agrobacterium* vectors to deliver fragments of viral genomes, or more specifically selected viral genes, for expression in transgenic plants. The study of the behavior of viruses in nonhost plants, where virus movement is normally restricted, can also be done easily through agroinfection. This technique acquires a particular interest for the mutational analysis of viral genomes, since mutant viral sequences will be present in every cell independently of whether they can generate active virus particles capable of normal systemic spread throughout the various tissues of the host or not.

a. Transient Expression

The mechanism in plant gene expression can be easily established by performing assay of transient expression of transforming DNA in isolated or regenerating plant protoplasts. The major limitation in this type of experiment conducted so far is that the amount of nucleic acid which is taken up and expressed in the recipient cell is very small. Even very sensitive tests for expression, such as the acetylation of radioactive chloramphenicol[154] or the expression of luciferase,[155] are not enough to detect the viral genes. The work has been made several times easier through agroinfection. Agroinfection allows the amplification of the incoming genetic information to occur by replication of viral sequences. This increases the quantity of assayable gene products several times. In agroinfection it is not necessary to prepare isolate protoplasts for the transfer of foreign genes, because agroinfection can transfer its T-DNA following the attachment with the recipient cell. This is particularly important in culture with cell suspensions which do not give a high planting efficiency after protoplasting.

b. Expression of Viral Genes in Host and Nonhost Plants

Viral genomes constitute one of the most readily available sources of plant gene regulatory sequences. Recent studies have revealed that the host range of *Agrobacterium* is much broader than that of individual kinds of viruses. Thus, it is possible to agroinfect even nonhost plants with viral sequences to understand the mechanism of expression of viral genes in different environments. The development of selectable marker genes for plants has been greatly facilitated by their use, since viral promoters are usually strong and constitutive.

When such CaMV regulatory sequences are appropriately linked to the coding sequence of bacterial neomycin phosphotransferase, for example, chimeric genes giving high levels of resistance to kanamycin in transgenic plants can be obtained.[37,95,96]

Shewmaker et al.[156] placed a CaMV genome in the T-DNA of a wild-type Ti plasmid, and the resulting strain of *Agrobacterium* was used to induce galls on a variety of host or nonhost plants. It was found from Northern blot analysis that the two previously identified[70,132,157] promoters were active when integrated into the nuclear genome of a variety of plants. These promoters directed transcription more efficiently in host than in nonhost plants. However, excision and autonomous replication of the CaMV DNA was not observed, but this is not surprising since a monomer broken by nonviral DNA in two positions (at the cloning site and at another site by an introduced selective marker) was used. It was not possible to determine the proportion of the cells transformed because the gall tissue was used for the analysis.

Agroinfection has been useful in studying the DNA sequences required for the activity of the CaMV 35S RNA promoter in (nonhost plant) transgenic plants.[132] Using a CaMV 35S RNA promoter-human growth hormone chimeric construction, a series of deletions in the 5′ end of the promoter was made and transferred into tobacco plants. Whereas sequences extending to −46 were sufficient for accurate transcription, sequences between −46 and −105 increased the level of transcription. No tissue specificity of expression was shown by the promoter.

The expression of TMV CP by a CaMV promoter in regenerated tobacco plants has also been reported. Using a TMV cDNA clone, a chimeric CaMV 35S promoter-TMV CP gene was made and leaf pieces were agroinfected.[37] Transgenic plants were selected on the basis of resistance to kanamycin, encoded by a CaMV 35S promoter—Tn5 hybrid gene. The level of coat protein synthesized from the nuclear gene was 6000 times less than that observed in acute viral infection. However, somewhat higher levels (500- to 100-fold) were obtained in independent experiments with a similar construct.[88] With the help of the same method *Brassica* tissue (a nonhost) was also agroinfected with the above-mentioned oligomers of PSTV which were agroinfectious on tomato,[144] but no PSTV RNA could be found in crown galls, suggesting that no replication of PSTV occurs in turnip when the viroid is introduced to the chromosome in the form of multimeric DNA.

c. Transgenic Plants Containing Oligomers of Viral Genomes

Plants transgenic for an oligomer of one component of a multicomponent viral genome which is by itself unable to produce infectious viruses can be agroinfected by another component of the virus. Grimsley and Bisaro[143] suggested three methods: (1) by inoculation or agroinfection of the plant with another component, (2) by transformation, producing plants transgenic for both components, or (3) by crossing to plants transgenic for another component. In all cases, the assay for functional complementation would be the replication of a virus containing all the components. Tomato golden mosaic virus (TGMV) belongs to the geminivirus group, whose members are characterized by their distinctive twin icosahedral particle morphology and genomes of circular single-stranded DNA.[76] The genome of TGMV is divided between two such DNA molecules, designated A and B, both of which are about 2.5 kb in length.[158,159] Viral DNA replication is thought to occur through the circular double-stranded replicative form of DNAs found in the nuclei of infected cells,[160,161] and the cloned double-stranded DNA components are infectious when both are released from the cloning vector and inoculated together on tobacco leaves.[162,163] Neither component alone produces symptoms or replicates to a detectable extent under these conditions.

Rogers et al.[57] separately introduced viral genome components into petunia cells using a Ti transformation-regeneration system. In these experiments both single and tandemly repeated inserts of each TGMV component in the Ti plasmid vector were constructed. These derivative plasmids were introduced into agrobacteria which were subsequently used in the leaf disk procedure[49] to obtain transformed regenerated petunia plants. All transformed plants were normal in morphology and most had only one or a few T-DNA insertions at a single chromosomal site. The transformed plants also flowered normally and were used in all possible combinations of crosses.[57] All progeny seedlings were normal in appearance with the exception of those which resulted from crosses of tandem A by B plants. One quarter of these seedlings displayed symptomatic A + B progeny. Plants revealed the presence of free, replicating single- and double-stranded viral DNA forms identical to those previously identified in TGMV-infected tobacco.[160] Both A and B components were represented in these DNA forms. Further, a similar analysis of nonsymptomatic parents showed that free viral DNA was also present in plants containing tandem insertions of the A component alone. These DNAs hybridized only with A-specific probes. Free viral DNA was not detected in plants transformed with a monomeric A component, or in any plants transgenic only for the monomeric or tandemly repeated B component.

Several conclusions can be drawn regarding the organization of the TGMV genome, in particular, and agroinfection, in general.[143] The results indicate that the TGMV A genome component encodes all viral functions necessary for the replication of viral DNA and, since plants which carry tandem A insertions do not show symptoms, that the B component must encode some function(s) essential to their production. Thus, the B component, itself incapable of replication, was shown to be activated by the independently replicating A component in progeny that results from the crossing of transformed A and B parent. Systemic infection also resulted at high frequency when transformed B plants were inoculated with free A component DNA.[143] These results clearly demonstrate that viral DNA can escape from the T-DNA after chromosomal integration. Free viral DNA could be detected in the progeny of agroinfected plants, while during natural infection TGMV is not known to be seed transmitted. Information concerning the mechanism by which TGMV DNA is released from the chromosomally integrated T-DNA is not yet available, although it might occur either by single-strand replication or by homologous recombination between tandemly repeated copies of the viral genome. The first alternative is analogous to the transcriptional mechanism observed in the case of CaMV.[79] If this method of escape proves general, it would be reasonable to suppose that vectors derived from viral replicons in T-DNA could indefinitely generate replicating extrachromosomal DNA in the host cell, since replication or transcription should not alter the chromosomal viral sequences. This could have important implications for the development of genetically transmissible high-copy-number plant expression vectors.

A variety of mutations can be produced in viral genomes by *in vitro* manipulations, but the majority of these are lethal, precluding subsequent *in vivo* analysis.[64,164,165] Agroinfection would allow integration of the mutated viral DNA, in oligomeric form, into the host plant nuclear DNA, as outlined schematically for caulimoviruses and geminiviruses in Figure 12. In this situation part of the viral life cycle may occur within each plant cell, allowing deductions to be made about the role of particular viral genes or nucleic acid sequences. This method offers an important advantage over the use of cultured plant cells for mutant analysis, since viral genes which affect symptom generation or some other aspect of the virus-host interaction can be studied. At present there are no reports on the use of this approach.

A combination of the *Agrobacterium*-based transformation system with viral vector systems can be made which draws together advantages of both types of systems.[143] A viral replicon that contains a cargo DNA of interest when integrated into the plant nuclear genome as a kind of "provirus" can provide a substrate for escape of free, extra, chromosomal DNAs carrying cargo DNA (Figure 12). Suitable constructions should allow the expression of viral and cargo DNA gene products to be regulated. Genetic studies outlined above should yield information concerning the viral functions required to initiate and maintain replication

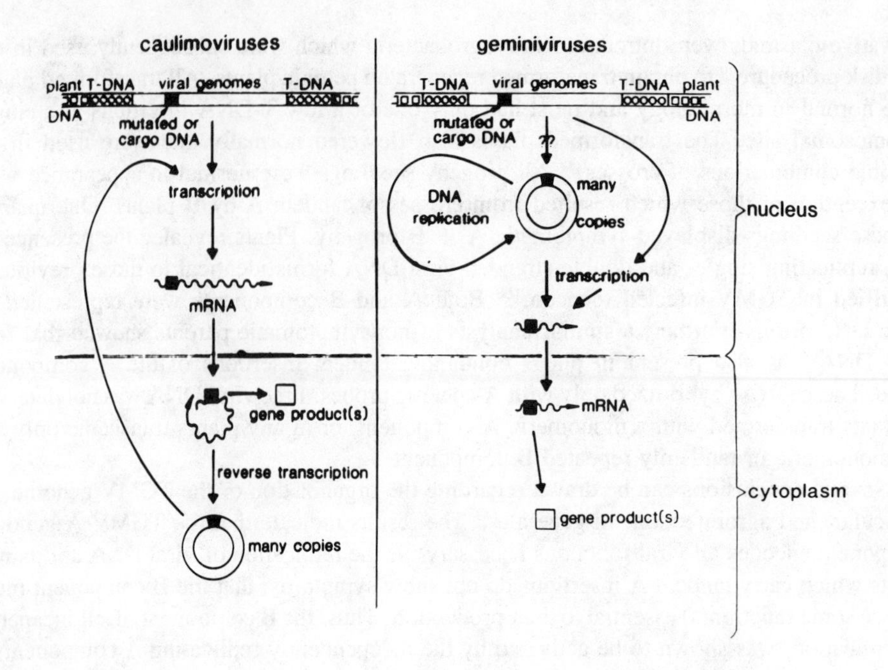

FIGURE 12. Plant transgenic for oligomers of viral genomes: analysis of viral mutant strains and expression of foreign genes. The viral functions necessary for RNA/DNA or DNA/DNA replication are incorporated into the nuclear DNA of every plant cell and may allow the production of a molecule which can replicate independently. Inclusion of appropriate control signals would allow the copy number of these molecules to be regulated, in a manner analogous to the induction of a provirus in a bacterial cell. Two systems are compared; caulimoviruses (left) are known to escape from transforming DNA by transcription/reverse transcription, whereas the mechanism by which geminiviruses (right) escape from DNA integrated in the plant genome is not yet understood. This approach might be used for analysis of *in vitro*-produced viral mutant strains which are lethal when inoculated as naked nucleic acid, or for the development of plant expression vectors. (From Grimsley, N. and Bisaro, D., *Plant DNA Infectious Agents*, Hohn, T. and Schell, J., Eds., Springer-Verlag, New York, 1987, 87. With permission.)

of the free nucleic acid sequences, and this knowledge would be of use in the development of these vector systems. While the nuclear DNA copy provides for stability and meiotic transmissibility of the cargo DNA, the many free copies of the cargo DNA in the viral replicon should allow for high level expression of the cargo DNA gene product.

d. Superinfection of Plants Transgenic for Viral Sequences

Plants transgenic for a fragment of a viral genome can be ''challenged'' with a different fragment of a viral genome in many of the following ways (Figure 13): (1) by inoculation of a particular part of a whole plant, for example, with viral DNA or by aphids carrying viral particles; (2) by regeneration of plants from agroinfected cells, producing plants transgenic for both fragments; or (3) by crossing to other agroinfected plants. In those cases where two or more defective parts of a viral genome are combined, systemic spread of the virus is an assay for recombination.

By agroinfection, with sequences important in the control of viral replication, for example, or by engineering sequences to produce antisense mRNA,[166] it may be possible to regenerate plants which are resistant to superinfection by related viruses. Plants transformed with viral components may also show resistance to viral-infected tobacco plants transgenic for a CaMV 35S RNA promoter-TMV coat protein-nopaline synthase 3′ end hybrid gene construction and have recently been reported to show significantly increased resistance to infection by TMV.[165]

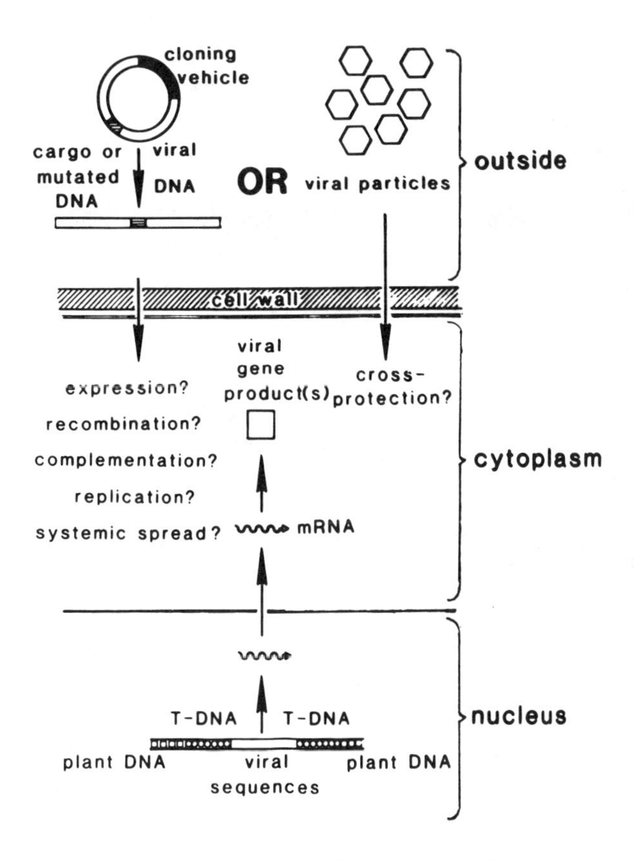

FIGURE 13. Superinfection of plants transgenic for viral sequences. A variety of interesting questions can be approached by studying the interaction between such transformed plants and a "challenge" inoculation. (From Grimsley, N. and Bisaro, D., *Plant DNA Infectious Agents,* Hohn, T. and Schell, J., Eds., Springer-Verlag, New York, 1987, 87. With permission.)

3. Satellite Defenses for Plants

Satellite RNAs are species of RNA associated with some strains of certain plant viruses, but not necessary for virus replication. Satellites are replicated upon and because they are packaged with viral genomic RNAs in the virus particle, they can accompany viruses released from diseased plant tissue to new infection sites. Although they are completely dependent on the virus for replication and transmission, their nucleotide sequence seems to be essentially unrelated to that of the viral genome. Why natural selection does not lead to their loss or elimination is not known, but it is clear that certain satellites reduce, sometimes markedly, the severity of the disease system resulting from viral infection where they are present.

Cucumber mosaic virus (CMV) particles contain three genomic single-stranded RNA molecules and a subgenomic species which acts as an mRNA for the CMV CP. Some isolates also contain a satellite RNA.[167] Recently, cDNA copies of the satellite have been made and, as monomers or tandem dimers, cloned in a plant expression vector.[168] Tobacco leaf disks were agroinfected with such constructs and transgenic plants were found to express the RNA species predicted from the positions of the promoter and the polyadenylation sites present in the expression vector, producing an RNA molecule much longer than the natural satellite RNA. Upon inoculation of parts of these transformed plants with a satellite-free isolate of CMV, satellite-length RNA molecules were found in the tissue 4 to 7 d later. Hybridization with cloned satellite sequences and passing of the CMV with its acquired satellite through another host plant confirmed its identity and biological activity. Thus, the CMV had acquired genetic information from the genome of its host plant. The authors suggest that it might be possible to protect crop plants from viral attack by transforming them with viral satellite sequences, since the presence of a satellite in a viral isolate sometimes leads to a reduction in severity of the viral symptoms.[169-171]

The ability to transform plant genomes with alien genetic material by the *Agrobacterium* Ti plasmid system has made it possible to test whether plant resistance to viruses can be increased by incorporating virus-related nucleotide sequences into the nuclear DNA of the host. Two groups have used this approach with satellite RNA associated with different viruses. Harrison et al.[172] used CMV and one of its satellites. It was established that normal satellite RNAs are produced in transgenic tobacco plants containing DNA versions of the satellite. However, the only satellite-free laboratory strain of CMV available is the one that caused mild symptoms. The new results of Harrison et al.[172] obtained with a more satellite-free strain isolated from the field (most field strains are satellite free) show that CMV replication and disease symptoms in the satellite-producing transgenic plant are attenuated. Tobacco plants transformed with a DNA copy of this satellite RNA produced large amounts of the satellite RNA on infection with a satellite-free inoculum of CMV, and, as a result, the satellite became transmissible as a component of the virus culture. CMV replication was greatly decreased and symptom development largely suppressed in these transgenic plants and their sexual progeny. The tomato aspermy virus (TAV), which is closely related to CMV, likewise induces satellite RNA synthesis accompanied by symptom suppression but with little or no decrease in TAV genome synthesis. These effects indicate that symptom suppression does not necessarily depend on a decrease in virus replication, and a genetically engineered protein by virus satellite nucleic acid is a feasible strategy for enhancing the virus resistance of crop plants.

Gerlach et al.,[173] on the other hand, carried out similar experiments with tobacco ringspot virus (TobRV, a member of the neovirus class) and its satellite (STobRV). It consists of 28-nm single-strand genomic RNAs of 4.8 and 7.2 kb. TobRV infects a wide range of dicotyledonous plants and is the causative agent of budblight disease of soybean. A small RNA which can replicate to high levels and be encapsidated by TobRV in infected plants has been found during serial passages of viral isolates. It is not required for virus propagation and has detectable sequence homology with the virus genomic RNAs. It is therefore termed the satellite RNA of TobRV. It can be considered a parasite of the virus and it ameliorates disease symptoms when present during infection. Gerlach et al.[173] reported the expression of forms of the STobRV sequence in transgenic tobacco plants. Plants which expressed full length of STobRV or its complementary sequence as RNA transcripts showed phenotypic resistance when infected with TobRV (Figure 14). This is correlated with the amplification of satelite RNA to high levels during virus infection of plants.

ACKNOWLEDGMENTS

The authors are thankful to Dr. N. Grimsley for supplying the literature on agroinfection and original figures for reproduction. Thanks are also due to Dr. W. L. Gerlach for supplying original photographs of the material for reproduction.

FIGURE 14. Symptom development on *N. tabacum* var. Samsun nontransgenic control plants and transgenic plants containing an active ST-gene construction at specific intervals after mechanical inoculation with TobRV. Leaf symptoms on nontransgenic control plants showing (A) individual necrotic local lesions, 1 week after inoculation; (B) strong systemic lesions and necrosis, 3 weeks after inoculation; (C) mottled, small systemically infected leaves, 6 weeks after inoculation; (D to F) leaf symptoms on transgenic plants containing ST+:D, local lesions, 1 week after inoculation; (E) symptomless leaf, 3 weeks after inoculation; (F) symptoms, ranging from weak systemic reactions without necrosis to symptomless, from leaves 6 weeks after inoculation; (G and H) growth of nontransgenic plants (marked C) compared with that of ST + transgenic plants (ST+), 3 weeks (G) and 6 weeks (H) after inoculation. (From Gerlach, W. L., Llewellyn, D., and Haseloff, J., *Nature,* 328, 802, 1987. With permission.)

REFERENCES

1. **Old, R. W. and Primrose, S. B.,** *Principles of Gene Manipulation,* 2nd ed., Blackwell Scientific, Oxford, 1981.
2. **Chilton, M. D., Saiki, R. K., Yadav, N., Gordon, M. P., and Quetier, R.,** T-DNA from *Agrobacterium* Ti plasmid is in the nuclear DNA fraction of crown gall tumor cells, *Proc. Natl. Acad. Sci. U.S.A.,* 77, 4060, 1980.
3. **Thomashow, M. F., Hugly, S., Buchholz, W., and Thomashow, L. S.,** Molecular basis for the auxin independent phenotype of crown gall tumor tissues, *Science,* 231, 616, 1986.
4. **Wang, K., Herrera-Estrella, L., Van Montagu, M., and Zambryski, P.,** Right 25-bp terminus of the nopaline T-DNA is essential for and determines direction of DNA transfer from *Agrobacterium* to the plant genome, *Cell,* 38, 455, 1984.
5. **Yadav, N. S., Vanderleyden, J., Bennett, D. R., Barnes, W. M., and Chilton, M. D.,** Short direct repeats flank the T-DNA on a nopaline Ti plasmid, *Proc. Natl. Acad. Sci. U.S.A.,* 79, 6322, 1982.
6. **Zambryski, P., Depicker, A., Kruger, D., and Goodman, H.,** Tumor induction by *Agrobacterium tumefaciens:* analysis of the boundaries of T-DNA, *J. Mol. Appl. Genet.,* 1, 361, 1982.
7. **Stachel, S. E. and Nester, E. W.,** The genetic and transcriptional organization of the vir region of the A6 Ti plasmid of *Agrobacterium tumefaciens, EMBO J.,* 5, 1445, 1986.
8. **Yanofsky, M., Porter, S., Young, C., Albright, L., Gordon, M., and Nester, E.,** The vir D operon of *Agrobacterium tumefaciens* encodes a site-specific endonuclease, *Cell,* 47, 471, 1986.
9. **Thomashow, L. S., Reeves, S., and Thomashow, M. F.,** Crown gall oncogenesis: evidence that a T-DNA gene from the *Agrobacterium* Ti plasmid pTiA6 encodes an enzyme that catalyzes synthesis of indoleacetic acid, *Proc. Natl. Acad. Sci. U.S.A.,* 81, 5071, 1984.
10. **Schroder, G., Waffenschmidt, S., Weiler, E. W., and Schroder, J.,** The T-region of Ti plasmids codes for an enzyme synthesizing indole-3-acetic acid, *Eur. J. Biochem.,* 138, 387, 1983.
11. **Akiyoshi, D. E., Klee, H., Amasino, R. M., Nester, E. W., and Gordon, M. P.,** T-DNA of *Agrobacterium tumefaciens* encodes an enzyme of cytokinin biosynthesis, *Proc. Natl. Acad. Sci. U.S.A.,* 81, 5994, 1984.
12. **Klee, H. J., White, F. F., Iyer, V. N., Gordon, M. P., and Nester, E. W.,** Mutational analysis of the virulence region of an *Agrobacterium tumefaciens* Ti plasmid, *J. Bacteriol.,* 153, 878, 1983.
13. **Stachel, S. E., Timmerman, B., and Zambryski, P.,** Generation of single-stranded T-DNA molecules during the initial stages of T-DNA transfer from *Agrobacterium tumefaciens* to plant cells, *Nature,* 322, 706, 1986.
14. **Horsch, R. B. and Klee, H. J.,** Rapid assay of foreign gene expression in leaf discs transformed by *Agrobacterium tumefaciens.* Role of T-DNA borders in the transfer process, *Proc. Natl. Acad. Sci. U.S.A.,* 83, 4428, 1986.
15. **Horsch, R. B., Klee, H. J., Stachel, S. E., Winans, S. C., and Nester, E. W.,** Analysis of *Agrobacterium tumefaciens* virulence mutants in leaf discs, *Proc. Natl. Acad. Sci., U.S.A.,* 83, 2571, 1986.
16. **Fromm, M. E., Taylor, L. P., and Walbot, V.,** Stable transformation of maize after gene transfer by electroporation, *Nature,* 319, 791, 1986.
17. **Hoekema, A., Hirsch, P. R., Hooykaas, P. J., and Schilperoort, R. A.,** A binary plant vector strategy based on separation of vir and T-region of the *Agrobacterium, Nature,* 303, 179, 1983.
18. **Stachel, S. E. and Zambryski, P.,** VirA and VirG control the plant-induced activation of the T-DNA transfer process of *Agrobacterium tumefaciens, Cell,* 46, 325, 1986.
19. **Stachel, S. E., Messens, E., Van Montagu, M., and Zambryski, P.,** Identification of the signal molecules produced by wounded plant cells that activate T-DNA transfer in *Agrobacterium tumefaciens, Nature,* 318, 624, 1985.
20. **Peralta, E. G., Hellmiss, R., and Ream, W.,** Overdrive, a T-DNA transmission enhancer on the *A. tumefaciens* tumor inducing plasmid, *EMBO J.,* 5, 1137, 1986.
21. **Zambryski, P., Joos, H., Genetello, C., Leemans, J., Van Montagu, M., and Schell, J.,** Ti plasmid vector for the introduction of DNA into plant cells without alteration of their normal regeneration capacity, *EMBO J.,* 2, 2143, 1983.
22. **Perani, L., Radke, S., Douglas, M. W., and Bosseat, M.,** Gene transfer methods for crop improvement: introduction of foreign DNA into plants, *Physiol. Plant.,* 68, 566, 1986.
23. **Fraley, R. T., Rogers, S. G., Horsch, R. B., Eichholtz, D., Flick, J. S., Fink, C. L., Hoffmann, N. L., and Sanders, P. R.,** The SEV system: a new disarmed Ti plasmid vector for plant transformation, *Biotechnology,* 3, 629, 1985.
24. **de Framond, A., Back, E., Chilton, W., Kayes, L., and Chilton, M. D.,** Two unlinked T-DNAs can transform the same tobacco plant cell and segregate in the E_1 generation, *Mol. Gen. Genet.,* 202, 125, 1986.

25. **Simpson, R. B., Spielmann, A., Margossian, L., and McKnight, T. D.,** A disarmed binary vector from *Agrobacterium tumefaciens* functions in *Agrobacterium rhizogenes, Plant Mol. Biol.*, 6, 403, 1986.
26. **Petit, A., Berkaloff, A., and Tempe, J.,** Multiple transformation of plant cells by *Agrobacterium* may be responsible for the complex organization of T-DNA in crown gall and hairy root, *Mol. Gen. Genet.*, 202, 388, 1986.
27. **McCormick, S., Niedermeyer, J., Fry, J., Barnason, A., Horsch, R., and Fraley, R.,** Leaf disc transformation of cultivated tomato *(L. esculentum)* using *Agrobacterium tumefaciens, Plant Cell Rep.*, 5, 81, 1986.
28. **Owens, L. D. and Cress, D. E.,** Genotypic variability of soybean response to *Agrobacterium* strains harboring Ti or Ri plasmids, *Plant Physiol.*, 77, 87, 1985.
29. **Sederoff, R., Stom, A. M., Chilton, W. S., and More, L. W.,** Gene transfer into lobolly pine by *Agrobacterium tumefaciens, Biotechnology,* 4, 647, 1986.
30. **Facciotti, D., O'Neal, J. K., Lee, S., and Shewmaker, C. K.,** Light-inducible expression of a chimaeric gene in soybean tissue transformed with *Agrobacterium, Biotechnology,* 3, 241, 1985.
31. **Klee, H., Horsch, R., and Rogers, S.,** *Agrobacterium* mediated plant transformation and its further application to plant biology, *Annu. Rev. Plant Physiol.*, 38, 467, 1987.
32. **Kyozuka, J., Hayashi, Y., and Shimamoto, K.,** High-frequency plant regeneration from rice protoplasts by novel nurse culture methods, *Mol. Gen. Genet.*, 206, 408, 1987.
33. **Vasil, V. and Vasil, I. K.,** Isolation and culture of cereal protoplasts. II. Embryogenesis and plantlet formation from protoplasts of *Pennisetum americanum, Theor. Appl. Genet.*, 56, 97, 1980.
34. **De Greve, H., Dhaese, P., Seurinck, J., Lemmess, H., Van Montague, M., and Schell, J.,** Nucleotide sequence and transcript map of the *Agrobacterium tumefaciens* Ti plasmid encoded octopine synthase gene, *J. Mol. Appl. Genet.*, 1, 499, 1982.
35. **De Greve, H., Leemans, J., Hernalsteens, J. P., Thia-Toong, L., De-Beuckeleer, M., Willmitzer, L., Otten, L., Van Montagu, M., and Schell, J.,** Regeneration of normal and fertile plants that express octopine synthase from tobacco crown galls after deletion of tumour controlling functions, *Nature,* 300, 752, 1982.
36. **Depicker, A., Stachel, S. E., Dhaese, P., Zambryski, P., and Goodman, H. M.,** Nopaline synthase: transcript mapping and DNA sequence, *J. Mol. Appl. Genet.*, 1, 561, 1982.
37. **Bevan, M., Barnes, W. M., and Chilton, M. D.,** Structure and transcription of the nopaline synthase gene region of T-DNA, *Nucleic Acids Res.*, 11, 369, 1983.
38. **Leemans, J., Deblaere, R., Willmitzer, L., De Greve, H., Hernalsteens, J. P., Van Montagu, M., and Schell, J.,** Genetic identification of functions of Ti-DNA transcript in octopine crown gall, *EMBO J.*, 1, 147, 1982.
39. **Klee, H. J., Gordon, M. P., and Nester, E. W.,** Complementation analysis of *Agrobacterium tumefaciens* Ti plasmid mutations affecting oncogenicity, *J. Bacteriol.*, 150, 327, 1982.
40. **Matze, A. J. M. and Chilton, M. D.,** Site-specific insertion of genes into T-DNA of the *Agrobacterium* tumor-inducing plasmid: an approach to genetic engineering of higher plant cells, *J. Mol. Appl. Genet.*, 1, 39, 1981.
41. **de Framond, A., Barton, K. A., and Chilton, M. D.,** Mini-Ti: a new vector strategy for plant genetic engineering, *Biotechnology,* 1, 262, 1983.
42. **Beck, E., Ludwig, G., Auerswald, E. A., Reiss, B., and Schaller, H.,** Nucleotide sequence and exact localization of the neomycin phosphotransferase gene from transposon Tn5, *Gene,* 19, 327, 1982.
43. **An, G., Watson, B. D., and Chiang, C. C.,** Transformation of tobacco, tomato, potato and *Arabidopsis thaliona* using binary T_1 vector system, *Plant Physiol.*, 8, 301, 1986.
44. **Bevan, M.,** Binary *Agrobacterium* vectors for plant transformation, *Nucleic Acids Res.*, 12, 8711, 1984.
45. **Bevan, M., Flavell, R. B., and Chilton, M. D.,** A chimaeric antibiotic resistance gene as a selectable marker for plant cell transformation, *Nature,* 304, 184, 1983.
46. **Fraley, R. T., Horsch, R. B., Matzke, A., Chilton, M. D., Chilton, W. S., and Sanders, P. R.,** In vitro transformation of *Petunia* cells by an improved method of co-cultivation with *A. tumefaciens* strains, *Plant Mol. Biol.*, 3, 371, 1984.
47. **Fraley, R. T., Rogers, S. C., Horsch, R. B., Sanders, P. R., Flick, J., Adams, S., Bittner, M., Brand, L., Fink, C., Fry, J., Galluppi, G., Goldberg, S., Hoffmann, N., and Woo, S.,** Expression of bacterial genes in plant cells, *Proc. Natl. Acad. Sci. U.S.A.*, 80, 4803, 1983.
48. **Horsch, R. B., Fraley, R. T., Rogers, S. G., Sanders, P. R., Lloyd, A. R., and Hoffmann, N. L.,** Inheritance of functional foreign genes in plants, *Science,* 223, 496, 1984.
49. **Horsch, R. B., Fry, J. E., Hoffmann, N. L., Wallroth, M., Eicholtz, D., Rogers, S. G., and Fraley, R. T.,** A simple and general method for transferring genes into plants, *Science,* 227, 1229, 1985.
50. **Klee, H. J., Yanofsky, M., and Nester, E. W.,** Vectors for transformation of higher plants, *Biotechnology,* 3, 637, 1985.
51. **Koncz, C. and Schell, J.,** The promoter of Ti-DNA gene 5 controls the tissue-specific expression of chimeric genes carried by a novel type of *Agrobacterium* binary vector, *Mol. Gen. Genet.*, 204, 383, 1986.

52. **Matzke, A. J. M. and Matzke, M. A.,** A set of novel Ti plasmid derived vectors for the production of transgenic plants, *Plant Mol. Biol.*, 7, 357, 1986.
53. **Van den Elzen, P., Lee, K. Y., Townsend, J., and Bedbrook, J.,** Simple binary vectors for DNA transfer to plant cells, *Plant Mol. Biol.*, 5, 149, 1985.
54. **Van den Elzen, P., Townsend, J., Lee, K. Y., and Bedbrook, J.,** A chimeric hygromycin resistance gene as a selectable marker in plant cells, *Plant Mol. Biol.*, 5, 299, 1985.
55. **Velten, J. and Schell, J.,** Selection-expression plasmid vectors for use in genetic transformation of higher plants, *Nucleic Acids Res.*, 13, 6981, 1985.
56. **Lloyd, A. M., Barnason, A. R., Rogers, S. G., Byrne, M. C., Fraley, R. T., and Horsch, R. B.,** Transformation of *Arabidopsis thaliana* with *Agrobacterium tumefaciens, Science,* 234, 464, 1986.
57. **Rogers, S. G., Bisaro, D. M., Horsch, R. B., Fraley, R. T., Hoffman, N. L., Brand, L., Elmer, J. S., and Lloyd, A. M.,** Tomato golden mosaic virus A component DNA replicates autonomously in transgenic plants, *Cell,* 45, 593, 1986.
58. **Waldron, C., Murphy, E. B., Roberts, J. L., Gustafson, G. D., Armour, S. L., and Malcolm, S. K.,** Resistance to hygromycin-B, *Plant Mol. Biol.*, 5, 103, 1985.
59. **Haughn, G. and Somerville, C.,** Sulfonylurea-resistant mutants of *Arabidopsis thaliana, Mol. Gen. Genet.*, 204, 430, 1986.
60. **Wullems, G., Molendijk, L., Ooms, G., and Schilperoort, R. A.,** Differential expression of crown gall tumor markers in transformants obtained after *in vitro Agrobacterium tumefaciens*-induced transformation of cell wall regenerating protoplasts derived from *Nicotiana tabacum, Proc. Natl. Acad. Sci. U.S.A.*, 78, 4344, 1981.
61. **Fry, J., Barnason, A., and Horsch, R. B.,** Transformations of *Brassica napus* with *Agrobacterium tumefaciens* based vectors, *Plant Cell Rep.*, 6, 321, 1987.
62. **Wallroth, M., Gerats, A. G. M., Rogers, S. G., Fraley, R. T., and Horsch, R. B.,** Chromosomal localization of foreign genes in *Petunia hybrida, Mol. Gen. Genet.*, 202, 6, 1986.
63. **Gardner, C. O., Melcher, V., Shockey, M. W., and Essenberg, R. C.,** Restriction enzyme cleavage maps of the DNA of cauliflower mosaic virus isolates, *Virology,* 103, 250, 1980.
64. **Howell, S. H.,** The molecular biology of plant DNA viruses, *CRC Crit. Rev. Plant Sci.*, 2, 287, 1985.
65. **Hull, R. and Covey, S. N.,** Does cauliflower mosaic virus replicate by reverse transcription?, *Trends Biochem. Sci.*, 8, 119, 1983.
66. **Pfeiffer, P. and Hohn, T.,** Involvement of reverse transcription in the replication of cauliflower mosaic virus: a detailed model and test of some aspects, *Cell,* 33, 781, 1983.
67. **Vanstogteren, G. M. S., Hooykaas, P. J. J., and Schilperoort, R. A.,** Expression of Ti plasmid genes in monocotyledonous plants infected with *Agrobacterium tumerfaciens, Nature,* 311, 763, 1984.
68. **Lebeurier, G., Hirth, L., Hohn, B., and Hohn, T.,** *In vivo* recombination of cauliflower mosaic virus DNA, *Proc. Natl. Acad. Sci. U.S.A.*, 79, 2932, 1982.
69. **Walden, R. M. and Howell, S. H.,** Uncut recombinant plasmids bearing nested cauliflower mosaic virus genomes infect plants by intergenomic recombination, *Plant Mol. Biol.*, 2, 927, 1983.
70. **Guilley, H., Dudley, R. K., Jonard, G., Balazs, E., and Richards, K. E.,** Transcription of cauliflower mosaic virus DNA: detection of promoter sequences, and characterisation of transcripts, *Cell,* 30, 763, 1982.
71. **Hohn, T., Hohn, B., and Pfeiffer, U. P.,** Reverse transcription in cauliflower mosaic virus, *Trends Biochem. Sci.*, 5, 205, 1985.
72. **Grimsley, N., Hohn, B., Hohn, T., and Walden, R. M.,** Agroinfection an alternative route for plant virus infection by using Ti plasmid, *Proc. Natl. Acad. Sci. U.S.A.*, 83, 3282, 1986.
73. **Haber, S., Tkegami, M., Bajet, M. B., and Goodman, R. M.,** Evidence for a divided genome in bean golden mosaic virus, a geminivirus, *Nature,* 289, 324, 1981.
74. **Hooykaas, P. J. J., Vanstogtaren, G. M. S., and Schilperoort, R. A.,** A Process for Incorporation of Foreign DNA into the Genome of Monocotyledonous Plants, European Patent Appl. EP 0159418, Publ. October 30, 1985.
75. **Hernalsteens, J. P., Thia-Toong, L., Schell, J., and Van Montagu, M.,** An *Agrobacterium*-transformed cell culture from the Monocot *Asperagus officinalis, EMBO J.*, 3, 3039, 1984.
76. **Stanley, D. M.,** The molecular biology of geminiviruses, *Adv. Virus Res.*, 30, 139, 1985.
77. **Grimsley, N., Hohn, T., Davies, J. W., and Hohn, B.,** *Agrobacterium*-mediated delivery of infectious maize streak virus into maize plants, *Nature,* 324, 177, 1987.
78. **Hohn, B., Hohn, T., Boulton, M. I., Davies, J. W., and Grimsley, N.,** Agroinfection of *Zea mays* with maize streak virus DNA, in *Plant Molecular Biology,* Von Wettstein, D. and Chua, N. H., Eds., Plenum Press, New York, 1987, 1.
79. **Grimsley, N., Hohn, T., and Hohn, B.,** Recombination in a plant virus: template-switching in cauliflower mosaic virus, *EMBO J.*, 5, 641, 1986.

80. **Leemans, J., Shaw, Ch., Deblaere, R., DeGreve, H., Hernalsteens, J. P., Maes, M., Van Montagu, M., and Schell, J.,** Site-specific mutagenesis of *Agrobacterium* Ti-plasmids and transfer of genes to plant cells, *J. Mol. Appl. Genet.*, 1, 39, 1981.
81. **Davey, M. R., Cocking E. C., Freeman, J., Pearce, N., and Tudor, I.,** Transformation of *Petunia* protoplasts by isolated *Agrobacterium* plasmids, *Plant Sci. Lett.*, 18, 307, 1980.
82. **Draper, J., Davey, M. R., Freeman, J. P., Cocking, E. C., and Cox, B. J.,** Ti plasmid homologous sequences present in tissues from *Agrobacterium* plasmid-transformed *Petunia* protoplasts, *Plant Cell Physiol.*, 23, 255, 1982.
83. **Krens, F. A., Molendijk, L., Wullems, G. J., and Schilperoort, R. A.,** In vitro transformation of plant protoplasts with Ti-plasmid DNA, *Nature*, 296, 72, 1982.
84. **Vaeck, M., Reynaerts, A., Hofte, H., Jensens, S., De Beukeleer, M., Dean, C., Zabeau, M., Van Montagu, M., and Leemans, J.,** Transgenic plants protected from insect attack, *Nature*, 327, 33, 1987.
85. **Watson, J. D., Hopkins, N. H., Roberts, J. W., Steitz, J. A., and Weiner, A. M.,** *The Molecular Biology of the Gene*, Benjamin/Cummings, Menlo Park, CA, 1987.
86. **Fischhoff, D. A., Bowdish, K. S., Perlak, F. J., Marrone, P. G., and McCormick, S. M.,** Insect tolerant transgenic tomato plants, *Biotechnology*, 5, 807, 1987.
87. **Anon.,** Rohm and Hass will field-test an engineered plant, *Chem. Week*, 139, 34, 1986.
88. **Abel, P. P., Nelson, R. S., De., Hoffmann, N., Rogers, S. G., Fraley, R. T., and Beachy, R. N.,** Delay of disease development in transgenic plants that express the tobacco mosaic virus coat protein gene, *Science*, 232, 738, 1986.
89. **Loesch-Fries, L. S., Merlo, D., Zinnen, T., Burhop, L., and Hill, K.,** Expression of alfalfa mosaic virus RNA in transgenic plants confer virus resistance, *EMBO J.*, 6, 1845, 1987.
90. **Nelson, R. S., McCormic, S. M., Delannay, X., Dube, P., Layton, J., Anderson, E. J., Koniewska, M., Proksch, R. K., Horsch, R. B., Rogers, S. G., Fraley, R. T., and Beachy, R. N.,** Virus tolerance, plant growth, and field performance of transgenic tomato plant expressing coat protein from tobacco mosaic virus, *Biotechnology*, 6, 403, 1988.
91. **De Block, M., Botterman, J., Vandewielve, M., Dockx, J., Theon, C., Gossele, V., Movva, N. R., Thompson, C., Vam Montagur, M., and Leemans, J.,** Engineering herbicide resistance in plants by expression of a detoxifying enzyme, *EMBO J.*, 6, 2513, 1987.
92. **Chalelf, R. S. and Ray, T. B.,** Herbicide resistant mutant from tobacco cell cultures, *Science*, 223, 1148, 1984.
93. **Comai, L., Facciotti, D., Haitt, W. R., Thompson, G., Rose, R. E., and Stalker, D. M.,** Expression in plants of a mutant aro A gene from *Salmonella typhimurium* confers tolerance to glyphosate, *Nature*, 317, 741, 1985.
94. **Shah, D., Horsch, R., Klee, H., Kishore, G., Winter, J., Turner, N., Hironaka, C., Sanders, P., Gasser, C., Aykent, S., Siegel, N., Rogers, S., and Fraley, R. T.,** Engineering herbicide tolerance in transgenic plants, *Science*, 233, 478, 1986.
95. **Crossway, A., Oakes, J. V., Irvine, J. M., Ward, B., Knauf, V. C., and Shewmaker, C. K.,** Integration of foreign DNA following micro-injection of tobacco mesophyll protoplasts, *Mol. Gen. Genet.*, 202, 179, 1986.
96. **Paszkowski, J., Shillito, R. D., Saul, M., Mandak, V., Hohn, T., Hohn, B., and Potrykus, I.,** Direct gene transfer to plants, *EMBO J.*, 3, 2717, 1984.
97. **Herrera-Estrella, L., Depicker, A., Van Montagu, M., and Schell, J.,** Expression of chimeric genes transferred into plant cells using Ti plasmid-derived vector, *Nature*, 303, 209, 1983.
98. **Comai, L., Sen, L., and Stalker, D.,** An altered aro A gene product confers resistance to the herbicide glyphosate, *Science*, 221, 370, 1983.
99. **Fillatti, J. J., Sellmer, J., McCown, B., Haissig, B., and Comai, L.,** *Agrobacterium* mediated transformation and regeneration of *Populus*, *Mol. Gen. Genet.*, 206, 192, 1987.
100. **Akinymiju, O. A., Iscbrands, J. G., Nelson, N. D., and Dickmann, D. I.,** Use of glyphosate in the establishment of *Populus* in short rotation intensive culture, in Proc. North American Poplar Council Meeting, Rhinelander, WI, Divison of Extension, Kansas State University, Manhattan, 1982.
101. **Hansen, E. A. and Netzer, D. A.,** Weed control using herbicides in short-rotation intensively cultured Poplar plantations, USDA Forest Service Research Paper NC-260, U.S. Department of Agriculture, 1985.
102. **Keen, P. J., Kerr, A., and New, P. B.,** Crown gall of stone fruit. II. Identification and nomenclature of *Agrobacterium* isolates, *Aust. J. Biol. Sci.*, 23, 585, 1970.
103. **Sciaky, D., Montoya, A. L., and Chilton, M. D.,** Fingerprints of *Agrobacterium* Ti plasmids, *Plasmid*, 1, 228, 1978.
104. **Parsons, T. J., Sinkar, V. P., Stettler, R. F., Nester, E. W., and Gordon, M. P.,** Transformation of poplar by *Agrobacterium tumefaciens*, *Biotechnology*, 4, 533, 1986.
105. **Anon.,** Combined research at Calgene, University of Wisconsin U.S. Forestry Service, herbicide resistance genetically engineered into poplars, *Bio Process Technol.*, 9, 6, 1987.

106. **White, D. W. R. and Wood, D.,** Transformation of forage legume *Trifolium repens* L. using binary *Agrobacterium* vectors, *Plant Mol. Biol.,* 8, 461, 1987.
107. **Umbeck, P. and Johnson, G.,** Genetically transformed cotton *(Gossypium hirsutum* L) plants, *Biotechnology,* 5, 263, 1987.
108. **Mueller, G. E. and Schieder, O.,** Interspecific T-DNA transfer through plant protoplast fusion, *Mol. Gen. Genet.,* 268, 235, 1987.
109. **Jorgensen, R., Snyder, C., and Jones, J. D. G.,** T-DNA is organized predominantly in inverted repeat structures in plants transformed with *Agrobacterium tumefaciens* C 58 derivatives, *Mol. Gen. Genet.,* 207, 471, 1987.
110. **Chyi, Y. S. and Phillips, G. C.,** High efficiency *Agrobacterium* mediated transformation of *Lycopersicon* based on conditions favourable for regeneration, *Plant Cell Rep.,* 6, 105, 1987.
111. **Mueller, A. J., Mendel, R. R., Schiewann, J., Simoens, C., and Inze, D.,** High meiotic stability of a foreign gene introduced into tobacco by *Agrobacterium* mediated transformation, *Mol. Gen. Genet.,* 207, 171, 1987.
112. **Ooms, G., Bains, A., Burrell, M., Karp, A., Twell, D., and Wilcox, E.,** Genetic manipulation of cultivars of oilseed rape *(Brassica napus)* using *Agrobacterium, Theor. Appl. Genet.,* 7, 325, 1985.
113. **Guerche, P., Joaanin, L., Tapfer, D., and Pelletier, G.,** Genetic transformation of oilseed rape *(Brassica napus)* by RiT-DNA of *Agrobacterium rhizogenes* and analysis of inheritance of the transformed phenotype, *Mol. Gen. Genet.,* 206, 382, 1987.
114. **Pua, E. C., Mehra-Palta, A., Nagy, F., and Chua, N.,** Transgenic plants of *Brassica napus* L., *Biotechnology,* 5, 815, 1987.
115. **Basiran, N., Armitage, P., Scott, R. J., and Draper, J.,** Genetic transformation of flax *(Linum usitatissimum)* by *Agrobacterium tumefaciens:* regeneration of transformed shoots via a callus phase, *Plant Cell Rep.,* 6, 396, 1987.
116. **Sukhapinda, K., Spivey, R., Simpson, R. B., and Shahin, E. A.,** Transgenic tomato *(Lycopersicon esculentum* L) transformed with binary vector in *Agrobacterium rhizogenes:* non-chimaeric origin of callus clone and low copy numbers of integrated vector T-DNA, *Mol. Gen. Genet.,* 206, 491, 1987.
117. **An, G., Watson, B. D., Stachel, S., Gordon, M. P., and Nester, E. W.,** New cloning vehicles for transformation of higher plants, *EMBO J.,* 4, 277, 1985.
118. **Ooms, C., Burrell, M. M., Karp, A., Bevan, A., and Hille, J.,** Genetic transformation in two potato cultivars with T-DNA from disarmed *Agrobacterium, Theor. Appl. Genet.,* 73, 744, 1987.
119. **Goldsbrough, P. B., Gelvin, S. B., and Larkins, B. A.,** Expression of maize zein genes in transformed sunflower cells, *Mol. Gen. Genet.,* 202, 374, 1986.
120. **Singleton, W. and Jones, D.,** Heritable characters of maize. XXXV. Male sterile, *J. Hered.,* 21, 266, 1930.
121. **Frankel, R.,** *Heterosis: Reappraisal of Theory and Practice,* Springer-Verlag, Berlin, 1983.
122. **Philouze, J.,** Genes marqueurs lies aux genes de sterilite made ms 35 et ms 32 chez la tomate, *Ann. Amelior. Plant.,* 24, 77, 1974.
123. **Tanksley, S.,** Molecular markers in plant breeding, *Plant Mol. Biol. Rep.,* 1, 3, 1983.
124. **Wiebe, G.,** A proposal for hybrid barley, *Agron. J.,* 52, 181, 1960.
125. **Jorgensen, R. A.,** A hybrid seed production method based on synthesis of novel linkages between marker and male sterile genes, *Crop Sci.,* 27, 8-6-810.
126. **DeBlock, M., Herrara-Estrella, L., Van Montagu, M., Schell, J., and Zambryski, P.,** Expression of foreign genes in regenerated plants and their progeny, *EMBO J.,* 3, 1681, 1984.
127. **Otten, L., DeGreve, H., Hernalsteens, J. P., Van Montagu, M., Schieder, O., Straub, J., and Schell, J.,** Mendelian transmission of genes introduced into plants by the Ti-plasmids of *Agrobacterium tumefaciens, Mol. Gen. Genet.,* 183, 209, 1981.
128. **Ream, W. and Gordon, M.,** Crown gall disease and prospects for genetic manipulation in plants, *Science,* 218, 854, 1982.
129. **Wallroth, M., Gerats, A., Rogers, S., Fraley, R., and Horsch, R.,** Chromosomal localization of foreign genes into *Petunia hybrida, Mol. Gen. Genet.,* 202, 6, 1986.
130. **Hille, J., Dekker, M., Luttighuis, H., van Kammen, A., and Zabel, P.,** Detection of T-DNA transfer to plant cells by *A. tumefaciens* virulence mutants using agroinfection, *Mol. Gen. Genet.,* 205, 411, 1986.
131. **Schardl, C. L., Byrd, A. D., Benzion, G., Altschuler, M. A., Hildeorand, D. F., and Hunt, A. G.,** Designing and construction of a varsatile system for the expression of foreign genes in plants, *Gene,* 61, 1, 1987.
132. **Odell, J. T., Dudley, R. K., and Howell, S. H.,** Structure of the 19S RNA transcript encoded by the cauliflower mosaic virus genome, *Virology,* 111, 377, 1981.
133. **Odell, J. T., Nagy, F., and Chua, N. H.,** Identification of DNA sequences required for activity of the cauliflower mosaic virus 35S promotor, *Nature,* 313, 810, 1985.

134. **Chyi, Y. S., Jorgensen, R. A., Goldstein, D., Tanksley, S. D. and Loaiza-Figueroa, F.,** Locations and stability of *Agrobacterium*-mediated T-DNA insertions in *Lycopersicon* genome, *Mol. Gen. Genet.,* 204, 64, 1987.
135. **Nagy, F., Morelli, G., Fraley, R. T., Rogers, S. G., and Chua, N. H.,** A short conserved sequence is involved in the light inducibility of a gene encoding ribulose 1,5-bis phosphate carboxylase small subunit of pea, *Nature,* 315, 200, 1985.
136. **Hildebrand, D. F., Altschuler, M., Bookjans, G., Benzoin, G., Hamilton-Kemp, T. R., Andersen, R. A., Rodriguezi, J. G., Polacco, J. C., Dahmer, M. L., Hunt, A. G., Wang, X., and Collins, G. B.,** Physiological and transformational analysis of lipoxygenases, in *The Metabolism, Structure and Function of Plant Lipids,* Stumpf, P., Mudd, J., and Nes, W., Eds., Plenum Press, New York, 1987, 715.
137. **Maiti, I. B., Hunt, A., and Wagner, G. J.,** Attempts to modify cadmium accumulation in plants through tissue-specific expression of the mouse metallothionin gene, in Abstr. 27th Annu. Meeting of the Phytochemical Society of North America, Paper 26, 1987.
138. **Stram, Y., Berger, P., and Pirone, T. P.,** Cloning and expression of the polyvirus helper component gene, in Abstr. 7th Int. Congr. Virol., Abstr. 704, 1987.
139. **Sanders, P. R., Winter, J. A., Barason, A. R., Rogers, S. G., and Fraley, R. T.,** Comparison of cauliflower mosaic virus 35S and nopaline synthase promotors in transgenic plants, *Nucl. Acids Res.,* 15, 1543, 1987.
140. **Gheysen, G., Dhaese, P., and Van Montagu, M.,** *Genetic Flux in Plants,* Hohn, B. and Dennis, E. S., Eds., Springer-Verlag, Vienna, 1985.
141. **Koukoliova-Nicola, Z., Shillito, R. D., Hohn, B., Wang, K., Van Montagu, M., and Zambryski, P.,** Involvement of circular intermediates in the transfer of T-DNA from *Agrobacterium tumefaciens* to plant cells, *Nature,* 313, 191, 1985.
142. TCCP (Tissues Culture for Crops Project), Progress Report, Colorado State University, Fort Collins, 1987.
143. **Grimsley, N. and Bisaro, D.,** Agroinfection, in *Plant DNA Infectious Agents,* Hohn, T. and Schell, J., Eds., Springer-Verlag, New York, 1987, 87.
144. **Gardner, R. C., Chanoles, K. R., and Owens, R. A.,** Potato spindle tuber viroid infections mediated by the Ti plasmid of *Agrobacterium tumefaciens, Plant Mol. Biol.,* 6, 221, 1986.
145. **Covey, S. N. and Hull, R.,** Advances in cauliflower mosaic virus, *Res. Oxford Surv. Plant Mol. Cell Biol.,* 2, 339, 1985.
146. **Howell, S. H., Walker, L. L., and Dudley, R. K.,** Cloned cauliflower mosaic virus DNA infects turnips *Brassica rapa, Science,* 208, 1265, 1980.
147. **Mullineauxl, P. M., Donson, J., Morris-Krisinich, B. A. M., Boulton, M. I., and Davies, J. W.,** The nucleotide sequence of maize streak virus DNA, *EMBO J.,* 3, 3062, 1984.
148. **Polson, A. and Von Wechmar, M. B.,** A novel way to transmit plant viruses, *J. Gen. Virol.,* 51, 179, 1980.
149. **Branch, A. D. and Robertson, H. D.,** A replication cycle for viroids and other small infectious RNAs, *Science,* 223, 450, 1984.
150. **Hutchins, C. J., Keese, P., Visavader, J. E., Rathjen, P. D., McInnes, J. L., and Symons, R. H.,** Comparison of multimeric plus and minus forms of viroids and virusoids, *Plant Mol. Biol.,* 4, 293, 1985.
151. **Cress, D. E., Kiefer, M. C., and Ownes, R. A.,** Construction of infectious potato spindle tuber viroid cDNA clones, *Nucleic Acids Res.,* 11, 6821, 1985.
152. **Tabler, M. and Sanger, H. L.,** Cloned single and double-stranded DNA copies of potato spindle tuber viroid (PSTV) RNA and co-inoculated subgenomic DNA fragments are infectious, *EMBO J.,* 3, 3055, 1984.
153. **Gardner, R. C. and Knauf, V. C.,** Transfer of *Agrobacterium* DNA to plants requires a T-DNA border but not the vir E locus, *Science,* 231, 725, 1986.
154. **Shaw, W.,** Chloramphenicol acetyltransferase from resistant bacteria, *Methods Enzymol.,* 53, 737, 1975.
155. **Ow, D. W., Wood, K. V., DeLuca, M., de Wet, J. R., Helsinki, D. R., and Howell, S. H.,** Transient and stable expression of the firefly luciferase gene in plant cells and transgenic plants, *Science,* 234, 856, 1986.
156. **Shewmaker, C. K., Caton, J. R., Houck, C. M., and Gardner, R. C.,** Transcription of cauliflower mosaic virus integrated into plant genomes, *Virology,* 140, 281, 1985.
157. **Covey, S. N., Lomonossoff, G. P., and Hull, R.,** Characterisation of cauliflower mosaic virus RNA sequences which encode major polyadenylated transcripts, *Nucleic Acids Res.,* 9, 6735, 1981.
158. **Bisaro, D. M., Hamilton, W. D. O., Coutts, R. H. A., and Buck, K. W.,** Molecular cloning and characterization of the two DNA components of tomato golden mosaic virus, *Nucleic Acids Res.,* 10, 4913, 1982.
159. **Hamilton, W. D. C., Stein, V. E., Coutts, R. H. A., and Buck, K. W.,** Complete nucleotide sequence of the infectious cloned DNA components to tomato golden mosaic virus: potential coding regions and regulatory sequences, *EMBO J.,* 3, 2197, 1984.

160. **Hamilton, W. D. O., Bisaro, D. M., and Buck, K. W.,** Identification of novel DNA forms in tomato golden mosaic virus infected tissue: evidence for a two component viral genome, *Nucleic Acids Res.*, 10, 4901, 1982.
161. **Coutts, R. H. A. and Buck, K. W.,** DNA and RNA polymerase activities of nuclei and hypotonic extracts of nuclei isolated from tomato golden mosaic virus infected tobacco leaves, *Nucleic Acids Res.*, 13, 7881, 1985.
162. **Hamilton, W. D. O., Sanders, R. C., Coutts, R. H. A., and Buck, K. W.,** Characterization of tomato golden mosaic virus as a geminivirus, *FEMS Microbiol. Lett.*, 11, 263, 1981.
163. **Hamilton, W. D. O., Bisaro, D. M., Coutts, R. H. A., and Buck, K. W.,** Demonstration of the bipartite nature of the genome of a single stranded plant virus by infection with the cloned DNA components, *Nucleic Acids Res.*, 11, 7387, 1983.
164. **Daubert, S., Shepherd, R. J., and Gardner, R. C.,** Insertional mutagenesis of cauliflower mosaic virus genome, *Gene*, 25, 201, 1983.
165. **Dixon, L. K., Koening, I., and Hohn, T.,** Mutagenesis of cauliflower mosaic virus, *Gene*, 25, 189, 1983.
166. **Ecker, J. R. and Davis, R. W.,** Inhibition of gene expression in plant cells by expression of antisense RNA, *Proc. Natl. Acad. Sci. U.S.A.*, 83, 5372, 1986.
167. **Kaper, J. M., Tousignant, M. E., and Lot, H.,** A low molecular weight replicating RNA associated with a divided genome plant virus: defective or satellite RNA, *Biochem. Biophys. Res. Commun.*, 72, 1237, 1976.
168. **Baulcombe, D. C., Saunders, G. R., Bevan, M. W., Mayo, M. A., and Harrison, B. D.,** Expression of biologically active viral satellite RNA from the nuclear genome of transformed plants, *Nature*, 321, 446, 1986.
169. **Bevan, M. W., Mason, S. E., and Goelet, P.,** Expression of tobacco mosaic virus coat protein by a cauliflower mosaic virus promoter in plants transformed by *Agrobacterium*, *EMBO J.*, 4, 1921, 1985.
170. **Waterworth, H.-E., Kaper, J. M., and Tousignant, M. E.,** CARNA 5, the small cucumber mosaic virus-dependent replicating RNA, regulates disease expression, *Science*, 204, 845, 1979.
171. **Mossop, D. W. and Francki, R. I. B.,** Comparative studies of two satellite RNAs of cucumber mosaic virus, *Virology*, 95, 395, 1979.
172. **Harrison, B. D., Mayo, M. A., and Baulcombe, C. D.,** Virus resistance in transgenic plants that express cucumber mosaic satellite RNA, *Nature*, 328, 799, 1987.
173. **Gerlach, W. L., Llewellyn, D., and Haseloff, J.,** Construction of a plant disease resistance gene from the satellite RNA of tobacco ringspot virus, *Nature*, 328, 802, 1987.
174. **Roca, W. M., Szabados, L., and Hussain, A.,** Annual Report, Biotechnology Research Unit, CIAT, Colombia, 1987.

Chapter 6

PLANT FROST INJURY AND ITS MANAGEMENT

I. INTRODUCTION

Injury of plants through frost is a serious problem to agriculturists. In the temperate zones, frost injury is one of the main limiting factors to crop production. Frost injury was also considered an unavoidable result of physical stress in plants.[1-3] In the U.S. alone, the losses due to frost injury are estimated at over $1 billion yearly.[4]

Mechanisms of frost injury to frost-sensitive plants which are damaged at temperatures warmer than 5°C have not been studied in detail.[1,3] Frost-sensitive plants are distinguished from frost-hardy plants by their relative inability to tolerate ice formation within their tissues.[2,3,5] Injury to plants through frost is mainly due to certain bacteria on the surface of leaves. These bacteria freeze water even at higher temperatures. This makes frost-sensitive plants more susceptible to damage.[6-11]

Most of the herbaceous annual plants, flowers of deciduous fruit trees, fruits of many plant species, and shoots and stems of certain forest trees such as *Eucalyptus* contain frost-sensitive tissue. Ice formed in or on frost-sensitive plants spreads rapidly both intercellularly and intracellularly, causing mechanical disruption of cell membranes.[5,12] Upon warming, the damage to plants is reflected by flaccidity and/or decoloration (Figure 1). It has been seen that there is no mechanism in plants to avoid such a frost injury.[5,12,13]

II. SUPERCOOLING OF WATER AND ICE FORMATION

It is an interesting feature of many liquids, including water, that do not invariably freeze at the melting point of the solid phase. These liquids can be supercooled (undercooled or subcooled) to several degrees centigrade below the melting point of the solid phase. The freezing of such liquids takes place only upon the spontaneous addition of a suitable catalyst for the liquid-solid phase transition. These catalysts for the water-ice phase transition are called ice nuclei.[14] Homogeneous ice nuclei catalyze water-ice phase transition at low temperatures, whereas heterogeneous ice nuclei are more important at temperatures approaching 0°C. Small volumes of pure water can be supercooled to approximately −40°C before the spontaneous homogeneous catalysis of ice formation occurs.[15] Even relatively large quantities of water readily supercool to −10 to −20°C.[15]

Ice nucleation of all heterogeneous ice nuclei is due to ordering of water molecules into an ice-like lattice, perhaps in the case of inorganic salts, by aggregation of water molecules onto the face of fractured crystals with lattice structures similar to ice.[16] The efficiency (defined by relatively warm threshold ice nucleation temperatures) of heterogeneous ice nuclei presumably increases with increasing numbers of water molecules oriented in a rigid ice-like array.[15] Most organic and inorganic materials such as dust particles nucleate ice only at temperatures lower than −10 to −15°C.[17,18] Dust particles, particularly certain mineral clays, have long been considered as primary sources of ice nuclei.[18,19] Mineral particles, particularly silver iodide, are the most common and the most thoroughly studied source of heterogeneous ice nuclei.[20] These mineral particles efficiently nucleate ice only at temperatures lower than −8 to −15°C.[20,21] Mineral particles of meteoric origin are also considered abundant atmospheric ice nuclei[22] and are active as ice nuclei primarily at temperatures colder than −15°C, and therefore are quite unlikely to account for ice nucleation at relatively warm subfreezing temperatures. Kaolinite is among the most active mineral ice nucleus sources, but it is active in ice nucleation only at temperatures below about −9°C.[17]

FIGURE 1. Frost injury. (A) Mature citrus trees shortly after a severe frost; (B) immature Bartlett pear fruit with internal discoloration typical of mild injury (right) compared with uninjured fruit; (C) newly emerged potato leaves after exposure to −4°C; (D) avocado foliage after a mild radiative frost. (From Lindow, S. E., *Plant Dis.*, 67, 327, 1983. With permission.)

Crystals of several organic compounds also have ice nucleation activity, including steroids,[23,24] amino acids,[25-27] proteins,[21] terpenes,[28] metaldehyde,[29] and phenazine.[30] Although these organic compounds are active in ice nucleation at relatively warm temperatures (warmer than −5°C), they are active as ice nuclei only in a crystalline form.[26] When solubilized, these compounds lose ice nucleation activity. The natural occurrence of the crystalline form of these organic compounds is likely to be small. Crystals of a number of inorganic compounds, however, are ice nuclei at temperatures warmer than −10°C.[16,26,31]

III. ICE NUCLEATION-ACTIVE BACTERIA ON PLANT SURFACE AND FROST DAMAGE

The ability of many frost-sensitive plants to supercool has been recognized for some time.[2,3,32] Ice nucleation activity in plants grown axenically appears to be very rare at temperatures above −5°C.[6,8-10] Significant ice nucleation activity is observed on greenhouse-grown plants only at temperatures lower than −8 to 10°C, because plant materials themselves are inefficient ice nuclei.[33-39] Modlibowska has shown flowers of small fruit trees to supercool only at −2°C before ice formation occurs.[40] Extensive supercooling has been reported for lemon, grapefruit, and other citrus species.[41-44] Variability in the degree of supercooling, which ranged from −2 to −14°C for a large number of different species, has also been reported.[33-39,45] Frost-sensitive plants, particularly when grown under greenhouse conditions, however, have the ability to supercool.

In most cases ice formation and subsequent frost damage under field conditions occur at temperatures warmer than the determined temperature limit for supercooling. This is due to the presence of heterogeneous ice nuclei on the surface of plants.[32] Ice nuclei are not uniformly distributed on a given leaf; and these nuclei vary in quantity both with maturity of leaves and among plant species.[33-37] Similar results have been presented by other scientists.[38,39] Earlier atmospheric ice nuclei were thought to be the sources of ice nuclei on plants. This was, however, ruled out by Marcellos and Single.[38]

The inability of plant tissues to supercool extensively in natural situations can be explained by the detection of up to 1000 ice nuclei per gram of plant tissue which are active at temperatures warmer than −5°C.[46] Most other organic and inorganic materials such as dust particles nucleate ice only at temperatures lower than −10°C. It is unlikely that these nuclei are important in limiting the supercooling of plant tissue at temperatures above −5°C, the temperature range at which most frost-sensitive plants are injured.[12] These observations indirectly indicate the presence of certain biological material on the surface of the plants which were later identified as large bacterial populations.

A. TYPES OF ICE NUCLEI-ACTIVE BACTERIA

Studies have shown that at least 95% ice nuclei on leaf surfaces active at −5°C or above are of bacterial origin.[7-9,46] The extent of frost damage at a given temperature (the chances of a given plant part freezing) increases with increasing populations of ice nucleation-active (INA) bacteria on that plant. Frost injury at a given temperature is more directly related to the numbers of actual bacterial ice nuclei on the plant at the time of freezing than to the population of INA bacteria.[47] Various species of INA bacteria have been demonstrated to be necessary and sufficient to account for the frost sensitivity of all frost-sensitive plants examined to date.

It has been shown that plant frost injury at temperatures only slightly below 0°C is due to the presence of various bacteria.[48-52] The incidence of frost injury to plants at a given temperature is also related directly to the logarithm of the numbers of bacterial ice nuclei and INA bacteria on plant surfaces.[8,14,48-51] Populations of INA bacteria on plants vary both temporarily and with plant species, but are generally found in the range from 10 to 10^7 bacteria per gram fresh weight of plant tissue.[53-55]

The concentration of ice nuclei in the atmosphere at a given location was observed to increase with increasing organic matter content of the soil at that location,[56] indicating the possibility of involvement of microbes in ice nucleation. Decaying vegetation was also found to be a source of abundant ice nuclei.[57,58] Finally, the bacterium *Pseudomonas syringae* van Hall, associated with decaying leaf material, was found to be responsible for ice nucleation.[58,59] This was followed by the identification of three species of *Pseudomonas* commonly found as epiphytes on leaf surfaces that act as catalysts for ice formation. Several pathovars[60] of *P. syringae* have now been found to be active in ice nucleation and are proving the most common INA bacteria ever found on plants in the U.S.[6,8,10,49,61,62] Certain other strains of both *Erwinia herbicola* (Lohnis) Dye and *P. fluorescens* Migula are also active in ice nucleation.[61-65] Approximately 50% of the many pathovars of *P. syringae* pv. *pisi* (Sadatt) and *P. syringae* pv. lachrymans are active in ice nucleation.[66-68] Ice nucleation phenotype has been suggested as a possible taxonomic tool in differentiating the many pathotypes of *P. syringae,* because most strains of a given pathovar of *P. syringae* tested for ice nucleation consistently yielded either a positive or a negative reaction.[67,68] Subsequent work established the involvement of several microbial strains in ice nucleation.[7-11] This was followed by the isolation of strains of *P. syringae* from natural nonsymptomatic plants.[7-11]

If populations of INA bacteria are established on the leaves of plants before they are cooled to −5°C, the plants freeze,[6,69] and if they are tender plants, they are killed. Several plants kept free from these bacteria can be maintained at below 0°C.[69] Several species of "tender" plants have been tested and frost injury was limited only to those plants that harbored populations of INA bacteria. All leaves on plants harboring high epiphytic populations of *P. syringae* were killed between −2 and −3°C; those with lower populations

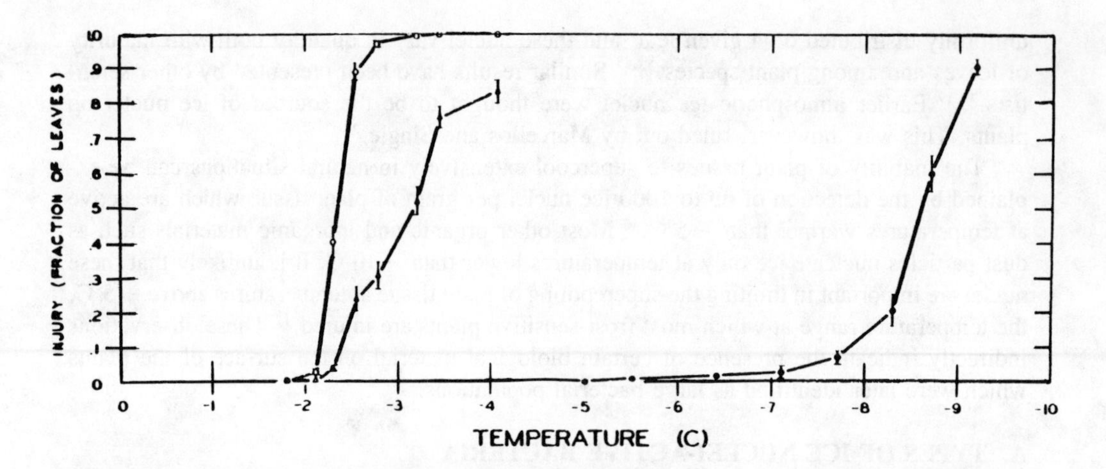

FIGURE 2. Frost damage as a function of temperature to corn seedlings with and without leaf populations of *P. syringae*. Plants were sprayed with *P. syringae* suspensions (≈0.5 ml/plant) of 3 × 10^8 cells per milliliter (○) or 3 × 10^5 cells per milliliter (△) and incubated in a mist chamber for 24 h. Other plants remained untreated (●). Then the plants were placed in a growth chamber at 0°C and cooled at 0.05°C/min. Groups of plants were removed from the chamber at the temperatures shown on the abscissa. (From Lindow, S. E., Arny, D. C., Barchet, W. R., and Upper, C. D., *Plant Cold Hardiness and Freezing Stress,* Li, P. H. and Sakai, A., Eds., Academic Press, New York, 1978, 249. With permission.)

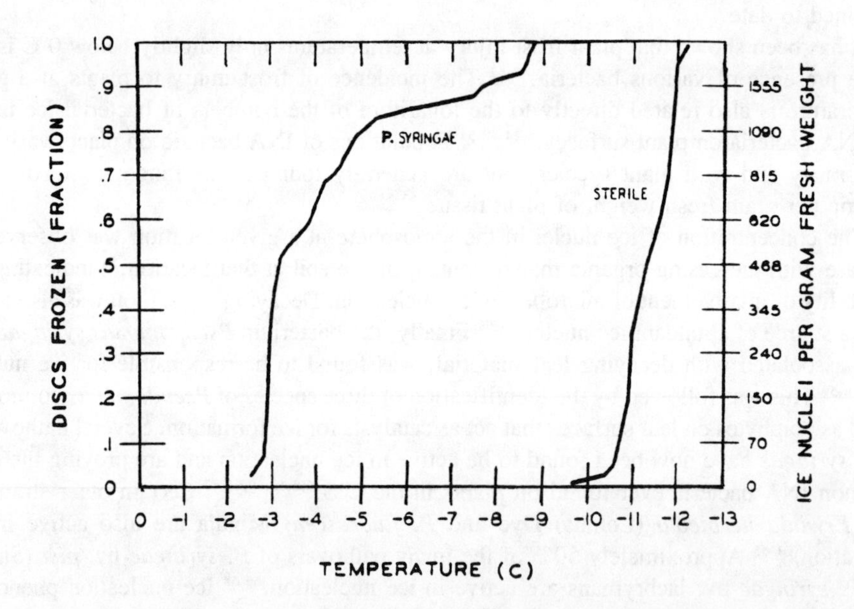

FIGURE 3. Comparison of ice nucleation activity of leaf disks from corn plants with leaf surface *P. syringae* population with those grown on sterile Hoagland's nutrient solution. (From Lindow, S. E., Arny, D. C., Barchet, W. R., and Upper, C. D., *Plant Cold Hardiness and Freezing Stress,* Li, P. H. and Sakai, A., Eds., Academic Press, New York, 1978, 249. With permission.)

of *P. syringae* were severally injured between −2 and −4°C, whereas those lacking populations of INA bacteria were not injured or killed until the temperature had reached −8 to −9°C (Figure 2).[8]

Lindow et al.[8] compared the ice nucleus content of axenic leaves of *Zea mays* with that of similar leaves on which *P. syringae* was established (Figure 3). Ice nuclei were not detected in axenic leaves at temperatures above −9 to −10°C. On the other hand, ice nuclei between −2 and −3°C were detected in disks harboring populations of *P. syringae*. Similar experiments with several other plant species yielded essentially the same results. Thus, leaves do not contain intrinsic ice nuclei active above −10°C, but ice nuclei active at higher temperatures are associated with leaves harboring INA bacteria.

It is believed that a single ice nucleus is sufficient to initiate ice formation and subsequent frost injury to an entire leaf, fruit, flower, or even groups of leaves or flowers, depending on the degree of restriction of ice propagation within a plant.[70] Since frost-sensitive plants must avoid ice formation to avoid frost damage, frost injury to these plants might be best considered a quantal response—either a plant part escapes ice formation or it does not.

With the progress in research in this field the number of new strains and species of INA bacteria is steadily increasing. Lindow[49] listed five different bacterial species which have the ability to serve as catalysts for ice formation both *in vitro* as well as on plant surfaces. Most isolates of approximately half of the described pathovars[60,71] of *P. syringae* van Hall are now known to exhibit ice nucleation activity *in vitro* at temperatures above −5°C.[6,59,66-68] Several isolates of *P. viridiflava* (Burkholder)[66] and *Xanthomonas campestris* pv. *translucens* Dye also exhibit ice nucleation activity at temperatures above −5°C.[14] Some isolates (probably less than ten of all strains) of *P. fluorescens* (Migula) and *E. herbicola* (Lohnis) Dye also possess ice nucleation activity at temperatures above −5°C.[21,64,65,72] These bacterial species appear to constitute the most abundant and most active catalysts for ice formation in nature.[72-74] None of the hundreds of other bacterial strains exhibit detectable ice nucleation activity.[6,8]

B. DISTRIBUTION OF INA BACTERIA ON PLANT SURFACES

Much work has not been done on the distribution of INA bacteria on plant surfaces. Epiphytic populations of various INA bacteria are common on most field-grown plants.[50,63,66,69,72] Nearly all of the 95 species of agricultural and wild plants, except for conifers and smooth-leaved crucifers sampled from several locations in North America, were found to contain epiphytic INA bacteria.[72] INA bacteria on plants have recently been reported from Israel and Japan.[63,65] Though there is no report, INA bacteria should be common under prevailing temperature conditions in several parts of India.

INA strains of both *P.s.* pv. *syringae* and *E. herbicola* can be detected on most plants but *P.s.* pv. *syringae* is the predominant species on the majority of plants investigated.[46] INA strains of *P. fluorescens* are only rarely found on plants in California. Because ice nucleation has only recently been reported in bacteria only a few laboratories have investigated the populations of INA bacteria on plants. However, several workers have studied leaf surface populations of phytophathogenic bacteria or their antagonists. Leaf surface populations of *P.s.* pv. *syringae* and *E. herbicola* have been reported on a variety of plants throughout the world and appear to be nearly ubiquitous epiphytes on nearly all plants studied.[75-93] Although the ubiquity of ice nucleation among strains of *E. herbicola* is as yet unknown, the observation is that at least half of the pathotypes of *P.s.* pv. *syringae* are active as ice nuclei, suggesting that INA bacteria have a worldwide distribution.[46] Similarly, strains of *P. fluorescens* are common soil and water inhabitants. Even an extremely low population of *P. fluorescens* strains is active in ice nucleation.

1. Factors Affecting the Distribution of INA Bacteria

The appearance and distribution of INA bacteria primarily depend upon the type of plant species available in a particular area. In addition, several physical, chemical, and biological factors influence the qualitative and quantitative distribution of INA bacteria.

a. Variation Due to Plant Species and Plant Parts

The numbers of INA bacteria on plant surfaces vary with plant species. Valencia and navel oranges (*Citrus* sp.) have been reported to harbor INA bacteria up to 100 cells per gram fresh weight of leaf tissue. However, English walnut (*Juglans regia* L.) and almond (*Prunus amygdalus* L.) were found to contain over 10^7 cells per gram fresh weight of leaf tissue on leaves. The population of INA bacteria is also dependent on the plant parts. For instance, as compared to leaves, large epiphytic populations of INA bacteria (principally *P.s.* pv. *syringae*) were present on emerging flowers and/or leaves of the plants.

b. Seasonal Variations

Large seasonal variations in the numbers of epiphytic INA bacteria on both annual and perennial plants have been observed.[3,46,47,74] Generally low populations (less than 100 cells per gram fresh weight of leaf on bud tissue) of INA bacteria are found on overwintering tissues of deciduous plants or on emerging cotyledons or leaves of annual plants.[8,74]

Populations of INA bacteria also varied on corn with the growing season.[8] Shortly after the corn plants emerged in early summer, both the total bacterial population and the population of INA bacteria were quite low. The abrupt increase in late July coincided with polination of the corn. The total population of bacteria stabilized nearly 10^8 cells per gram tissue weight in early August, but the population of INA bacteria continued to increase and was nearly 10^6 cells per gram tissue weight at the time the plants were killed by frost in late September. Bacterial populations found on healthy pear (*Pyrus communis* L.) flowers and leaves under California growing conditions are also typical of this variation.[47] A 1000-fold increase in bacterial populations occurred on pear during the 3-week period immediately following bud break. Populations of INA bacteria decreased after May, finally declining with the onset of hot, dry weather to less than 100 cells per gram by late summer.[47]

Epiphytic populations of ice nucleation strains of *Pseudomonas syringae* of up to 10^6 cells per gram fresh weight were found on healthy tissues of commercially managed almond (*Prunus dulcis*) orchards in California.[14] Leaf bacteria accounted for 99% of the ice nuclei active at a temperature higher than −5°C on almond. Large seasonal variations in population of INA bacteria and ice nuclei were observed with maximum populations found shortly after full bloom (Figure 4).

C. ACTIVITY OF INA BACTERIA

The ice nucleation activity of bacteria is variable and depends on the type of bacterium active as ice nucleus, its population density, type of host plant, temperature, etc.

1. Variation among Different INA Bacteria

The strains of *Pseudomonas syringae* and *E. herbicola* studied to date are the most active naturally occurring ice nuclei[94,95] yet discovered. These bacteria catalyze ice formation at a temperature as warm as −1°C (Figure 5). Every cell of *P. syringae, E. herbicola,* or *P. fluorescens* is not active as an ice nucleus at a given time.[9,50,55,58,96,97] Many isolates of *P. syringae* exhibit significant ice nucleation activity at temperatures lower than −4°C.[68] However, other isolates exhibit a reduced frequency of expression of ice nucleation at this temperature.[67,68] The frequency of ice nucleation among cells of INA strains of *E. herbicola* examined to date are approximately 10^4-fold lower than the most active strains of *P. syringae* or *P. fluorescens* at −5°C and about 100-fold lower at −9°C when grown under similar *in vitro* cultural conditions.[10,11,55,63]

The ice nucleation in corn leaf disks harboring *P. syringae* occurs at the same temperatures as in suspensions of *P. syringae*.[8] Most of the nuclei in this suspension were active between −2 and −5°C, (temperature range) in which frost injury to tender plants occurs. *E. herbicola* is somewhat less efficient than *P. syringae* as an ice nucleus.[69] However, suspensions of *E. herbicola* also contain substantial numbers of ice nuclei active between −2 and −5°C (Figure 6).[8]

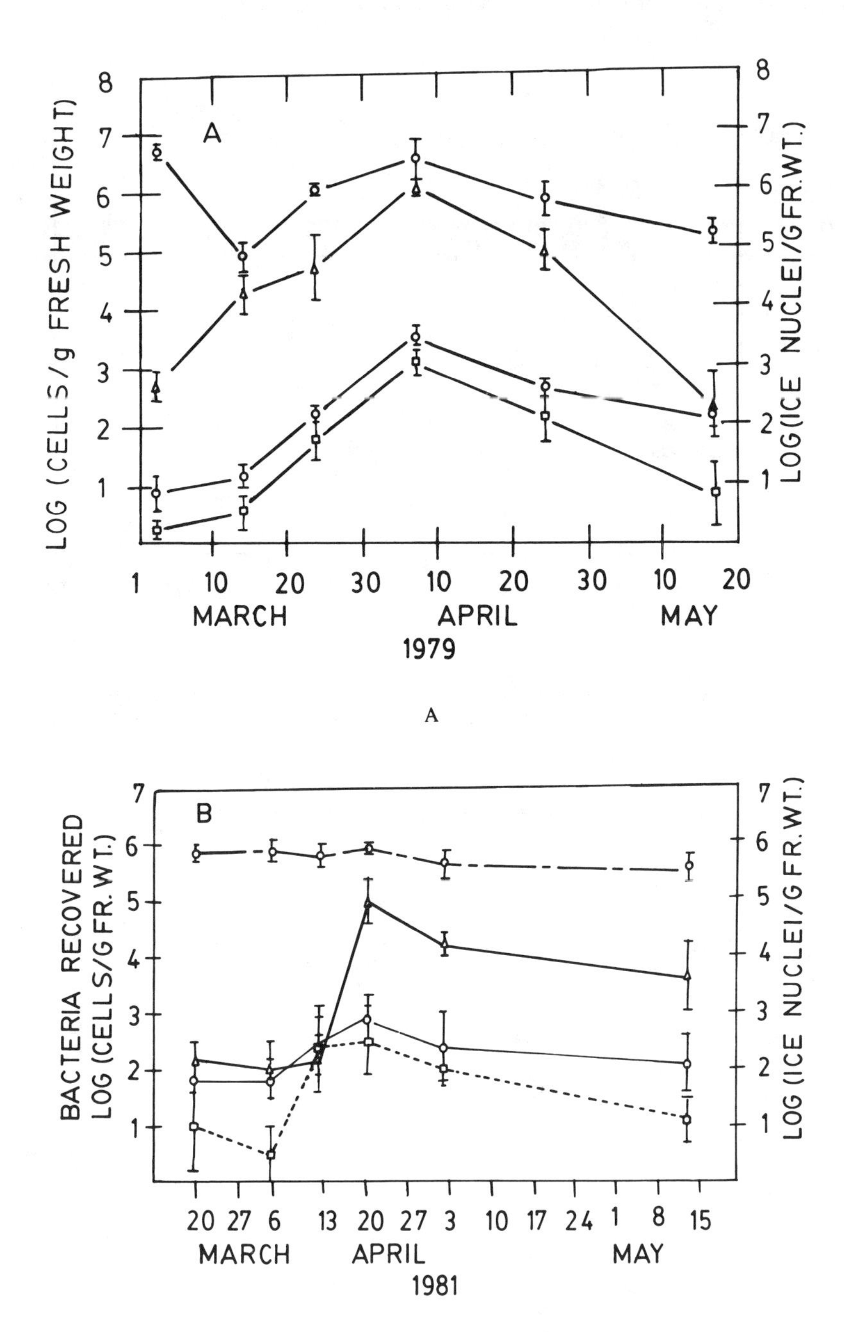

A

B

FIGURE 4. Total bacteria (○), INA bacteria (△), and ice nuclei active at −5°C (□) or −9°C (◇) on leaves, flowers, and young nuts of almond grown near Turlock, CA in 1979 (A) and 1981 (B). (From Lindow, S. E. and Connell, J. H., *J. Am. Soc. Hortic. Sci.*, 109, 48, 1984. With permission.)

2. Variation Due to Temperature

The number of cells active as ice nuclei is primarily dependent on the temperature. The fraction of cells of both *P. syringae* and *E. herbicola* on leaf surfaces expressed as ice nuclei increases sharply with decreasing temperature from −1 to −6°C.[46] At the warmest temperature of −1°C only a very low fraction (less than one cell in 10^8) is expressed as an ice

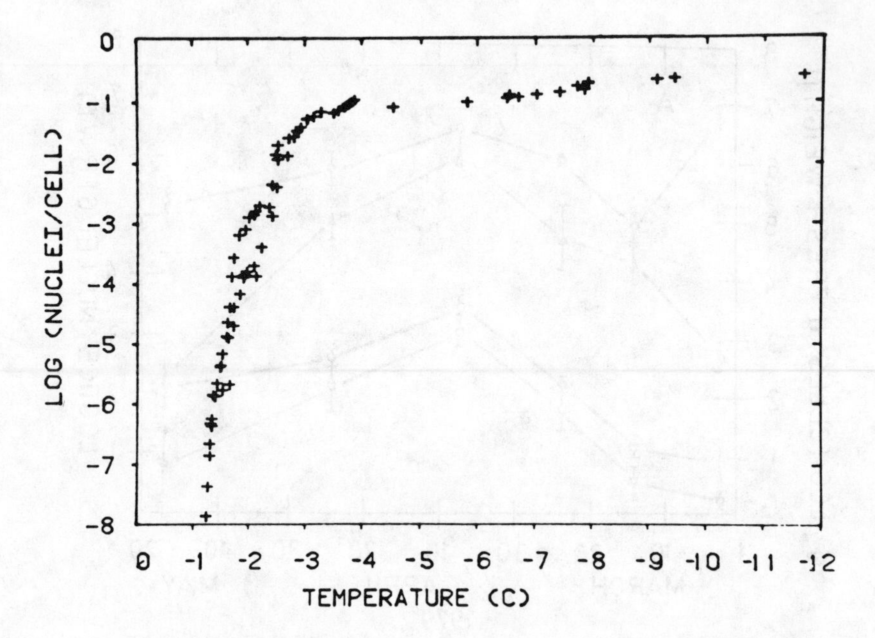

FIGURE 5. Ice nucleation activity of *P. syringae* (logarithm of fraction of cells active in ice nucleation) as a function of temperature. (From Lindow, S. E., *Plant Dis.*, 67, 327, 1983. With permission.)

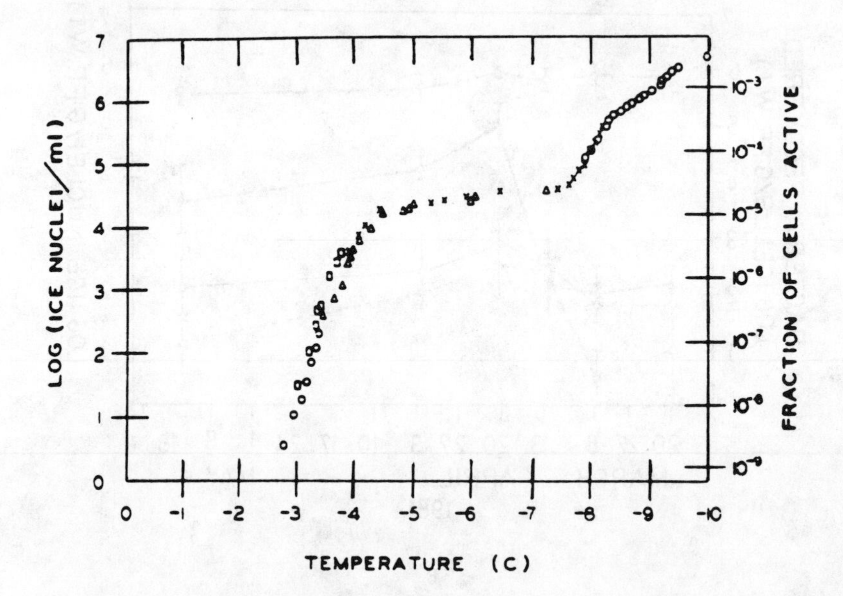

FIGURE 6. Cumulative ice nucleus concentration and fraction of cells in *E. herbicola* cell suspensions that were active in ice nucleation as a function of temperature. The different symbols represent determinations from different dilutions. (From Lindow, S. E., Arny, D. C., Barchet, W. R., and Upper, C. D., *Plant Cold Hardiness and Freezing Stress*, Li, P. H. and Sakai, A., Eds., Academic Press, New York, 1978, 249. With permission.)

nucleus; however, at −5°C up to 10% of the cells of *P. syringae* on leaf surfaces are active in ice nucleation.[46] Although many isolates of *P. syringae, P. fluorescens,* and *E. herbicola* are active in ice nucleation at temperature above −2°C, ice nucleation activity is detectable in other bacterial strains only at temperatures approaching −10°C.[65-68] Ice nucleation activity also does not appear to be a phenotypic characteristic expressed at a given time by every cell of a bacterial isolate capable of ice nucleation. Over 1000 bacterial ice nuclei active at temperatures above $-5°C$ per gram of leaf tissue have been measured on plant surfaces.[46] Thus, the epiphytic habit of the phytopathogenic bacterium *P. syringae* is a major source of ice nuclei active at small levels of supercooling on leaf surfaces as well as a reservoir of inoculum for disease initiation.

3. Population Density of INA Bacteria

The population size of epiphytic INA bacteria on plants determines the likelihood of freezing injury to plants at temperatures above -5 to $-6°C$. Nearly all plant species investigated do not contain intrinsic ice nuclei active at temperatures above -5 to $-6°C$.[37-39,55,98] The difficulty of obtaining axenically grown woody plant tissues makes the origin of ice nuclei (reported in field and greenhouse-grown woody tissues, such as peach) more difficult to determine.[99] The majority of plant species tested, however, has been shown to readily supercool below $-5°C$. Exogenous ice-nucleating agents are required to initiate damaging ice formation in frost-sensitive plant species at above $-5°C$. Since all INA bacterial species, particularly *P. syringae,* are efficient products of ice nuclei active at temperatures above $-5°C$, they have been shown to influence the supercooling point of plant tissues.[9,10,46,47,55,100] The incidence of frost damage to several plant species, both in greenhouse and in field conditions, is related directly with the logarithm of population size of INA bacteria on plants.[9,14,46,47] The average frequency of ice nucleation activity on plants appears to be approximately 10^{-3} per cell.[14,55] Since not every bacterial cell is active in ice nucleation, the incidence of frost damage to plants (incidence of ice-nucleating events on plants) is related directly with the logarithm of the number of bacterial ice nuclei per unit mass.[8]

Large reductions in the population size of INA bacteria on leaf surfaces are necessary to achieve a substantial reduction in the incidence of frost damage to plants, which is related directly to the logarithm of bacterial population size (with a relatively small value of the regression coefficient).[8] Small reductions in numbers of epiphytic bacteria, i.e., less than fivefold, do not substantially reduce the incidence of frost damage to plants.[8] It has been estimated that epiphytic populations of INA bacteria must be reduced over 50-fold to achieve a substantial reduction in the incidence of frost damage at a given temperature.

The supercooling point of individual plant parts (the lowest temperature that can be achieved before ice nucleation and, therefore, frost injury occur) is related directly with the logarithm of bacterial population size. At temperatures above $-5°C$, the probability of a given bacterial cell being active as an ice nucleus decreases logarithmically with increasing temperature. Therefore, as the number of INA bacteria on an experimental sample, such as a leaf, increases, the probability of that sample containing a higher temperature ice nucleus also increases. The percentage of leaves that freeze at a given temperature is an estimate of the frequency with which leaves contain a sufficiently large population of INA bacteria to contribute an ice nucleus.[101] One bacterial ice nucleus per leaf is sufficient to cause ice formation in the entire leaf and cause frost damage to that leaf. As shown Figure 7 for *P. syringae* strain Cit 7 (rifampicin resistant), a significant log-linear relationship between the freezing temperature of individual leaves and the population size of strain Cit 7 is observed. As the mean log bacterial population size associated with a given leaf decreases, the freezing temperature also decreases.[102] It is also clear from Figure 7, however, that large reductions in the epiphytic population size of INA bacteria are required to achieve significant (greater than 1°C) reductions in freezing temperatures of leaves.

4. Miscellaneous Conditions

In addition to temperature, type, and density of INA bacteria several *in vitro* cultural conditions including medium composition, solid vs. liquid growth medium, and growth temperature were found to effect profoundly the ice nucleation efficiency of cells of many

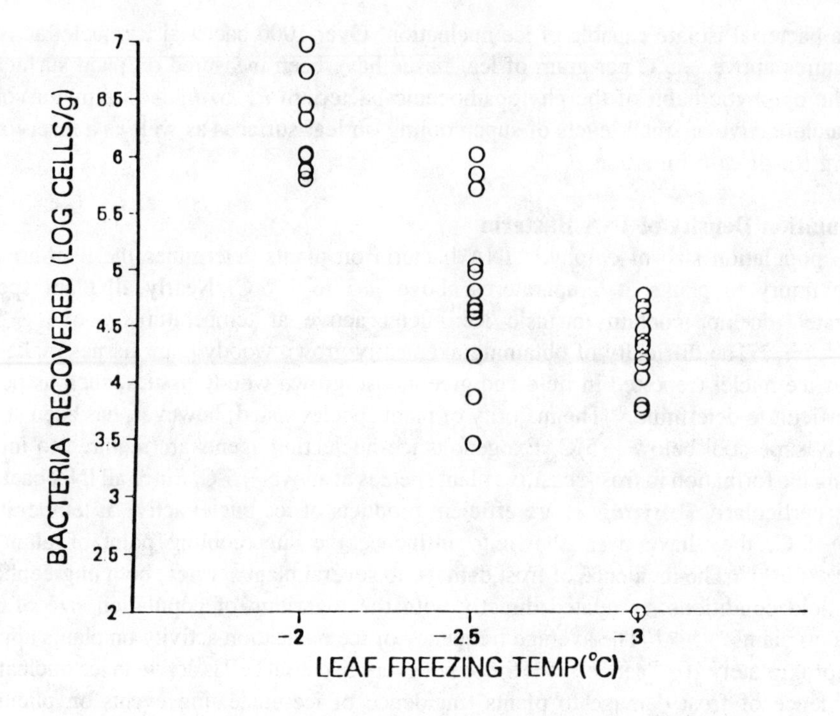

FIGURE 7. Relationship between the population size of *P. syringae* strain Cit 7 on individual bean *(Phaseolus vulgaris)* leaves and the supercooling point of these leaves. Individual bean leaves were immersed in a test tube containing 10 ml of ice nuclei-free water and cooled slowly. Leaves that froze at each of the temperatures shown on the abscissa were warmed to 20°C, sonicated in an ultrasonic cleaning bath at 20°C for 9 min, and the population size of *P. syringae* strain Cit 7 determined by dilution plating of leaf washings on King's medium B containing 100 μg rifampicin ml^{-1}. (From Lindow, S. E., *Microbiology of Phylosphere,* Fokkema, N. J. and Oorden, A. J., Eds., Cambridge University Press, New York, 1986, 293. With permission.)

INA strains of *P. syringae* and *E. herbicola*. These conditions also affect the temperature at which ice nucleation activity is expressed in these cells.[46,47,103-105] However, Maki et al.[59] reported that the cell-to-ice nucleus ratio of their *P. syringae* isolate was constant under different growth conditions.

D. QUANTITATIVE ESTIMATION OF INA BACTERIA

Development of techniques for quantitative and qualitative measurement of INA bacteria has gone a long way to investigate the bacteria responsible for plant damage. In the beginning only measurement of total number of INA bacteria at a given temperature was possible.[8-10,46,47,104] Subsequent development of a droplet freezing assay[106,107] formed the basis for most measurements of bacterial ice nuclei. This technique gives an estimate of the number of freezing nuclei, defined as those heterogeneous ice nuclei that are active when suspended in water. The major benefit of this technique is that it gives reliable and repeatable quantitative assessment of ice nuclei contributed by bacteria in almost all situations. An accurate measurement of the temperature dependence of ice nucleation has been obtained using ice nucleus spectrometers, in which the freezing temperatures of a large collection of droplets are slowly cooled.[8-10,72,104,107] A modification of this method in which aqueous suspensions of bacteria are placed in calibrated capillary tubes has been reported to increase slightly the accuracy of determination of ice nucleation temperatures.[65]

The activity of bacteria as contact ice nuclei, in which dry bacterial cells contact and nucleate supercooled water droplets suspended in an isothermal cloud chamber, has also been reported.[64,108] Determination of the supercooling point of individual leaves gives a rapid estimate of minimum populations of INA bacteria present on plant samples known to contain only bacterial ice nuclei.[62] The speed of this procedure may be useful in estimating the distribution of bacterial populations on plants, which has recently been shown to be log-normal,[62] although the accuracy of this method of estimating the bacterial population of INA bacteria is low.

The development of a new technique known as the rapid freezing technique by Lindow et al.[9] facilitated the assessment of quantitative and nonselective estimates of populations of INA bacteria. This technique has proven valuable for the quantitative assessment of the ecological impact of INA bacteria. Hundreds of individual bacterial colonies isolated on nonselective media can be scored rapidly for the potential to produce ice nuclei and INA strains which can be isolated directly from colonies held at constant subfreezing temperatures.[72]

IV. CONTROL OF PLANT FROST INJURY

A. TRADITIONAL METHODS

Protection from frost injury in plants requires physical warming of plant tissue to at least 0°C to avoid internal ice formation.[109] In physical methods of frost prevention the use of stationary wind machines or helicopters to mix the cold layer of air nearest the ground with warmer air aloft during radiation has been used.[109] In certain cases heaters have also been employed to heat the air in the vicinity of plants in need of protection.[109] During the period of cold temperature, water applied to soil by sprinklers or by furrow irrigation has been used to heat the air. Radiative heat losses can also be reduced by the application of artificially generated fogs or foam-like insulation to cover the plants. These methods reduce loss of heat from plants and retain heat otherwise lost from the soil. Another commonly used method of frost management is the application of water directly to plant parts during periods of freezing temperatures.[12] Any ice formed during such a process is limited to the exterior of the plant. Frost damage does not result as long as additional water is applied to ice-covered plant parts during the entire period, maintaining the air temperature below 0°C. The latent heat of fusion released when water freezes to form ice warms the ice-water mixture on leaves to 0°C. This mixture will remain at 0°C as long as water is continuously available to freeze on plant surfaces.[109] In addition, plants also try to resist the formation of ice and the freezing point of the plant tissue is slightly lower than 0°C.[12]

The available physical methods of frost prevention protect the plants from frost damage at ambient air temperatures of −3°C or above also pose certain problems and are not free from limitations. Application of water may result in an accumulation of ice on some plants that can cause mechanical breakage of limbs and other plant parts. Sprinkler irrigation of leaves for frost control requires large amounts of water and this method is ineffective because wind or poor sprinkler coverage prevents continuous wetting of the plants.[109] In addition, many physical methods are expensive and require a large amount of water for better implementation. Burning of large quantities of fossil fuels can deteriorate environmental quality and artificially generated fogs can create safety hazards to motorists.

B. BACTERICIDES

The physical methods as described earlier are not appealing and highly expensive. Thus, it was necessary to develop new methods for effective control. The extensive work by Lindow and associates led to the development of chemical and biological methods for the control of INA bacteria.

The first alternative method which was made available for this purpose was the use of bactericides. This method is quite effective for the control of INA bacteria. Large (100- to

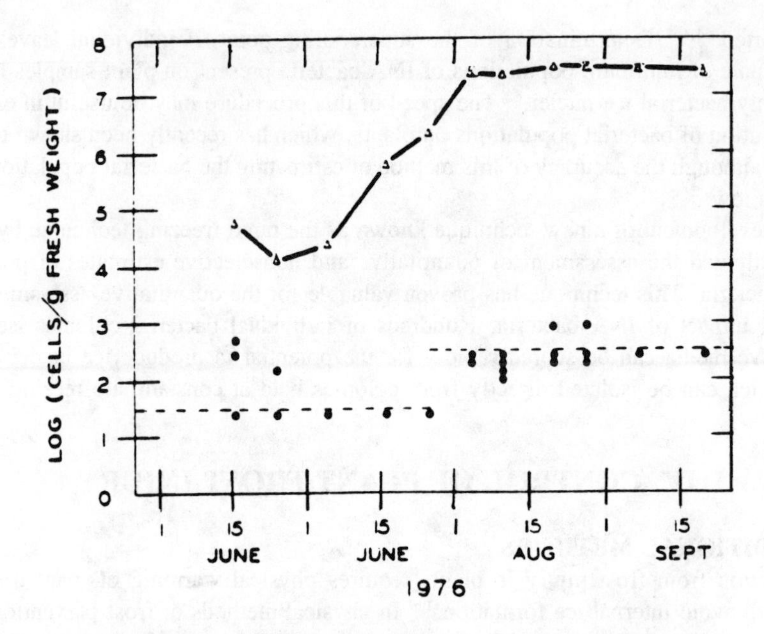

FIGURE 8. Total (△) and INA (●) bacterial populations on corn leaves sprayed with streptomycin throughout the 1976 growing season. Each individual determination of INA bacteria is represented (●). The dashed line represents the estimated limit of detection of INA bacteria. Values plotted below the detection limit represent samples from which INA bacteria were not detected. Vertical bars represent the standard error of the mean log population. (From Lindow, S. E., Arny, D. C., Barchet, W. R., and Upper, C. D., *Plant Cold Hardiness and Freezing Stress,* Li, P. H. and Sakai, A., Eds., Academic Press, New York, 1978, 249. With permission.)

1000-fold) reductions in populations of epiphytic INA bacteria are observed following protectant bactericide applications.[8-10,46,96,103,104] The number of ice nuclei on bactericide-treated plants was also significantly lower than on untreated plants,[46,47,72,104] thereby reducing the chances of frost injury to a given plant part at temperatures above −5°C. Significant frost control has been achieved with experimental applications of bactericides on several different crops such as corn (*Zea mays* L.), beans (*Phaseolus vulgaris* L.), potatoes (*Solanum tuberosum* L.), squash (*Cucurbita* spp.), and tomatoes (*Lycopersicon esculentum* Mill.).[8,46,96,97,103]

Plants treated with streptomycin (Figure 8) had about the same total bacterial population as the untreated controls,[8] but had substantially lower numbers of INA bacteria.[8] Treatment with cupric hydroxide also decreased the numbers of INA bacteria as effectively as did treatment with streptomycin, and also caused about a tenfold decrease in the total population of bacteria compared to the untreated controls.

A single ice nucleus is sufficient to cause ice formation and thus frost injury to an entire leaf, fruit, or flowers. Figure 9 shows ice propagation in a field-grown navel orange leaf exposed to natural radiative frost injury at about −4°C.[94] Typical symptoms of frost injury are apparent. Ice formation was observed initially only on an isolated water-soaked area and had not yet propagated to the periphery of the leaf as evidenced by ice formation in dew droplets only in the vicinity of the water-soaked area. The water-soaked lesions, presumably initiated by a single nucleus, expanded over a period of several minutes. The water-soaked lesions were common on leaves of untreated trees and were nearly absent on leaves of trees treated with one of the several bactericides. Fruit on such trees also escaped ice formation (Figure 10).

Some problems also exist in the use of bactericides for frost prevention. Most bactericides

FIGURE 9. Formation of ice in a citrus leaf during a natural radiative frost of approximately −4°C near Exeter, CA in December 1978. (From Lindow, S. E., *Plant Dis.*, 67, 327, 1983. With permission.)

FIGURE 10. Undamaged navel orange fruit (top) was treated with a cupric hydroxide formulation 3 weeks before a radiative frost of approximately −6°C. Untreated frost-damaged fruit (bottom) shows dryness and crystalline inclusions in segment membrane. (From Lindow, S. E., *Plant Dis.*, 67, 327, 1983. With permission.)

such as streptomycin kill growing INA bacteria rapidly on contact in culture, but these bacteria lose their ability to nucleate ice very slowly *in vitro*,[8] although exceptions have been reported.[11,63] A similar phenomenon may operate on leaf surfaces. Strategies to minimize the potential for development of resistance to INA bacteria by bactericides must also be developed if agents such as streptomycin are to be used frequently for frost management.[94] Without proper management the development of resistance to effective bactericides such as streptomycin or oxytetracycline, as seen in other phytopathogenic bacteria, may quickly preclude the use of such materials for either disease or frost management.[71,84,110,111]

Though the application of bactericides is appealing and appears to have the potential for effective frost control, much work is needed before using bactericides on a particular plant. The history of the crop, extent of damage, type of INA bacteria involved, effective doses of bactericides, nature and type of bactericides, and their persistence are some of the important parameters which must be worked out before using a chemical as bactericide. In addition, sufficient knowledge of the effects of these chemicals on bacteria other than INA bacteria should be gathered.

Lindow and Connell[14] have been successful in working out the application schedule of bactericides for the control of INA bacteria in commercially managed almond *(Prunus dulcis)*. They observed large epiphytic populations of INA strains of *Pseudomonas syringae* (up to 10^6 cell per gram fresh weight) on healthy tissues of the plants. Leaf bacteria accounted for over 99% of the ice nuclei active at temperatures higher than -5°C on almond. Large seasonal variations in populations of INA bacteria and ice nuclei on almond were observed, with maximum populations found shortly after full bloom. The populations were reduced from 10- to 100-fold by three weekly applications of bactericides starting at bud break (Figure 11), or a single application at 10% bloom of a non-INA antagonistic bacterium isolated from an almond leaf surface. The frost sensitivity of trees treated with antagonistic bacterium T7-3 was lower than untreated trees, but higher than trees treated with the bactericide cupric hydroxide or a mixture of streptomycin and oxytetracycline. The numbers of ice nuclei active at -5°C on trees treated once with a mixture of urea and sodium carbonate were reduced significantly compared to untreated controls for about 10 d following treatment. However, neither the numbers of ice nuclei active at -9°C nor the populations of INA bacteria were reduced by treatment with urea and sodium carbonate. Frost injury to detached almond spurs cooled to -3°C was reduced by all treatments of urea and sodium carbonate that reduced the numbers of bacterial ice nuclei on almond tissue. These studies clearly indicated that bactericides, ice nucleation inhibitors, and antagonistic bacteria have all the promise for reducing the ice nucleus contents of almond tissue. In addition, these studies also revealed that dynamics of distribution and stage of sensitivity of the INA bacteria, the time of the season when chemicals should be applied, must be worked out for effective control.

C. ANTAGONISTIC BACTERIA

Only a small fraction of the total bacteria found on leaf surfaces are active catalysts for ice formation. Lindow[51] reported that around 0.1 to 10.0% of the total bacteria are involved directly in ice nucleation and frost injury. In natural environments INA bacteria are present on leaf surfaces along with several other bacteria and their activity is thus dependent on these organisms. Most of the bacteria other than INA bacteria compete for space and food with INA bacteria. However, this natural competition alone is not enough to keep a check on the population of INA bacteria and to reduce frost damage. The knowledge gained on the mechanism of competition among INA and non-INA bacteria has been exploited to a large extent to reduce the population of INA bacteria. The non-INA bacteria which are the natural competitors of INA bacteria act as antagonists, and their release on the surface of the leaves can avoid the appearance of INA bacteria right from the beginning.

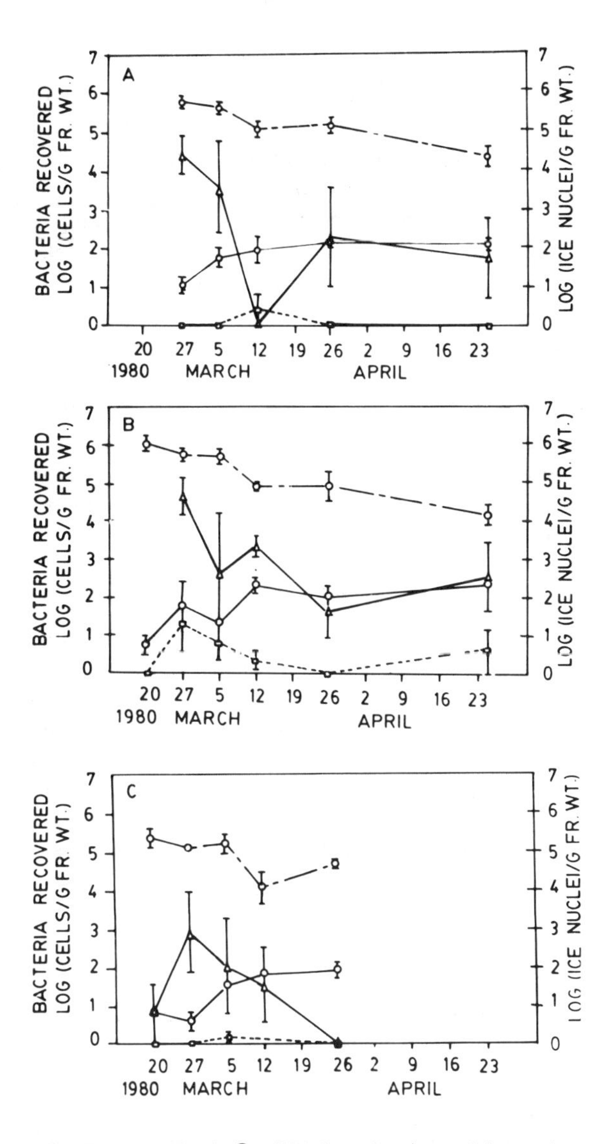

FIGURE 11. Total (○), INA bacteria (△), and ice nuclear active at −5°C (□) or −9° C (◇) on leaves, flowers, and young nuts of untreated almond grown near Turlock, CA. (A) Starting at bud break and treated with a mixture of streptomycin and oxytetracyline; (B) treated with cupric hydroxide; (C) grown near Fresno, CA and treated at 10% bloom in 1981 with antagonistic bacterium T 7-3. (From Lindow, S. E. and Connell, J. H., *J. Am. Soc. Hortic. Sci.*, 109, 48, 1984. With permission.)

The utilization of this technique for the control of INA bacteria was not so easy, and a lot of research work was done by Lindow and co-workers before implementing this method for commercial purposes.[94,96,97] The essential information which is necessary for the implementation of this method is the time of appearance of INA bacteria on a particular crop and population fluctuation between INA bacteria and antagonists. In addition, information regarding the number of competitors which should be released on the surface of the plant for effective control is also required.

Lindow and co-workers selected bacterial competitors on the basis of their antibiosis against INA bacteria *in vitro* and/or their effective colonization on leaf surfaces.[8,69,102,112]

They identified these antagonists and cultured them *in vitro* in order to raise populations large enough to be released or sprayed on the surface of the plants. The antagonists were established by foliar application of plant parts prior to colonization of these plants by INA bacteria.[69,103] In this way they succeeded in reducing significantly the population of INA bacteria,[94] and thus frost damage to the crop was reduced.

Encouraged by their initial success Lindow and co-workers employed this method to control the populations of INA bacteria on corn leaf.[8,112] A non-INA strain of *Erwinia herbicola* (M232A) competes successfully with INA strains of *E. herbicola* and *P. syringae*.[8] Populations of M232A were established on corn by spraying plants several times with suspensions of M232A during the growing season. As a result, treatments with M232A suspensions elevated the total bacterial population (both INA and non-INA bacteria) relative to untreated controls during the early part of the growing season. However, by late July, the total population was essentially the same on treated as on untreated leaves. The number of INA bacteria was substantially lower in the presence of M232A than on untreated controls throughout the growing season. Indeed, no INA bacteria were found on leaves of plants treated with M232A until late July.

Changes in frost sensitivity to field-grown leaves throughout the growing season were measured by subjecting detached leaves to $-5°C$ in a growth chamber.[8] The number of untreated corn leaves injured at $-5°C$ was relatively low early in the season when INA bacterial populations were low, and increased throughout the season. Frost injury to leaves treated with a bactericide (RH6401) that had decreased neither the numbers of INA bacteria nor the numbers of ice nuclei per gram was not different from that to control leaves. However, frost injury to leaves treated with the competing bacterium M232A, or with the bactericides that did decrease numbers of INA bacteria was substantially less than that of control leaves. The injury to detached corn leaves at $-5°C$ was directly proportional to the logarithm of the population of INA bacteria on the leaves (Figure 12). Significant reduction in colonization of *P. syringae* occurred on plants colonized by non-INA bacterium compared with plants not already colonized.[8]

1. Factors Affecting the Activity of Antagonistic Bacteria

Several biological and nonbiological factors are known to affect the efficiency of antagonists in reducing the populations of INA bacteria.

a. Time of Inoculation of Plants with Antagonists

Though antagonists reduce the populations of INA bacteria, the efficiency of the technique primarily depends on the stage of the plant at which the inoculum is applied. Sufficient reduction in frost damage was noticed only when antagonistic bacteria were applied to plants prior to or at the same time to challenge inoculations (INA bacteria).[96] If antagonistic bacteria were applied to plants after the establishment of an INA bacterial population, there was no reduction in INA bacteria. Thus, the relative time of inoculation of plants with INA bacteria greatly determines the extent of reduction in damage or the population size of INA bacteria applied as a challenge inoculum. Naturally occurring strains of *P. fluorescens* and *P. putida* applied to pear trees under field conditions reduced the population size of naturally occurring INA bacteria compared to untreated trees only when applied prior to the establishment of large population sizes of INA bacteria.[51,52] The population size of INA bacteria on treated as well as untreated pear trees was similar for several days following inoculation of trees with non-INA bacteria.[52,96,97] However, the population size of INA bacteria on untreated trees increased relative to those on treated trees after inoculation for up to 30 d.[73,96] When antagonistic bacteria were applied to pear trees only after large population sizes of *P. syringae* had developed, no subsequent differences in population size of INA bacteria *(P. syringae)* between treated and untreated trees were observed.

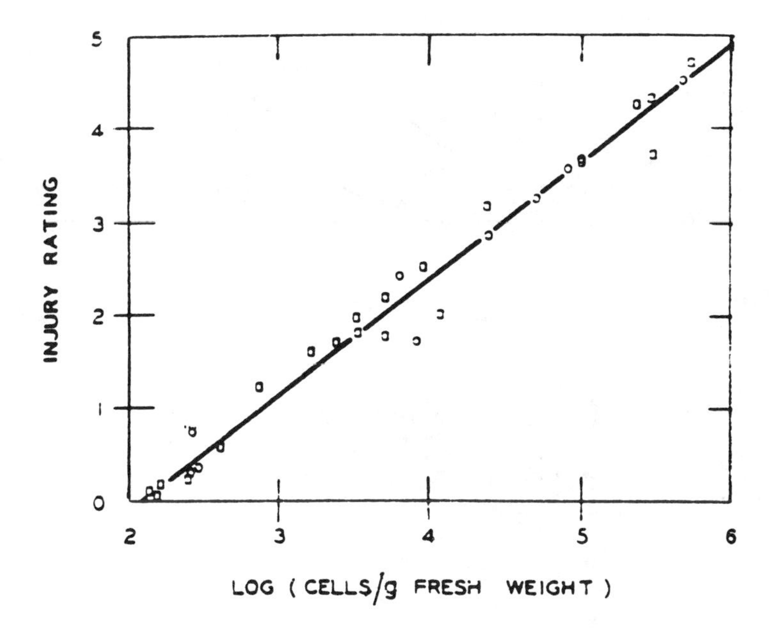

FIGURE 12. Relationship between log INA bacterial population and frost injury at −4.5 to −5.0°C. The mean number of detached corn leaves per five-leaf bundle which froze at −4.5 to −5.0°C from unsprayed control plots (○) or from plants receiving bactericides or antagonistic bacteria treatments (□). The line drawn represents the linear regression $Y = 1.24 \times -2.54$; $r = 0.983$, $p < 0.01$ for data from untreated leaves. (From Lindow, S. E., Arny, D. C., Barchet, W. R., and Upper, C. D., *Plant Cold Hardiness and Freezing Stress,* Li, P. H. and Sakai, A., Eds., Academic Press, New York, 1978, 249. With permission.)

Treatment of corn (*Z. mays* L.) seedlings in a growth chamber with an *E. herbicola* isolate (M232A) that was not active in ice nucleation decreased the amount of frost damage incited by INA isolate of *P. syringae* and *E. herbicola* at −5°C. The percentage of leaves damaged by frost decreased significantly if M232A was applied at any time before and up to 12 h after application of an INA *E. herbicola* 26 (EH26) isolate.

Populations of an INA *E. herbicola* isolate and a streptomycin-resistant mutant of M232A (M232ASR11) were estimated on maize seedlings treated with M232ASR11 before and after treatment with the INA *E. herbicola.*[96] Total populations of bacteria at the time of freezing were nearly constant (~10^7 cfu/g fresh weight), but the fraction that was INA decreased with increasing time of pretreatment with M232ASR11 (Figures 13 and 14). A significant linear correlation was found between the logarithm of INA populations of *E. herbicola* present on leaves at the time of freezing and frost injury to these leaves. M232A significantly reduced frost incited by six different *E. herbicola* and two different *P. syringae* isolates.

b. Effect of Inoculum Size of Non-INA Bacteria (Antagonists)

The effectiveness of selected non-INA bacteria to exclude INA bacteria during subsequent challenge inoculations increased with increasing relative population size of the antagonistic bacteria. The maximum epiphytic bacterial population size observed on symptomless plants is often approximately 10^7 cells g^{-1} when epiphytic populations are measured 1 or more days after inoculation.[48,73,77,96] Several physical or chemical factors probably limit the total number of these bacteria on plant surfaces.

An inverse relationship between the population size of antagonistic bacteria and that of coexisting applied INA bacteria has been observed both under greenhouse[96] and field conditions.[14] When a non-INA *E. herbicola* isolate (M232A) was applied to plants prior to INA

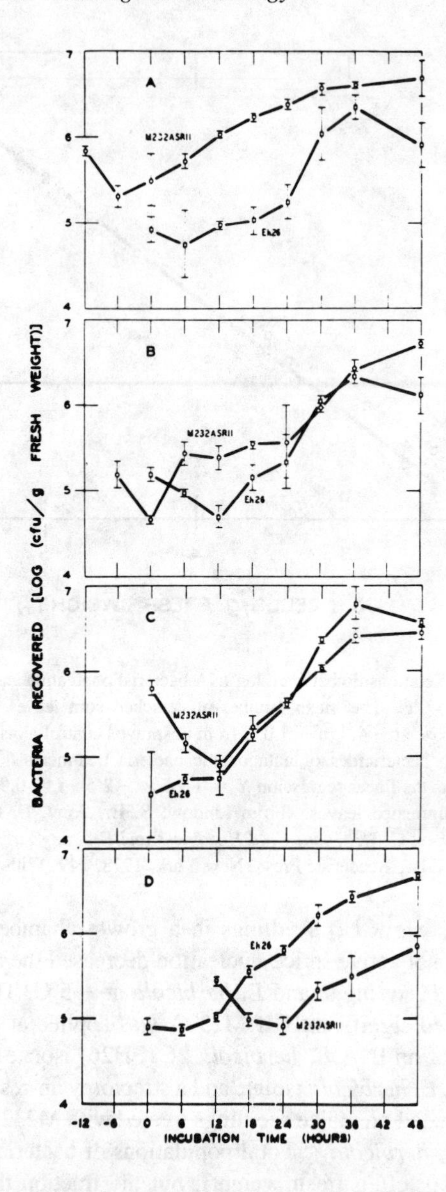

FIGURE 13. Growth of mixed populations of *E. herbicola* isolates M232ASR11 and 26 on corn seedlings and the effect of time of application of M232ASR11 relative to the application of Eh 26. Eh 26 (□, 1.2 × 10^8 cfu/ml) in phosphate buffer was applied to all plants at incubation time marked "O". Plants were sprayed with M232ASR11 (9 × 10^7 cfu/ml) in phosphate buffer at the relative times indicated by the first data point for each curve (i.e., 12 h before Eh 26 in the set [A] 6 h before Eh 26 in [B], etc.) and incubated in a mist chamber. Each value represents the mean log of three determinations of bacterial population. The vertical bars represent the standard errors. (From Lindow, S. E., Arny, D. C., and Upper, C. D., *Phytopathology*, 73, 1097, 1983. With permission.)

strains of this species, it grew rapidly and comprised over 90% of the total bacteria found on plants.[96] Similarly, when the INA strain of *E. herbicola* was inoculated onto plants prior to a non-INA strain of this species, it grew rapidly and comprised the majority of bacterial cells recoverable from plant surfaces.[96] Frost damage to corn seedlings decreased with increased densities of non-INA *E. herbicola* isolate M232A (10^5 to 10^9 cfu/ml) applied 24

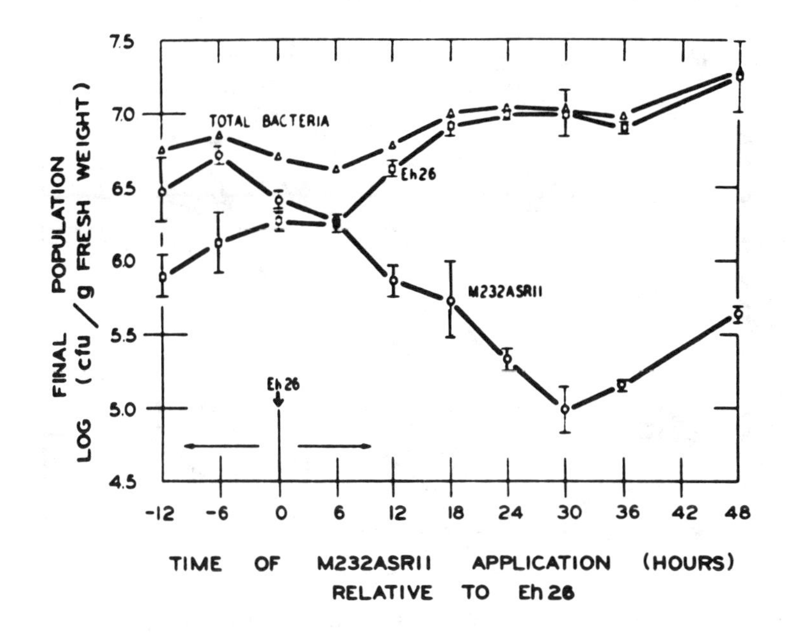

FIGURE 14. Effect of time of application of the antagonist (*E. herbicola* M232ASR11) relative to the INA isolate (*E. herbicola* 26) on final bacterial population (48 h after application of Eh 26) in the experiment described in Figure 13. After Eh 26 had been on the plants for 48 h, populations of M232ASR11 (○), Eh 26 (□), and total bacteria (△) were determined. The vertical bars represent the standard errors. (From Lindow, S. E., Arny, D. C., and Upper, C. D., *Phytopathology*, 73, 1097, 1983. With permission.)

h before challenging with an INA *E. herbicola* (8×10^5 or 8×10^6 cfu/ml). The cell densities of INA *E. herbicola* isolate 26 (Eh26) also affected the damage. Thus, although frost damage was significantly reduced by M232A at all cell densities of Eh26, the percent reduction in injury provided by M232A compared with Eh26 alone decreased as the density of Eh26, which was applied, increased. M232A was found to reduce frost damage incited by INA pseudomonas 31 (Ps 31) as well as Eh 26 (Figure 15). The largest reduction in damage to plants challenged with Ps31 occurred when less than $\sim 10^6$ cfu of Ps per millimeter was used. The M232A-mediated reduction of frost damage was considerably less with challenges of Ps31 at $\sim 10^6$ cfu/ml. Accordingly, a greater degree of reduction of the population of *P. syringae* by M232A would be required to achieve a comparable reduction to frost injury. Similarly, frost damage to plants treated with a single application of an INA *E. herbicola* strain decreased as the concentration of non-INA *E. herbicola* cells applied prior to the INA strain increased up to 10^8 cells ml^{-1}. Therefore, it appears that modification of the leaf surface microflora by preemptive exclusion rather than displacement mechanisms is required, in addition to the development of a sufficiently large population size of the antagonistic bacterium on the plant surface.[96] Thus, if both antagonistic and target organisms have similar growth rates on plant surfaces, the arrival of antagonistic organisms on plants must precede that of target organisms, so that a large population size will be achieved prior to any multiplication of target organisms. Early and rapid epiphytic colonization of leaves also appears to be a prerequisite for biological control of frost injury and probably for modifications of leaf surface bacterial populations.

c. Time of Inoculation in Relation to Host Plant

Lindow et al.[96,97] have emphasized the need to use bacterial antagonists on newly exposed tissue of perennial plants. The treatment at this stage is most effective when applied at the

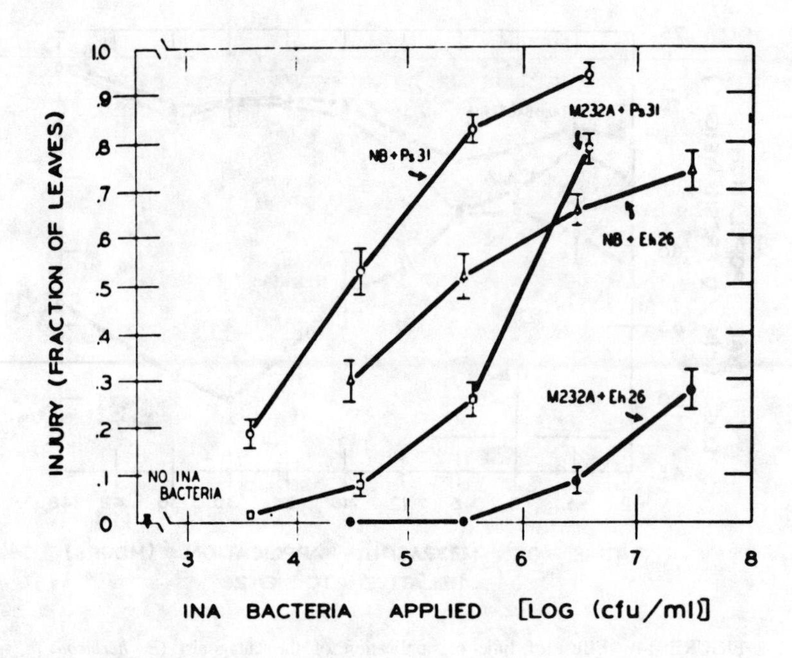

FIGURE 15. Frost damage to corn seedlings pretreated with the antagonist (*E. herbicola* M232A) and nutrient broth (NB) challenged with a range of cell densities of the INA isolates *E. herbicola* 26 and *P. syringae* 31. Three days before exposure to −4.5°C, three sets of seedlings were sprayed with M232A (3 × 10^8 cfu/ml) in nutrient broth (●, □, and ▲) and two sets with nutrient broth alone (○ and △) and then incubated in a mist chamber. One day later plants were sprayed with Ps 31 (○ and □) or Eh 26 (● and △) and in phosphate buffer and were returned to the mist chamber. Controls (▲) were not treated with either Eh 26 or Ps 31. The vertical bars represent the standard errors. (From Lindow, S. E., Arny, D. C., and Upper, C. D., *Phytopathology*, 73, 1097, 1983. With permission.)

onset of favorable environmental conditions (cool, moist weather) before significant colonization and/or multiplication of INA bacteria has occurred on these plants. Lindow[94] demonstrated that antagonistic bacterium A506, when applied to pear trees as a foliar spray at approximately 20% bloom during the first week of April 1979, reduced significantly the appearance of INA bacteria. The population of INA bacteria was over 1000-fold greater on expanding untreated leaves than on treated ones. The primary effect of treatment was prevention of the increase of INA bacteria active at −5°C, which would have otherwise appeared on untreated plants during April. The treatment with antagonistic bacteria also reduced the number of INA bacteria active at −9°C.

The populations of INA bacteria are low shortly after the emergence of the annual plants, and subsequently large epiphytic populations on untreated plants appear.[8,46,47,51,52] Antagonistic bacteria have been applied to seeds of annual plants or as a foliar spray shortly after emergence of these plants. Under field conditions, populations of INA bacteria have been found to decrease from 10- to 10,000-fold on such plants treated with antagonistic bacteria.[8,69,103,104] Thus, antagonistic bacterial treatment at a particular time prevents the subsequent increase of INA bacteria that otherwise would appear in large numbers. Reductions in frost damage to treated plants had varied in magnitude, but was related directly to reductions in populations of INA bacteria.[46,47,104] Efficient bacterial antagonists effectively colonize the mature tissues for a relatively long period of time (1 to 4 months) following a single foliar application.[51] Most antagonistic bacteria thus appear to influence frost sensitivity of the plants on which they reside, much like a protectant bactericide, by limiting the populations of INA bacteria on leaf surfaces throughout a period of freezing conditions.

Some bacterial strains readily colonize many different hosts, whereas other strains effectively colonize only the host from which they are originally isolated. Thus, some degree of host specificity is also observed among antagonistic bacterial strains as demonstrated by differences in colonization of a given host.

D. CONSTRUCTION AND USE OF ICE$^-$ BACTERIA FOR THE CONTROL OF INA BACTERIA

1. Construction of Ice Nucleation-Deficient Bacteria

Until recently, the nature of the substance(s) responsible for ice nucleation activity by INA bacteria was not known. However, ice nucleation activity is associated with the bacterial cell (i.e., nondiffusible) evidently localized on the outer membrane. One or more proteins are involved, directly or indirectly, in its expression.[113] Mutants deficient in ice nucleation activity have been isolated by Lindow and Staskawicz.[104] Lindow and co-workers have been able to successfully clone the ice nucleation genes from *P. syringae* and *Erwinia herbicola*. The most detailed genetic analysis of cloned ice genes has been made of a clone from *P. syringae* Cit 7.[113] Initially, Orser et al.[114] used several bacterial strains and plasmids. The initial cloning and subcloning and other vector constructions are shown in Figure 16. Representative of the plasmids isolated from *Escherichia coli* HB101 Ice$^+$ clone were $_p$ICE3 and $_p$ICE7, originating from genomic libraries of *P. syringae* Cit7R1 and *E. herbicola* 26SR6-2, respectively.[115] The fragments that carried the ice genes were subsequently identified by subcloning and Tn5 insertion analysis and by transfer of Ice$^-$ mutants of *P. syringae* and was found to be a small fragment of chromosomal DNA.Southern blot hybridization of total DNA from Cit7R1 digested with various restriction enzymes and probed with 32p-labeled DNA from subcloned ice inserts demonstrated colinearity of the cloned fragment with the chromosomal ice region of this strain. The ice region from *P. syringae* Cit7R1 was further characterized by combination of subcloning, deletion, and Tn5 insertion analysis. The ice region spanned 3.5 to 4.0 kilobases (kb) and was continuous over the region in *P. syringae* Cit7R1. The ice region of *E. herbicola* spanned approximately 4.5 kb.[114]

Ice genes from *P. syringae* and *E. herbicola* completely reproduced the ice phenotype in *E. coli*. Ice$^+$ transductants carrying cosmid clones such as $_p$ICE1 of $_p$ICE7, and expressed ice nucleation activity. The ice nucleation activity of *E. coli* strains containing these and two other cloned ice genes reflected the characteristic of ice nucleation activity of the DNA source strains when grown at their respective optimum temperature for expression of ice nucleation activity, but the optimum temperatures differed slightly (Figures 17 and 18). *E. coli* strains containing cloned ice genes from *P. syringae* and *E. herbicola* expressed maximum ice nucleation at a temperature about 3°C lower than the respective DNA source strain. In view of the low number of cells expressing ice nucleation activity in both the *E. coli* recombinants and DNA source strains, cloning the ice genes in high copy number vector or downstream from plasmid promotors was used to increase their level of expression. Plasmids PICE1.1, carrying a 9.5-kb ice insert from strain Cit7R1, cloned into $_p$BR325, whereas plasmids carrying the 4.5-kb ice insert cloned into the high-copy-number vector $_p$BR322, $_p$UC8, or $_p$UC9 were cloned into $_p$ICE1.2, $_p$ICE1.8, and $_p$ICE1.9 (Figure 19). The plasmids were introduced into strains HB101 and JM83 by transformation. Substantial increases in nucleation activity of both *E. coli* and *P. syringae* were obtained by subcloning DNA fragments on multicopy plasmid vectors.

There was a substantial humology between the ice regions of *P. syringae* and *E. herbicola,* although individual restriction sites with *ice* region differed between species. DNA sequences, internal to the clonal *P. syringae* Cit7RI ice region when hybridized to total genomic DNA from *E. herbicola,*[114] produced a lower intensity hybridization signal in *E. herbicola* DNA than that observed with DNA from the homologous *P. syringae* strain. The number and location of restriction sites within or near the ice region of these two species

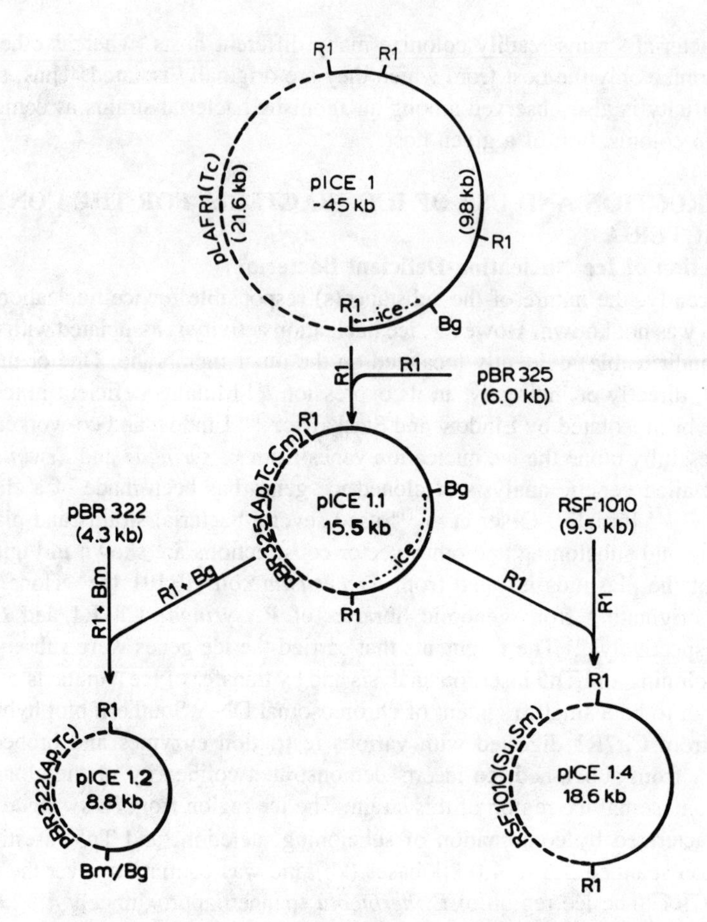

FIGURE 16. Pedigrees of some plasmids carrying the *P. syringae* Cit7R1 *ice* region. Abbreviations: *S*, SalI; RI, EcoRI; Bg, *Bgl*II; Bm, BamHI; Tc, tetracycline; Ap, ampicillin; Cm, chloramphenicol; Sm, streptomycin; Su, sulfonamide; Bm/Bg, BamHI-*Bgl*II fusion site. The enzymes used to digest each plasmid are indicated on the arrows. The Ice phenotype of each plasmid is indicated in Figure 19 or discussed in the text. (From Orser, C. S., Staskawicz, B. J., Panopoulos, N. J., Dahlbeck, D., and Lindow, S. E., *J. Bacteriol.*, 164, 359, 1985. With permission.)

also differed considerably. DNA conferring the Ice$^+$ phenotype was localized on different-sized EcoR1 restriction fragments in *P. syringae* and *E. herbicola*. In addition, the ice region of *E. herbicola* did not contain restriction sites for SalI or PvuII, whereas these enzymes produced multiple restriction fragments of the *P. syringae* Cit7RI *ice* region.

The species of INA bacteria discovered so far exhibit at least some DNA sequence homology with the *P. syringae* Cit7 ice gene. However, the degree of homology varies intraspecifically and interspecifically. Lindow[74] found that *P. syringae* TLP2 (isolated from a healthy potato leaf surface at Tulekake, CA) had a high degree of DNA sequence homology with *P. syringae* Cit7 (isolated from a healthy citrus leaf surface near Exeter, CA). In these two species, however, cloned ice genes of both of these *P. syringae* strains contain at least two internal SalI restriction sites. The positions of restriction endonuclease sites internal to the ice gene differed considerably.[74]

Lindow and associates were able to isolate the ice region of *P. syringae* strains Cit7 and TLP2.[74,114] This permitted the construction of site-directed deletion mutants within this gene in both of these strains. Deletions internal to the ice region of *P. syringae* Cit7 and

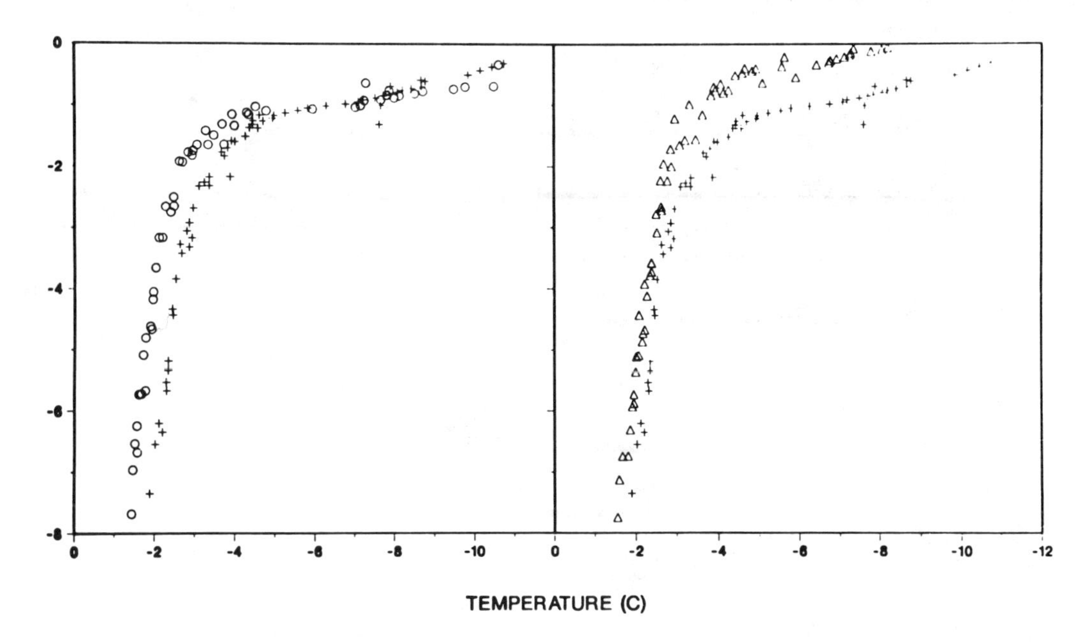

FIGURE 17. Ice nucleation of *P. syringae* Cit7R1 (+), *E. coli* HB101 (pICH1) (○), and *E. coli* HB101 (pICE1.1) (△). Forty 10-μl droplets of appropriate serial dilutions of bacterial cultures were placed on an ice nucleation spectrometer. The rate of decrease in temperature, measured with a Digitec Thermistor thermometer, was 0.1°C/min. The freezing of individual droplets was recorded as the temperature decreased. (From Orser, C., Staskawicz, B. J., Panopoulos, N. J., Dahlbeck, D., and Lindow, S. E., *J. Bacteriol.*, 164, 359, 1985. With permission.)

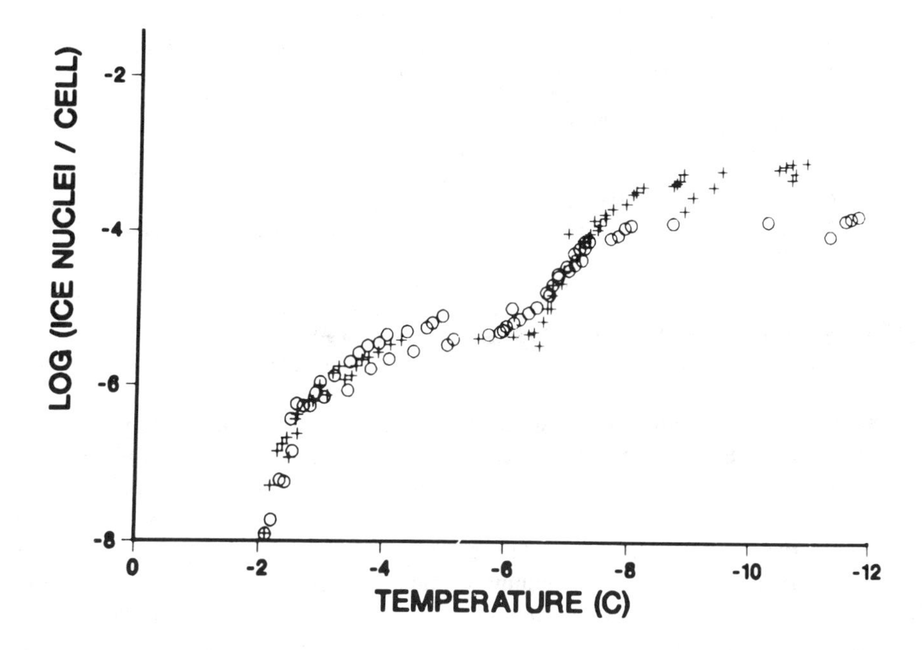

FIGURE 18. Ice nucleation spectra of *E. herbicola* 26SR6-2 (+) and *E. coli* HB101 (pICE7) (○). Assay and growth conditions were as described in the legend to Figure 17. (From Orser, C., Staskawicz, B. J., Panopoulos, N. J., Dahlbeck, D., and Lindow, S. E., *J. Bacteriol.*, 164, 359, 1985. With permission.)

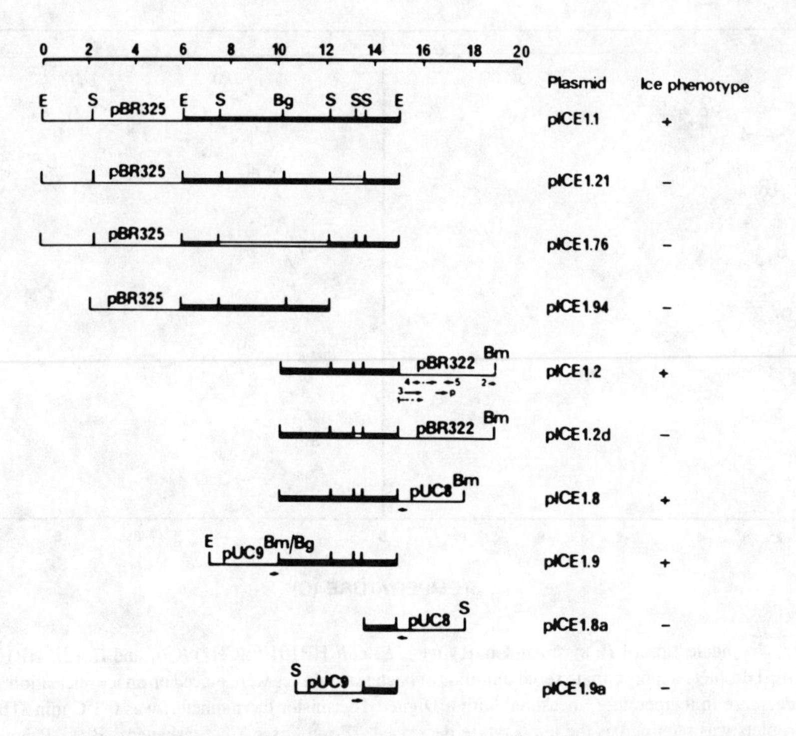

FIGURE 19. Deletion and subcloning analysis of the *ice* region of *P. syringae* Cit7R1. Only some restriction endonuclease sites are shown. Each site is designated once in the map of the plasmid in which it appears first, and all vertically aligned sites are the same. All plasmids are shown as linearized molecules aligned at the right-most EcoR1 end of their ice inserts (except pICE1.94, in which it is deleted). Deletions are shown as open boxes and vectors as thin lines. Abbreviations: E, *Eco*R1; S, *Sal*I; Bm, *Bam*HI; Bg, *Bgl*II; Bm/Bg, *Bam*HI-*Bgl*II fusion site. Arrows indicate the location, direction, and approximate range of influence (downstream) of various vector promoters. Promoters 1, 2, 3, 4, 5, and p on pICE1.2 correspond to P1-P2, P3, P4, P5, Pp, respectively in pBR322. The lactose promoter on pUC8- and pUC9-derived plasmids is shown by an arrow below each of the maps. Dashed-line arrows indicate the read-through activity of the P4 and the P1 and P2 promoters of pBR322, which were assumed to be deleted during the construction of p1CE1.2 and p1CE1.2d. The deletion in pICE1.94 includes the pBR325 segment between the EcoR1 and Sal1 sites in the Tc^r gene. (From Orser, C., Staskawicz, B. J., Panopoulos, N. J., Dahlbeck, D., and Lindow, S. E., *J. Bacteriol.*, 164, 359, 1985. With permission.)

TLP2 were constructed *in vitro* by digestion with SalI endonuclease followed by religation. Hybrid plasmids containing DNA sequences from the ice region, but lacking one SalI restriction fragment internal to the ice region (and therefore incapable in conferring ice nucleation), were obtained and propagated in *E. coli*. Site-directed mutagensis of the homologous wild-type chromosome was obtained by Lindow[74] by reciprocal exchange of the appropriate plasmid-borne deletion-containing ice region according to the method described by Ruvkun and Ausubel.[115] A recombination event between regions of homology of the chromosome and partial ice region carried by the plasmid $_p$BR325, which does not replicate in *P. syringae*, caused integration of the vector and deletion-containing ice region into the chromosome of the recipient strain.[74] Growth of a collection of such *cis*-merodiploid strains of *P. syringae* Cit7 and TLP2 (in the absence of selection for tetracycline resistance conferred by the integrated plasmid vector) was performed to allow the accumulation of Ice$^-$ haploid exocisants within the population of cells due to a recombinational event distal to the end of the *P. syringae* ice region at which the first recombination event had occurred. Such excisant

P. syringae strains were detected by their reversion to tetracycline sensitivity upon loss of the nonreplicative plasmid vector $_{p}$BR325. Southern blot analysis of wild-type *cis*-merodiploid, and fully haploid excisant strains of *P. syringae,* obtained after resolution of *cis*-merodiploid strains by using as a hybridization probe, confirmed the expected genomic organization of the ice region. Progeny isolated after site-directed mutagenesis either had reverted to a wild-type ice gene organization and Ice$^+$ phenotype or had retained the deletion-containing ice region of the appropriate homologous strain and were phenotypically Ice$^-$.

The gene conferring ice nucleation activity was thus first cloned from *P. syringae* strain Cit12.[114] Warren et al.[116] identified the corresponding genes by DNA hybridization with cloned Cit7 gene of plasmid $_{p}$ICE.1, in naturally occurring isolates of *P. syringae* (S203) and *P. fluorescens* (MS1650). In each case the hybridizing sequence was shown to be present as a single copy per genome. They constructed the deletion mutants of *P. syringae* and *P. fluorescens in vitro* and introduced into the bacteria by marker exchange. In other diagnostic and colonization tests the mutant strains were indistinguishable from their wild-type progenitors.

2. Use of Ice$^-$ Site-Directed Deletion Mutant of *Pseudomonas syringae* for Frost Control

Seedling of corn, bean, and potato plants colonized by *Ice*-deletion mutants of two *P. syringae* strains can be cooled to $-5°C$ before significant frost injury results.[74] This site-directed Ice$^-$ deletion mutant of *P. syringae* Cit7 and TLP2 does not cause frost damage to plants. In addition, the frost damage to corn plants treated with Ice$^-$ deletion mutants of strains Cit7 or TLP2 does not differ significantly from that to untreated greenhouse-grown plants at all temperatures above $-7°C$.

Site-directed Ice$^-$ deletion mutants of *P. syringae* TLP2 (such as TLP2 dell) were found to be highly antagonistic to *Ice$^+$ P. syringae* strains on leaf surfaces. *P. syringae* TLP2 dell grew logarithmically on bean leaf surfaces and achieved a maximum population size in excess of 10^6 cells per gram fresh weight.[74] INA *P. syringae* B728A did not exceed a population size of approximately 10^2 cells per gram fresh weight when inoculated onto plants colonized by strain TLP2 dell. However, when B728A was applied to a subset of plants not previously colonized by strain TLP2 dell, it grew logarithmically and reached a population size in excess of 10^4 cells per gram fresh weight within 2 d. Thus, site-directed Ice$^-$ deletion mutants of *P. syringae* such as TLP2 dell appear to be effective in competitive exclusion of other *P. syringae* strains on plant surface when high populations of these Ice$^-$ strains can be established on plants.

The main aim of production of Ice$^-$ bacteria as projected by Lindow and co-workers is to release this bacterium for the control of INA bacteria without causing any change in ecology of the microbial population on leaf surface. Lindow[74] also determined influences of several variables including physical, chemical, and biological environmental conditions on such Ice$^-$ bacteria. In addition, interaction of Ice$^-$ bacteria with normal bacterial populations under field conditions was also studied. Lindow[74] proposed a field test of *P. syringae* TLP2 dell and an Ice$^-$ mutant of *P. syringae* Cit7 (Cit7 dellB) on potato plants at the University of California Agricultural Experiment Station at Tulelake. The proposed field trial was to be conducted in approximately 0.16 ha of potato fields which were treated with nine different chemical and bacterial treatments including Ice$^-$ *P. syringae* Cit7 dellB and TLP2 dell (with and without subsequent foliar sprays of the homologous respective Ice$^+$ wild-type strain). Additional treatments included the use of two Ice$^+$ *P. syringae* strains, a naturally occurring non-INA strain of *P. fluorescens,* and a bactericide. All treatments were to be replicated four times in a randomized complete block design. The plot area was surrounded on all sides by at least 10 m of soil which was kept free from all other plant species. The presence of Ice$^-$ mutants and wild-type strains both inside and outside of the plot area was to be monitored periodically by isolating bacteria from plant and insect samples on selective antibiotic-containing media. The presence of INA strains both within and outside of the plot area was also monitored by a replica freezing technique after their isolation on a nonselective medium such as King's medium B. The population size of applied bacteria and the predominant microbial species on plants within and outside the plot area was determined both as an indication of the safety of this experiment as well as to provide the desired information on the specificity with which interactions of bacteria occurred on plant surfaces.

The legal aspects of using ice nucleation-deficient bacteria for frost protection are responsible for the attention accorded to this research by the popular media. The novel interest here was that the field tests proposed would have involved the first deliberate release of genetically engineered bacteria. In the U.S., the National Institute of Health (NIH) already had established a Recombinant Advisory Committee (RAC) to consider the hazards of accidental release. The RAC guidelines show that the accidental release of engineered *P. syringae* is not considered a serious problem, since the construction of such a strain is permitted under low levels of physical containment. The key difference between the accidental and deliberate release is in the number of organisms likely to be introduced into the environment. Thus, the main question was whether the dispersal of large numbers of engineered bacteria would cause significant environmental impact beyond the field plot where they were to be applied.

Lindow and Panopoulos applied to RAC for permission to release ice nucleation-deficient bacteria; permission was denied because of their initial plan to replace the ice nucleation gene with an element conferring resistance to kanamycin. Their second application was approved but legal constrains prevented execution in the field test. In 1984 the Environmental Protection Agency (EPA) determined that ice nucleation-deficient bacteria should be considered a form of pesticide. The classification will probably apply to most bacterial products with agricultural applications, but not to other types of recombinant organisms. The existing procedures for regulation of pesticides dictated that an experimental permit (EUP) would be required for field testing of bacteria. The EPA is responsible for granting an EUP, basing their decision of a review of submitted data.

With the development of the Ice$^-$ genetically engineered microbe, the first authorized field trials of this organism for the control of frost injury to plants began during spring of 1987, making this year a watershed year for biotechnology.[48] The proposed field trial by Lindow and co-workers, though delayed by 3 years, was launched for field tests of genetically modified ''ice minus'' bacteria that are supposed to prevent frost damage to crop plants. Additional requests for approval of studies that require the deliberate introduction of genetically engineered microorganisms into the environment are in the regulatory pipeline, a technology taking the first step to what may eventually be a large-scale application outside. The release of ice minus bacteria in to the field through the continuous efforts of Lindow and co-workers has given new spark to an issue that has dogged the 15-year history of recombinant DNA technology, especially with regard to its use in producing genetically altered microbes.

Are such microbes safe or dangerous in some way to man or other species? No indication of the hazard has emerged during years of laboratory experiments with recombinant DNA, but the environmental release of recombinant microbes has raised concerns that the organisms might disrupt the ecosystem, perhaps displacing the indigenous bacterial species or otherwise causing harm to plants or animals. Previous experiences with pests such as the gypsy moth and Africanized ''killer'' bee are frequently cited examples of what can go wrong when new species are introduced into the environment.

The current approach of determining the introduction of a particular genetically engineered microbe into the environment (constitutes a possible hazard) focuses on finding the answers to the following:

1. Will the released organism survive?
2. Will it multiply?
3. Will it spread beyond its original area of application?
4. Can it transfer its genetic material to other organisms?
5. Will the original organism or any of those that might pick up its genes prove harmful?

If one looks into the work done on ice minus bacteria before their release into the field by Lindow and associates, one feels convinced that sufficient data have been collected to provide satisfactory answers to the first three questions raised. They found that ice minus bacteria survived and multiplied in the environment after their release. The possibilities of this bacterium spreading beyond its original area of application were ruled out. However, answers to the remaining two questions are not easy because the nature of microbes is so complex that they can even do things which are not expected. The seriousness of the situation will, however, depend upon the type of genetically engineered organism to be released. Thus, a general method cannot be devised that can work with all such organisms.

Extensive work was done by Lindow and associates on the utilization of carbon and nitrogen sources and antibiotic sensitivity profiles of ice minus bacteria.[94,114] In order to satisfy questions of their identity, the specificity of genetic modifications and safety to applicators as well as to the environment as a whole are required. Biochemical reactions were performed for both Ice$^-$ mutants and wild-type Ice$^+$ strains of *P. syringae* proposed for use in field experiments (Table 1). While the Ice$^+$ strains Cit7 and TLP2 differed from each other in certain characteristics such as carbon sources utilization, both clearly matched with the description of *P. syringae* van Hall. The Ice$^-$ site-directed deletion mutants of Cit7 and TLP2 did not differ from their parental strains in any of the characteristics tested.[74]

Most bacterial isolates designated as *P. syringae* are isolated from symptomatic plants and are pathogenic on one or more plant species. These strains are usually designated as pathovars[60] within this species on the basis of such host responses. However, strains Cit7 and TLP2 were isolated from asymptomatic plants, and neither they nor their deletion mutants exhibited pathogenicity to any of an extensive list of plants tested as potential hosts for these strains. Lindow[74] inoculated several plants by different methods with wild-type Ice$^+$ and Ice$^-$ deletion mutants of *P. syringae* Cit7 and TLP2. The major plants inoculated included *(Avena sativa)* cv. cayuse, barley *(Hordeum vulgare)* cv. Steptoe, tomato *(L. esculentum)* cv. Bonny Best. potato *(S. tuberosum)* cv. Russet Burbank, corn *(Z. mays)*, wheat *(Triticum aestivum)* cv. Twin, citrus *(Citrus sinensis)*, almond *(Prunus dulcis)* cv. Ne Plus, pear *(Pyrus cummunis)* cv. Bartlett, cherry *(Prunus avium)* cv. Bing, bean *(Phaseolus vulgaris)* cv. Kidney, cowpea *(Vigna unguiculata)*, horseradish *(Armoracia rusticana)*, alfalfa *(Medicago sativa)*, onion *(Allium cepa)*, tobacco *(Nicotiana tabacum)* cv. Turk, *Gomphrena globosa, Chenopodium amerantacolor,* mustard *(Discurainia sophia)*, wild oats *(Avena fatua)*, dandelion *(Taraxacum offcinale)*, pepper *(Capsicum annum)* cv. R10, pepper *(C. annum)* cv. Cal Wonder, quack grass *(Agropyron repens)*, black nightshade *(S. nigrum)*, American black nightshade *(S. nodiflorum)*, petunia *(Petunia × hybrida)* cv. Purple Cascade, barley *(H. vulgare)* cv. Larker, wheat *(Triticum aestivum)* cv. Modoc, mallow *(Malva rotundifolia)*, pea *(Pisum sativum)* cv. Tinga, *Bromus orcuttianus, H. glaucum, B. tectorium* var. *glabratus, Asclepias fascicularis,* purslane *(Portulaca oleracea)*, oats *(Avena sativa)* cv. Sierra, perennial ryegrass *(Lolium perenne)*, hairy vetch *(Vicia villosa)*, petunia *(Petunia × hybrida)* cv. Mitchell, buttercup *(Ranunculus* sp.), red clover *(Trifolium pratense)*, and *Chenopodium guinea*. Neither wild-type nor Ice$^-$ deletion mutants were pathogenic on any of the plants tested. *Pseudomonas syringae* Cit7, TLP^2, Cit7 dell, and TLP2 dellB gave a hypersensitive

TABLE 1
Characteristics of the Wild-Type and Ice⁻ Deletion Mutants of *P. Syringae*[a]

Characteristic	Cit7	Cit7 dellB	TLP2	TLP2 dell
Ice nucleation	+	−	+	−
Respiration	Strict aerobe	Strict aerobe	Strict aerobe	Strict aerobe
Oxidase	−	−	−	−
Arginine dihyrdolase	−	−	−	−
Pectolytic enzymes	−	−	−	−
Levan production	+	+	−	−
Tobacco hypersensitivity	+	+	+	+
Gram stain	−	−	−	−
Poly-β-hydroxybutyrate	−	−	−	−
Fluorescence	+	+	+	+
Potato rot	−	−	−	−
Urease	−	−	−	−
β-Galactosidase	−	−	−	−
Lysine decarboxylase	−	−	−	−
Ornithine decarboxylase	−	−	−	−
Denitrification	−	−	−	−
Motility	+	+	+	+
Color	White	White	White	White
Maximum growth temp	35°C	35°C	35°C	35°C
Bacteriocin group	None	None	12	12
Utilization of				
Acetamide	+	+	+	
Acetate	+	+	+	+
cis-Aconitate	+	+	+	
Adipate	−	−	−	
Adonitol	−	−	−	
D-Alanine	+	+	+	
L-Alanine	+	+	+	
β-Alanine	+	+	+	
Allantoate	−	−	−	
p-Aminobenzoate	−	−	−	
D-Aminobutyrate	−	−	−	
L-Aminobutyrate	−	−	−	
α-Aminobutyrate	+	+	+	
α-Aminoethanol	−	−	−	
α-Aminovalcrate	+	+	+	
N-Amylamine	−	−	−	
Anthranilate	−	−	−	
D-Arabinose	−	−	−	−
L-Arabinose	+	+	+	
D-Arabitol	−	−	−	
L-Arabitol	−	−	−	
D-Arginine	−	−	−	
L-Arginine	+	+	+	+
L-Ascorbate	−	−	−	
D-Asparagine	−	−	−	
L-Asparagine	−	−	−	
D-Aspartate	−	−	−	
L-Aspartate	+	+	+	+
Azelate	−	−	−	
Benzoate	−	−	−	
Bentaine	+	+	+	
2,3-Butanediol	−	−	−	
Butanol	−	−	−	
Butyrate	+	+	+	
Caprate	+	+	+	

TABLE 1 (continued)
Characteristics of the Wild-Type and Ice⁻ Deletion Mutants of *P. Syringae*[a]

Characteristic	Cit7	Cit7 dellB	TLP2	TLP2 dell
Caproate	+	+	+	
Caprylate	+	+	+	
Choline	+	+	+	
Citraconate	–	–	–	
Citrate	+	+	+	+
Citronelate	–	–	–	
L-Citrulline	–	–	–	
L-Cysteine	–	–	–	
Dulcitol	+	+	+	
Erythritol	+	+	+	
Ethanol	+	+	+	
Fructose	+	+	+	+
α-L-Fucose		–	–	
Fumarate	+	+	+	
D-Galactose	+	+	+ +	
α-D-Galacturonate	+	+	+	
Gallate	–	–	–	
Gentisate	–	–	–	
D-Glacenate	+	+	+	
Glucose	+	+	+	+
d-Glucuronate	+	+	+	
L-Glutamate	+	+	+	+
L-Glutamine	+	+	+	
Glutarate	+	+	–	
Glutathione	+	+	+ +	
D-Glycerate	+	+	+	
Glycerol	+	+	+	+
Glycine	+	+	+	
Glycolate	–	–	+	
Heptonate	+	+	+	
Hesperatin	–	–	–	
Hippurate	–	–	–	
DL-Homoserine	–	–	–	
L-Homoserine	–	–	–	
p-Hydroxybutyrate	–	–	–	
p-Hydroxybutyrate	+	+	+	
α-Hydroxybutyrate	–	–	–	
β-Hydroxymethyl glutarate	+	+	+	
Inositol	+	+	+	+
Inulin	–	–	–	
β-Isoleucine	–	–	–	
Isovalerine	–	–	–	
Itaconitate	–	–	–	
α-Ketoglutamine	+	+	+	
Kynurenate	–	–	–	
D-Lactate	+	+	+	
L-Lactate	+	+	+	
α-Lactose	–	–	–	
Laurate	+	+	+	
Laurylalcohol	–	–	–	
L-Leucine	+	+	+	
Levulinate	+	+	–	
L-Lysine	+	+	+	
D-Lyvose	+	+	+	
D-Malate	+	+	+	
L-Malate	+	+	+	

TABLE 1 (continued)
Characteristics of the Wild-Type and Ice$^-$ Deletion Mutants of *P. Syringae*[a]

Characteristic	Cit7	Cit7 dellB	TLP2	TLP2 dell
Malonate	+	+	+	
p-Hydroxybenzoate	+	+	+	
Mannitol	+	+	+	
Mannose	+	+	+	
D-Melezitose	−	−	−	
Mesaconate	−	−	− −	
α-Methylglucoside	−	−	−	
D-Methionine	−	−	−	
Mucate	+	+	+	
D-Norvaline	−	−	−	
L-Norvaline	−	−	−	
Oleate	+	+	−	
D-Ornithine	−	−	−	
L-Ornithine	−	−	−	
Oxalate	−	−	−	
Palatinose	−	−	−	
Palmitate	−	−	−	
D-Pantothenate	−	−	−	
Perseitol	−	−	−	
Phenylacetate	−	−	−	
D-Phenylalanine	−	−	−	
L-Phenylalanine	−	−	−	−
Phytate	−	−	−	
Picolenate	−	−	−	
Pimelate	−	−	−	
Maltose	−	−	−	
DL-Pipecolate	+	+	+	
L-Pipecolate	+	+	+	
Polygalacturonate	−	−	−	
D-Proline	+	+	−	
L-Proline	+	+	+	+
β-Proponal	−	−	−	
Pullalan	−	−	−	
Putracene	+	+	−	
Quinate	+	+	+	
D-(+)-Raffinose	−	−	−	
L-(−)-Rhamnose	−	−	−	
L-(+)-Rhamnose	−	−	−	
D–Ribose	+	+	+	
Rutin	+	+	+	
Saccharate	+	+	+	
Salicin	−	−	−	
Saligenin	−	−	−	
Sarcosine	+	+	+	
D-Serine	+	+	−	
L-Serine	+	+	+	
Shikimate	+	+	+	
Sorbitol	+	+	+	
L-Sorbose	−	−	−	
Sorbate	−	−	−	
Sterate	−	−	−	
Stearylate	−	−	−	
Suberate	−	−	−	
Succinate	+	+	+	
Sucrose	+	+	+	+
D-Tartrate	+	+	−	

TABLE 1 (continued)
Characteristics of the Wild-Type and Ice⁻ Deletion Mutants of *P. Syringae*[a]

Characteristic	Cit7	Cit7 dellB	TLP2	TLP2 dell
L-Tartrate	–	–	–	
m-Tartrate	+	+	+	+
Trehalose	–	–	–	
L-Threonine	–	–	–	
Triacetin	–	–	–	
Tryonellin	+	+	+	
Tryptophan	+	+	–	
D-Tyrosine	–	–	–	
L-Tyrosine	–	–	–	–
n-Valerate	–	–	–	
D-Valine	–	–	–	
L-Valine	–	–	–	
Xylitol	–	–	–	
Xylar	–	–	–	
Susceptibility to				
Sulfathiazole	S	S	S	S
Nalidixic acid	S	S	S	S
Neomycin	S	S	S	S
Bacitracin	R	R	R	R
Nitrofurantoin	R	R	R	R
Penicillin	R	R	R	R
Lincomycin	R	R	R	R
Streptomycin	S	S	S	S
Erythromycin	S	S	S	S
Novobiocin	R	R	R	R
Polymyxin B	S	S	S	S
Chloramphenicol	S	S	S	S
Ampicillin	R	R	R	R
Gentamicin	S	S	S	S
Tetracycline	S	S	S	S
Colistin	S	S	S	S
Vancomycin	R	R	R	RR
Kanamycin	S	S	S	S
Rifampin	R	R	R	R
Carbenicillin	S	S	S	S
Chloretetracycline	S	S	S	S
Cephaloridine	R	R	R	R
Oxytetracycline	S	S	S	S

Note: Carbon and nitrogen source utilization tests were performed in a minimal medium; +, utilization of compound; –, nonutilization. Antibiotic susceptibility tests were performed on King's medium B at 30°C: S, susceptible to antibiotic tested; R, resistant to antibiotic tested.

From Lindow, S. E., *Engineered Organisms in the Environment; Scientific Issues,* Halvorson, H. O., Pramer, D., and Rogui, M., Eds., American Society of Microbiology, Washington, D.C., 1985, 23. With permission.

response (Table 2) when injected into tobacco leaves, which is typical of other phytopathogenic members of *P. syringae* and other plant pathogens.[74] While at least one potential host for these bacteria cannot be excluded, all major agricultural crops in northern California, as well as many native plants and introduced weeds common in northern California, are not potential hosts for these bacterial strains.

Lindow and associates also worked out the behavior of parental Ice⁺ and Ice⁻ deletion mutants of *P. syringae*. These bacteria did not differ from each other in their growth and survival on plant surfaces.[74,114] Survival in soil, or survival under stress of freezing and

TABLE 2
Colonization and Survival on a Variety of Plants by Ice$^+$ *P. Syringae* Strains on Ice$^-$ Deletion Mutants

Plant tested	*P. syringae* populations[a] (log [cfu/g fresh weight])			
	TLP2	TLP2 dell	Cit7	Cit7 dellB
Wet incubation				
Wild oats *(Avena fatua)*	4.30 ab IJK	3.99 b HIJ	3.94 b FG	4.67 ab BC
Oats *(A. sativa)* cv. Cayuce	4.06 bc KL	3.27 c J	3.47 c FG	3.70 bc DEF
Pea *(Pisum sativum)* cv. Tinga	4.66 ab GHIJK	3.76 b IJ	5.07 a BCD	4.40 ab BCD
Black nightshade *(Solanum nigrum)*	6.08 a ABCD	5.73 ab CDE	5.28 bc BC	6.07 ab A
American black nightshade *(S. nodiflorum)*	5.46 a DEFG	4.88 a FG	4.59 a BCDE	3.42 a EF
Petunia *(Petunia × hybrida)* cv. Purple Cascade	4.02 a KL	4.36 a GHI	3.45 ab FG	3.04 abc F
Petunia *(P. × hybrida)* cv. Mitchell	3.32 ab L	4.33 a GHI	3.26 ab G	1.73 bc G
Mallow *(Malva rotundifolia)*	4.55 ab HIJK	5.73 a CDE	4.30 abc DEFG	3.10 b F
Descurainia sophia	5.41 ab DEFGH	5.21 b K	4.86 b BCD	5.11 b BC
Sisymbrium orientale	5.17 b EFGHI	6.99 ab A	4.46 bc BCDEF	4.80 b BC
Horseradish *(Armoracia rusticana)*	4.24 abc JK	4.69 ab G	4.65 ab BCDE	3.09 c F
Asclepias fascicularis	4.30 a IJK	5.76 a CD	4.15 a DEFG	4.65 a BC
Wheat *(Triticum aestivum)* cv. Modoc	4.52 a IJK	3.62 ab J	3.46 ab FG	3.68 ab DEF
Wheat *(T. aestivum)* cv. Twin	5.02 a GHIJK	5.04 a EFG	4.25 bc CDEFG	4.70 ab BC
Barley *(Hordeum vulgare)* cv. Steptoe	3.94 a KL	3.86 a HIJ	1.69 b H	0.08 b H
Barley *(H. vulgare)* cv. Larker	4.65 a CDEF	4.49 bc GH	4.22 bcd CDEFG	4.24 bcd CDE
Pepper *(Capsicum annuum)* cv. Cal Wonder	5.06 abc FGHIJ	5.49 ab DEF	4.76 bc BCDE	6.10 a A
Pepper *(C. annuum)* cv. R10	5.94 a BCDE	5.83 a BCD	5.18 ab BCD	4.92 bc BC
Quack grass *(Agropyron repens)*	6.86 a A	6.52 ab AB	5.42 c B	6.14 bc A
Dandelion *(Taraxacum officinale)*	6.64 a AB	5.69 a CDE	4.73 bc EFG	5.19 ab B
Potato *(Solanum tuberosum)* cv. Russett Burbank	6.32 ab ABC	6.04 b BCD	6.71 a A	6.51 a A
Bean *(Phaseolus vulgaris)* cv. Eagle	6.81 a AB	6.72 a A	6.32 a A	6.41 a A
Corn *(Zea mays)*	6.41 a ABC	6.30 a ABC	6.51 a A	6.02 a A
Dry incubation				
Wild oats *(Avena fatua)*	4.62 ab AB	4.43 abc DE	4.37 ab AB	4.69 a A
Oats *(A. sativa)* cv. Cayuce	4.16 bc ABC	4.51 ab BCDE	5.25 ab A	4.16 bc AB
Pea *(Pisum sativum)* cv. Tinga	4.68 ab AB	4.41 ab CDE	4.93 a A	4.80 ab A
Black nightshade *(S. nigrum)*	3.81 d BCD	4.09 d DEGF	3.78 d ABC	4.81 cd A
American black nightshade *(S. nodiflorum)*	4.11 a ABC	3.94 a DEFG	4.07 a AB	4.75 a A

Petunia *(Petunia × hybrida)* cv. Purple Cascade	1.82 bc E	3.05 abc GH	3.41 ab ABC	1.32 c CDE
Petunia *(P. × hybrida)* cv. Mitchell	2.75 abc DE	2.44 abc H	0.74 c E	0.90 bc E
Mallow *(M. rotundifolia)*	3.75 b EFG	3.71 b EFG	3.16 b ABCD	1.14 c DE
Descurainia sophia	3.69 c BCD	5.38 ab A	3.40 c ABC	3.56 c ABCD
Sisymbrium orientale	4.55 bc AB	5.66 ab AB	2.93 c ABCDE	3.91 bc AB
Horseradish *(Armoracia rusticana)*	3.87 bc ABCD	5.46 a ABC	4.14 abc AB	3.59 bc ABCD
Asclepias fascicularis	4.09 a ABC	4.09 a DEFG	2.23 b BCDE	2.01 b BCDE
Wheat *(Triticum aestivum)* cv. Modoc	2.93 b CDE	3.11 ab FGH	1.09 c DE	2.91 bc ABCDE
Wheat *(T. aestivum)* cv. Twin	4.00 c ABCD	4.05 c DEFG	3.42 d ABC	3.79 cd ABC
Barley *(H. vulgare)* cv. Steptoe	2.99 ab CDE	3.30 a EFGH	1.59 b CDE	3.34 a ABCDE
Barley *(H. vulgare)* cv. Larker	4.67 b AB	3.54 d EFGH	3.77 cd ABC	4.34 bcd AB
Pepper *(C. annuum)* cv. Cal Wonder	4.12 c ABC	4.36 bc CDE	4.26 c AB	4.12 c AB
Pepper *(C. annum)* cv. R10	4.36 bc AB	4.33 bc CDEF	4.45 bc AB	4.04 c AB
Quack grass *(Agropyron repens)*	5.21 c A	5.11 c BCD	3.93 d AB	4.15 d AB
Dandelion *(Taraxacum officinale)*	3.89 bc ABCD	3.38 c EFGH	0.94 d E	1.16 d DE
Potato *(Solanum tuberosum)* cv. Russett Burbank	—	—	—	—
Bean *(Phaseolus vulgaris)* cv. Eagle	—	—	—	—
Corn *(Z. mays)*	—	—	—	—

[a] Before treatment, all plants were grown under greenhouse conditions that minimized leaf wetting and therefore epiphytic bacterial populations. Plants were spray inoculated with an aqueous bacterial suspension of ca. 10^5 CFU/ml and placed in a mist chamber at 24°C for 2 d. Leaves were then harvested from half of the plants, and the population of applied *P. syringae* strains was determined by dilution plating of leaf washing on King's medium B plus 100 μg of rifampin per milliliter. The remaining plants were placed in a growth chamber and held at 30°C and 30% relative humidity with a 16-h photoperiod for 2 additional days; then the bacterial populations were measured as above. The values reported are the mean of three determinations of population size for each bacterium-plant combination. Mean values in each row followed by the same lowercase letter do not differ significantly ($p = 0.05$) as determined by Duncan's multiple range test.

From Lindow, S. E., *Engineered Organisms in the Environment; Scientific Issues,* Halvorson, H. O., Pramer, D., and Rogui, M., Eds., American Society of Microbiology, Washington, D.C., 1985, 23. With permission.

thawing was also the same.[74] *P. syringae* TLP2 and TLP2 dell also grew logarithmically at the same rate and achieved the same maximum population size on potato leaf surfaces. Similarly, while the population size of strains TLP2 and Cit7 differed from each other on a given plant species, population sizes of respective Ice$^-$ deletion mutants did not differ appreciably from their parental strain on that plant species (Table 2). There were considerable difference of strains TLP2 and Cit7 in final population size on different plant species. However, the population size of the respective deletion mutants did not differ greatly from that of their parental types under either environmental treatment (Table 2). The population sizes of strains TLP2 and Cit7 or respective Ice$^-$ deletion strains were generally higher on potatoes and beans than they were on other cultivated or wild plants on which they would not be applied in the proposed field experiment.[74] The viable populations of *P. syringae* TLP2 and Cit7 and respective Ice$^-$ deletion mutants of these strains decreased from more than 10^6 cells per gram immediately after incorporation into nonsterile soil collected at Tulelake, CA, to undetectable levels (fewer than three cells per gram) 17 d after incubation in soil at 20°C. The behavior of wild-type and Ice$^-$ deletion mutants of strains TLP2 and Cit7 did not differ significantly from their respective parental strain in this natural soil. The poor survival of these *P. syringae* strains is typical of the survival of this species in soil. The population sizes or the wild type or Ice$^-$ deletion mutants of these strains in the soil would be expected to be low in years subsequent to their application in field trials. Although the members of viable cells of *P. syringae* TLP2 and Cit7 each decreased with increasing numbers of freezing and thawing cycles of 15-min duration, the viability of respective Ice$^-$ deletion mutants of these strains did not differ from that of the parental type. Thus, site-directed deletions internal to the ice region of these strains do not appear to affect significantly the tolerance of these strains[74] to physical stresses such as freezing, and thus would not confer a selective advantage for the Ice$^-$ phenotype in natural situations.

Lindow also found lack of expected significant ecological impact of these strains following their application in small field trials in a remote area of northern California.[74] Studies of the ecological habitat and interactions of the wild-type and Ice$^-$ mutant strains of *P. syringae,* as well as a comparison of the *in vitro* and *in vivo* characteristics of these strains did not reveal any major difference. A detailed study on characteristics of the wild-type and Ice$^-$ mutant *in vitro* and *in vivo* strains also did not reveal any unexpected differences between these strains. Since wild-type *P. syringae* Cit7 and TLP2 are natural components of the leaf surface ecology of plants in northern California, introductions of novel genotypes into this area will not occur in the proposed experiments. Lindow reported a significant alteration of the bacterial composition on the surfaces of leaves treated directly with Ice$^-$ mutants of *P. syringae.* Since a population size-dependent antagonism phenomenon has been documented for *P. syringae,* small changes in leaf surface populations of this species are not expected to significantly alter other components of the leaf surface microflora. Therefore, significant impacts on the microbial ecology of the treated area will almost certainly be restricted to the actual plot area to be treated.

Ice nucleation-deficient mutants of *Pseudomonas* strains derived by site-directed mutagenesis were recently tested for their efficacy as biological control agents of frost injury on blossoms of greenhouse-grown strawberry plants.[117] Inhibition of one bacterial strain by its near-isogenic counterpart was dose dependent rather than strain dependent. The INA-deletion mutants of *P. syringae* and *P. fluorescens* biovar I, inoculated at 10^7 cfu per blossom, inhibited growth of their ice nucleation-active (INA$^+$) parental strains inoculated at 10^2 cfu per blossom. The INA$^+$ parental strains inhibited INA$^-$ derivatives when inoculum dose was reversed. Inhibition was incomplete unless doses differed by 10^5-fold. No inhibition occurred when two strains were inoculated simultaneously at equal doses (either 10^2 or 10^7 cfu per blossom). The INA$^-$ *P. syringae* strain protected blossom against freezing by other *P. syringae* strains, but did not inhibit or protect against INA$^+$ *P. fluorescens*. The INA$^-$

P. fluorescens strain was much more effective as an inhibitor of *P. syringae* strains than the other *P. fluorescens* strains.

Generally, an organism with the same trait as ice minus bacterium can be obtained simply by screening naturally occurring strains or by conventional mutagenesis in which the parent organism is submitted to a mutation-inducing agent such as ultraviolet light, and the progeny are subsequently surveyed for the desired mutation. The Ice$^-$ bacteria which are genetically engineered forms of *P. syringae* or *P. fluorescens* are only such bacteria. The parent organism contributes to frost damage of plants by secreting a protein that acts as a nucleating center for ice crystal formation on leaves. The genetically engineered Ice$^-$ strains were produced by recombinant DNA methods to remove the gene that encodes the ice nucleation protein, in hopes of producing bacteria that could grow on leaves in place of the parent strain and thereby prevent ice formation. Ice$^-$ strains of bacteria are also common in nature. It is estimated that roughly half of the *P. syringae* strains collected from around the world lack the ability to nucleate ice formation. The genetic basis for this lack in the naturally occurring strains are unknown, whereas it is crystal clear for those produced by recombinant DNA technology.[74,102]

In addition, frost injury is an important problem and its target microorganisms are well known and can be well quantified based on their phenotype of ice nucleation activity. Frost injury on the leaf surface can now be easily quantified. Even in the absence of frost injury, information gained on the ecology and control of INA pathovars of *P. syringae* could be exploited to achieve management of the disease initiated by these and other bacteria by reduction of epiphytic inoculum sources on host plants. Young vegetative tissues have been generally found to harbor low populations of INA bacteria.[8]

INA$^-$ epiphytic colonizers occur naturally and can also be derived from INA$^+$ strains by chemical mutagenesis. Strains of both origins have been demonstrated to be capable of providing some degree of frost protection.[46] Why then it is necessary to construct competitors by genetic engineering? One rather cynical misconception is that genetic engineering can improve only the patentability of the microorganisms.[115]

It seems unlikely that naturally occurring INA$^-$ bacteria would occupy the exact same niches as INA$^+$ strains, even though they may be epiphytic colonizers of the same plant species.[116] There is little direct evidence for this idea and one view would be by determining whether most naturally occurring INA$^-$ and INA$^+$ bacteria constitute genetically distinct populations. Since niche exclusion is probably essential for effective competition, a naturally occurring INA$^-$ bacterium is therefore likely to be inferior to one that is isogenic with an INA$^+$ strain, except at the locus controlling INA. Chemical mutagenesis is not a satisfactory way to obtain such isogenic strains. For practical reasons it is common to use chemical mutagens at concentrations which are likely to cause multiple mutations; the additional mutations which do not affect INA are likely to be undetectable in the laboratory, but affect fitness in the field. The possibility of unknown mutations being present will always complicate experiments with chemically induced mutants. Genetic engineering offers the means to obtain truly isogenic INA$^-$ mutants and thus perform more straightforward experiments with increased chance of success. In addition, deletions through genetic engineering will be nonrevertible; this is considered a desirable trait in an organism which is scheduled for deliberate release into the environment.

The deletion of the genetic sequences that code for ice nuclei had no demonstrable effect on competence for epiphytic colonization. Thus, there is no evidence to suggest that the release of genetically engineered INA$^-$ deletion mutants would alter the population of INA$^-$ and non-INA bacteria in the natural environment. In a field test, it is hoped that the large-dose advantage given to the INA$^-$ bacteria by the applicator would be sufficient to allow successful biological control at the test site.[117] However, once the INA$^-$ strains were dispersed beyond the experimental plot, they would have no dose advantage; in the environment at large, natural strains would outnumber.

E. CONTROL OF ALMOND, PEAR, TOMATO, AND POTATO

Many factors are involved in the control of INA bacteria and in certain situations neither chemical control nor use of antagonists alone is effective in controlling the frost damage. Lindow and associates suggested the use of the integrated approach to control the INA bacteria.[46-49] They were able to demonstrate this approach by controlling fire blight and frost injury to pear, almond, and other crops.[51] Antagonistic non-INA bacteria applied at 10% bloom to pear trees colonized pear flowers and leaves for over 3 months and reduced significantly the epiphytic populations of *P. syringae* and *Erwinia amylovora*. The incidence of both frost injury and, later, fire blight was reduced significantly compared to untreated trees.[51] The control of frost injury and fire blight from a single application of antagonistic bacteria was nearly as good as from weekly applications of a mixture of streptomycin and oxytetracycline or cupric hydroxide.

Because antagonistic bacteria are effective only in exclusion of INA bacteria, a knowledge of the population dynamics of target bacterial species on such plants was also thought necessary to implement at integrated pest management. The population size of INA bacteria on untreated field-grown potato, citrus, pear, and tomato plants varies greatly during the growing season.[8-10,14,46,51] The population size of INA bacteria on potato and tomato was smallest shortly after emergence of these annual plants from seeds.[47] The population size of INA bacteria increased by a factor of over 10^3 and 10^4 within 2 to 5 weeks after emergence of potato and tomato, respectively. The population size of INA bacteria on pear was smallest shortly after the emergence of leaves and flowers from overwintering buds in early spring.[10,47] The INA bacterial population size increased by a factor of over 10^4 within 4 weeks after the appearance of young vegetative tissues during spring. The population size of INA bacteria declined approximately 10 weeks after emergence of vegetative tissue on pear with the onset of hot, dry weather typical of the Mediterranean climate of California.[14,74,94-97] No INA bacteria are detected on navel orange until the beginning of October, when maximum air temperatures begin to decline and free moisture first appears on leaves due to rain, and increasingly heavy dew at night.[47] The total population size of INA bacteria on navel orange reaches a maximum in early January, coinciding with the period of maximum frost hazard to this plant.[46,47] Antagonistic bacteria could be applied to navel oranges upon the onset of favorable environmental conditions in October, and prior to the colonization of foliage by INA bacteria. For other crops favorable conditions for bacterial growth occur when natural a population of INA bacteria are small following seedling or bud emergence. Thus, early spring periods exist in which environmental conditions are favorable for the development of epiphytic populations of INA bacteria, and in which the colonization of young, emerging tissues of potato, tomato, and pear is incomplete. Applications of antagonistic bacteria to these crops at this time prevent subsequent increase in population size of INA bacteria.[14,46,47,74]

Large seasonal variations in the population size of INA bacteria on untreated almond trees were observed during the years 1982 to 1985. Population dynamics of INA bacteria on almond were similar to those seen on pear.[102] Relatively small populations of INA bacteria were seen on young, vegetative, or reproductive tissues in early spring. Large increases in population size of INA bacteria were seen in the years 1982 to 1984, and more modest increases in 1985, within 2 to 4 weeks following flowering of trees. The maximum population size of INA bacteria on almond trees during these years was dependent on the frequency and abundance of rainfall during the months of February, March, and April when newly expanded almond leaf and flower tissue was available for colonization by INA bacteria. Small total populations of INA bacteria in 1985 reflect an almost total absence of rainfall during the early spring of that year. Antagonistic bacteria (used as competitors to exclude subsequent increases in INA bacterial populations sizes on almond trees) must be inoculated onto trees shortly after flowering in these trees, to precede the development of populations of the target species. Antagonistic bacteria applied to plants with the purpose of biological control of INA bacteria became the dominant bacterial species recovered over a period from several days to more than 2 months. The absolute population size of antagonistic bacteria, genetically marked by spontaneous mutations conferring resistance to the antibiotic rifampycin to allow their unambiguous identification and enumeration, varied between different plant species to which they were applied.[14,46,47,73,74,96] However, their population sizes as a proportion of the total population size of bacteria recoverable on King's medium B was often larger than 90%. The population size of antagonistic bacteria applied to pear trees increased for 2 to 3 weeks following application and constituted between 90 to 95% of the total bacteria recoverable from leaf surfaces of treated pear trees.[73,74,96,102] An inverse relationship between the population size of antagonistic bacteria applied to plants and the population size of INA bacteria was seen under field conditions on treated almond trees.[14]

F. MECHANISMS OF ANTAGONISM

Not all bacteria or fungi applied to plant foliage as biological control agents of plant disease exhibit *in vitro* antibiosis toward the foliar pathogen. However, fungal and bacterial plant pathogens have been shown to be inhibited on plants as well as *in vitro* by certain antagonistic bacteria used as biological control agents. Properties of only a few non-INA bacterial antagonists have been studied so far.[118,119] In some cases non-INA bacteria antagonists to INA bacteria on plants are also antagonistic to these bacteria *in vitro*. Similarly, some, but not all bacteria that reduced the incidence of fire blight of pear were inhibitory to *E. amylovora* in culture.[46,47] Antibiosis which relates to the production of certain antibiotic types of compounds by antagonists may be responsible for reducing the populations of INA *P. syringae*. Although antibiosis has been reported to be important in interactions of microorganisms in the rhizosphere and antibiotic production is widely used as a prerequisite for testing of potential biological control agents of both foliar and root diseases, little is known of the importance of antibiosis in the interactions of microbes on leaves. Knowledge of the primary mechanisms of such interactions would be important in designing selection procedures for identification of potential antagonistic bacteria as biological control agents.

Antibiosis plays a minor role in antagonism of *P. syringae* on leaves by other bacteria.[46] For instance, nearly all mutants of antibiotic-producing antagonistic bacteria deficient in production of antibiotics *in vitro* did not differ from the parental strain in antagonism of *P. syringae* on plants.[46] Similarly, antagonistic bacteria that controlled both *P. syringae* and *E. amylovora* were not inhibitory, *in vitro,* to both these above-mentioned species.[46,47] Therefore, antagonism on leaf surfaces may be a general phenomenon so that nonspecific control of more than one target microorganism may be possible.

The study by Lindemann and Suslow[117] demonstrated what has already been shown by others in that biological control may be effective in the absence of antibiosis. The apparent growth inhibition could be the result of nutrient depletion and/or the saturation of colonizable sites as the maximum population capacity of the blossoms is reached. The first alternative could be investigated using near-isogenic lines with altered substrate utilization spectra. Since the interspecific competition between *P. syringae* and *P. fluorescens* was not reciprocal, antibiosis cannot be ruled out as a part of the competitive mechanism.

The reduction in population of INA bacteria and subsequent reduction to frost injury through the use of non-INA bacteria on plants may be due to two factors: (1) the non-INA bacteria produce certain chemicals which may prove harmful/toxic/lethal to INA bacteria, thereby either reducing the population of existing INA bacteria or simply not giving any chance for them to colonize the plant parts; (2) alternatively, non-INA bacteria can colonize and grow in sufficient numbers prior to the arrival of INA bacteria on plant surfaces. Due to competition for surface and food the population of INA bacteria may not be able to appear on the plants.

The control of INA bacteria through foliar spray of antagonists has gained a significant importance in recent years. As in other habitats antibiosis is a commonly assigned mechanism for the interaction of bacterial species on leaf surfaces.[120-127] *In vitro* antibiosis is common in many bacterial strains which exhibit biological control of either fungal or bacterial plant pathogens. Several studies have revealed a poor correlation between antibiosis *(in vitro)* and *in vivo* antagonism.[74,122,128,129] There is also limited information on the nature of antibiosis in relation to interactions of bacteria on leaf surfaces. There are reports that antibiotics including bacteriocin-like compounds and siderophores are inhibitory and responsible for biological control of certain foliar diseases. It is not certain whether the production of these compounds is necessary for the biological control,[122,125] and the experimental evidences for such a necessity are also lacking. There are also reports indicating that bacterial siderophores are not produced on leaf surfaces.[105] Lindow[74] found that antibiosis was not necessary for the preemptive exclusion of INA bacteria by non-INA bacteria on plant surfaces.[74] Out of 88 bacteria antagonists to *P. syringae* on the leaf surfaces only 58 produced compounds which were inhibitory to *P. syringae*.[74] Many strains inhibitory to *P. syringae in vivo* were not inhibitory *in vitro*.[102] It was interesting to note that antibiosis in antibiotic-producing strains was shown not to be necessary for exclusion of *P. syringae* on leaf surfaces using antibiosis-deficient mutants. *P. syringae* populations on plants pretreated with antibiosis-deficient mutants or with antibiosis-positive antagonistic bacteria did not differ at the time of assay.[74] These results indicate that antibiosis is not necessary for successful antagonism among bacteria on leaf surfaces either when antagonistic bacteria are allowed to colonize plants prior to challenge inoculation with INA bacteria or when coinoculated with challenge strains.

The second alternative, which then appears to be the possible mechanism of antagonisms is competition for limiting environmental resources.[102] This possibility has been supported from the experiment using both near-isogenic (chemically induced mutations) or fully isogenic (site-directed *in vitro* mutagenesis) bacterial strains. Near-isogenic strains were obtained following mutagenesis with ethylmethane sulfonate and other mutagens. Treatment of leaves with ice parental strains of these bacteria reduced both the population size of Ice$^+$ *P. syringae* strains and frost damage to leaves at $-5°C$ compared to plants treated only with Ice$^+$ *P. syringae* strains.[74] The population size of Ice$^-$ mutants of *P. syringae* did not differ from that of Ice$^+$ parental strains when each of them was inoculated singly onto the surfaces of corn or bean plants.

Construction of fully isogenic Ice$^+$ and Ice$^-$ strains of *P. syringae* has further made it possible to test the efficiency of competition on leaf surfaces as a mechanism of biological control of frost injury. Lindow and co-workers have finally been successful in identifying the genes for ice nucleation activity both in *P. syringae* and *E. herbicola*. Ice$^+$ phenotype in recipient *Escherichia coli* cells has been expressed.[113-115,130-132] The gene determining ice nucleation in *P. syringae* is rather large (approximately 4.2 kb) and is under the control of its own *cis*-acting regulatory region.[115] Deletions within this ice region have been constructed *in vitro* by molecular genetic techniques.[114,115] Reciprocal exchange of the deletion containing ice gene constructed *in vitro* (carried on a plasmid such as $_p$BR325 which does not replicate within *P. syringae*) for the homologous functional ice gene in recipient *P. syringae* strain occurs following conjugal transfer of such a plasmid to recipient strains.[114] These studies revealed that Ice$^-$ deletion mutants of *P. syringae* constructed *in vitro* do not differ from the Ice$^+$ parental strain in their growth rate on plant surfaces. Not only this, their ability to establish a particular population size following inoculation of any of 25 different plant species, survival during repeated freeze/thaw cycles, or survival in soil or in any of over 100 biochemical and physiological tests does not differ from Ice$^+$ parental strain.[130-133] The Ice$^-$ deletion mutants of *P. syringae* constructed *in vitro* when applied to corn, bean, tomato, potato, or strawberry plants reduced the population size of Ice$^+$ *P. syringae* strains applied as challenge inoculations compared to plants not treated with Ice$^-$ bacteria prior to such challenge inoculations.[74,117,132] Reductions in populations of Ice$^+$ *P. syringae* strains on plants preinoculated with Ice$^-$ strains were inversely proportional to the population size of Ice$^-$ *P. syringae* strains at the time of challenge inoculations. A similar reciprocal reduction in population size of isogenic strains of tomato was observed when either of two isogenic strains was used alternatively as an "antogonist" or "challenge" strain.[132] Thus, competition among isogenic strains on plants appears to be a common mechanism of interaction and one which can lead to large changes in the balance of microbial population sizes on leaves when modified by artificial inoculations of antagonist strains.

The limiting resources for which competition may occur on leaf surfaces can include specific localized sites on leaf surfaces. Limitations of nutrients and other factors will certainly not allow the new population of bacteria to appear at such sites. Nonrandom distribution of bacterial cells on leaf surfaces has been reported by several workers[92,134] and there is every possibility for these to occur in greater number at certain sites such as near veins or abaxial epidermal cell junctions. However, all such localities are not colonized by bacteria at all times and their distribution may depend upon several factors.

The mechanism of antagonism can be best explained from the knowledge gained on the distribution of INA and non-INA bacteria on leaf surfaces. The distribution of bacterial population sizes among individual leaves of a given plant species has been shown to be lognormal.[62,135] The lognormal distribution of population sizes of bacteria is also observed even if the area within an examined leaf is decreased.[102] Thus, no matter at what scale population size is being described, certain leaves, certain portions of leaves, or even collections of a few epidermal cells support the growth and/or survival of many more bacteria than do the average leaf or subsection of leaf. This alone cannot explain such lognormal distributions of bacterial population sizes of bacterial inoculum. In the beginning a normal distribution of bacterial population sizes among treated leaves is observed in leaves inoculated with bacteria, but it quickly develops a lognormal distribution of population sizes.[135] Based on these observations Lindow[102] hypothesized that the environment, either chemical or physical, at a few such localized sites may be far more conductive for the growth and/or survival of bacterial cells on the leaf surface than most other sites. Most competition among bacterial cells may, therefore, be expected to occur at such localized areas. A clear understanding of factors limiting bacterial population sizes on leaf surfaces should lead not only to a better understanding of the mechanisms of interaction between bacteria on leaf surfaces, but also to selection of more efficient antagonistic bacteria. The application of nutrients on leaf surfaces at times does not increase significantly the bacterial populations,[94-97,136] which indicates a relative lack of habitable sites on the leaf surface. Nutrients applied to uninhabitable sites on leaves, if not mobile on the leaf surfaces, would not become available to resident bacteria. Similarly, a lack of habitable sites would restrict population sizes.

G. MOLECULAR MECHANISM OF ICE NUCLEATION

Genes encoding ice nucleation activity have been cloned from *Pseudomonas syringae*[114] and *Pseudomonas fluorescence*.[137] Molecular analysis of these strains was facilitated considerably by the demonstration that a single small region of the chromosome cloned from *P. syringae* could confer ice nucleation activity on *Escherichia coli*.[114] Sequence analysis of ice nucleation gene from *P. syringae* showed 122 imperfect repeats of a consensus 24 nucleotide motif.[138] The inaW gene of *P. fluorescence* was found to possess a single long open reading frame of inaZ gene of *P. syringae*.[139] The two genes have diverged by many amino acid substitutions and have effectively randomized the third bases of homologous codons.

The genes conferring ice nucleation in *P. syringae* and *P. fluorescence* each encode the production of a single protein of approximately 150 and 180 kDa, respectively.[137,140] The unique large protein produced by the cloned ice gene of *P. syringae* is sufficient to confer expression of the ice phenotype in all heterologous genetic background tested.[114,139] The bacterial ice nucleation phenotype is very sensitive to heat (>40°C), pH changes, proteases, 2-mercaptoethanol, urea, and sodium dodecyl sulfate, indicating that protein involved in ice nucleation. Ice nucleation activity of *P. syringae* and *Erwinia herbicola* is also partially sensitive to other membrane pertubants such as borate compounds and lectins.

The ice nucleation genes of two pseudomonas have been found to encode homologous proteins[114,139] and strains of *E. coli* transformed with these genes exhibited Ina^+ phenotype. Much of the amino acid sequence deduced from either gene is composed of interleaved repetition of 8, 16, and 48 amino acids indicating that the function of the ice nucleating proteins is the formation of an ice crystal template.[138] The protein product of the gene (inaZ) responsible for ice nucleation by *P. syringae* has been identified and purified after over-expression in *E. coli.*[140] The amino acid composition and the N-terminal sequence of the purified, denatured protein corresponded well with that predicted from the sequence of the inaZ gene. The protein does not appear to be independently involved in the formation of ice crystals but its activity is associated with lipids. Delipidation of partially purified outer membranes of *P. syringae* by various delipidating agents results in a significant loss of ice nucleation activity associated with the cell envelops of *P. syringae* and other ice nucleation active bacteria.[141] This activity was restored by reconstitution with various phospholipids in a cholate dialysis procedure. However, lipid classes differ in their ability to restore ice nucleation activity. This suggests that hydrophobic environment provided either by lipids or certain detergent micelles is required for proper assembly and structural organisation of an oligomeric ice protein complex enabling its expression as ice nucleus.

ACKNOWLEDGMENTS

We are thankful to Dr. S. E. Lindow of University of California, Berkeley, for providing literature on ice nucleation. This chapter is mainly the compilation of his extensive research work on this subject.

REFERENCES

1. **Cary, J. W. and Mayland, H. F.,** Factors influencing freezing of supercooled water in tender plants, *Agron. J.*, 62, 715, 1970.
2. **Chandler, W. H.,** Cold resistance in horticultural plants, a review, *Proc. Am. Soc. Hortic. Sci.*, 64, 552, 1958.
3. **Mayland, H. F. and Cary, J. W.,** Chilling and freezing injury to growing plants, *Adv. Agron.*, 22, 203, 1970.
4. **White, G. F. and Haas, J. E.,** *Assessment of Research on Natural Hazards,* MIT Press, Cambridge, 1975, 304.
5. **Burke, M. J., Gusta, L. A., Quamme, H. A., Weiser, C. J., and Li, P. H.,** Freezing and injury to plants, *Annu. Rev. Plant Physiol.*, 27, 507, 1976.
6. **Arny, D. C., Lindow, S. E., and Upper, C. D.,** Frost sensitivity of *Zea mays* increased by application of *Pseudomonas syringae, Nature,* 262, 282, 1976.
7. **Lindow, S. E., Arny, D. C., Barchet, W. R., and Upper, C. D.,** Bacterial ice nuclei as incitants of warm temperature frost damage, *Am. Phytopathol. Soc. Proc.*, 3 (Abstr.), 224, 1976.
8. **Lindow, S. E., Arny, D. C., Barchet, W. R., and Upper, C. D.,** The role of bacterial ice nuclei in frost injury to sensitive plants, in *Plant Cold Hardiness and Freezing Stress,* Li, P. H. and Sakai, A., Eds., Academic Press, New York, 1978, 249.
9. **Lindow, S. E., Arny, D. C., Barchet, W. R., and Upper, C. D.,** Bacterial ice nucleation inhibitors and reduction of frost damage to plants, *Phytopathol. News,* 12 (Abstr.), 138, 1978.

10. **Lindow, S. E., Arny, D. C., Barchet, W. R., Baker, L. S., and Upper, C. D.,** Protection of beans against frost injury by modification of populations of epiphytic ice nucleation active bacteria, *Proc. 3rd Int. Congr. Plant Pathol.*, Abstr., p. 75, 1978.
11. **Yankofsky, S. A., Levin, Z., and Moshe, A.,** Association with citrus of ice-nucleating bacteria and their possible role as causative agents of frost damage, *Curr. Microbiol.*, 5, 213, 1981.
12. **Levitt, J.,** *Responses of Plants to Environmental Stresses,* Academic Press, New York, 1972, 306.
13. **Mazur, P.,** Freezing injury to plants, *Annu. Rev. Plant Physiol.*, 20, 419, 1969.
14. **Lindow, S. E. and Connell, J. H.,** Reduction of frost injury to almond by control of ice nucleation active bacteria, *J. Am. Soc. Hortic. Sci.*, 109, 48, 1984.
15. **Bigg, E. K.,** The supercooling of water, *Proc. Phys. Soc. London Sect. B*, 66, 688, 1953.
16. **Camp, P. R.,** The formation of ice at water-solid interfaces, *Ann. N.Y. Acad. Sci.*, 125, 317, 1965.
17. **Mason, B. J.,** Ice-nucleating properties of clay minerals and stony meteories, *Q. J. R. Meteorol. Soc.*, 84, 553, 1960.
18. **Mason, B. J. and Hallett, J.,** Ice-forming nuclei, *Nature (London)*, 197, 357, 1957.
19. **Schnell, R. C. and Vali, G.,** World-wide source of leave-derived freezing nuclei, *Nature (London)*, 246, 212, 1973.
20. **Vonnegut, B.,** Nucleation of supercooled water clouds by silver iodide smokes, *Chem. Rev.*, 44, 277, 1949.
21. **Zettlemoyer, A. C., Tcheurekdjian, N., and Chessick, J. J.,** Surface properties of silver iodide, *Nature (London)*, 192, 653, 1961.
22. **Schnell, R. C.,** Bacteria acting as natural ice nucleants at temperatures approaching −1°C, *Bull. Am. Meteorol. Soc.*, 57, 1356, 1976.
23. **Fukuta, N. and Mason, B. J.,** Epitaxial growth of ice on organic crystals, *J. Phys. Chem. Solids*, 24, 715, 1963.
24. **Feeney, R. E. and Yeh, Y.,** Antifreeze proteins from fish blood, *Adv. Protein Chem.*, 32, 19, 1978.
25. **Barthakur, N. and Maybank, J.,** Anomalous behavior of some amino-acids as ice nucleators, *Nature (London)*, 200, 866, 1963.
26. **Parungo, F. P. and Lodge, J. P., Jr.,** Molecular structure and ice nucleation of some organics, *J. Atmos. Sci.*, 22, 309, 1965.
27. **Parungo, F. P. and Lodge, J. P., Jr.,** Amino acids as ice nucleators, *J. Atmos. Sci.*, 24, 274, 1967.
28. **Rosinski, J. and Parungo, F.,** Terpeneiodine compounds as ice nuclei, *J. Appl. Meterol.*, 5, 119, 1966.
29. **Fukuta, N.,** Ice nucleation by metaldehyde, *Nature (London)*, 199, 475, 1963.
30. **Head, R. B.,** Ice nucleation by phenazine, *Nature (London)*, 196, 736, 1962.
31. **Garten, V. A. and Head, R. B.,** A theoretical basis of ice nucleation by organic crystals, *Nature*, 205, 160, 1965.
32. **Lucas, J. W.,** Subcooling and ice nucleation in lemon, *Plant Physiol.*, 29, 245, 1954.
33. **Kaku, S.,** Changes in supercooling in growing leaves of some evergreen plants, *Bot. Mag.*, 77, 283, 1966.
34. **Kaku, S.,** Changes in supercooling and freezing processes accompanying leaf maturation in *Buxus*, *Plant Cell Physiol.*, 12, 147, 1971.
35. **Kaku, S.,** Changes in supercooling in growing leaves of some evergreen plants and their relation to intercellular space, osmotic value and water content, *Bot. Mag.*, 79, 90, 1966.
36. **Kaku, S.,** High ice nucleating ability in plant leaves, *Plant Cell Physiol.*, 14, 1035, 1973.
37. **Kaku, S.,** Analysis of freezing temperatures distribution in plants, *Cryobiology*, 12, 154, 1975.
38. **Marcellos, H. W. and Single, W. V.,** Ice nucleation on wheat, *Agric. Meteorol.*, 16, 125, 1976.
39. **Marcellos, H. W. and Single, W. V.,** Supercooling and heterogeneous nucleation of freezing in tissues of tender plants, *Cryobiology*, 16, 74, 1979.
40. **Modlibowska, I.,** Some factors affecting supercooling of fruit blossoms, *J. Hortic. Sci.*, 37, 249, 1962.
41. **Hendershoot, C. H.,** The response of orange trees and fruit to freezing temperatures, *Proc. Am. Soc. Hortic. Sci.*, 80, 247, 1962.
42. **Yelenosky, G.,** Cold hardening in citrus stems, *Plant Physiol.*, 56, 540, 1975.
43. **Yelenosky, G. and Horanic, G.,** Sub-cooling in wood of citrus seedlings, *Cryobiology*, 5, 281, 1969.
44. **Young, R. H.,** Freezing points and lethal temperatures of citrus leaves, *Proc. Am. Soc. Hortic. Sci.*, 88, 272, 1966.
45. **Proebsting, E. L., Andrews, P. K., and Gross, D.,** Supercooling young developing fruit and floral buds in deciduous orchards, *Hortic. Sci.*, 17, 67, 1982.
46. **Lindow, S. E.,** Population dynamics of epiphytic ice nucleation active bacteria on frost sensitive plants and frost control by means of antagonistic bacteria, in *Plant Cold Hardiness*, Li, P. H. and Sakai, A., Eds., Academic Press, New York, 395, 1982.
47. **Lindow, S. E.,** Epiphytic ice nucleation active bacteria as incitants of frost injury and plant disease, in *Phytopathogenic Prokaryotes*, Lacy, G. and Mount, M., Eds., Academic Press, New York,, 1982, 344.
48. **Anon.,** Assessing the risks of microbial release, *Science*, 237, 1413, 1987.

49. **Lindow, S. E.,** The role of bacterial ice nucleation in frost injury to plants, *Annu. Rev. Phytopathol.*, 21, 363, 1983.
50. **Lindow, S. E., Arny, D. C., and Upper, C. D.,** Increased frost sensitivity of maize in the presence of *Pseudomonas syringae, Proc. Am. Phytopathol. Soc.*, 2 (Abstr.), 57, 1975.
51. **Lindow, S. E.,** Role of antibiosis in antagonism against ice nucleation active bacteria by epiphytic bacteria, *Phytopathology,* 72 (Abstr.), 986, 1982.
52. **Lindow, S. E.,** Integrated control of frost injury and fire blight of pear with antagonistic epiphytic bacteria, *Phytopathology,* 72 (Abstr.), 946, 1982.
53. **Gross, D. C., Cody, Y. S., Proebsting, E. L., Radamaker, G. K., and Spotts, R. A.,** Distribution, population dynamics, and characteristics of ice nucleation-active bacteria in deciduous fruit tree orchards, *Appl. Environ. Microbiol.*, 46, 1370, 1983.
54. **Hirano, S. S. and Upper, C. D.,** Ecology and epidemiology of foliar plant pathogens, *Annu. Rev. Phytopathol.*, 21, 243, 1983.
55. **Lindow, S. E., Arny, D. C., and Upper, C. S.,** *Erwinia herbicola,* an active ice nucleus incites frost damage to maize, *Phytopathology,* 68, 523, 1978.
56. **Vali, G.,** Ice nucleation relevant to the formation of hail, Stormy Weather Group, McGill University Scientific Rep. MW-58, 1968, 58.
57. **Schneel, R. C. and Vali, G.,** Atmospheric ice nuclei from decomposing vegetation, *Nature,* 236, 163, 1972.
58. **Vali, G., Christensen, M., Fresh, R. W., Galyon, E. L., Maki, L. R., and Schnell, R. C.,** Biogenic ice nuclei. II. Bacterial sources, *J. Atmos. Sci.*, 33, 1565, 1976.
59. **Maki, L. R., Galyon, E. L., Chang-Chien, M., and Caldwell, D. R.,** Ice nucleation induced by *Pseudomonas syringae, Appl. Microbiol.*, 28, 456, 1974.
60. **Dye, D. W., Bradbury, J. F., Goto, M., Hayward, A. C., Lelliott, R. A., and Schroth, M. N.,** International standards for naming pathovars of phytopathogenic bacteria and a list of pathovar names and pathotype strains, *Rev. Plant Pathol.*, 59, 153, 1980.
61. **Hirano, S. S., Nordheim, E. V., Arny, D. C., and Upper, C. D.,** Lognormal distribution of epiphytic bacterial populations on leaf surfaces, *Appl. Environ. Microbiol.*, 44, 695, 1982.
62. **Hirano, S. S., Rouse, D. I., and Upper, C. D.,** Frequency of bacterial ice nuclei on snap bean *(Phaseolus vulgaris* L.) leaflets as a predictor of bacterial brown spot, *Phytopathology,* 72 (Abstr.), 1006, 1982.
63. **Yankofsky, S. A., Levin, Z., Bertold, T., and Sandlerman, N.,** Some basic characteristics of bacterial freezing nuclei, *J. Appl. Meteorol.*, 20, 1013, 1981.
64. **Maki, L. R. and Willoughby, K. J.,** Bacteria as biogenic sources of freezing nuclei, *J. Appl. Meteorol.*, 17, 1049, 1978.
65. **Makino, T.,** Micropipette method: a new technique for detecting ice nucleation activity of bacteria and its application, *Annu. Phytopathol. Soc. J.*, 48, 452, 1982.
66. **Paulin, J. P. and Luisetti, J.,** Ice nucleation activity among phytopathogenic bacteria, *Proc. Int. Conf. Plant Pathol. Bacteriol. 4th (Angiers France),* 2, 725, 1978.
67. **Hirano, S. S., Maher, E. A., Lindow, S. E., Kelman, A., and Upper, C. D.,** Types of ice nucleation activity among plant pathogenic fluorescent pseudomonads, *Proc. Am. Phytopathol. Soc.*, 12, 176, 1978.
68. **Hirano, S. S., Maher, E. A., Lindow, S. E., Kelman, A., and Upper, C. D.,** Ice nucleation activity of fluorescent plant pathogenic pseudomonads, *Phytobacteriologie,* 32, 717, 1978.
69. **Lindow, S. E., Arny, D. C., and Upper, C. D.,** Control or frost damage to corn in the field by a bacterium antagonistic to ice nucleation active bacteria, *Proc. Am. Phytopathol. Soc.*, 4(Abstr.), 169, 1977.
70. **Single, W. V. and Olien, C. R.,** Freezing processes in wheat stems, *Aust. J. Biol. Sci.*, 20, 1025, 1967.
71. **Dye, D. W.,** Development of streptomycin-resistant variants of *Pseudomonas syringae* van Hall in culture and in peach seedlings, *N. Z. J. Agric. Res.*, 1, 44, 1958.
72. **Lindow, S. E., Arny, D. C., and Upper, C. D.,** Distribution of ice nucleation-active bacteria on plants in nature, *Appl. Environ. Microbiol.*, 36, 831, 1978.
73. **Lindow, S. E.,** Integrated control and role of antibiotics in biological control of fireblight and frost injury, in *Biological Control on the Phylloplane,* Windels, C. and Lindow, S. E., Eds., American Phytopathological Society Press, Minneapolis, 1985, 83.
74. **Lindow, S. E.,** Ecology of *Pseudomonas syringae* relevant to the field use of Ice$^-$ deletion mutants constructed in vitro for plant frost control, in *Engineered Organisms in the Environment; Scientific Issues,* Halvorson, H. O., Pramer, D., and Rogui, M., Eds., American Society of Microbiology, Washington, D.C., 1985, 23.
75. **Billing, E. and Baker, L. A. E.,** Characteristics of *Erwinia*-like organisms found in plant material, *J. Appl. Bacteriol.*, 26, 58, 1963.
76. **Cameron, J. R.,** *Pseudomonas* content of cherry trees, *Phytopathology,* 60, 1343, 1970.
77. **Crosse, J. E.,** Interactions between saprophytic and pathogenic bacteria in plant disease, in *Ecology of Leaf Surface Microorganisms,* Preece, T. F. and Dickinson, C. H., Eds., Academic Press, London, 1971, 283.

78. **Crosse, J. E.,** Epidemiological relations of the pseudomonad pathogens of deciduous fruit trees, *Annu. Rev. Phytopathol.,* 4, 291, 1966.
79. **Crosse, J. E.,** Bacterial canker of stone fruits. VI. Inhibition of leaf-scar infection of cherry by a saprophytic bacterium from the leaf surfaces, *Ann. Appl. Biol.,* 56, 149, 1965.
80. **Crosse, J. E.,** Bacterial canker of stonefruits. A comparison of leaf-surface populations of *Pseudomonas mors-prunorum* in autumn on two cherry varieties, *Ann. Appl. Biol.,* 52, 97, 1963.
81. **Crosse, J. E.,** Bacterial canker of stonefruits. IV. Investigation of a method for measuring the inoculum potential of cherry trees, *Ann. Appl. Biol.,* 47, 306, 1959.
82. **Dowler, W. M. and Weaver, D. J.,** Isolation and characterization of fluorescent pseudomonads from apparently healthy peach trees, *Phytopathology,* 65, 233, 1975.
83. **English, H. and Davis, J. R.,** The source of inoculum for bacterial canker and blast of stone trees, *Phytopathology,* 50(Abstr.), 634, 1960.
84. **Ercolani, G. L.,** Epiphytic survival of *Pseudomonas morsprunorum* Wormald, from cherry and *P. syringae* van Hall from pear on the host and on the non-host plant, *Phytopathol. Mediterr.,* 8, 197, 1969.
85. **Ercolanik, G. L., Hagedorn, D. J., Kelman, A., and Rand, R. E.,** Epiphytic survival of *Pseudomonas syringae* on hairy vetch in relation to epidemiology of bacterial brown spot of bean in Wisconsin, *Phytopathology,* 64, 1330, 1974.
86. **Freigoun, Sp. O. and Crosse, J. E.,** Host relations and distribution of a physiological and pathological variant of *Pseudomonas mors-prunorum, Ann. Appl. Biol.,* 81, 317, 1975.
87. **Garden, L., Prunier, J. P., and Luisetti, J.,** Studies on bacterial diseases of fruit trees. IV. Research and study on variation of *P. mors-prunorum* f. sp. persicae on peach trees, *Ann. Phytopathol.,* 4, 229, 1972.
88. **Gibbins, L. N.,** *Erwinia herbicola:* a review and perspective, *Proc. Int. Conf. Plant Pathol. Bacteriol. 4th (Angiers France),* 2, 403, 1978.
89. **Laurence, J. A. and Kennedy, B. W.,** Population changes of *Pseudomonas glycinae* on germinating soybean seeds, *Phytopathology,* 64, 1470, 1974.
90. **Leben, C. and Daft, G. C.,** Influence of an epiphytic bacterium on cucumber anthracnose, early blight of tomato, and northern leaf blight of corn, *Phytopathology,* 55, 760, 1965.
91. **Leben, C., Schroth, M. N., and Hildebrand, D. C.,** Colonization and movement of *Pseudomonas syringae* on healthy bean seedlings, *Phytopathology,* 60, 677, 1970.
92. **Leben, C. and Miller, T. D.,** A pathogenic Pseudomonad from healthy field-grown soybean plants, *Phytopathology,* 63, 1464, 1973.
93. **Mew, T. W. and Kennedy, B. W.,** Growth of *Pseudomonas glycinea* on the surface of soybean leaves, *Phytopathology,* 61, 715, 1971.
94. **Lindow, S. E.,** Methods of preventing frost injury through control of epiphytic ice nucleation active bacteria, *Plant Dis.,* 67, 327, 1983.
95. **Schnell, R. C., Miller, S. W., and Allee, P. A.,** Ice nucleation studies on bacterial serosols, Proc. 5th Conf. Agric. For. Meteorol., 16th Conf. Biometeorology, American Meteorological Society, Anatheim, CA, 1981.
96. **Lindow, S. E., Arny, D. C., and Upper, C. D.,** Biological control of frost injury. I. An isolate of *Erwinia herbicola* antagonistic to ice nucleation active bacteria, *Phytopathology,* 73, 1097, 1983.
97. **Lindow, S. E., Arny, D. C., and Upper, C. D.,** Biological control of frost injury. II. Establishment and effects of an antagonistic *Erwinia herbicola* isolate on corn in the field, *Phytopathology,* 73, 1103, 1983.
98. **Rajashekar, C. B., Li, P. H., and Carter, J. V.,** Frost injury and heterogeneous ice nucleation in leaves of tuber-bearing *Solanum* species, *Plant Physiol.,* 71, 749, 1983.
99. **Ashworth, E. N. and Davis, G. A.,** Ice nucleation within peach trees, *J. Am. Soc. Hortic. Sci.,* 109, 198, 1984.
100. **Andersen, G. L. and Lindow, S. E.,** Local differences in epiphytic bacterial population size and supercooling point of citrus correlated with type of surrounding vegetation and rate of bacterial immigration, *Phytopathology,* 75(Abstr.), 1321, 1985.
101. **Hirano, S. S., Baker, L. S., and Upper, C. D.,** Ice nucleation temperature of individual leaves in relation to population sizes of ice nucleation active bacteria and frost injury, *Plant Physiol.,* 77, 259, 1985.
102. **Lindow, S. E.,** Strategies and practice of biological control of ice nucleation active bacteria on plants, in *Microbiology of Phylosphere,* Fokkema, N. J. and Oorden, A. J., Eds., Cambridge University Press, New York, 1986, 293.
103. **Lindow, S. E.,** Frost damage to potato reduced by bacteria antagonistic to ice nucleation active bacteria, *Phytopathology,* 69(Abstr.), 1036, 1979.
104. **Lindow, S. E. and Staskawicz, B. J.,** Isolation of ice nucleation deficient mutants of *Pseudomonas syringae* and *Erwinia herbicola* and their transformation with plasmid DNA, *Phytopathology,* 71, 237, 1981.
105. **Lindow, S. E., Loper, J. E., and Schroth, M. N.,** Lack of evidence for in situ fluorescent pigment production by *P.s. syringae* on leaf surfaces, *Phytopathology,* 74, 825, 1984.

106. **Vali, G. and Stansbury, E. J.,** Time-dependent characteristics of the heterogeneous nucleation of ice, *Can. J. Phys.,* 44, 477, 1966.
107. **Vali, G.,** Quantitative evaluation of experimental results on the heterogeneous freezing nucleation of supercooled liquids, *J. Atmos. Sci.,* 28, 402, 1971.
108. **Schnell, R. C. and Vali, G.,** Biogenic ice nuclei. I. Terrestrial and marine sources, *J. Atmos. Sci.,* 33, 1554, 1976.
109. **Blanc, M. L.,** Protection against frost damage, World Meteorl. Org. Tech. Note 51, Geneva, 1969.
110. **Moller, W. J., Schroth, M. N., and Thomson, S. V.,** The scenario of fire blight and streptomycin resistance, *Plant Dis.,* 65, 563, 1981.
111. **Schroth, M. N., Thomson, S. V., and Moller, W. J.,** Streptomycin resistance in *Erwinia amylovora, Phytopathology,* 69, 565, 1979.
112. **Cary, J. W. and Lindow, S. E.,** The effect of leaf water variable on ice nucleating, *Pseudomonas syringae* in beans, *Hortic. Sci.,* 21, 1417, 1986.
113. **Orser, C., Staskawicz, B. J., Loper, J., Panopoulos, N. J., Dahlbeck, D., Lindow, S. E., and Schroth, M. N.,** Cloning of genes involved in bacterial ice nucleation and fluorescence pigment/siderophore production, in *Molecular Genetics of the Bacteria-Plant Interaction,* Puhler, A., Ed., Springer-Verlag, Berlin, 1983, 353.
114. **Orser, C. S., Staskawicz, B. J., Panopoulos, N. J., Dahlbeck, D., and Lindow, S. E.,** Cloning and expression of bacterial ice nucleation genes in *Escherichia coli, J. Bacteriol.,* 164, 359, 1985.
115. **Ruvkun, G. B. and Ausubel, F. M.,** A general method for site-directed mutagenesis in prokaryotes, *Nature (London),* 289, 85, 1981.
116. **Warren, G. J., Lindemann, J., Suslow, T. V., and Green, R. L.,** Ice nucleation deficient bacteria as frost protection agents, in *Biotechnology in Agricultural Chemistry,* ACS Symp. Ser. 334, LeBaron, H. M., Mumma, R. O., Honeycutt, R. C., and Duesing, J. H., Eds., American Chemical Society Books, Washington, D.C., 1987, 215.
117. **Lindemann, J. and Suslow, T. V.,** Competition between ice nucleation wild type and ice nucleation deficient deletion mutant strains of *Pseudomonas syringae* and *P. fluorescence* biovar and biological control of first frost injury on strawberry blossoms, *Phytopathology,* 77, 882, 1987.
118. **Lindow, S. E., Arny, D. C., and Upper, C. D.,** Bacterial ice nucleation: a factor in frost injury to plants, *Plant Physiol.,* 70, 1084, 1982.
119. **Lindow, S. E., Hirano, S. S., Arny, D. C., and Upper, C. D.,** The relationship between ice nucleation frequency of bacteria and frost injury, *Plant Physiol.,* 70, 1090, 1982.
120. **Teliz-Ortiz, M. and Burkholder, W. H.,** A strain of *Pseudomonas fluorescens* antagonistic to *Pseudomonas phaseolicola* and other bacterial plant pathogens, *Phytopathology,* 50, 119, 1960.
121. **Goodman, R. N.,** In vitro and in vivo interactions between components of mixed bacterial cultures isolated from apple bud, *Phytopathology,* 55, 217, 1965.
122. **Leben, C.,** Epiphytic microorganisms in relation to plant disease, *Annu. Rev. Phytopathol.,* 3, 209, 1965.
123. **Leben, C. and Daft, G. C.,** Migration of bacteria on seedling plants, *Can. J. Microbiol.,* 12, 1119, 1966.
124. **Chakravarti, B. P., Leben, C., and Daft, G. C.,** Numbers and antagonistic properties of bacteria from buds of field-grown soybean plants, *Can. J. Microbiol.,* 18, 696, 1972.
125. **Beer, S. V., Norelli, J. L., Rundle, J. R., Hodges, S. S., Palmer, J. R., Stein, J. I., and Aldwinckle, H. S.,** Control of fireblight by non-pathogenic bacteria, *Phytopathology,* 70(Abstr.), 459, 1980.
126. **Thomson, S. V., Schroth, M. N., Moller, W. J., and Reil, W. C.,** Efficacy of bactericides and saprophytic bacteria in reducing colonization and infection of pear flowers by *Erwinia amylovora, Phytopathology,* 66, 1457, 1976.
127. **Spurr, H. W.,** Experiments on foliar disease control using bacterial antagonists, in *Microbial Ecology of the Phylloplane,* Blakeman, J. P., Ed., Academic Press, London, 1981, 369.
128. **Hsu, S. C. and Lockwood, J. L.,** Mechanisms of inhibition of fungi in agar by Streptomycetes, *J. Gen. Microbiol.,* 57, 149, 1969.
129. **Anderews, J. H.,** Strategies for selecting antagonistic micoorganisms from the phylloplane, in *Biological Control on the Phylloplane,* Windels, C. E. and Lindow, S. E., Eds., American Phytopathological Society, St. Paul, 1985, 31.
130. **Lindemann, J., Arny, D. C., and Upper, C. D.,** Epiphytic population of *Pseudomonas syringae* pv. *syringae* on snap bean and nonhost plants and the incidence of bacterial brown spot disease in relation to cropping patterns, *Phytopathology,* 74, 1329, 1984.
131. **Lindemann, J. and Suslow, T. V.,** Characteristics relevant to the question of environmental fate of genetically engineered INA^- deletion mutant strains of *Pseudomonas,* in Proc. 6th Int. Conf. on Plant Pathogenic Bacteria, Civerolo, E., Ed., U.S. Department of Agriculture, College Park, MD, 1988, 1005.
132. **Lindemann, J.,** Genetic manipulation of microorganisms for biological control, in *Biological Control on the Phylloplane,* Windels, C. E. and Lindow, S. E., Eds., American Phytopathological Society, St. Paul, 1985, 116.

133. **Kozloff, L., Lute, M., and Westaway, D.,** Phosphatidylinositol as a component of the ice nucleating site of *Pseudomonas syringae* and *Erwinia herbicola, Science,* 226, 845, 1984.
134. **Leben, C.,** Survival of plant pathogenic bacterial Ohio Agric. Res. Dev. Cent. Wooster Spec. Circ. 100, 1974.
135. **Haefele, D. M. and Lindow, S. E.,** Changes in leaf surface characteristics influence the mean, variance, and nucleation frequency of epiphytic ice nucleation active bacterial populations, *Phytopathology,* 74(Abstr.), 882, 1984.
136. **Morris, C. E. and Rouse, D. I.,** Role of nutrients in regulating epiphytic bacterial populations, in *Biological Control on the Phylloplane,* Windels, C. E. and Lindow, S. E., Eds., American Phytopathological Society, St. Paul, 1985, 63.
137. **Corotto, L. V., Wolber, P. K., and Warren, G. J.,** Ice nucleation activity of *Pseudomonas fluorescens:* mutagenesis, complementation analysis and identification of a gene product, *EMBO J.,* 5, 231, 1986.
138. **Warren, G., Corotto, L., and Wolber, P.,** Conserved repeats in diverged ice nucleation structural genes from two species of *Pseudomonas, Nucleic Acids Res.,* 14, 8047, 1986.
139. **Green, R. L. and Warren, G. J.,** Physical and functional repetition in a bacterial ice nucleation gene, *Nature (London),* 317, 645, 1985.
140. **Wolber, P. K., Deininger, C. A., Southworth, M. W., Vandekerckhove, J., van Motagu, M., and Warren, G. J.,** Identification and purification of a bacterial ice nucleation protein, *Proc. Natl. Acad. Sci. U.S.A.,* 83, 7256, 1986.
141. **Govindarajan, A. G. and Lindow, S. E.,** Phospholipid requirement for expression of ice nuclei in *Pseudomonas syringae* and *in vitro, J. Biol. Chem.,* 263, 9333, 1988.

INDEX

A

Abscisic acid (ABA), 209
Acetohydroxy acid synthase, 262
Adenosine triphosphatase (ATPase), 172
Adh, see Alcohol dehydrogenase
Adventitious shoots, 76—77
Agar cultures, 129
Agglutination of protoplasts, 131
Agrobacterium Ti plasmid system, 255—263
  limitations in, 260—261
  T-DNA and, 256—258, 263, 265
Agrobacterium tumefaciens, 145
Agroinfection, 280—290
  viral storage and, 281—285
  viral transmission and, 281—285
Albino gene complementation, 133
Albino species, 133
Alcohol dehydrogenase (Adh), 6, 34, 44, 70
Alfalfa
  disease resistance in, 52
  frost injury in, 325
  long-term high-frequency regeneration of, 230, 233, 235
  protoplasts of, 126
  somaclonal variation in, 12—14, 25, 29, 38, 42, 46
Allium
  cepa, see Onion
  sativum, see Garlic
Alloplasmic male sterility, 164, 169, 172
Alloplasmic plants, 165
Almond, 325
Aluminum resistance, 49
Ambelmoschus esculentus, 105
Amino acid analog resistance mutants, 133
Aminoethyl cystine, 151
Aminoglycoside antibiotics, 151
4-Amino-3,5,6-trichloropicolinic acid (picloram, TCP), 51, 84, 238
Amitrol, 51
Amphidiploid plants, 165
β-Amylase, 6
Andigenum spp., 97
Androgenesis, 161
Aneuploidy, 7, 26, 29
Antagonistic bacteria, 312—319, see also specific types
Anthracnose, 22
Anthranilate synthase, 237
Antibiotics, 151, 169, see also specific types
Apical shoots, 74—76
Apium graveolens, see Celery
Apple, 91, 98, 176
Arabusta spp., see Coffee
Arachis spp., 118
Asparagus, 89, 98, 118
Asymmetric nuclear hybrids, 167
Asymmetric somatic hybrids, 167
AT, 87
ATPase, see Adenosine triphosphatase
Atrazine, 169, 244
Autotetraploidy, 19
Auxillary shoots, 74—76
Auxin, 75, 77
Avena
  fatua, see Wild oats
  sativa, see Oats
  sterilis, see Wild oats
Azetidine-2-carboxylic acid, 235

B

BA, see 6-Benzyladenine
BAA, see 6-Benzylaminipurine
Bacteria, 255, 256, see also specific types
  antagonistic, 312—319
  ice nucleation-active, see Ice nucleation-active (INA) bacteria
  ice nucleation-deficient, 319—325, 329, 332—333
Bacterial phosphotransferase, 262
Banana, 98
BAP, see Benzylaminopurine
Barley, 44, 54, 58, 233, 238, 325
Beans, 310, 325
Begonia spp., 76, 89
Bentazone, 244
6-Benzyladenine (BA), 140
6-Benzylaminipurine (BAA), 83
Benzylaminopurine (BAP), 75, 85, 86, 101, 201, 204, 207, 213
Betula spp., 82
Bialaphos, 274
Binary vectors, 260, 277
Brassica spp., 46, 58, 118, 127, 140
Breeding goals, 223
Brome grass, 128

C

Calcium, phosphate-DNA coprecipitate, 145
Calliclones, 1
Callus cultures, 77—80, 216
Callus-derived protoplasts, 118
Calystegia spp., 118
Carbenicillin, 263
Carnation, 58
Carrot
  genetic transfer in, 158
  glyphosate resistance in, 215
  long-term high-frequency regeneration of, 232
  mass propagation of, 101
  protoplasts of, 139, 149
  somaclonal variation in, 46, 58
Cassava, 98, 101
CAT, see Chloramphenicol transferase
Caulimovirus system of plant gene vectors, 263—267

cDNA, 282, 289
CDS, see Chlorophyll-deficient sectors
Celery, 7, 20, 29, 30, 42, 46, 101, 158
Cell cultures, see Cultures
Cell division, 131, 195
Cellobiose, 128
Cell selection, 214—223
  in vitro, 217—218
  methods in, 218—223
  plated cells and, 222—223
  screening for, 216—217
  strategies for, 214—218
  suspension culture and, 220—222
Cell sorting system, 136
Cell suspensions, 118
Cellulase, 118, 126
Cellulase RS, 126
Cellulose, 126
Cellulysin, 126
Cell wall-degrading enzymes, 117
Cell wall regeneration, 131
Centrifugation, 128
Cheimantha spp., 89
Chemically mediated methods of protoplast
    transformation, 145—149
Chemotherapy, 92—96, 111
Cherry, 145, 325
Chickpea, 230
Chloramphenicol, 275
Chloramphenicol transacetylase, 275
Chloramphenicol transferase (CAT), 149, 150
4-Chlorophenoxyacetic acid (CPA), 84
p-Chlorophenoxyacetic acid (PCPA), 140
Chlorophyll deficiency, 169
Chlorophyll-deficient mutants, 153
Chlorophyll-deficient sectors (CDS), 37
Chloroplast DNA (cpDNA), 36, 159, 165, 171, 178
Chloroplasts, 244
Chlorsulfuron, 31
Chondriomes, 162—164, 172—173
Chromosomes
  changes in number of, 20, 22, 24—30
  rearrangements of, 35
  structural changes in, 30—31
Chrysanthemum, 58, 78
Cinchona spp., 82
Circular RNA, 282
Citrus
  aurantinum, see Sour orange
  limon, see Lemon
  paradisi, see Grapefruit
  reticulata, see Mandarin orange
  sinensis, see Orange
Citrus fruit, 76, 77, 80, 91, 92, 106, 110, see also
    specific types
  frost injury in, 325
  protoplasts of, 130, 176
  regeneration of, 152
Cloning, 73, 99—101, 103, 105—108, 195
Cloning vectors, 264—269
Clover, 46, 58, 236, 325
CMS, see Cytoplasmic male sterility
Coconut, 80
Cocos nucifera, see Coconut
Cocultivation, 262
Coffee, 77, 79, 106
  cultures of, 82
  germplasm and, 98
  protoplasts of, 145, 176
Cointegrating vectors, 258—260
Complementation methods, 133—139
Corn, see Maize
Cowpea, 325
CPA, see 4-Chlorophenoxyacetic acid
cpDNA, see Chloroplast DNA
Crepis capillaris, 6, 30
Cryopreservation, 98, 99
Cryptomeria japonica, 76
CTT, see Cytoplasmic triazine tolerance
Cucumis spp., 173
Cultures, 73—111, 196—213, see also specific types
  agar, 129
  applications of, see specific applications
  callus, 77—80, 216
  cassava, 101
  in chemotherapy, 92—96, 111
  coffee, 82
  defined, 195
  in disease elimination, 96—98
  environment for, 87—89
  fruit, 91
  future prospects for, 108—111
  genotypes and, 89—90
  germplasm and, 98—99
  liquid droplet, 129
  mass propagation and, 99—101, 103, 105—108
  media for, 83—86, 128
  meristem, 90—92, 98
  oat, 78, 89
  onion, 87
  in pathogen-free plant production, 90—98
  proso millet, 79
  of protoplasts, 128—131
  protoplasts of, 126
  rice, 200—203
  seed production and, 99—101, 103, 105—108
  shoot induction and, 74—80
  shoot proliferation and, 74—80
  small-scale cloning and, 99—101, 103, 105—108
  suspension, 196, 220—222
  in thermotherapy, 92—96, 99, 111
  tomato, 78
Cunnignhamia lanceolata, 76
Cupric hydroxide, 310, 312
Cybrids, 168
Cynara scolymus, 107
Cytogenes, 36—37
Cytokinin isopentenyl adenosine, 256
Cytokinins, 75, 77, 103, 129
Cytoplasm, 173

Cytoplasmic genetic traits, 161—165
Cytoplasmic hybrids, 162
Cytoplasmic male fertility, 163
Cytoplasmic male sterility (CMS), 156, 157, 162, 163, 165, 172, 175, 177
Cytoplasmic triazine tolerance (CTT), 156, 157

D

2,4-D, see 2,4-Dichlorophenoxy acetic acid
Dandelion, 325
Datura spp., 20, 118, 133
Daucus spp., 133, see also Carrot
DCMU, 87
Decay curve-type responses, 228
Deplasmolysis, 117
DHFR, see Dihydrofolate reductase
2,4-Dichlorophenoxy acetic acid (2,4-D), 52, 84, 85, 101, 110, 140
  long-term regeneration and, 201, 203, 205, 207, 209, 210, 212, 213
  protoplasts of, 118, 129, 145
  regeneration and, 199, 238, 244
Dihydrofolate reductase (DHFR), 262
6-(γ,γ-Dimethylallyamino)-purine (2iP), 75, 83, 129, 140
Dioscorea alata, 83, 103
Direct methods of protoplast transformation, 145—149
Disarmed vectors, 258
Disease elimination, 96—98
Disease resistance, 18, 19, 52—53, 195, 238—242
DNA, 5, 8, 176, see also specific types
  chloroplast, 36, 159, 171, 178
  mitochondrial, 36, 159, 163—165, 172, 177, 178
  pure, 145
  Ti-plasmid, 150
  transferred, see Transferred DNA (T-DNA)
  transformation of, 148
  uptake of, 255
  viral, 281
DNA viruses, 255, see also specific types
Donor plants, 81—83
Donor-recipient protoplast fusion, 168—170
Donor-recipient technique, 162
Drechslera maydis, 37
Driselase, 118
Droplet freezing assay, 308

E

Eggplant, 140
Elaeis guineansis, see Oil palm
Electrophoresis, 137
Electroporation, 149—151
ELISA, see Enzyme-linked immunosorbent assay
Embryogenesis, 103
Embryogenic callus, 78
Embryoids, 77
EMS, see Methane sulfonic acid ethyl ester
Enolpyruvylshikimate 3-phosphate (EPSP), 274
5-Enol-pyruvylshikimate-3-phosphate synthase, 273
Enzyme-linked immunosorbent assay (ELISA), 19, 91
Enzymes, 117, 120—122, 126—127, see also specific types
EPSP, see Enolpyruvylshikimate 3-phosphate
Erwinia carotovara, 10
Ethionine (ETH), 33, 50
Ethionine resistance, 50—51
Eucalyptus spp., 82, 107
Explants, 81—83, 196—200

F

Feeder layers, 130—131, 151
Filteration-centrifugation, 128
Flax, 126, 230
Flow cytometry, 136
Fluorescein isothiocyanate, 136
Fluorescence microscopy, 136
Fluorescent cell sorting, 135—137
Fluorescent probes, 135
Foreign genes, 149—151, 279—280
Freesia spp., 78
Frost injury, 299—338
  antagonistic bacteria in control of, 312—319
  bactericides in control of, 309—312
  ice formation and, 299—300
  ice nucleation-bacteria and, see Ice nucleation-active (INA) bacteria
  ice nucleation-deficient bacteria in control of, 319—325, 329, 332—333
  supercooling of water and, 299—300
  traditional methods in control of, 309
Fruit crops, 91, see also specific types
Fungal toxins, 169, 171
6-Furfurylaminopurine, 83
Fusion, 133, 168, see Protoplast fusion

G

GA, 101, 103, 209, 210
Gametoclonal variants, 32, 47
Gamma-rays, 130, 162
Garlic, 30, 46
Gemini virus system, 267—269
Genes
  amplification of, 34—35
  deamplification of, 34—35
  expression of, 34—35, 279—280
  foreign, 149—151, 279—280
  organelle-based, 3
  RNA, 8
  transfer of, 149—151
Genetically transformed protoplasts, 137—138
Genetic analysis, 37—44
Genetic bases of somaclonal variation, 20—44
  chromosome number changes and, 20, 22, 24—30
  chromosome structural changes and, 30—31

genetic analysis and, 37—44
point mutations and, 31—37
Genotypes, 89—90
Geranium, 1, 26, 46, 58, 91
Germplasm, 98—99
Globe artichoke, 106
Glucose, 127, 128
Glutamine synthase (GS), 34
Glycine max, see Soybean
Glyphosate, 51, 52, 215, 273, 275
Gossypium hirsutum, 142
Grape, 98
Grapefruit, 144
Growth, 195
GS, see Glutamine synthase

H

Haploid protoplasts, 140
Haploid tobacco, 151
Haplopappus gracilis, 5, 20, 30, 126
Haworthia spp., 29, 54
Helicase, 119
Helonipsis orientalis, 87
Hemicellulose, 126
Herbage grasses, 98
Herbicides, see also specific types
plant gene vectors and, 273—276
resistance to, 49, 157, 169, 237—238, 273—276
tolerance of, 51—52
triazine, 156, 157, 244
Heterokaryotic fusion, 133
Heteroplastomic state, 169
Homokaryotic fusion, 133
Hordeum spp., 2, 20, 30, 42, 54
Hormones, 195, see also specific types
Horn worm resistance, 12
Horseradish, 325
Host-specific pathotoxins resistance, 49
Hyacinthus, 77
Hydroponics system, 230
Hydroxyurea, 51
Hyoscyamus muticus, 139

I

IAA, see Indole-3-acetic acid
IBA, see Indole-3-butyric acid
Ice formation, 299—300
Ice nucleation, 337—338
Ice nucleation-active (INA) bacteria, 300—309
activity of, 304—308
distribution of on plant surfaces, 303—304
measurement of total number of, 308
population density of, 307
quantitative estimation of, 308—309
types of, 301—303
variation among, 304—307
Ice nucleation-deficient bacteria, 319—325, 329, 332—333
INA, see Ice nucleation-active bacteria
Indoleacetamide, 256
Indole-3-acetic acid (IAA), 83, 84, 201, 204, 206, 207, 209
Indole-3-butyric acid (IBA), 84, 86, 209, 210
Induced mutations, 32—34
Insertional transposon mutagenesis, 256
Intergeneric hybrid regeneration, 158—161
Intergenic regions (IR), 264
Interspecific hybrid regeneration, 152—157
In vitro cell selection, 217—218
Iodoacetate prefusion treatment, 170
2iP, see 6-(γ,γ-Dimethylallyamino)-purine; 2-Isopentyl adenine
IPC, see Isopropyl *N*-phenyl carbonate
Ipomoea spp., 118
IR, see Intergenic regions
Irradiation, 130—131, 151, 163
Isoleucine, 237
Isopentenyl transferase, 256
2-Isopentyl adenine (2iP), 75, 83, 129, 140
Isopropyl *N*-phenyl carbonate (IPC), 151

K

Kalachoe blossfeldiana, 77
Kanamycin, 134, 137, 138, 145, 151, 175, 262
Kinetins (KIN), 85, 201, 203, 206, 212

L

Leaf-derived protoplasts, 131
Leaf disk transformation, 262—263
Leaf mesophyll protoplasts, 119, 125
Legumes, 143, see also specific types
Lemon, 144
Lettuce, 1, 18, 42, 46
Leucojum spp., 77
LIH, see Limited internal homology
Lilium spp., 78, 81
Lilly, 58
Limited internal homology (LIH), 258
Lincomycin resistance, 163, 164, 169
Linolenic acids, 160
α-Linolenic acids, 160
γ-Linolenic acids, 160
Liquidambar spp., 79, 82, 110
Liquid droplet culture, 129
Lolium spp., 1, 30, 42
Long-term high-frequency regeneration, 195—246
of alfalfa, 235
applications of, see specific applications
of barley, 238
cell selection in, see Cell selection
2,4-D and, 199, 201, 244
disease resistance and, 238—242
explant source and, 196—200
future prospects for, 243—246
herbicide resistance and, 237—238
of maize, 206—207, 237, 240
of millet, 205—206
mineral stress tolerance and, 235—237

of oats, 198, 207
of pigeon pea, 212—213
of proso millet, 199
of rice, 200—203, 234
salt tolerance and, 223—235
of sorghum, 198, 205
of soybean, 207—212, 244
of tobacco, 235, 238, 239
of tomato, 238, 239
of wheat, 203—205
LS, 85
Lycopersicon spp., see Tomato
Lysine, 237

M

Macerase, 126
Macerozyme R-10, 119
Maize
ATPase in, 172
disease resistance in, 53
frost injury in, 310, 325
genetic analysis of, 37
long-term high-frequency regeneration of, 206—207, 233, 237, 240
protoplasts of, 149, 150
somaclonal variation in, 7, 17—18, 30, 36, 42, 44, 58
Malaxis pludosa, see Orchid
Male sterility, 156, 157, 162—165, 169, 172, 175, 177
Mandarin orange, 144
Mangifera, see Mango
Mango, 77
Maninot esculenta, see Cassava
Mannitol, 127, 128
Marker genes, 276—279
Mass propagation, 99—101, 103, 105—108
Mecerozyme R-10, 118
Medicago spp., 26, 51, 118
Meicelase, 118, 126
Meristem cultures, 90—92, 98
Mesophyll cells, 118
Mesophyll protoplasts, 127, 129, 130, 133, 135
Metabolic complementation, 134
Methane sulfonic acid ethyl ester (EMS), 33, 50
Methionine, 237
Methionine sulfoximine (MSO), 151, 241, 242
Methionine tRNA, 264
Methotrexate, 262
Microinjection methods of protoplast transformation, 145—149
Microisolation, 135—137
Micropropagation, 103, 106
Millet, 75, 109, 205—206
Minerals, 195, see also specific types
Mineral stress, 235—237
Mitochondrial DNA (mtDNA), 36, 37, 159, 163—165, 172, 177, 178
Mitotic crossing-over, 35—36
Molecular mechanism of ice nucleation, 337—338
Monophenylamines, 86
Monosomics, 28
Monosomy, 7
Morphogenesis, 80—90
donor plants and, 81—83
explants and, 81—83
Morus indica, 107
MSO, see Methionine sulfoximine
mtDNA, see Mitochondrial DNA
Mustard, 325
Mutagenesis, 256
Mutations, see also specific types
amino acid analog resistance, 133
chemical induction of, 222
chlorophyll-deficient, 153
defined, 214
induced, 32—34
kanamycin-resistant, 175
nitrate reductase-deficient, 133, 134, 137
nonallelic albino, 133
nutritional, 176
point, 7, 31—37
regeneration of, 176
selection of, 151—152
spontaneous, 31—32
streptomycin-resistant, 134, 137, 160, 175

N

NAA, see Naphthylene-1 acetic acid
Naphthalene acetamide (NAM), 279
Naphthylene-1 acetic acid (NAA), 110, 129, 212, 213
Narcissus spp., 77
Neomycin phosphotransferase (NPT), 262, 274
Nerine spp., 77
New variety development, 173—176
Nicotiana spp., see Tobacco
Nitrate reductase, 138
Nitrate reductase-deficient mutants, 133, 134, 137
N-Nitroso-M-methylurea (NMU), 164
Nonallelic albino mutants, 133
Nondisarmed vectors, 259—260
Nonembryogenic callus, 78
Nopalines, 255, 256
Nopaline synthase, 263
Nopaline synthase promotors, 148
NPT, see Neomycin phosphotransferase
Nucellar embryoids, 77
Nuclear-controlled traits, 169
Nuclear division in protoplasts, 170
Nuclear genetic transfer, 165—168
Nuclear hybrids, 167
Nutrients, 195, see also specific types
Nutritional mutants, 176

O

Oats, 30
cultures of, 78, 89
frost injury in, 325
long-term high-frequency regeneration of, 198, 207

somaclonal variation in, 29, 42, 54
wild, 89, 325
Octopines, 255, 256, see also specific types
Oil palm, 80, 103, 144
Okra, 105
Onion, 87, 325
Onozuka R-10, 118
Opines, 255, 256, see also specific types
Orange, 144, 310
Orchid, 77
Organelle-based genes, 5
Organelle trait transfer, 168—173
Organelle transfer, 161—165
Organogenesis, 131
Osmotic stress, 231
Osmoticum, 127
Oxytetracycline, 312

P

PAGE, 91
Panicum miliaceum, see Proso millet
Paraquat, 52, 238, 244
PAT, see Phosphinothricin acetyltransferase
Pathogen-free plant production, 90—98
PCPA, see *p*-Chlorophenoxyacetic acid
Pear, 325
Peas, 325
Pecinase, 118
Pectin, 126
Pectinase, 118, 126
Pectolyase Y-23, 118, 126
PEG, see Polyethylene glycol
Pelargonium spp., see Geranium
Peppers, 325
Pest resistance, 270—273, see also specific types
Petal + cell culture fusion products, 135
Petal + leaf fusion products, 135
Petunia, 77, 133
cocultivation of, 262
frost injury in, 325
genetic transfer in, 158
herbicide resistance in, 273
long-term high-frequency regeneration of, 236
marker genes and, 278
protoplasts of, 139, 149, 150, 152, 176
somaclonal variation in, 7
Phaseolus vulgaris, 6
Phenolics, 129
Phenovariants, 1
Phermedipham, 244
Phloroglucinol, 86, 109
Phoma lingam, 31
Phosphinothricin, 52
L-Phosphinothricin, 34
Phosphinothricin (PPT), 274
Phosphinothricin acetyltransferase (PAT), 274
Physcomitrella spp., 134
Phytotoxic compounds, 152
Picea spp., 82
Picloram (4-amino-3,5,6-trichloropicolinic acid, TCP), 51, 84, 238
Pigeon pea, 212—213
Pigmentation deficiency, 169
Pineapple, 46
Pinus spp., 82
Plant gene vectors, 255—269
Agrobacterium Ti plasmid system as, see under *Agrobacterium* Ti plasmid system
agroinfection and, see Agroinfection
applications of, see specific applications
caulimovirus system of, 263—267
defined, 255
foreign gene expression and, 279—280
gemini virus system and, 267—269
herbicide resistance and, 273—276
limitations in, 260—261
marker genes and, 276—279
pest resistance and, 270—273
Plasmolysis, 117
Plastomes, 162, 164, 169, 171—172
Plated cells, 222—223
PLO, see Poly-L-ornithine
Point mutations, 7, 31—37
Polembryony, 77
Polyethylene glycol (PEG), 131, 133, 145, 148, 149, 152, 220, 225, 234, 235
Polyethylene glycol (PEG)-electroporation, 150
Poly-L-ornithine (PLO), 145
Polyploid, 26
Polyvinyl alcohol (PVA), 145
Polyvinyl polypyrrolidone, 109
Populus spp., 79, 110
Potato, 10, 27, 30
diseases of, 97
frost injury in, 310, 325
genetic transfer in, 158
germplasm and, 98
protoplasts of, 125, 127, 131
somaclonal variation in, 9—11, 29, 58
Potato-tomato somatic hybrid plants, 167
PPT, see Phosphinothricin
Proliferation rate, 80—90
donor plants and, 81—83
explants and, 81—83
for shoots, 74—80
Propagation, mass, 99—101, 103, 105—108
Proso millet, 79, 199
Protoclones, 1, 26, 31
Protoplast fusion, 131—138, 153, 161—165, 168—170
Protoplasts, 117—179, see also specific types
agglutination of, 131
applications of, see specific applications
callus-derived, 118
cell culture, 126
culture of, 129—131
culture-derived, 131
culture media for, 128
defined, 117
DNA transformation of, 148
electroporation of, 149—151
fusion of, see Protopast fusion
future prospects for, 176—179

genetically transformed, 137—138
growth of, 122—125
haploid, 140
irradiation of, 130—131
isolation of, 117—127, 131—138
leaf-derived, 131
leaf mesophyll, 119, 125
mesophyll, 127, 129, 130, 133, 135
new variety development and, 173—176
nuclear division in, 170
organelle trait transfer and, 168—173
plating of, 130—131
potato, 127
purification of, 127—129
in quantitative assays, 152
regeneration and, see Regeneration
suspension, 135
tobacco, see Tobacco protoplasts
tomato, 127, 128
totipotent, 118
transformation of, 145—149
visual selection of, 134—135
X-irradiated, 151
Prunus spp., 79, 91, see also Sweet cherry
Pseudotsuga spp., 76, 82
Pure DNA, 145
PVA, see Polyvinyl alcohol

Q

Quack grass, 325
Quantitative assay of phytotoxic compounds, 152

R

Rape, 42
Rapid freezing technique, 309
Red cedar, 238
Red clover, 236, 325
Regeneration, 152—168
cell wall, 131
of citrus fruit, 152
cytoplasmic genetic traits and, 161—165
frequency of, 196
intergeneric hybrids and, 158—161
interspecific hybrids and, 152—157
long-term high-frequency, see Long-term high-frequency regeneration
of mutants, 176
nuclear genetic transfer and, 165—168
organelle transfer and, 161—165
protoplast fusions and, 161—165
from protoplasts, 139—145
of somatic hybrids, 131—138
whole plant, 131
Research, 53—55
Resistance, 10, see also specific types
to aluminum, 49
to antibiotics, 160, 169
to disease, 18, 19, 52—53, 195, 238—242
to ethionine, 50—51
to fungal toxin, 169
to glyphosate, 215, 275
to herbicides, 49, 157, 169, 237—238, 273—276
to horn worm, 12
host-specific pathotoxins, 49
pathotoxins, 49
to pests, see Pest resistance
to streptomycin, see under Streptomycin
to stress, 195
to tentoxin, 163, 169
Rhodamine-6G (R6G), 164
Rhozyme, 118
Ribulose-1,5-bisphosphate carboxylase/oxygenase, 163
Rice
genetic analysis of, 37
long-term high-frequency regeneration of, 200—203, 234
protoplasts of, 126, 143, 150
somaclonal variation in, 7, 16—17, 42
RNA, 264, 282, 289, 290, see also specific types
RNA genes, 8
Robertsonian chromosome fissions and fusions, 3
rRNA, 162
RUBPcase, 171
Rye, 8
Ryegrass, 29, 325

S

Saccharum hybrida, 26
Saintpaulia ionantha, 88
Salt tolerance, 51—52, 223—235
Satellite defenses, 289—290
Satellite-length RNA, 289
Satellite RNA, 290
Seed production, 99—101, 103, 105—108
Sequoadendron, 76
Shoots
adventitious, 76—77
apical, 74—76
auxillary, 74—76
induction of, 74—80
proliferation of, 74—80
Small-scale cloning, 99—101, 103, 105—108
Snapdragon, 238
Solanum
accule, 27
brevidens, 10
chacoense, 131
demissum, 27
fernandezianum, 10
melongena, see Eggplant
stoloniferum, 27
tuberosum, see Potato
Somaclonal variation, 1—58, 161, 215, 227, 230, 241, 242
in alcohol dehydrogenase, 7
in alfalfa, 12—14, 25, 29, 38, 42, 46
applications of, see Somaclonal variation applications
in barley, 44, 54, 58
in *Brassica*, 46, 58

in carnation, 58
in carrot, 46, 58
in celery, 7, 20, 29, 42, 46
in chrysanthemum, 58
in clover, 46, 58
defined, 1
experimental approach to, 6—9
future prospects for, 55—58
in garlic, 46
genetic bases of, see Genetic bases of somaclonal variation
in geranium, 46
in *Haworthia* spp., 29, 54
in *Hordeum jubatum,* 54
in *Hordeum* spp., 42
in *Hordeum vulgare,* 54
in lettuce, 18, 42, 46
in lilly, 58
in *Lolium* spp., 42
in maize, 7, 17—18, 30, 36, 42, 44, 58
methodology in, 2—9
in oats, 29, 42, 54
in *Pelarogonium,* 58
in *Petunia* species, 7
in pineapple, 46
in potato, 9—11, 29, 58
potential applications of, 1
protoplasts of, 140
in rape, 42
in rice, 7, 16—17, 42
in rye, 8
in ryegrass, 29
in sorghum, 44, 58
in strawberry, 19—20, 46
in sugarcane, 18—19, 58
in sweet potato, 46
terminology in, 1—2
in tobacco, 7, 11—12, 42, 44, 58
in tomato, 7, 19, 42, 46
in triticale, 29, 54
in *Triticum,* 30
in wheat, 7, 9, 14—16, 29, 42, 44, 54
Somaclonal variation applications, 44—55
disease resistance and, 52—53
in research, 53—55
somatic cell selection as, 50—52
strategies in, 46—50
Somaclones, 6, 10
Somatic cells, 50—52, 103, 140
Somatic fusion, 168
Somatic hybridization, 176
Somatic hybrids, 131—138, 161, 165—168, see also specific types
Somatic recombination, 7
Sorbitol, 127, 128
Sorghum, 44, 58, 150, 198, 205
Sour orange, 144
Soybean, 126, 149—151, 207—212, 228, 236, 244
Spectinomycin resistance, 169
Spectrophotometric estimation, 3
Sphaerocarpus spp., 134
Spheroplasts, 145
Spontaneous mutations, 31—32
Squash, 310
SRI, see Streptomycin-resistant mutations
Sterility, 156, 157, 162—165, 172, 175, 177, see also specific types
Steviarebandiana, 107
Strawberry, 19—20, 46, 98
Streptomycin, 137, 162—164, 169, 312
Streptomycin-resistant (SRI) mutations, 134, 137, 160, 175
Stress, see also specific types
mineral, 235—237
osmotic, 231
resistance to, 195
tolerance of, 195, 196, 216, 217, 220
Stylosanthes spp., 22, 24, 25, 135
Subcellular organelles, 138
Sucrose, 127, 128
Sugarcane, 18—19, 58, 98
Sulfonylurea, 51, 273
Sunflower, 238
Supercooling of water, 299—300
Superinfection, 288
Suspension cultures, 196, 220—222
Suspension protoplasts, 135
Sweet cherry, 145
Sweet potato, 46
Symplasm, 97

T

2,4,5-T, see 2,4,5-Trichlorophenoxyacctic acid
Tabtoxin, 240, 241
Taxodium distichum, 76
TCP, see 4-Amino-3,5,6-trichloropicolinic acid
T-DNA, see Transferred DNA
Tentoxin, 163, 169, 171
Thermotherapy, 92—96, 99, 111
Threonine, 237
Ti, see Tumor-inducing
Tissue cultures, see Cultures
Tissues, 196—200, see also specific types
Tobacco, 19, 26, 31, 34, 36
cell division of, 162
cocultivation of, 262
frost injury in, 325
genetic analysis of, 37
genetic transfer in, 158
genotypes and, 90
haploid, 151
herbicide resistance in, 275
long-term high-frequency regeneration of, 230, 234, 235, 238, 239
marker genes and, 276
pest resistance in, 272
protoplasts of, see Tobacco protoplasts
somaclonal variation in, 7, 11—12, 42, 44, 58
somatic hybrids of, 166

X-irradiated, 163
Tobacco protoplasts, 119, 126, 149, 153, 169, 170, 176
fusion and, 177, 178
mutants of, 134, 137—139, 148, 151
new varieties of, 174, 175
sources of, 118
Tomato, 26, 32, 119
cultures of, 78
frost injury in, 310, 325
genetic transfer in, 158
long-term high-frequency regeneration of, 233, 238, 239
protoplasts of, 127, 128, 142
somaclonal variation in, 7, 19, 42, 46
Totipotency of cells, 214
Totipotent protoplasts, 118
Transferred DNA (T-DNA), 256—258, 263, 265, 275, 277—280, 282—285
agroinfection and, 280, 282—285
transfer of, 284—285
viral genome release from, 283
Transposable elements, 7, 35—36, 255
Triazine herbicides, 156, 157, 244
Tricale, 31
2,4,5-Trichlorophenoxyacetic acid (2,4,5-T), 84
Trichoderma viride, 118
1,3,5-Trihydroxybenzene (phloroglucinol), 86, 109
Trisomics, 28
Trisomy, 7
Triticale, 29, 30, 54
Triticum spp., 15, 30, see also Wheat
tRNA, 162, 264
Tryptophan monooxygenase, 256
T-toxin, 242
Tulipa spp., 81, 82
Tumor-inducing (Ti)-plasmid DNA, 150
Tumor-inducing (Ti) plasmids, 255, see also *Agrobacterium* Ti plasmid system

U

Ulmus, 79

V

Valine, 151
Variants, defined, 214
Vectors, see also specific types
binary, 260, 277
cloning, 264—269
cointegrating, 260
disarmed, 258
nondisarmed, 259—260
plant gene, see Plant gene vectors
Verticillium, 26
Vetch, 325
Vicia rosea, 145
Vigna aconitifolia, 142
Viruses, 90, 97, see also specific types
DNA, 255
DNA and, 281
gemini, 267—269
genetic information and, 285—288
RNA and, 264
storage of, 281—285
transmission of, 281—285
Visual selection of protoplasts, 134—135
Vitamins, 195, see also specific types
Vitrification, 109

W

Watersoaking, 109
Wheat
frost injury in, 325
gene expression in, 35
long-term high-frequency regeneration of, 203—205
protoplasts of, 150
somaclonal variation in, 7, 9, 14—16, 29, 30, 42, 44, 54
Whole plant regeneration, 131
Wild oats, 89, 325

X

X-irradiation, 151, 163

Z

Zea mays, see Maize
Zeatin, 75, 209, 210